Lecture Notes in Computer Science   8851

Commenced Publication in 1973
Founding and Former Series Editors:
Gerhard Goos, Juris Hartmanis, and Jan van Leeuwen

T0212706

Luca Maria Aiello   Daniel McFarland (Eds.)

Social Informatics

6th International Conference, SocInfo 2014
Barcelona, Spain, November 11-13, 2014
Proceedings

 Springer

Volume Editors

Luca Maria Aiello
Yahoo Labs
Avinguda Diagonal 177, 08018 Barcelona, Spain
E-mail: alucca@yahoo-inc.com

Daniel McFarland
Stanford Graduate School of Education
485 Lasuen Mall, Stanford, CA 94305-3096, USA
E-mail: dmcfarla@stanford.edu

ISSN 0302-9743   e-ISSN 1611-3349
ISBN 978-3-319-13733-9   e-ISBN 978-3-319-13734-6
DOI 10.1007/978-3-319-13734-6
Springer Cham Heidelberg New York Dordrecht London

Library of Congress Control Number: 2014955457

LNCS Sublibrary: SL 3 – Information Systems and Application, incl. Internet/Web
and HCI

Typesetting: Camera-ready by author, data conversion by Scientific Publishing Services, Chennai, India

Printed on acid-free paper

Springer is part of Springer Science+Business Media (www.springer.com)

Preface

This volume contains the papers presented at SocInfo 2014, the 6th International Conference on Social Informatics, held during November 11–13, 2014, in Barcelona, Spain. After the conferences in Warsaw, Poland, in 2009, Laxenburg, Austria, in 2010, Singapore in 2011, Lausanne, Switzerland, in 2012, and Kyoto, Japan, in 2013, the International Conference on Social Informatics returned to Europe.

SocInfo is an interdisciplinary venue for researchers from computer science, informatics, social sciences, and management sciences to share ideas and opinions, and present original research work on studying the interplay between socially centric platforms and social phenomena. The ultimate goal of social informatics is to create a better understanding of socially centric platforms not just as a technology, but also as a set of social phenomena. To that end, we have invited interdisciplinary papers, on applying information technology in the study of social phenomena, on applying social concepts in the design of information systems, on applying methods from the social sciences in the study of social computing and information systems, on applying computational algorithms to facilitate the study of social systems and human social dynamics, and on designing information and communication technologies that consider social context.

This year's special purpose of the conference was to to bridge the gap between the social sciences and computer science. We see the challenges of this as at least twofold. On the one hand, social scientific research is still largely overlooked and under-utilized in computational arenas. On the other, social scientists seldom take advantage of computational instruments and the richness of online-generated data. Our ambition is to make SocInfo a conference that is equally attractive to computer scientists and social scientists alike, by putting emphasis on the methodology needed in the field of computational social science to reach long-term research objectives. We have envisioned SocInfo as a venue that attracts open-minded researchers who relax the methodological boundaries between informatics and social sciences so to identify common tools, research questions, and goals.

We were delighted to present a strong technical program at the conference as a result of the hard work of the authors, reviewers, and conference organizers. We received a record of 147 submissions (124 long papers, 23 short papers, not including incomplete or withdrawn submissions). From these, 28 of the 123 long papers were accepted (23%), with another 11 accepted as short paper (9%); three of the 23 short papers were accepted (13%), for a total of 28 long papers, and 14 short papers. We also allowed the authors of accepted papers to opt for a "presentation only" mode with no inclusion in the proceedings: The authors of three papers decided to go for that option. Every paper was reviewed by a group of at least three reviewers, among which at least one had a background in social or organizational sciences. We were also pleased to invite Duncan Watts, Michael Macy, Lada Adamic, Daniele Quercia, Dirk Helbing, and Bruno Goncalves to give exciting keynote talks.

This year SocInfo 2014 included nine satellite workshops: the City Labs Workshop, the Workshop on Criminal Network Analysis and Mining (CRIMENET), the Workshop on Interaction and Exchange in Social Media (DYAD), the Workshop on Exploration of Games and Gamers (EGG), the Workshop on HistoInformatics, the Workshop on Socio-Economic Dynamics, Networks and Agent-Based Models (SEDNAM), the Workshop on Social Influence (SI), the Workshop on Social Scientists Working with Start-Ups, and the Workshop on Social Media in Crowdsourcing and Human Computation (SoHuman).

We would like to thank the authors of submitted papers and presenters as well as the participants for making the conference and the workshops a success. We express our gratitude to the senior and regular Program Committee members and reviewers for their hard and dedicated work. We are extremely grateful to the program co-chairs Ingmar Weber, Kristina Lerman, and Fabio Rojas for their great work in putting together a high-quality program and for directing the activity of the Program Committee. We owe special thanks to Estefania Ricart and Natalia Pou, our local co-chairs, who had a vital role in all the stages of the organization. We thank our publicity chairs Paolo Boldi, Tsuyoshi Murata, Emilio Ferrara, Barbara Poblete, Symeon Papadopoulos, and our Web chair Michele Trevisiol. Also, last but not least we are grateful to Adam Wierzbicki for his continuous support.

Lastly, this conference would not be possible without the generous help of our sponsors and supporters: Microsoft Research, Facebook, Yahoo, the Stanford Center for Computational Social Science, Barcelona Media, and the FP7 EU project SocialSensor.

October 2014   Luca Maria Aiello
                                                                 Daniel McFarland

Organization

Organizing Committee

Local Co-chairs

Estefania Ricart   Yahoo Labs, Barcelona, Spain
Natalia Pou   Administrative Coordinator, Yahoo Labs, Barcelona, Spain

General Co-chairs

Luca Maria Aiello   Yahoo Labs, Barcelona, Spain
Daniel McFarland   Stanford, Palo Alto, CA, USA

Honrary General Co-chairs

Alejandro Jaimes   Yahoo Labs, New York, NY, USA
Dirk Helbing   ETH, Zurich, Switzerland

Program Co-chairs

Ingmar Weber   Qatar Computing Research Institute, Doha, Qatar
Kristina Lerman   University of Southern California, Los Angeles, CA, USA
Fabio Rojas   Indiana University, Bloomington, IN, USA

Publicity Co-chairs

Paolo Boldi   University of Milan, Milan, Italy
Tsuyoshi Murata   Tokyo Institute of Technology, Tokyo, Japan
Emilio Ferrara   Indiana University, Bloomington, IN, USA
Symeon Papadopoulos   CERTH-ITI, Thessaloniki, Greece
Barbara Poblete   University of Chile, Santiago, Chile

Steering Chair

Adam Wierzbicki   Polish-Japanese Institute of Information Technology, Warsaw, Poland

Web Chairs

Michele Trevisiol   Universitat Pompeu Fabra, Barcelona, Spain

Senior Program Committee

Ulrik Brandes	Universität Konstanz, Germany
Elizabeth Bruch	University of Michigan, USA
Carlos Castillo	QCRI, Qatar
Meeyoung Cha	KAIST, South Korea
Jacob Cheadle	UCLA, USA
Munmun De Choundry	Georgia Tech, USA
Andreas Flache	University of Groningen, The Netherlands
Bruno Gonçalves	Aix-Marseille Université, France
Haewoon Kwak	Telefonica Research, Spain
Mounia Lalmas	Yahoo Labs, UK
Ee Peng Lim	Singapore University, Singapore
Ka-yuet Liu	UCLA, USA
Gloria Mark	University of California Irvine, USA
Porter Mason	Oxford University, UK
Alice Oh	KAIST, South Korea
Elizabeth Pontikes	University of Chicago-Booth, USA
Daniele Quercia	Yahoo Labs, Spain
Beate Volker	Utrecht University, The Netherlands
Ed Walker	UCLA, USA
Lynne Zucket	UCLA, USA

Regular Program Committee

Palakorn Achananuparp Aek	LARC, Singapore
Yong-Yeol Ahn	Indiana University, USA
Leman Akoglu	Stony Brook University, USA
Frederic Amblard	University of Tolouse, France
Jisun An	Cambridge University, UK
Yasushito Asano	Kyoto University, Japan
Brandy Aven	Carnegie Mellon University, USA
Nick Beauchamp	Northeastern University, USA
Michal Bojanowski	University of Warsaw, Poland
Piotr Bródka	Wroclaw University, Poland
Matthias Brust	University of Central Florida, USA
James Caverlee	Texas A&M University, USA
Fabio Celli	University of Trento, Italy
Lu Chen	Ohio State, USA
Freddy Chua	Singapore University, Singapore
Giovanni Luca Ciampaglia	Indiana University, USA
David Corney	City University, UK
Rense Corten	Utrecht University, The Netherlands
Michele Coscia	Harvard University, USA
Andrew Crooks	George Mason, USA

Guillaume Deffuant   Laboratoire d'Ingénierie pour les Systémes
                                    Complexes, France
Bruce Desmarais   UMass, USA
Jana Diesner   University of Illinois, USA
Victor Eguiluz   IFISC, Spain
Emilio Ferrara   Indiana University, USA
Fabrizio Ferraro   IESE Business School, Spain
Fabian Flck   GESIS, University of Koblenz, Germany
Santo Fortunato   Aalto University, Finland
Vanessa Frias-Martinez   University of Maryland, USA
Wai-Tat Fu   University of Illinois, USA
Aram Galstyan   University of Southern California, USA
Ruth Garcia-Gavilanes   Universitat Pompeu Fabra, Spain
Manuel Garcia-Herranz   Universitad Autónoma de Madrid, Spain
Armando Geller   Scensei, Switzerland
Maria Giatsoglou   Aristotle University, Greece
Sandra Gonzalez-Bailon   UPenn, USA
Andrea Gorbatai   Berkeley/Haas, USA
Christophe Guéret   Data Archiving & Net Services,
                                    The Netherlands
Ido Guy   IBM Research, Israel
Przemyslaw Grabowicz   MPI, Germany
Nir Grinberg   Cornell University, USA
Alex Hanna   University of Wisconsin-Madison, USA
Sharique Hasan   Stanford University, USA
Purohit Hemant   Knoesis, USA
Stephan Humer   Berlin University of the Arts, Germany
Adam Jatowt   Kyoto University, Japan
Marco Alberto Javarone   University of Sassari and University of Cagliari,
                                    Italy
Andreas Kaltenbrunner   Barcelona Media, Spain
Przemyslaw Kazienko   Wroclaw University, Poland
Kazama Kazuhiro   Wakayama University, Japan
Brian Keegan   Northeastern University, USA
Nakata Keiichi   University of Reading, UK
Adam Kleinbaum   Dartmouth College, USA
Andreas Koch   University of Salzburg, Austria
Konstantinos Konstantinidis   CERTH, Greece
Farshad Kooti   University of Southern California, USA
Nicolas Kourtellis   Yahoo Labs, Spain
Wonjae Lee   KAIST, South Korea
Janette Lehmann   Universitat Pompeu Fabra, Spain
Sune Lehmann   Technical University of Denmark, Denmark
Ilias Leontiadis   Telefonica Research, Spain

Yefeng Liu   Baidu, China
Matteo Magnani   Uppsala University, Sweden
Marco Janssen   Arizona State University, USA
Drew Margolin   Cornell University, USA
Carlos Martin-Dancausa   Robert Gordon University, UK
Winter Mason   Facebook, USA
Karissa McKelvey   Google, USA
Yelena Mejova   QCRI, Qatar
Takis Metaxas   Wellesley College, USA
Patrick Meyer   QCRI, Qatar
Stasa Milojević   Indiana University, USA
Giovanna Miritello   Telefonica Research, Spain
John Mohr   University of California Santa Barbara, USA
Yamir Moreno   BIFI, Spain
Mikolaj Morzy   Poznan University of Technology, Poland
Tsuyoshi Murata   Tokyo Institute of Technology, Japan
Shinsuke Nakajima   Kyoto Sangyo University, Japan
Anne-Marie Oostveen   ISI Foundation, Italy
André Panisson   ISI Foundation, Italy
Mario Paolucci   Institute of Cognitive Sciences and
                               Technologies, Italy
Symeon Papadopoulos   CERTH-ITI, Greece
Paolo Parigi   Stanford University, USA
Ruggero Pensa   University of Turin, Italy
Giorgos Petkos   CERTH-ITI, Greece
Gregor Petric   University of Ljubljana, Slovenia
Jo-Ellen Pozner   University of California Berkeley, USA
Michal Ptaszynski   Hokkai-Gakuen University, Japan
Jose Ramasco   IFISC, Spain
Georgios Rizos   CERTH-ITI, Greece
Luca Rossi   University of Copenhagen, Denmark
Giancarlo Ruffo   University of Turin, Italy
Diego Saez-Trumper   Yahoo Labs, Spain
Mostafa Salehi   University of Bologna, Italy
Claudio Schifanella   University of Turin, Italy
Rossano Schifanella   University of Turin, Italy
Frank Schwitzer   ETH, Switzerland
Xiaoling Shu   University of California Davis, USA
Alex Shulz   TU Darmstadt, Germany
Rok Sosic   Stanford University, USA
Emma Spiro   University of Washington Seattle, USA

Sponsors

Microsoft Research (research.microsoft.com)
Facebook (www.facebook.com)
Yahoo (labs.yahoo.com)
Stanford, Center for Computational Social Science
(css-center.stanford.edu)
Barcelona Media (www.barcelonamedia.org)
SocialSensor (www.socialsensor.eu)
IEEE Special Technical Community on Social Networking
(stcsn.ieee.net)

Table of Contents

Networks, Communities, and Crowds

Interpersonal Links and Gender Biases

News, Credibility, and Opinion Formation

Science and Technology

Organizations, Society, and Social Good

On Joint Modeling of Topical Communities and Personal Interest in Microblogs

Tuan-Anh Hoang and Ee-Peng Lim

Living Analytics Research Centre
Singapore Management University
{tahoang.2011,eplim}@smu.edu.sg

Abstract. In this paper, we propose the *Topical Communities and Personal Interest* (**TCPI**) model for simultaneously modeling topics, topical communities, and users' topical interests in microblogging data. **TCPI** considers different topical communities while differentiating users' personal topical interests from those of topical communities, and learning the dependence of each user on the affiliated communities to generate content. This makes **TCPI** different from existing models that either do not consider the existence of multiple topical communities, or do not differentiate between personal and community's topical interests. Our experiments on two Twitter datasets show that **TCPI** can effectively mine the representative topics for each topical community. We also demonstrate that **TCPI** significantly outperforms other state-of-the-art topic models in the modeling tweet generation task.

Keywords: Social media, Microblogs, Topic modeling, User modeling.

1 Introduction

Microblogging sites such as Twitter[1] and Weibo[2] allow users to publish short messages, which are called *tweets*, sharing their current status, opinion, and other information. Embedded in these tweets is a wide range of topics. Empirical and user studies on microblog usage have showed that users may tweet about either their personal topics or background topics [11,29,13]. The former covers individual interests of the users. The latter is the interest shared by users in topical communities and they emerge when users in the same community tweet about common interests [6]. Background topics are thus the results of interests of the topical communities.

There are previous works on modeling background topics in social media as well as in general document corpuses, e.g., [30,23,10]. However, most of these works model a single background topic or a distribution of background topics. In this work, we instead consider the existence of multiple topical communities, each with a different background topic distribution. Examples of such communities include IT professionals, political groups, entertainment fans, etc.. The IT

---

[1] www.twitter.com
[2] http://www.weibo.com

L.M. Aiello and D. McFarland (Eds.): SocInfo 2014, LNCS 8851, pp. 1–16, 2014.

community covers topics such as technology, science, etc.. The political community covers topics such as welfare, budget, etc.. A user who is associated with a topical community will therefore adopt topics from the interest of the community. The members of these communities may not be socially connected to one another. Hence, when modeling users on social media, we have to consider both the user's personal interests and his topical communities.

In this work, we aim to model topical communities as well as users' topical interests in microblogging data. We want to consider different topical communities, and also to learn topical interests of each user and her dependence on the topical communities to generate content.

A simple way to identify the topical communities is first performing topic modeling on the set of tweets using one of existing models (e.g., LDA [4]) to find out topical interests of the users, then assign the most common topics of all the users to be the topical communities' topics. Such an approach however does not allow us to distinguish between multiple topical communities, nor allow each topical community to have multiple topics. It also does not allow us to quantify, for each user, the degree in which the user depends on topical communities in generating content. We therefore propose to jointly model user topical interests and topical communities' interests in a same framework where each user has a parameter controling her bias towards generating content based on her own interests or based on the topical communities.

Our main contributions in this work consist of the following.

- We propose a probabilistic graphical model, called *Topical Communities and Personal Interest* model (abbreviated as **TCPI**), for modeling topics and topical communities, as well as modeling users' topical interests and their dependency on the topical communities in generating content.
- We develop a sampling method to infer the model's parameters. We further develop a regularization technique to bias the model to learn more semantically clear topical communities.
- We apply **TCPI** model on two Twitter datasets and show that it significantly outperforms other state-of-the-art models in modeling tweet generation task.
- An empirical analysis of topics and topical communities for the two datasets has been conducted to demonstrate the efficacy of the **TCPI** model.

The rest of the paper is organized as follows. We first discuss the related works on modeling topics in social media in Section 2. We then present our proposed model in detail in Section 3. Next, we describe two experimental datasets and report results of experiments in applying the proposed model on the two dataset in Section 4. Finally, we give our conclusions and discuss future work in Section 5.

2 Related Work

In this section, we review previous works that are closely related to our work. These works fall into two categories: (i) the works on analyzing topics in microblogs, and (ii) works on analyzing communities in social networks.

2.1   Topic Analysis

Michelson *et. al.* first examined topical interests of Twitter users by analyzing the named entities mentioned in their tweets [16]. Hong *et. al.* then conducted an empirical study on different ways of performing topic modeling on tweets using the original LDA model [9] and Author-topic model [21]. They found that topic learnt from documents formed by aggregating tweets posted by the same users may help to significantly improve some user profiling tasks. Similarly, Mehrotra *et. al.* investigated different ways of forming documents from tweets in order to improve the performance of LDA model for microblogging data [15]. They found that grouping the tweets containing the same hashtags may lead to a significant improvement. Using the same approach, Ramage *et. al* proposed to use Supervised LDA model [20] to model topics of tweets where each tweet is labeled based on linguistic elements (e.g., hashtags, emoticons, and question marks, etc.) contained in the tweet; and Qiu *et. al.* proposed to jointly modeling topics of tweets and their associated posting behaviors (i.e., tweet, retweet, or reply) [19]. Lastly, the work by Zhao *et. al.* [30] is particularly close to our work. In this work, the authors proposed **TwitterLDA** topic model, which is considered as state-of-the-art topic model for microblogging data. **TwitterLDA** is a variant of LDA, in which: (i) documents are formed by aggregating tweets posted by the same users; (ii) a single background topic is assumed; (iii) there is only one common topic for all words in each tweet; and (iv), each word in a tweet is generated from either the background topic or the user's topic. The plate notation of **TwitterLDA** model is shown in Figure 1 (a), and it's generative process is as follows.

- Sample the background topic $\phi_B \sim Dirichlet(\beta)$
- For each $k = 1, \cdots, K$, sample the k-th topic $\phi_k \sim Dirichlet(\beta)$
- Sample the dependence on background topic $\mu \sim Beta(\rho)$
- For each user u, sample u's topic distribution $\theta_u \sim Dirichlet(\alpha)$
- Generate tweets for the user u: for each tweet t that u posts:
  1. Sample topic for the tweet $z_t \sim Multinomial(\theta_u)$
  2. Sample the tweet's words: for each word $w_{t,n}$ at slot n:
     - Sample $y_{t,n} \sim Bernoulli(\mu)$
     - If $y_{t,n} = 0$, sample from background topic: $w_{t,n} \sim Multinomial(\phi_B)$; else ($y_{t,n} = 1$), sample from topic z_t: $w_{t,n} \sim Multinomial(\phi_{z_t})$

TwitterLDA model however does not consider multiple background topics, and impractically assume that all users have the same dependency on the unique background topic (as the paramter μ is common for all the users).

It is important to note that our work is similar but not exactly the same with works on finding global topics (e.g., [10,23]). Global topics are shared by all the users and not specific for any community. On the other hand, topics of each topical community is specific for the community, and are shared mostly by users within the community.

2.2 Community Analysis

Most of the early works on community analysis in social networks are finding social communities based on social links among the users. For example, Newman proposed to discover social communities by finding a network partition that maximizes a measure of "compactness" in community structure called *modularity* [18]; Airoldi *et. al.* proposed a statistical mixed membership model [1]. There are also works on finding topical communities based on user generated content (e.g., [31,23]), and users' attributes and interest affiliations (e.g., [24,26,27]). Ding *et. al.* conducted an empirical study showing that social community structure of a social network may significantly be different from topical communities discovered from the same network [5]. Moreover, most of existing works on analyzing topical community do not differentiate users' personal interests from those of topical communities. They assume that a user's topical interests is determined purely based on her topical communities' interests. This assumption is not practical when applying for microblogging users since they express interest in a vast variety of topics of daily life, and their interests are therefore not always determined by their topical communities.

Lastly, it is also important to note that our work is different from works on finding topical interests of social communities (e.g., [28,22]). Topical interests of each social community includes most common topics shared by users within the community, and hence may not specific for the community, i.e., two different social communities may have the same topical interests. On the other hand, each topical community is uniquely determined based on its topical interests: different topical communities have significantly different topical interests.

3 Topical Community and Personal Interest Model

3.1 Assumptions

Our model relies on the assumptions that: (i) users generate content topically; and (ii) users generate content according either to their personal interests or some topical communities. The first assumption suggests that, for each user, there is always an underlying topic explaining content of the every tweet she posts. The second assumption suggests that, while different users generally have different personal topical interests, their generated content also share some common topics of the topical communities the users belong to. For example, most of the users tend to tweet about daily activities and entertainment although these topics may not represent their real personal interest. During an election campaign, a user who is not personally interested in politics, may still tweet more about political topics as she follows the prevalent topical community of interests in political topics. Hence, to model users' content accurately, it is important to determine topical communities as well as their own personal interests.

3.2 Generative Process

Based on the assumptions as presented above, we propose the **TCPI** model to model user generated tweet from a vocabulary $\mathcal{V}$. The **TCPI** model has K latent

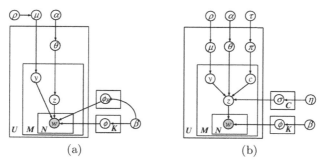

Fig. 1. Plate notation for: (a) **TwitterLDA** and (b) **TCPI** models

topics, where each topic k has a multinomial distribution ϕ_k over the vocabulary $\mathcal{V}$. To capture the topical communities, the **TCPI** model assumes that there are C topical communities, where each community c has a multinomial distribution σ_c over the K topics. Each user u also has a personal topic distribution θ_u over the K topics and a community distribution π_u over the C topical communities. Moreover, each user has a dependence distribution μ_u which is a Bernoulli distribution indicating how likely the user tweets based on her own personal interests (μ_u^0) or based on the topical communities ($\mu_u^1 = 1 - \mu_u^0$). Lastly, we assume that θ_u, π_u, σ, and ϕ have Dirichlet priors α, τ, η, and β respectively, while μ_u has Beta prior ρ.

In **TCPI** model, we assume the following generative process for all the posted tweets. To generate a tweet t for user u, we first flip a biased coin $y_{u,t}$ (whose bias to head up is μ_u^0) to decide if the tweet is based on u's personal interests, or based on one of the topical communities u belongs to. If $y_{u,t} = 0$, we then choose the topic z_t for the tweet according to u's topic distribution θ_u. Otherwise, $y_{u,t} = 1$, we first choose a topical community c according to u's community distribution π_u, then we choose z_t according to the chosen community's topic distribution σ_c. As tweets are short with no more than 140 characters, we assume that each tweet has only one topic. Once the topic z_t is chosen, words in t are then chosen according to the topic's word distribution ϕ_{z_t}. In summary, the **TCPI** model has the plate notation as shown in Figure 1 (b) and the generative process as follows.

- For each $k = 1, \cdots, K$, sample the k-th topic $\phi_k \sim Dirichlet(\beta)$
- For each $c = 1, \cdots, C$, sample the c-th community's topic distribution $\sigma_c \sim Dirichlet(\eta)$
- For each user u
  1. Sample u's topic distribution $\theta_u \sim Dirichlet(\alpha)$
  2. Sample u's community distribution $\pi_u \sim Dirichlet(\tau)$
  3. Sample u's dependence distribution $\mu_u \sim Beta(\rho)$
- Generate tweets for the user u: for each tweet t that u posts:
  1. Sample $y_{u,t} \sim Bernoulli(\mu_u)$
  2. Sample topic for the tweet: if $y_{u,t} = 0$, sample $z \sim Multinomial(\theta_u)$; if $y_{u,t} = 1$, sample a community $c \sim Multinomial(\pi_u)$, then sample $z_t \sim Multinomial(\sigma_c)$

$$p(y_j^i = 0 | \mathcal{T}, \mathcal{Y}_{-t_j^i}, \mathcal{C}_{-t_j^i}, \mathcal{Z}, \alpha, \beta, \tau, \eta, \rho) \propto$$

$$\propto \frac{\mathbf{n_y}(0, u_i, \mathcal{Y}_{-t_j^i}) + \rho_0}{\sum\limits_{y=0}^{1} \left(\mathbf{n_y}(y, u_i, \mathcal{Y}_{-t_j^i}) + \rho_y \right)} \cdot \frac{\mathbf{n_{zu}}(z_j^i, u_i, \mathcal{Z}_{-t_j^i}) + \alpha_{z_j^i}}{\sum\limits_{k=1}^{K} \left(\mathbf{n_{zu}}(k, u_i, \mathcal{Z}_{-t_j^i}) + \alpha_k \right)} \qquad (1)$$

$$p(y_j^i = 1, c_j^i = c | \mathcal{T}, \mathcal{Y}_{-t_j^i}, \mathcal{C}_{-t_j^i}, \mathcal{Z}, \alpha, \beta, \tau, \eta, \rho) \propto$$

$$\propto \frac{\mathbf{n_y}(1, u_i, \mathcal{Y}_{-t_j^i}) + \rho_1}{\sum\limits_{y=0}^{1} \left(\mathbf{n_y}(y, u_i, \mathcal{Y}_{-t_j^i}) + \rho_y \right)} \cdot \frac{\mathbf{n_{cu}}(c, u_i, \mathcal{C}_{-t_j^i}) + \tau_c}{\sum\limits_{c=1}^{C} \left(\mathbf{n_{cu}}(c, u_i, \mathcal{C}_{-t_j^i}) + \tau_c \right)} \cdot \frac{\mathbf{n_{zc}}(z_j^i, c, \mathcal{Z}_{-t_j^i}, \mathcal{C}_{-t_j^i}) + \eta_{cz_j^i}}{\sum\limits_{k=1}^{K} \left(\mathbf{n_{zc}}(k, c, \mathcal{Z}_{-t_j^i}, \mathcal{R}_{-t_j^i}) + \eta_{ck} \right)}$$

$$(2)$$

Fig. 2. Probabilities used in **jointly sampling coin and topical community** for tweet t_j^i without regularization

3. Sample the tweet's words: for each word slot n, sample the word $w_{t,n} \sim Multinomial(\phi_{z_t})$

3.3   Model Learning

Consider a set of microblogging users together with their posted tweets, we now present the algorithm for performing inference in the **TCPI** model. We use U to denote the number of users and use W to denote the number of words in the tweet vocabulary $\mathcal{V}$. We denote the set of all posted tweets in the dataset by $\mathcal{T}$. For each user u_i, we denote her j-th tweet by t_j^i. For each posted tweet t_j^i, we denote N_{ij} words in the tweet by $w_1^{ij}, \cdots, w_{N_{ij}}^{ij}$ respectively, and we denote the tweet's topic, coin, and topical community (if exists) by z_j^i, y_j^i, and c_j^i respectively. Lastly, we denote the bag-of-topics, bag-of-coins, and bag-of-topical communities of all the posted tweets in the dataset by $\mathcal{Z}$, $\mathcal{Y}$, and $\mathcal{C}$ respectively.

Due to the intractability of LDA-based models [4], we make use of sampling method in learning and estimating the parameters in the **TCPI** model. More exactly, we use a collapsed Gibbs sampler ([14]) to iteratively and jointly sample the latent coin and latent topical community, and sample latent topic of every posted tweet as follows.

For each posted tweet t_j^i, the j-th tweet posted by user u_i, we use $\mathcal{Y}_{-t_j^i}$, $\mathcal{C}_{-t_j^i}$, $\mathcal{Z}_{-t_j^i}$ to denote the bag-of-coins, bag-of-topical communities and bag-of-topics, respectively, of all other posted tweets in the dataset except the tweet t_i^j. Then the coin y_j^i and the topical community c_j^i of t_j^i are jointly sampled according to equations in Figure 2, and the topic z_j^i of t_j^i is sampled according to equations in Figure 3. Note that when $y_j^i = 0$, we do not have to sample c_j^i, and the current c_j^i (if exists) will be discarded. In these equations, $\mathbf{n_y}(c, u, \mathcal{C})$ records the number of times the coin y is observed in the set of tweets of user u for the bag-of-coins $\mathcal{Y}$.

$$p(z_j^i = z|y_j^i = 0, \mathcal{T}, \mathcal{Y}_{-t_j^i}, \mathcal{C}, \mathcal{Z}_{-t_j^i}, \alpha, \beta, \tau, \eta, \rho) \propto$$

$$\propto \frac{\mathbf{n_{zu}}(z, u_i, \mathcal{Z}_{-t_j^i}) + \alpha_z}{\sum_{k=1}^{K} \left(\mathbf{n_{zu}}(k, u_i, \mathcal{Z}_{-t_j^i}) + \alpha_k\right)} \cdot \prod_{n=1}^{N_{ij}} \frac{\mathbf{n_w}(w_n^{ij}, z, \mathcal{Z}_{-t_j^i}) + \beta_{zw_n^{ij}}}{\sum_{v=1}^{W} \left(\mathbf{n_w}(v, z, \mathcal{Z}_{-t_j^i}) + \beta_{zv}\right)} \quad (3)$$

$$p(z_j^i = z|y_j^i = 1, \mathcal{T}, \mathcal{Y}_{-t_j^i}, \mathcal{C}, \mathcal{Z}_{-t_j^i}, \alpha, \beta, \tau, \eta, \rho) \propto$$

$$\propto \frac{\mathbf{n_{zc}}(z, c_j^i, \mathcal{Z}_{-t_j^i}, \mathcal{C}_{-t_j^i}) + \eta_{c_j^i z}}{\sum_{k=1}^{K} \left(\mathbf{n_{zc}}(k, c_j^i, \mathcal{Z}_{-t_j^i}, \mathcal{C}_{-t_j^i}) + \alpha_{c_j^i k}\right)} \cdot \prod_{n=1}^{N_{ij}} \frac{\mathbf{n_w}(w_n^{ij}, z, \mathcal{T}_{-t_j^i}, \mathcal{Z}_{-t_j^i}) + \beta_{zw_n^{ij}}}{\sum_{v=1}^{W} \left(\mathbf{n_w}(v, z, \mathcal{T}_{-t_j^i}, \mathcal{Z}_{-t_j^i}) + \beta_{zv}\right)} \quad (4)$$

Fig. 3. Probabilities used in **sampling topic** for tweet t_j^i without regularization

Similarly, $\mathbf{n_{zu}}(z, u, \mathcal{Z})$ records the number of times the topic z is observed in the set of tweets of user u for the bag of topics $\mathcal{Z}$; $\mathbf{n_{zc}}(z, c, \mathcal{Z}, \mathcal{C})$ records the number of times the topic z is observed in the set of tweets that are tweeted based on the topical community c by any user for the bag-of-topics $\mathcal{Z}$ and the bag-of-topical communities $\mathcal{C}$; $\mathbf{n_{cu}}(c, u, \mathcal{C})$ records the number of times the topical community c is observed in the set of tweets of user u; and $\mathbf{n_w}(w, z, \mathcal{T}, \mathcal{Z})$ records the number of times the word w is observed in the topic z for the set of tweets $\mathcal{T}$ and the bag-of-topics $\mathcal{Z}$.

In the right hand side of Equation 1: (i) the first term is proportional to the probability that the coin 0 is generated given the priors and (current) values of all other latent variables (i.e., the coins, topical communities (if exist), and topics of all other tweets); and (ii) the second term is proportional to the probability that the (current) topic z_j^i is generated given the priors, (current) values of all other latent variables, and the chosen coin. Similarly, in the right hand side of Equation 2: (i) the first term is proportional to the probability that the coin 1 is generated given the priors and (current) values of all other latent variables; (ii) the second term is proportional to the probability that the topical community c is generated given the priors, (current) values of all other latent variables, and the chosen coin; and (iii) the third term is proportional to the probability that the (current) topic z_j^i is generated given the priors, (current) values of all other latent variables, and the chosen coin as well as the chosen community.

The terms in the right hand side of Equations 3 and 4 respectively have the similar meaning with those of Equations 1 and 2.

3.4 Sparsity Regularization

As we want to differentiate users' tweets based on personal interests from topical communities and to differentiate one topical community from the others, we would prefer a clear distinction among these latent factors. In other words, we want topical communities' topic distributions and users' topic distributions to be skewed on different topics, and topical communities' topic distribution to be also skewed on different topics. More exactly, in estimating parameters in the **TCPI** model, we need to obtain sparsity in the following distribution.

- Topic specific coin distribution $p(y|z)$ where y is a coin and z is a topic: the sparsity in this distribution is to ensure that each topic z is mostly covered by either users' personal interests or topical communities.
- Topic specific topical community distribution $p(c|z)$ where c is a topical community and z is a topic: the sparsity in this distribution is to ensure that each topic z is mostly covered by one or only a few topical communities.

To obtain the sparsity mentioned above, we use the *pseudo-observed variable* based regularization technique proposed by Balasubramanyan *et. al.* [2] as follows.

Topic Specific Coin Distribution Regularization. Since the topic specific coin distributions are determined by both coin and community joint sampling and topic sampling steps, we regularize both these two steps to bias the distributions to expected sparsity.

In Coin and Topical Community Joint Sampling Steps. In each coin and topical community sampling step for the tweet t_j^i, we multiply the right hand side of equations in Figure 2 with a corresponding regularization term $\mathcal{R}_{\text{topCoin-C\&C}}(y|z_j^i)$ which is computed based on empirical entropy of $p(y|z_j^i)$ as in Equation 5.

$$\mathcal{R}_{\text{topCoin-C\&C}}(y|z_j^i) = exp\left(- \frac{\left(H_{y_j^i=y}\left(p(y'|z_j^i)\right) - \mu_{\text{topCoin}}\right)^2}{2\sigma_{\text{topCoin}}^2}\right) \qquad (5)$$

Fig. 4. Topic specific coin distribution regularization terms used in sampling coin and/or **topical community** for tweet t_j^i

In Topic Sampling Steps. In each topic sampling step for the tweet t_j^i, we multiply the right hand side of equations in Figure 3 with a corresponding regularization term $\mathcal{R}_{\text{topCoin-Topic}}(z|t_j^i)$ which is computed based on empirical entropy of $p(y|z)$ as in Equation 6.

In Equations 5, $H_{y_j^i=y}\left(p(y'|z_j^i)\right)$ is the empirical entropy of $p(y'|z_j^i)$ when $y_j^i = y$. Similarly, in Equations 6, for each topic z', $H_{z_j^i=z}\left(p(y|z')\right)$ is the empirical entropy of $p(y|z')$ when $z_j^i = z$. The two parameters μ_{topCoin} and σ_{topCoin} is respectively the expected mean and expected variance of the entropy of $p(y|z)$. These expected mean and expected variances are pre-defined parameters. Obviously, with a low expected mean μ_{topCoin}, these regularization terms (1) increase weight for values of y, c, and z that give lower empirical entropy of $p(y|z)$, and hence increasing the sparsity of these distributions; but (2) decrease weight for

$$\mathcal{R}_{\text{topCoin-Topic}}(z|t_j^i) = exp\left(-\sum_{z'=1}^{K} \left[\frac{\left(H_{z_j^i=z}\big(p(y|z')\big) - \mu_{\text{topCoin}}\right)^2}{2\sigma_{\text{topCoin}}^2} \right] \right) \qquad (6)$$

Fig. 5. Topic specific coin distribution regularization terms used in sampling **topic** for tweet t_j^i

$$\mathcal{R}_{\text{topComm-C\&C}}(y, c|z_j^i) = exp\left(-\frac{\left(H_{y_j^i=y,c_j^i=c}\big(p(c'|z_j^i)\big) - \mu_{\text{topComm}}\right)^2}{2\sigma_{\text{topComm}}^2} \right) \qquad (7)$$

Fig. 6. Topic specific topical community distribution regularization terms used in sampling **coin** and/or **topical community** for tweet t_j^i

$$\mathcal{R}_{\text{topComm-Topic}}(z|t_j^i) = exp\left(-\sum_{z'=1}^{K} \left[\frac{\left(H_{z_j^i=z}\big(p(c|z')\big) - \mu_{\text{topComm}}\right)^2}{2\sigma_{\text{topComm}}^2} \right] \right) \qquad (8)$$

Fig. 7. Topic specific topical community distribution regularization terms used in sampling **topic** for tweet t_j^i

values of y, c, and z that give higher empirical entropy of $p(y|z)$, and hence decreasing the sparsity of these distributions.

Topic Specific Topical Community Distribution Regularization. Similarly, since the topic specific topical community distributions are determined by both coin and topical community joint sampling and topic sampling steps, we regularize both these two steps to bias the distributions to expected sparsity.

In Coin and Topical Community Joint Sampling Steps. In each coin and topical community sampling step for the tweet t_j^i, we also multiply the right hand side of equations in Figure 2 with a corresponding regularization term $\mathcal{R}_{\text{topComm-C\&C}}(y, c|z_j^i)$ which is computed based on empirical entropy of $p(c'|z_j^i)$ as in Equation 7.

In Topic Sampling Steps. In each topic sampling step for the tweet t_j^i, we also multiply the right hand side of equations in Figure 3 with a corresponding regularization term $\mathcal{R}_{\text{topComm-Topic}}(z|t_j^i)$ which is computed based on empirical entropy of $p(c|z)$ as in Equation 8.

In Equations 7, $H_{y_j^i=y,c_j^i=c}\big(p(c'|z_j^i)\big)$ is the empirical entropy of $p(c'|z_j^i)$ when $y_j^i = y$ and $c_j^i = c$. Similarly, in Equations 8, for each topic z', $H_{z_j^i=z}\big(p(c|z')\big)$ is the empirical entropy of $p(c|z')$ when $z_j^i = z$. The two parameters μ_{topComm} and σ_{topComm} is respectively the expected mean and expected variance of the entropy of $p(c|z)$. These expected mean and expected variances are pre-defined

Table 1. Statistics of the experimental datasets

Dataset	SE	Two-Week
#user	14,595	24,046
#tweets	3,030,734	3,181,583

parameters. Obviously, with a low expected mean μ_{topComm}, these regularization terms (1) increase weight for values of y, c, and z that give lower empirical entropy of $p(c|z)$, and hence increasing the sparsity of these distributions; but (2) decrease weight for values of y, c, and z that give higher empirical entropy of $p(c|z)$, and hence decreasing the sparsity of these distributions.

In our experiments, we used sampling method with the above regularization setting $\mu_{\text{topCoin}} = \mu_{\text{topComm}} = 0$, $\sigma_{\text{topCoin}} = 0.3$, $\sigma_{\text{topComm}} = 0.5$. We also used symmetric Dirichlet hyperparameters with $\alpha = 50/K$, $\beta = 0.01$, $\rho = 2$, $\tau = 1/C$, and $\eta = 50/K$. Given the input dataset, we train the model with 600 iterations of Gibbs sampling. We took 25 samples with a gap of 20 iterations in the last 500 iterations to estimate all the hidden variables.

4   Experimental Evaluation

4.1   Datasets

Using snowball sampling, we collected the following two datasets for evaluating the **TCPI** model.

SE Dataset. This dataset is collected from a set of Twitter users who are interested in technology, and particularly in software development. To construct this dataset, we first utilized 100 most influential software developers in Twitter provided in [12] as the seed users. These are highly-followed users who actively tweet about software engineering topics, e.g., *Jeff Atwood*[3], *Jason Fried*[4], and *John Resig*[5]. We further expanded the user set by adding all users following at least five seed users. Lastly, we took all tweets posted by these users from August 1st to October 31st, 2011 to form the first dataset, called **SE** dataset.

Two-Week Dataset. The second dataset is a large corpus of tweets collected just before the 2012 US presidential election. To construct this corpus, we first manually selected a set of 56 *seed users*. These are highly-followed and politics savy Twitter users, including major US politicians, e.g., Barack Obama, Mitt Romney, and Newt Gingrich; well known political bloggers, e.g., America Blog, Red State, and Daily Kos; and political desks of US news media, e.g., CNN Politics, and Huffington Post Politics. The set of users was then expanded by adding all users following at least three seed users. Lastly, we used all the tweets posted

---

[3] http://en.wikipedia.org/wiki/Jeff_Atwood
[4] http://www.hanselman.com/blog/AboutMe.aspx
[5] http://en.wikipedia.org/wiki/John_Resig

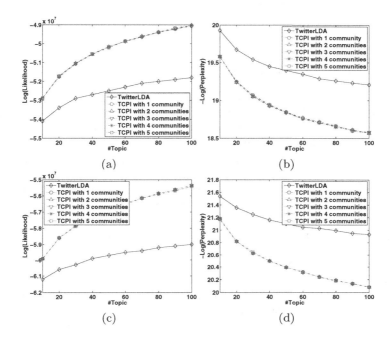

Fig. 8. Loglikelihood and Perplexity of **TwitterLDA** and **TCPI** in: ((a) and (b)) **SE**, and ((c) and (d)) **Two-Week** datasets

by these users during the two week duration from August 25th to September 7th, 2012 to form the second dataset, known as the **Two-Week** dataset.

We employed the following preprocessing steps to clean both datasets. We first removed stopwords from the tweets and filtered out tweets with less than 3 non stop-words. Next, we excluded users with less than 50 (remaining) tweets. This minimum thesholds are necessary so that, for each user, we have enough number of tweet observations for learning both influence of the user's personal interests and that of the topical communities in tweet generation.

Table 1 shows the statistics of the two datasets after the preprocessing steps. As shown in the table, the two datasets after filtering are still large, with about 200 tweets per user in **SE** dataset; and 120 tweets per user in **Two-Week** dataset. This allows us to learn the latent factors accurately.

4.2   Evaluation Metrics

To examine the ability of **TCPI** model in modeling tweet generation, we compare **TCPI** with **TwitterLDA** model. We adopt *likelihood* and *perplexity* for evaluating the two models. For each user, we randomly selected 90% of tweets of the user to form a training set, and use the remaining 10% of the tweets as the test set. We then learn the **TCPI** and **TwitterLDA** models using the training set, and using the learnt models to generate the test set. Lastly, for each model, we compute the likelihood of the training set and perplexity of the test set. The model with a higher likelihood, or lower perplexity is considered better for the task.

4.3 Performance Comparison

Figures 8 (a) and (b) show the performance of **TwitterLDA** and **TCPI** models in topic modeling on **SE** dataset. Figures 8 (c) and (d) show the performance of the models on **Two-Week** dataset. As expected, larger number of topics K gives larger likelihood and smaller perplexity, and the amount of improvement diminishes as K increases. The figures show that: (1) **TCPI** significantly outperforms **TwitterLDA** in topic modeling task; and (2) **TCPI** is robust against the number of topical communities as its performance does not significantly change as we increase the number of the communities from 1 to 5.

4.4 Background Topics and Topical Communities Analysis

We now examine the background topics and topical communities found by the **TwitterLDA** and **TCPI** models respectively. Considering both time and space complexities, and since it is not practical to expect a large number of topics falling in topical communities, we set the number of the topical communities in **TCPI** model to 3, and set the number of topics in both models to 80.

Table 2. Top words of **background topic** found in **SE** dataset by **TwitterLDA** model

life,making,video,blog,change,reading,job,home,thought,line team,power,game,business,money,friends,talking,starting,month,company

Table 3. Top topics of **topical communities** found in **SE** dataset by **TCPI** model

Community Id	Community Label	Top topics		
		Topic Id	Topic Label	Probability
0	Daily life	61	Daily stuffs	0.535
		79	Traveling	0.086
		25	Food and drinks	0.063
1	Apple's product	50	iOS	0.274
		74	Networking services	0.146
		37	iPhone and iPad	0.091
2	Software development	24	Programming	0.614
		9	Conference and meeting	0.105
		15	Operating systems	0.056

Table 2 shows the top words of the background topic found by **TwitterLDA** model in **SE** dataset, and Table 3 shows the top topics of each topical community found by **TCPI**. Note that, other than background topic, the labels of

Table 4. Top words of topics found in **SE** dataset by **TCPI** model

9	Conference and meeting	conference,meeting,team,weekend,code,session,home event,book,friends,friday,coffee,room,folks lunch,presentation,job,slides,minutes,beer
15	Operating sytems	windows,linux,mac,laptop,ubuntu,server,machine desktop,running,computer,systems,usb,ssd,lion software,#linux,apple,macbook,installing,win8,power
24	Programming	code,javascript,git,ruby,java,github,rails,data,api,server,tests php,node,python,language,blog,simple,programming,testing,files
25	Food and drinks	coffee,eating,chicken,dinner,cream,ice,lunch,beer cheese,bacon,chocolate,breakfast,recipe,delicious pizza,salad,wine,pumpkin,bread,butter
37	iPhone and iPad	iphone,apple,ipad,event,ipod,video,ios,retina,macbook,#apple,screen #iphone5,mac,battery,lightning,camera,connector,imac,nano,price
50	iOS	mac,ios,iphone,windows,chrome,apple,lion,ipad,google,screen,mountain android,text,safari,version,browser,itunes,desktop,keyboard,tweetbot
61	Daily stuffs	home,kids,house,#fb,life,coffee,dog,car,wife,room bed,thought,cat,playing,wearing,making,music,baby,friends,weekend
74	Networking services	email,facebook,google,spam,emails,page,blog,service,link,gmail password,mail,users,linkedin,api,inbox,client,links,message,user
79	Traveling	home,train,san,city,ride,bike,airport,weather,car,bus,rain weekend,francisco,traffic,london,road,minutes,heading,#fb,plane

other topics are manually assigned after examining the topics' top words (shown in Tables 4) and top tweets. For each topic, the topic's top words are the words having the highest likelihoods given the topic, and the topic's top tweets are the tweets having the lowest perplexities given the topic. The label of each topical community is also manually assigned based on examining the community's top topics. The tables show that: (i) the background topic found by **TwitterLDA** model is not sematically clear; and (ii) the topical communitiess and their extreme topics found by **TCPI** model are both semantically clear and reasonable. In **SE** dataset, other than *Daily life* community as reported in [11], it is expected that professional communities *Software Development* and *Apple's product* exist in the dataset as most of its users are working in IT industry. This agrees with the findings by Zhao *et. al.* [29] that people also use Twitter for gathering and sharing useful information relevant to their profession.

Table 5. Top words of **background topic** found in **Two-Week** dataset by **TwitterLDA** model

life,making,home,america,called,house,change,thought,video,talking line,american,money,country,job,obama,friends,fact,lost,hell

Similarly, Table 5 shows the top words of the background topic found by **TwitterLDA** model in **Two-Week** dataset, and Table 6 shows the top topics of each topical community found by **TCPI** model. Again, the topics' labels are manually assigned after examining the topics' top words (shown in Tables 7)

Table 6. Top topics of **topical communities** found in **Two-Week** dataset by **TCPI** model

Community Id	Community Label	Top topics		
		Topic Id	Topic Label	Probability
0	Daily life	1	Daily stuffs	0.622
		32	Happenings in DNC and RNC 2012	0.062
		25	Food and drinks	0.052
1	Republicans' activities	10	Republican candidates	0.210
		32	Happenings in DNC and RNC 2012	0.196
		0	Presidential candidates' speeches	0.066
2	Campaigning speeches	0	Presidential candidates' speeches	0.203
		18	Speeches at DNC 2012	0.175
		16	Goverment and people	0.108

Table 7. Top words of topics found in **Two-Week** dataset by **TCPI** model

0	Presidential candidates speeches	obama,romney,gop,media,lies,speech,party,ryan fact,#dnc2012,#tcot,convention,dems,truth republicans,facts,mitt,democrats,campaign,liberal
1	Daily stuffs	life,home,kids,class,mom,house,car,bed,god,friends,room thought,baby,weekend,friend,person,family,hair,game,dog
10	Republican candidates	romney,#gop2012,mitt,#rnc2012,speech,#rnc,ann,ryan christie,america,paul,obama,president,chris rubio,#romneyryan2012,#tcot,american,convention,condi
16	Goverment and people	america,obama,government,god,party,country,american #tcot,freedom,rights,democrats,#dnc2012,americans,gop gop,constitution,liberty,nation,war,power,states
18	Speeches at DNC 2012	obama,#dnc2012,#tcot,biden,#dnc,joe,dnc clinton,#dncin4words,speech,#p2,america,president,bill bill,romney,god,michelle,barack,chair,dems
25	Food and drinks	coffee,chicken,ice,cream,eating,dinner,cheese,beer,lunch,bacon chocolate,breakfast,pizza,wine,#dnc,salad,milk,ate,making,home
32	Hapenning in DNC and RNC 2012	#dnc2012,#gop2012,convention,#rnc2012,speech,#rnc romney,obama,rnc,#dnc,dnc,tampa,ryan gop,stage,mitt,biden,charlotte,paul,music

and top tweets; and the communities' labels are also manually assigned based on examining the communities' top topics. Also, the tables show that: (i) the background topic found by **TwitterLDA** model is not sematically clear; and (ii) the topical communities and their extreme topics found by **TCPI** model are both semantically clear and reasonable. In **Two-Week** dataset, other than *Daily life*, it is expected that political communities *Republicans' activities*, and *Campaining speeches* exist in the dataset as it was collected during a politically active period with many political events related to the American 2012 presidential election, e.g., the national conventions of both democratic (DNC 2012[6]) and republican (RNC 2012[7]) parties.

5   Conclusion

In this paper, we propose a novel topic model called **TCPI** for simultaneously modeling topical communities and users' topical interests in microblogging data. Our model differentiates users' personal interests from their topical communities while learning both the two set of latent factors at the same time. We also report experiments on two Twitter datasets showing the effectiveness of the proposed model. **TCPI** is shown to outperform TwitterLDA, another state-of-the-art topic model for modeling tweet generation.

In the future, we would like to consider the scalability of the proposed model. Possible solutions for scaling up the model are approximated and distributed implementations of Gibbs sampling procedures [17], and stale synchronous parallel implementation of variational inference procedures [7]. Moreover, it is potentially helpful to incorporate prior knowledge into the proposed model. Examples of the prior knowledge are topic indicative features [3], and groundtruth community labels for some users [25,8].

Acknowledgements. This research is supported by the Singapore National Research Foundation under its International Research Centre @ Singapore Funding Initiative and administered by the IDM Programme Office, Media Development Authority (MDA).

References

1. Airoldi, E.M., Blei, D.M., Fienberg, S.E., Xing, E.P.: Mixed membership stochastic blockmodels. J. Mach. Learn. Res. 9 (2008)
2. Balasubramanyan, R., Cohen, W.W.: Regularization of latent variable models to obtain sparsity. In: SDM 2013 (2013)
3. Balasubramanyan, R., Dalvi, B., Cohen, W.W.: From topic models to semi-supervised learning: Biasing mixed-membership models to exploit topic-indicative features in entity clustering. In: Blockeel, H., Kersting, K., Nijssen, S., Železný, F. (eds.) ECML PKDD 2013, Part II. LNCS, vol. 8189, pp. 628–642. Springer, Heidelberg (2013)
4. Blei, D.M., Ng, A.Y., Jordan, M.I.: Latent dirichlet allocation. J. Mach. Learn. Res. (2003)
5. Ding, Y.: Community detection: Topological vs. topical. Journal of Informetrics 5(4), 498–514 (2011)
6. Grabowicz, P.A., Aiello, L.M., Eguiluz, V.M., Jaimes, A.: Distinguishing topical and social groups based on common identity and bond theory. In: WSDM (2013)
7. Ho, Q., Xing, E., et al.: More effective distributed ml via a stale synchronous parallel parameter server. In: NIPS (2013)
8. Hoang, T.A., Cohen, W.W., Lim, E.P.: On modeling community behaviors and sentiments in microblogging. In: SDM 2014 (2014)

---

[6] http://en.wikipedia.org/wiki/2012_Democratic_National_Convention
[7] http://en.wikipedia.org/wiki/2012_Republican_National_Convention

9. Hong, L., Davison, B.D.: Empirical study of topic modeling in twitter. In: SOMA 2010 (2010)
10. Hong, L., Dom, B., Gurumurthy, S., Tsioutsiouliklis, K.: A time-dependent topic model for multiple text streams. In: KDD 2011 (2011)
11. Java, A., Song, X., Finin, T., Tseng, B.: Why we twitter: An analysis of a microblogging community. In: Zhang, H., Spiliopoulou, M., Mobasher, B., Giles, C.L., McCallum, A., Nasraoui, O., Srivastava, J., Yen, J. (eds.) WebKDD 2007. LNCS, vol. 5439, pp. 118–138. Springer, Heidelberg (2009)
12. Jurgen, A.: Twitter top 100 for software developers (2009), http://www.noop.nl/2009/02/twitter-top-100-for-software-developers.html
13. Kooti, F., Yang, H., Cha, M., Gummadi, P.K., Mason, W.A.: The emergence of conventions in online social networks. In: ICWSM 2012 (2012)
14. Liu, J.S.: The collapsed gibbs sampler in bayesian computations with applications to a gene regulation problem. J. Amer. Stat. Assoc. (1994)
15. Mehrotra, R., Sanner, S., Buntine, W., Xie, L.: Improving lda topic models for microblogs via tweet pooling and automatic labeling. In: SIGIR (2013)
16. Michelson, M., Macskassy, S.A.: Discovering users' topics of interest on twitter: A first look. In: AND 2010 (2010)
17. Newman, D., Asuncion, A., Smyth, P., Welling, M.: Distributed algorithms for topic models. The Journal of Machine Learning Research 10, 1801–1828 (2009)
18. Newman, M.E.J.: Modularity and community structure in networks. PNAS (2006)
19. Qiu, M., Jiang, J., Zhu, F.: It is not just what we say, but how we say them: Lda-based behavior-topic model. In: SDM 2013 (2013)
20. Ramage, D., Hall, D., Nallapati, R., Manning, C.D.: Labeled lda: A supervised topic model for credit attribution in multi-labeled corpora. In: ECML (2009)
21. Rosen-Zvi, M., Griffiths, T., Steyvers, M., Smyth, P.: The author-topic model for authors and documents. In: UAI (2004)
22. Sachan, M., Dubey, A., Srivastava, S., Xing, E.P., Hovy, E.: Spatial compactness meets topical consistency: Jointly modeling links and content for community detection. In: WSDM (2014)
23. Xie, P., Xing, E.P.: Integrating documet clustering and topic modeling. In: UAI (2013)
24. Yang, J., Leskovec, J.: Community-affiliation graph model for overlapping network community detection. In: ICDM (2012)
25. Yang, J., Leskovec, J.: Defining and evaluating network communities based on ground-truth. In: ICDM (2012)
26. Yang, J., McAuley, J., Leskovec, J.: Community detection in networks with node attributes. In: ICDM (2013)
27. Yang, J., McAuley, J., Leskovec, J.: Detecting cohesive and 2-mode communities indirected and undirected networks. In: WSDM (2014)
28. Yin, Z., Cao, L., Gu, Q., Han, J.: Latent community topic analysis: Integration of community discovery with topic modeling. ACM TIST (2012)
29. Zhao, D., Rosson, M.B.: How and why people twitter: The role that micro-blogging plays in informal communication at work. In: GROUP 2009 (2009)
30. Zhao, W.X., Jiang, J., Weng, J., He, J., Lim, E.-P., Yan, H., Li, X.: Comparing twitter and traditional media using topic models. In: Clough, P., Foley, C., Gurrin, C., Jones, G.J.F., Kraaij, W., Lee, H., Mudoch, V. (eds.) ECIR 2011. LNCS, vol. 6611, pp. 338–349. Springer, Heidelberg (2011)
31. Zhou, D., Manavoglu, E., Li, J., Giles, C.L., Zha, H.: Probabilistic models for discovering e-communities. In: WWW 2006 (2006)

Bridging Social Network Analysis and Judgment Aggregation

Silvano Colombo Tosatto and Marc van Zee

University of Luxembourg, Luxembourg, Luxembourg

Abstract. Judgment aggregation investigates the problem of how to aggregate several individuals' judgments on some logically connected propositions into a consistent collective judgment. The majority of work in judgment aggregation is devoted to studying impossibility results, but the relationship between the (social) dependencies that may exist between voters and the outcome of the voting process is traditionally not studied. In this paper, we use techniques from social network analysis to characterize the relations between the individuals participating in a judgment aggregation problem by analysing the similarity between their judgments in terms of social networks. We obtain a correspondence between a voting rule in judgment aggregation and a centrality measure from social network analysis and we motivate our claims by an empirical analysis. We also show how large social networks can be simplified by grouping individuals with the same voting behavior.

1 Introduction

Social choice theory studies the problem how to reach collective consent between a group of people in the area of economic theory. It includes among others voting theory, preference aggregation and judgment aggregation. Judgment aggregation is the most recent formal theory of social choice, which investigates how to aggregate individual judgments on logically related propositions to a group judgment on those propositions. Examples of groups that need to aggregate individual judgments are expert panels, legal courts, boards, and councils. The problem of aggregating judgments gained popularity in the last ten years, since it has been shown to be general in the sense that both voting theory and preference aggregation are subsumed by it [16]. The majority of work in judgment aggregation is devoted to studying impossibility results similar to the work in preference aggregation by Arrow [1,15], leading to the development of a large number of aggregation rules such as majority outcome, premise-based aggregation, and conclusion-based aggregation [16]. These rules are all concerned with the general problem of selecting outputs that are *consistent* or *compatible* with individual judgments [11]. However, the relation between the (social) dependencies that may exist between the voters and the outcome of the voting scenario is traditionally not studied.

Arguably, there may exist (social) relationships between the individuals that can have an influence on their individual judgments, and consequently on the

L.M. Aiello and D. McFarland (Eds.): SocInfo 2014, LNCS 8851, pp. 17–33, 2014.

aggregated outcome. For instance, a subgroup of the individuals can be close friends and therefore vote alike, or an individual in the group may be a dominant person and thus may influence the voting behavior of other voters. A representation of the social structure of a judgment aggregation problem makes it possible to identify influential voters in the entire group or in a subgroup of voters. This information can be useful for different purposes. Firstly, it can be used to determine the outcome of the voting process, simply by looking at what voters have a central position in the voting process and deriving the outcome from these voters. Secondly, it can be used to detect cartels in voting scenarios. A cartel is a formal, explicit agreement among competing firms. It is a formal organization of producers and manufacturers that agree to fix prices, marketing, and production. Finally, the social dependencies may allow one to simplify the voting problem by reducing the number of voters to the most important ones.

It does not seem obvious to extract such information from a judgment aggregation scenario, merely by relying on the tools that judgment aggregation offers. However, we believe that a possible natural solution to this problem can be provided by using techniques from social network analysis (SNA) to derive dependencies between voters. SNA views social relationships in terms of graph theory, consisting of nodes (representing individuals within the network) and ties (which represent relationships between the individuals, such as friendship, kinship, organisational position, sexual relationships, etc.) [21,4]. These networks are often depicted in a social network diagram, where nodes are represented as points and ties are represented as lines. The centrality of vertices, or the identification of which vertices are more "central" than others, is a key issue in network analysis.

In this paper, we explore the possibility to apply SNA to judgment aggregation by systematically translating a judgment aggregation problem to a social network. This social network reflects the agreement between voters derived from their judgments on the issues in the judgment aggregation problem. We analyse this network using the degree centrality measure, which is arguably the most well-known measure of node centrality from SNA. We formally prove an equivalence between the "average voter" voting rule in judgment aggregation and the degree centrality measure, showing that the social network can be used as an instrument to decide on a consistent and compatible outcome in the voting process. We motivate our claim with an experimental analysis, indicating that by varying the parameters of the centrality measure, we are able to fine-tune the outcome of the voting process. Finally, we show that large networks with many voters and few issues can be simplified significantly by clustering individuals that vote the same.

The paper is organised as follows: We start by discussing related work in Section 2. In Section 3 we introduce the basic notions of judgment aggregation and two voting rules, and we introduce basic terminology from social network analysis in Section 4. In Section 5 we show how we can systematically obtain a social network from a judgment aggregation problem using simple matrix operations. We use this method in Section 6, where we show a correspondence between a centrality measure on the graph and a voting rule in judgment aggregation. Empirical results are discussed in Section 7 and we show how to simplify the social networks in Section 8.

2   Related Work

There is substantial research in social science showing that social dependencies exist between voters and that this can have an influence on the outcome of the voting process. For instance, Gerber *et al.* [9] performed a large-scale field experiment involving several hundred thousand registered voters, demonstrating the profound importance of social pressure as an inducement to political participation. Nickerson [19] performs two field experiments showing that within households, 60% of the propensity to vote is passed onto the other member of the household. This suggests a mechanism by which civic participation norms are adopted and couples grow more similar over time. Kenny [12] uses survey responses from the 1984 South Bend study to model the relationship between political discussion partners. Again, the evidence indicates that certain types of both individually based and socially based participation are affected by those in the immediate social environment.

Possibly caused by the recent popularity of online social networks such as Facebook, Twitter, LinkIn, Pinterest, and others, most recent research combining social choice theory with social network analysis pursues in the opposite direction from ours. Social networks are taken as the starting point and one investigates to what extent fair and consistent voting can be implemented on such networks. For instance, both Salehi-Abari and Boutilier [22] as well as Boldi *et al.* [2] study how members of a social network derive utility based on both their own preferences and the satisfaction of their neighbors. Here, users can only express their preferences for one among the people they are explicitly connected with, and this preferences can be propagated transitively. Both Lerman and Galstyan [13] and Lerman and Ghosh [14] study the role of social networks in promoting content on Digg, a social news aggregator that allows users to submit links to and vote on news stories. Their results suggest that pattern of the spread of interest in a story on the network is indicative of how popular the story will become.

There is significantly less work trying to obtain social networks from social choice problems. Endriss and Grandi [5] investigate the problem of graph aggregation, where individuals do not give a judgment over alternatives, but instead provide a directed graph over a common set of vertices. Judgment aggregation reduces then to computing a single graph that best represents the information inherent in this profile of individual graphs. This is considerably different from our work, since we obtain a graph from the dependencies between voters, assuming that voters give a judgment over alternatives.

3   Judgment Aggregation

In this section we recall the framework of judgment aggregation [16,24]. The problem is formulated as *binary aggregation with integrity constraints*, which is equivalent to judgment aggregation when the individual judgments are complete and consistent [10]. We also define several voting rules that we use throughout the paper.

3.1   Basic Definitions

A judgment aggregation problem consists of a set of individuals having to aggregate their preferences over a set of issues. The preferences of each individual are expressed by saying either *yes* or *no* for each of the issues proposed.

Let $\mathcal{N} = \{1, 2, \ldots, n\}$ be a finite set of individuals, and let $\mathcal{I} = \{1, 2, \ldots, m\}$ be a finite set of *issues*. We want to model collective decision making problems where the group of individuals $\mathcal{N}$ have to jointly decide for which issues in $\mathcal{I}$ to choose "yes" and for which to choose "no". A ballot $B \in \{0, 1\}^m$ associates either 0 ("no") or 1 ("yes") with each issue in $\mathcal{I}$. We write B_j for the jth element of B. Thus, $B_j = 1$ denotes that the individual has accepted the jth issue, and $B_j = 0$ denotes that the individual has rejected it.

In general, not every possible ballot might be a *feasible* or *rational* choice. For instance, if the issues are tasks that are to be executed by a group of people, then a task constraint might mean that deciding to execute certain tasks makes it impossible to execute other tasks.

Formally, let $PS = \{p_1, \ldots, p_m\}$ be a set of propositional symbols, one for each issue $\mathcal{I}$. An *integrity constraint* is a formula $IC \in \mathcal{L}_{PS}$, where $\mathcal{L}_{ps}$ is obtained from PS by closing under the standard propositional connectives. Let $Mod(IC) \subseteq \{0, 1\}^m$ denote the set of models of IC, i.e. the set of rational ballots satisfying IC.

A *profile* is a vector of rational ballots $\mathbf{B} = (B_1, \ldots, B_n) \in Mod(IC)^n$, containing one ballot for each individual. We write B_{ij} to denote the ith individual's choice about the jth issue, i.e. the jth choice of ballot B_i. Since ballots are vectors themselves, we can consider $\mathbf{B}$ as a matrix of size $n \times m$. The *support* of a profile $\mathbf{B} = (B_1, \ldots, B_n)$ is the set of all ballots that occur at least once within $\mathbf{B}$:

$$\text{SUPP}(\mathbf{B}) = \{B_1\} \cup \ldots \cup \{B_n\}.$$

A *Voting Rule* $F : \{0, 1\}^{m \times n} \to 2^{\{0,1\}^m}$ is a function that maps each profile $\mathbf{B}$ to a set of ballots. This means that an aggregation rule can have one or multiple outcomes, also called an *irresolute voting rule*. A voting rule is called *collectively rational* when all outcomes satisfy the integrity constraints.

One of the most well-known voting rules is the (weak) *majority rule*, which accepts an issue if a weak majority accepts it:

$$Maj(\mathbf{B})_j = 1 \text{ iff } |\{i \in \mathcal{N} \mid \mathbf{B}_{ij} = 1\}| \geq \left\lceil \frac{n}{2} \right\rceil.$$

Example 1. Suppose the following judgment aggregation scenario consisting of six individuals (a, b, c, d, e, f) voting on an agenda composed of four issues (p, q, r, z). The agenda is subject to the following integrity constraint: $IC = (p \land q \land r) \Leftrightarrow z$. The majority outcome is depicted in the last row.

Issue:	p	q	r	z
a	0	1	1	0
b	1	0	0	0
c	1	1	1	1
d	1	0	0	0
e	1	0	1	0
f	0	0	1	0
Maj	1	0	1	0

The Hamming distance between two ballots $B = (B_1, \ldots, B_m)$ and $B' = (B'_1, \ldots, B'_m)$ is defined as the sum of the amount of issues on which they differ:

$$H(B, B') = |\{j \in \mathcal{I} \mid B_j \neq B'_j\}|$$

For example, $H((1, 0, 0), (1, 1, 1)) = 2$. The Hamming distance between a ballot B and a profile $\mathbf{B}$ is the sum of the Hamming distances between B and the ballots in $\mathbf{B}$:

$$\mathcal{H}(B, \mathbf{B}) = \sum_{i \in \mathcal{N}} H(B, B_i)$$

3.2   The Average Voter Rule

Endriss and Grandi [6] recently proposed the average voter rule, which reduces the space of the possible outcomes to the ballots proposed by the voters. In this way, the consistency of the outcome of the voting process is guaranteed, given that all voters vote consistently. It was later shown by Grandi and Pigozzi [11] that this rule satisfies several desirable properties.

Definition 1 (AVR). *The average voter rule (AVR) is the voting rule that selects those individual ballots that minimise the Hamming distance to the profile:*

$$AVR(\boldsymbol{B}) = \underset{B \in \text{SUPP}(\boldsymbol{B})}{\arg\min} \ \mathcal{H}(B, \boldsymbol{B})$$

4   Social Network Analysis

A social network usually is represented as a graph. The vertices are the individuals, and the edges represent the social connections. In this paper, we consider the symmetric case where social networks are represented by undirected graphs. An edge which joins a vertex to itself is called a *loop*. The number of edges that are incident to a vertex is called the *degree* of a vertex. The *neighborhood* of a vertex v is the set of all vertices adjacent to v.

We denote a weighted network (or weighted graph) with $G = (V, E, W)$ with the vertex set $V(G) = \{v_1, \ldots, v_n\}$, edge set E, and weight matrix W, where each edge $e = (v_i, v_j)$ is labeled with a weight w_{ij}. We assume that if two vertices are not connected, then there exists an edge of weight 0 connecting them. Since we only consider undirected networks, $w_{ij} = w_{ji}$. We define the *sum-weight* s_i of a vertex v_i with $s_i = \sum_{j=1}^{n} w_{ij} = \sum_{u \in N(v_i)} w_{v_i u}$, where $N(v_i)$ is the neighborhood of v_i. We denote the degree k_i of a vertex v_i with $k_i = |N(v_i)|$, i.e. k_i denotes the number of neighbors of v_i.

The centrality of vertices, identifying which vertices are more "central" than others, has been a key issue in network analysis. Freeman [8] originally formalized three different measures of vertex centrality: degree, closeness, and betweenness. In this paper, we will only consider the degree centrality. Degree is the number of vertices that a focal vertex is connected to, and measures the local involvement of the vertex in the network. This measure is originally formalised for binary graphs [8], but we will consider recent proposal [20] that uses a tuning parameter α to control the relative importance of number of edges compared to the weights on the edges.

The degree centrality measure is defined as the product of the number of vertices that a focal vertex is connected to, and the average weight to these vertices adjusted by the tuning parameter. The degree centrality for a vertex i is computed as follows:

$$C_D^{W\alpha}(i) = k_i \times \left(\frac{s_i}{k_i}\right)^{\alpha} = k_i^{(1-\alpha)} \times s_i^{\alpha} \tag{1}$$

where W is the weight matrix of graph, α is a positive tuning parameter, k_i is the size of the neighborhood of vertex i and s_i the sum of the weights of the incident edges. If α is between 0 and 1, then having a high degree is favorable over weights, whereas if it is set above 1, a low degree is favorable over weights. In Section 6 we elaborate on different levels of α for degree centrality.

5   Towards a Social Network

In this section we will bridge the two problem domains that we introduced above using a technique introduced in social theory by Breiger [3]. This technique is originally used to analyse membership of people to groups, however we use it to represent agreement between voters.

5.1   Matrix Translation

We use the following transformations to obtain a social network from a voting profile.

Definition 2 (Similarity matrix). *Given a profile matrix B. The* similarity *matrix $\mathcal{B}$ is obtained from B as follows:*

$$\mathcal{B}_{ij} = \begin{cases} 1 & \text{if } B_{ij} = 1 \\ -1 & \text{if } B_{ij} = 0 \end{cases}$$

Definition 3 (Voter-to-voter matrix). *Given a similarity matrix $\mathcal{B}$ of size $n \times m$ and $V^* = \mathcal{B}(\mathcal{B}^T)$, where multiplication is ordinary (inner product) matrix multiplication. The* voter-to-voter matrix V *of V^* is constructed as follows:*

$$V_{ij} = \frac{V_{ij}^* + m}{2}$$

The following theorem states the main result of this section, showing that the voter-to-voter matrix counts the equal elements between each two rows of the original profile matrix.

Theorem 1. *Let $\mathbf{B}$ be a profile matrix of size $n \times m$, $\mathcal{B}$ the similarity matrix of $\mathbf{B}$, and V the corresponding voter-to-voter matrix of $\mathcal{B}$. V_{ij} contains the amount of equal elements in row i and j of $\mathbf{B}$, i.e.:*

$$V_{ij} = |\{\mathbf{B}_{ik} \mid \mathbf{B}_{ik} = \mathbf{B}_{jk}, 1 \leq k \leq m\}|$$

Thus, V_{ij} denotes the number of times that both voters i and j voted "yes" or they both voted "no" for the same issue.

Example 2 (Continued). We can translate the matrix $\mathbf{B}$ of Example 1 that corresponds to this voting profile to a similarity matrix (Figure 1a). Next, we calculate V^* and obtain the the voter-to-voter matrix V after normalising the result (Figure 1b).

	p	q	r	z
a	-1	1	1	-1
b	1	-1	-1	-1
c	1	1	1	1
d	1	-1	-1	-1
e	1	-1	1	-1
f	-1	-1	1	-1

(a) Voting Profile **B**

	a	b	c	d	e	f
a	4	1	2	1	2	3
b	1	4	1	4	3	2
c	2	1	4	1	2	1
d	1	4	1	4	3	2
e	2	3	2	3	4	3
f	3	2	1	2	3	4

(b) Voter-to-voter matrix (V)

Fig. 1. Transforming the voting profile to a voter-to-voter matrix

The voter-to-voter matrix is *symmetric* with respect to its main diagonal: If some voter i agrees with a voter j on some issues, then j agrees with i on the same issues as well. This implies *reflexivity*: a voter always agrees with itself over every issue, and similarly for any issue. Therefore, the main diagonal of the voter matrix is always equal to the number of issues.

5.2   Relational Graphs

The voter-to-voter matrix V can be represented as an undirected, weighted graph. In such a graph, a voter is represented by a node, and an edge represents the agreement between two voters. Formally, an edge (i, j) connects two vertices i and j if the matrix entry V_{ij} has a value larger than 0. We denote the obtained graphs with $G_V = (V_V, E_V, W_V)$. Notice that the matrix V is equivalent to the weight matrix of the corresponding graph, i.e. $V = W_V$. We call the graph G_V as the *voter graph*.

Example 3 (Continued). Figure 2 shows the voting graph resulting from the voter-to-voter matrix depicted in Figure 1b. We can see that the strongest connection is between the individuals b and d, representing the fact that their ballots are equivalent. Differently, c can be considered an outlier due to its weak connections with the other individuals. Note that for the sake of readability, the edges with a weight of 1 have not been labeled in Figure 2 and reflexive edges have been omitted.

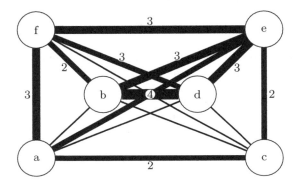

Fig. 2. Voter graph (G_V)

The voter-to-voter graph expresses the agreement between the individuals through the weighted edges. In the following section we show that the individuals in the voter graph that are most central according to the degree centrality measure corresponds to the voters that are selected by the average voter rule in the judgment aggregation profile.

6   Theoretical Analysis

We start out with a straightforward equivalence between the Hamming distance between two voters and the edge that connects the two voters in the corresponding voting graph.

Lemma 1. *The Hamming distance between two ballots B_i and B_j is equal to $m - w_{ij}$ in the corresponding voter graph G_V, i.e. $H(B_i, B_j) = m - w_{ij}$.*

We use this lemma to obtain an equivalence between the Hamming distance to a profile and the total weight of the corresponding node in the voter graph.

Lemma 2. *The Hamming distance between a ballot B_i and a profile $\boldsymbol{B}$ is equal to $mn - s_i$, where s_i is the sum of the weights of the incident edges of vertex i in the voter graph constructed from $\boldsymbol{B}$:*

$$H(B_i, \boldsymbol{B}) = mn - s_i$$

Example 4 (Continued). In Example 1, we have $H(a, b) = 3$ and $H(a, \boldsymbol{B}) = 11$. In the corresponding graph in Figure 2 we have that $w_{ab} = 1$ and thus $m - w_{ab} = 4 - 1 = 3$, which corresponds to the Hamming distance between a and b. Moreover, $s_a = 13$ (including the reflexive weight of 4), so $mn - s_a = 24 - 13 = 11$, which corresponds to the Hamming distance between a and the profile $\boldsymbol{B}$.

Since the average voter rule selects the voter that minimises the distance with the profile, we can obtain the following equivalence:

Lemma 3. *The average voter rule (AVR) (Definition 1) selects the voters corresponding to the maximum total weight vertices in the voter graph, i.e.:*

$$AVR(\boldsymbol{B}) = \underset{i \in V_V}{\operatorname{argmax}}\, s_i.$$

Next, we obtain that the average voter rule corresponds to the node with the highest degree centrality when the tuning parameter $\alpha = 1$:

Theorem 2. *The AVR selects those individual ballots that have the maximal degree centrality value when $\alpha = 1$. Suppose $\alpha = 1$:*

$$AVR(\boldsymbol{B}) = \underset{i \in V_V}{\operatorname{argmax}}\, C_D^{W\alpha}(i)$$

Proof. Follows directly from Eq.(1) and Lemma 3

Example 5. Consider the voting profile in Figure 3a, where a set of four voters have to decide on five issues. We can see in the bottom part of the table that the average voter rule corresponds to the set of individuals $\{a, c, d\}$. The voter graph of this voting profile is shown in Figure 3b. Recall that the degree centrality score for the nodes when $\alpha = 1$ can be calculated by summing up the weights of all incident edges. Thus, the value of node a, c, and d are all 6 while the value of node b is 4. Therefore, the set of voters selected using the degree centrality measure is $\{a, c, d\}$ as well, which is in line with Theorem 2.

7   Empirical Analysis

In this section we analyse the effect of varying the tuning parameter α in the degree centrality measure by comparing the outcomes with those of the average voter rule, taking the majority voting rule as our base measure. We first provide an intuitive discussion on the effect of varying the tuning parameter, followed by an empirical analysis.

	1	2	3	4	5
a	0	1	1	1	1
b	1	0	0	0	0
c	0	1	1	0	0
d	0	0	0	1	1
AVR:	0	1	1	1	1
	0	1	1	0	0
	0	0	0	1	1

(a) Profile

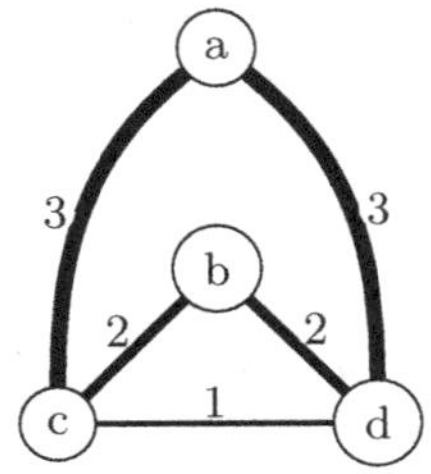

(b) Voter graph

Fig. 3. Example judgment aggregation profile with voter-to-voter graph

7.1 Varying the Tuning Parameter

Reconsider the voting scenario together with the majority outcomes and the average voter outcomes of Figure 3. The outcome of the degree centrality for varying α are depicted in Figure 4. As can be seen from the table, for $\alpha = 1$, the degree centrality measure corresponds to the s_i measure, which measures the sum of the weights of the edges connected to that node. Therefore the nodes a, c and d are all chosen as the average voter because of their greater degree centrality measure. When $\alpha < 1$ the amount of edges play are larger role and only c and d are chosen as the most central because of their three connected edges against the two of nodes a and b. Contrast this with $\alpha > 1$, when the weight of the edges play the prominent role in deciding the most central node. In this case a is picked as the most central node by having the edges with the largest weights connected to it. This analysis suggests that in some cases, by using different values for the tuning parameter α to compute the most central node in a graph, it is possible to obtain a more fine-grained voting rule than the result of the average voter rule.

The outcomes obtained using the degree centrality measure can be compared with the vector of "average votes" $(\frac{1}{4}, \frac{2}{4}, \frac{2}{4}, \frac{2}{4}, \frac{2}{4})$, showing for each issue the proportion of voters who chose 1 rather than 0. We can see that only the first issue is uncontroversial, while the no unique decision on the other issues is possible. Having multiple available outcomes is not uncommon for voting rules such as the majority rule and AVR. However, in these cases fine tuning the α parameter may lead to more resolute outcomes by exploiting the structure of the voter graph.

7.2 Experimental Setup

The setup of the empirical analysis performed[1] consists of a judgment aggregation problem with 100 voters and 4 issues, with no integrity constraints. We have chosen for relatively many voters because the degree centrality measure is based

---

[1] The experiment has been coded in Java and can be found on the web:
http://icr.uni.lu/marc/code/socinfo2014/src.zip

Vertex	s_i	$C_D^{W\alpha}$ when $\alpha =$			
		0	0.5	1	1.5
a	6	2	3.46	6	10.39
b	4	2	2.83	4	5.66
c	6	3	4.24	6	8.48
d	6	3	4.24	6	8.48

Fig. 4. Degree centrality scores when different values of α are used

on graph theory whereby these measures are more effective on large graphs due to the more dependencies and similarities between the individuals. We leave out the integrity constraints since logical constraints on the issues are not the focus of our work. The votes are generated pseudo-randomly such that all votes are complete, meaning that each voters votes either "yes" or "no" for each issue.

In order to compare the different measures we use the majority rule as the *base measure*. The majority rule is generally considered to be the most well-known voting rule, and is most likely also one of the most used rules. We compare the outcome of the average voter rule and the degree centrality measure for different values of α with the base measure by computing the *Hamming Distance*.

The experiment is reiterated 5000 times for each value of the tuning parameter α. If a measure produces multiple outcomes, we measure the distance to the base measure for each result. All these distances are stored in a list L_M for each measure M. For each value of the tuning parameter we use L_M to compute the mean, the standard deviation σ and the average number of outcomes per benchmark O_{avg}, i.e. $O_{avg} = \frac{|L_M|}{5000}$ for the measure M. The value O_{avg} can be seen as a measure for resoluteness: The closer this number is to 1, the more resolute the voting rule is, which means that the number of outcomes is effectively smaller.

7.3   Results and Analysis

Figure 5 shows the results of the experiment. From the figure it can be seen that, as shown in Section 6, the average voter rule corresponds to the degree centrality measure when the tuning parameter $\alpha = 1$. When the value of α increases from 1 to 3, the average distance to the base measure slowly increases, meaning that the result of the degree centrality measure is further away from the majority based rule. On the other hand, the average number of outcomes also decreases, which seems to suggest that while the voting rule becomes more resolute (i.e. results in less outcomes) when α increases, it also becomes less precise. The results for $\alpha = 0$ are somewhat surprising, since the average distance to the base measure is very high compared to the average distance of the average voter rule, and the average number of outcomes is very large as well. A possible explanation for this deviation may have to do with the density of the graph. When α is 0, the weights on the edges is completely disregarded and the degree centrality value of a node is solely determined by the number of other nodes connected to it (see Equation 1). The networks that we obtain are usually rather dense, so it seems that selecting an outcome merely based on the number of ties is not precise enough, which might explain the large number of outcomes for $\alpha = 0$. For α values smaller than 0 or larger than 3, the results remained more or less constant.

Voting rule	α	mean	σ	O_{avg}
MRV	-	0.06	0.24	1.9
Degree	0.0	1.16	0.88	7.2
	0.5	0.07	0.26	1.88
	1.0	0.06	0.24	1.9
	1.5	0.07	0.24	1.83
	2.0	0.07	0.26	1.83
	2.5	0.08	0.28	1.81
	3.0	0.10	0.33	1.77

Fig. 5. Benchmarking results showing Hamming distances from majority based rule

8   Simplifying the Social Network

As we mentioned previously, it can be the case that a judgment aggregation problem features a big set of voters having to decide over a small set of issues. In this case it is inevitable that many of the voters involved in the voting process will have identical votes. Consider for instance a group of 100 voters that has to decide over 4 issues. The number of possible voting profiles (assuming no integrity constraints) is $2^4 = 16$, meaning that there will be at least 84 non-identical voters, so at most 16% of the voting profiles are unique.

A group of individuals voting the same way is represented in the voter graph as a strongly connected component of the graph where each of the connections among the nodes in the component has weight equal to the amount of issues in the voting scenario. In addition to the connections among the tightly connected components, the nodes are also connected to other nodes using edges with variable weights depending on the amount of agreement between the voters represented by the nodes as it is shown in Figure 6.

By having a high number of nodes in the graph it is necessary to calculate the degree centrality measure of each one of them in order to decide the most central one(s). Moreover the representation of the voter graph would be cluttered by all the edges being the graph almost completely connected. Additionally, when considering the most central nodes in such graphs, all nodes belonging to the same strongly connected component have the same degree centrality value, hence if one of them is the most central, then each of them is.

Both problems, the cluttered graph and redundant calculations of the degree centrality, can be solved by reducing the amount of node represented in the graph itself. Because the nodes that belong to the same strongly connected components have the same properties in term of centrality, we can represent each strongly connected component as a single node in the graph and connect it

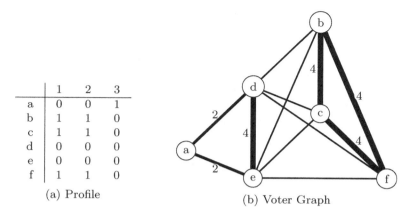

	1	2	3
a	0	0	1
b	1	1	0
c	1	1	0
d	0	0	0
e	0	0	0
f	1	1	0

(a) Profile   (b) Voter Graph

Fig. 6. Non-Simplified Translation

to the other strongly connected components (also represented by a single node) using edges weighted according to the agreement. To keep track of the size of the strongly connected components reduced to nodes, a weight equivalent to the cardinality of the component is associated to the node. The simplified voting graph of Figure 6b is shown in Figure 7.

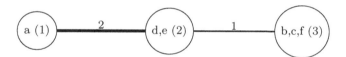

Fig. 7. Simplified Voter Graph

The degree centrality on the simplified graph can be calculated for each of the nodes using the following equations, which produces a result equivalent to the one that would have been obtained by calculating it on a node belonging to the strongly connected component in the non simplified graph. The two equations allow to compute in the simplified graph the size of the neighbourhood k_i and the size of sum of the weights of the connected edges s_i.

$$s_n = i \cdot (c_n - 1) + \sum_{(n,m) \in E} w_{nm} \cdot c_m \qquad (2)$$

$$k_n = (c_n - 1) + \sum_{(n,m) \in E} c_m \qquad (3)$$

Where c_i is the size of the strongly connected component to which node i belongs, i is the number of issues, and w_{nm} is the weight of the edge between n and another node m in the voter graph.

9   Conclusions and Future Work

In the present paper we show that deciding the average voter in a judgment aggregation problem corresponds to selecting the most central voter in a social network where the strength of the ties in the network follows from similar voting behavior of two individuals in the judgment aggregation problem.

To the best of our knowledge, this is the first attempt to correlate the two areas by showing that by remodelling the problems, classic techniques as the centrality measure used by Breiger [3] to analyse people membership to groups, is comparable to use the average voter rule, proposed by Grandi and Pigozzi [11], to solve a judgment aggregation problem.

The connection between the two fields shown in this work hints that some of the techniques used in one of the areas could be indeed adapted and reused in the other to solve some of the problems. As we show in our empirical analysis, by varying the tuning parameter α used to compute the centrality measure, the results obtained change. As discussed the parameter α switches the emphasis between the weights and the number of edges connecting the nodes, hence using a different tuning parameter than $\alpha = 1$ already corresponds in some sense to a different voting rule, however whether these new rules can be useful is still to be decided.

Additionally, we propose a way to simplify a tightly connected graph where some strongly connected components are present. By collapsing the strongly connected components in a single node, we avoid to represent a cluttered and unreadable graph in addition to have to calculate the degree centrality measures for the collapsed nodes instead for the whole strongly connected component. For future work, we would like to compare this method against other graph sparsification methods such as [23,7,18,17] to find out whether our approach can be optimized or extended. For instance, a more general treatment of this network simplification technique might refer to community detection, which is not restricted to cliques with the same weight but can define groups of nodes in other ways.[2] Reducing the number of ties will also be useful because SNA methods are often conceived with sparse graphs in mind, while our approach often produces very dense graphs. In addition, the impact of the tuning parameter α seems to be related to these missing edges only and requires future study.

Lastly, consistency is one of the main objects of study in judgment aggregation. The fact that the proposed framework does not consider constraints seems to represent a significant limitation. In particular it does not seem obvious how the connection between traditional social network analysis (SNA) measures and voting rules can be maintained. We leave this to future studies.

Acknowledgments. Silvano Colombo Tosatto and Marc van Zee are supported by the National Research Fund, Luxembourg.

---

[2] This was suggested by an anonymous reviewer.

References

1. Arrow, K.J.: Social Choice and Individual Values. John Wiley and Sons, New York (1951)
2. Boldi, P., Bonchi, F., Castillo, C., Vigna, S.: Voting in social networks. In: Cheung, D.W.-L., Song, I.-Y., Chu, W.W., Hu, X., Lin, J.J. (eds.) CIKM, pp. 777–786. ACM (2009)
3. Breiger, R.L.: The duality of persons and groups. Social forces (1974)
4. D'Andrea, A., Ferri, F., Grifoni, P.: An overview of methods for virtual social networks analysis. In: Computational Social Network Analysis. Computer Communications and Networks, ch. 1, pp. 3–25. Springer, London (2010)
5. Endriss, U., Grandi, U.: Graph aggregation. In: Proceedings of the 4th International Workshop on Computational Social Choice (COMSOC-2012) (September 2012)
6. Endriss, U., Grandi, U.: Binary aggregation by selection of the most representative voter. In: Proceedings of the 7th Multidisciplinary Workshop on Advances in Preference Handling (August 2013)
7. Foti, N.J., Hughes, J.M., Rockmore, D.N.: Nonparametric sparsification of complex multiscale networks. PloS one 6(2), e16431 (2011)
8. Freeman, L.C.: Centrality in social networks conceptual clarification. Social Networks, 215 (1978)
9. Gerber, A.S., Green, D.P., Larimer, C.W.: Social pressure and voter turnout: Evidence from a large-scale field experiment. American Political Science Review 102, 33–48 (2008)
10. Grandi, U., Endriss, U.: Binary aggregation with integrity constraints. In: Proceedings of the 22nd International Joint Conference on Artificial Intelligence (July 2011)
11. Grandi, U., Pigozzi, G.: On compatible multi-issue group decisions. In: Proceedings of the 10th Conference on Logic and the Foundations of Game and Decision Theory (2012)
12. Kenny, C.B.: Political participation and effects from the social environment. American Journal of Political Science, 259–267 (1992)
13. Lerman, K., Galstyan, A.: Analysis of social voting patterns on digg. In: Proceedings of the First Workshop on Online Social Networks, WOSN 2008, pp. 7–12. ACM, New York (2008)
14. Lerman, K., Ghosh, R.: Information contagion: an empirical study of the spread of news on digg and twitter social networks. CoRR, abs/1003.2664 (2010)
15. List, C., Puppe, C.: Judgment aggregation: A survey. Handbook of Rational and Social Choice (2009)
16. List, C., Polak, B.: Introduction to judgment aggregation. Journal of Economic Theory 145(2), 441–466 (2010)
17. Macdonald, P.J., Almaas, E., Barabási, A.-L.: Minimum spanning trees of weighted scale-free networks. EPL (Europhysics Letters) 72(2), 308 (2005)
18. Mathioudakis, M., Bonchi, F., Castillo, C., Gionis, A., Ukkonen, A.: Sparsification of influence networks. In: Proceedings of the 17th ACM SIGKDD International Conference on Knowledge Discovery and Data Mining, pp. 529–537. ACM (2011)
19. David, W.: Nickerson. Is voting contagious? evidence from two field experiments. American Political Science Review 102(01), 49–57 (2008)
20. Opsahl, T., Agneessens, F., Skvoretz, J.: Node centrality in weighted networks: Generalizing degree and shortest paths. Social Networks 32(3), 245–251 (2010)
21. Pinheiro, C.A.R.: Social Network Analysis in Telecommunications. Wiley and SAS Business Series. Wiley and SAS Business Seriee (2011)

22. Salehi-Abari, A., Boutilier, C.: Empathetic social choice on social networks. In: Fourth International Workshop on Computational Social Choice (2012)
23. Ángeles Serrano, M., Boguñá, M., Vespignani, A.: Extracting the multiscale backbone of complex weighted networks. Proceedings of the National Academy of Sciences 106(16), 6483–6488 (2009)
24. Slavkovik, M.: Judgment aggregation for multiagent systems. PhD thesis, Univerity of Luxembourg (2012)

A   Appendix: Proofs

Theorem 1. *Let B be a profile matrix of size $n \times m$, $\mathcal{B}$ the similarity matrix of B, and O the corresponding normalised matrix of $\mathcal{B}$. O_{ij} contains the amount of equal elements in row i and j of B, i.e.:*

$$O_{ij} = |\{\boldsymbol{B}_{ik} \mid \boldsymbol{B}_{ik} = \boldsymbol{B}_{jk}, 1 \leq k \leq m\}|$$

Proof. We prove this theorem directly.

1. Suppose arbitrary rows $\mathbf{B}_i, \mathbf{B}_j$ of some profile matrix $\mathbf{B}$, where $y = |\{\mathbf{B}_{ik} \mid \mathbf{B}_{ik} = \mathbf{B}_{jk}, 1 \leq k \leq m\}|$ and $x = m - y$.
2. From 1., it follows that y is the amount of equal elements between rows i and j in $\mathbf{B}$, and x is the amount of elements that are unequal.
3. Let $\mathcal{B}$ the similarity matrix of $\mathbf{B}$ and O the normalized matrix of $A = \mathcal{B}(\mathcal{B}^T)$ according to Definition 3.
4. From 3. and the definition of inner product multiplication, it follows that each cell of the matrix A is calculated as follows: $A_{ij} = \sum_{k=1}^{m} \mathcal{B}_{ik}\mathcal{B}_{jk}$.
5. From Definition 2 it follows that if $\mathcal{B}_{ik} = \mathcal{B}_{jk}$, then $\mathcal{B}_{ik}\mathcal{B}_{jk} = 1$, and otherwise $\mathcal{B}_{ik}\mathcal{B}_{jk} = -1$.
6. From 4. and 5., it follows that $\sum_{k=1}^{m} \mathcal{B}_{ik}\mathcal{B}_{jk} = y - x$. Therefore $A_{ij} = y - x$.
7. From 6. and Definition 3, it follows that $O_{ij} = \frac{A_{ij}+m}{2} = \frac{y-x+m}{2} = \frac{y-x+x+y}{2} = y$.
8. From 2. and 7., it follows that O_{ij} is the amount of equal elements between rows i and j in $\mathbf{B}$.

Lemma 1. *The Hamming distance between two ballots B_i and B_j is equal to $m - w_{ij}$ in the corresponding voter graph G_V, i.e. $H(B_i, B_j) = m - w_{ij}$.*

Proof. We prove this lemma directly.

1. Suppose two ballots B_i and B_j containing m issues, and y to be the amount of issues on which the voters i and j agree.
2. From 1. and Theorem 1 it follows that the voter-to-voter normalised matrix V, constructed from a profile $\mathbf{B}$ containing B_i and B_j, has $V_{ij} = y$.
3. From 2. and the construction of the voter-to-voter matrix, it follows that G_V is the voter graph constructed from V and the weight of the edge between the vertices i and j in G_V, written w_{ij}, is y.
4. From 3. and Hamming distance definition, it follows that the Hamming distance $H(B_i, B_j) = m - y$,
5. From 4. and 2., it follows that $H(B_i, B_j) = m - w_{ij}$.

Lemma 2. *The Hamming distance between a ballot B_i and a profile $\boldsymbol{B}$ is equal to $mn - s_i$, where s_i is the sum of the weights of the incident edges of vertex i in the voter graph constructed from $\boldsymbol{B}$:*

$$H(B_i, \boldsymbol{B}) = mn - s_i$$

Proof. Suppose some profile $\boldsymbol{B}$, a ballot $B_i \in \boldsymbol{B}$ and a voter graph G_V constructed from $\boldsymbol{B}$. The Hamming distance between B_i and $\boldsymbol{B}$ is

$$\sum_{j \in \mathcal{N}} H(B_i, B_j)$$

$$= \sum_{j \in \mathcal{N}} m - w_{ij} \qquad \text{(Lemma 1)}$$

$$= mn - \sum_{j \in \mathcal{N}} w_{ij}$$

$$= mn - s_i$$

Lemma 3. *The average voter rule AVR (Definition 1) selects the voters corresponding to the maximum total weight vertices in the voter graph, i.e.:*

$$AVR(\boldsymbol{B}) = \operatorname*{argmax}_{i \in V_V} s_i.$$

Proof.

$$AVR(\mathbf{B}) = \operatorname*{argmin}_{B \in \textsc{Supp}(B)} \mathcal{H}(B, \mathbf{B}) \qquad \text{(Definition 1)}$$

$$= \operatorname*{argmin}_{i \in V_V}(mn - s_i) \qquad \text{(Lemma 2)}$$

$$= \operatorname*{argmax}_{i \in V_V} s_i$$

Friend Grouping Algorithms for Online Social Networks: Preference, Bias, and Implications

Motahhare Eslami, Amirhossein Aleyasen, Roshanak Zilouchian Moghaddam,
and Karrie Karahalios

University of Illinois at Urbana-Champaing,
Computer Science Department,
Urbana, IL, US
{eslamim2,aleyase2,rzilouc2,kkarahal}@illinois.edu

Abstract. Managing friendship relationships in social media is challenging due to the growing number of people in online social networks (OSNs). To deal with this challenge, OSNs' users may rely on manually grouping friends with personally meaningful labels. However, manual grouping can become burdensome when users have to create multiple groups for various purposes such as privacy control, selective sharing, and filtering of content. More recently, recommendation-based grouping tools such as Facebook smart lists have been proposed to address this concern. In these tools, users must verify every single friend suggestion. This can hinder users' adoption when creating large content sharing groups. In this paper, we proposed an automated friend grouping tool that applies three clustering algorithms on a Facebook friendship network to create groups of friends. Our goal was to uncover which algorithms were better suited for social network groupings and how these algorithms could be integrated into a grouping interface. In a series of semi-structured interviews, we asked people to evaluate and modify the groupings created by each algorithm in our interface. We observed an overwhelming consensus among the participants in preferring this automated grouping approach to existing recommendation-based techniques such as Facebook smart lists. We also discovered that the automation created a significant bias in the final modified groups. Finally, we found that existing group scoring metrics do not translate well to OSN groupings–new metrics are needed. Based on these findings, we conclude with several design recommendations to improve automated friend grouping approaches in OSNs.

Keywords: Automated Grouping, Clustering Algorithms, Online Social Networks.

1   Introduction

Mailing lists, chat groups, Facebook lists, and Google+ circles are a few examples of tools that facilitate group creation in social media. We create groups to help us manage large amounts of information, in this case people. By creating a mailing list for an alumni group, we no longer need to memorize a long list of names. Instead, we can recall the group name and use it for exchanging messages [9]. In the context of OSNs, in 2007, Facebook introduced friend lists, manually created lists of Facebook friends,

L.M. Aiello and D. McFarland (Eds.): SocInfo 2014, LNCS 8851, pp. 34–49, 2014.

for the purpose of selectively sharing and reading content [24]. Twitter introduced lists in late 2009 for filtering content from one's network [25]. In 2011, Google+ introduced circles that enable selective sharing and filtering of posts on the site. Recent studies have emphasized the desire and feasibility of grouping for privacy control, sharing, and filtering [14,16,15,27]. These studies found that people desired groupings or clusters of members in their community. However, due to the high cost of creating groups manually, the majority of manual group creation mechanisms remained underused. A case in point was the 2010 Facebook announcement that only about 5% of Facebook users had created at least one Facebook list [7].

Given the significant burden of manual grouping, later work in OSN group creation proposed automating group creation while allowing users to modify the created groups [14,16]. Following this philosophy, in 2011, Facebook introduced smart lists. Smart lists differ from the original Facebook lists in that they use a recommender system to automatically assign friends to different groups. Example groups include close friends, acquaintances, family, and others [23]. Similarly, recommendation-based tools such as FeedMe [4] and ReGroup [2] suggest recipients for a post based on prior sharing patterns and the content. Such automated recommendation-based techniques can be helpful in social media systems such as email. However, when applied to large, public OSNs such as Facebook, Google+ and Twitter, these techniques put a relatively high burden on users to verify friend suggestions–for every contact individually. If one user sends ten messages on an OSN, this requires verifying all of the recipients for all ten messages.

But automating group creation and allowing user modification need not to be limited to recommender systems. One can utilize clustering algorithms to create populated groups from the onset, and then allow for personal curation. While the feasibility of structural network clustering for group creation in social networks has been investigated before [14], less is known about the benefits and drawbacks of using various automated clustering algorithms for grouping people within a social media interface. This work is a first step in that direction.

In this paper, we present a grouping tool that automatically creates groups within Facebook using three different clustering algorithms: Markov Clustering, OSLOM, and Louvain. The interface then enables the users to modify the groupings as needed. To verify the usefulness of our tool and to compare the effectiveness of the three algorithms, we evaluated our tool using both human perception and traditional clustering evaluation metrics. The following summarizes our three major findings:

– We found that users preferred automated groupings with the proposed graphical tool over existing manual or recommendation-based grouping tools such as Facebook smart lists. In addition, two of the three clustering algorithms we evaluated (Markov Clustering and Louvain), performed significantly well in terms of human satisfaction and traditional clustering evaluation metrics. These algorithms are appropriate candidates for automated friend grouping applications.

– Comparing the final groupings from different algorithms created by each participant, we found a significant difference between these groupings (14%). This relatively high difference illustrates a bias resulting from the automation in users' final groupings. We argue this bias arises primarily from (1) being influenced by the

algorithmic groupings, (2) the existing hierarchical structure in social relationships, (3) having friends with multiple roles and (4) the user's uncertainty when grouping.
 – We explored group composition before and after modification based on two validated and efficient metrics that assess the quality of groups in the absence of ground truth: *Conductance* and *Triad-Participation Ratio (TPR)*. We found four categories of groups that did not fit the traditional definition of a group assumed by these metrics. We posit that such groups which exist in social network sites such as Facebook, therefore, require different group quality assessment metrics.

In the following section, we begin by reviewing previous studies on friend grouping in OSNs. Then, we introduce group detection in networks and the three clustering algorithms we used to build our automated friend grouping tool. After explaining our mixed-methods study, we discuss the results of our study using both quantitative and qualitative evaluations. We conclude by suggesting future directions for friend grouping algorithms and interfaces.

2   Literature Review

Selective sharing, filtering of content and privacy control are cited as major motivators for the creation of groups on OSNs. Early work exploring group creation focused primarily on privacy control interfaces [14,16]. In this domain, manual creation and annotation of groups was costly in terms of time and frustration due to unintuitive interfaces. This approach resulted in a lack of use of personalized, curated privacy settings [13]. While these studies emphasize privacy, the implications extend to information filtering and selective sharing [15]. Studies on group creation demonstrate that people are not willing to use current grouping techniques in OSNs as they were intended. For example, a study on Facebook lists at 2010 showed that only 20% of participants' friends were included in Facebook lists and none of the participants used these lists for controlling privacy [14]. In a related study, Kelly et al. asserted that participants using Facebook lists to create groups included few friends [16]. A 2012 study of Google+ notes that although users perceive grouping friends on OSNs positively, Google+ circles were only moderately used to selectively post to groups and filter incoming content [27].

Jones and O'Neill's suggest that existing list and grouping tools have not met expectations [14]. They conducted a study asking people to create groups of their Facebook friends to apply group privacy settings. They discovered that organizing contacts into groups required too much time and effort; therefore, users were unwilling to group in this manner. Similarly, Kelley et al. [16] conducted a study asking users to apply four manual strategies (card sorting, grid tagging, file hierarchy and Facebook friend lists) to create groups in Facebook. They suggested that assistance through automation in creating and modifying friend groups could be enormously helpful for OSNs users.

In 2010, FeedMe [4], a content sharing web plug-in for Google Reader, has been proposed to recommend friends who might be interested in receiving a message about a topic. In this vein, in 2011, Facebook launched smart lists, human assisted lists through automation. Example lists include close friends and family groups. The interface includes a recommendation system whereby additional friends are suggested for given groups [24]. Similarly, Katango, a start-up now acquired by Google+ [18], launched a

Facebook mobile application to automatically sort friends in groups with minimal user assistance [17]. Subsequently, Amershi et al. [2] presented ReGroup, an interactive machine learning system that suggests members for the groups. In the context of private messaging systems, SocialFlows [21], an email-based application created friend groups based on the history of email communication.

Many of the mentioned automated grouping techniques employ clustering or group detection algorithms to discover friend groups in OSNs. In spite of many existing clustering algorithms, there is no gold standard for grouping members in social networks [10]. A main reason is the lack of a '*ground-truth*' or gold standard template for a group. Most current evaluation metrics for clustering algorithms rely on a pre-existing ground-truth for comparison to a derived group. While some clustering algorithms such as Markov Clustering perform very well analyzing protein-protein interaction network [6], finding meaningful relationships for grouping in social networks is not straightforward. With dynamically changing relationships and networks in social media, it is not clear a single ground truth exist at any point in time.

Despite this, researchers approximate ground-truth to explore the nature of groups. Jones and O'Neill [14] applied a clustering algorithm on Facebook. They then used manually created groups as ground-truth for comparison with the automated grouping results. This approach assumes users know and can identify real groups within their structural social networks. In a similar vein, a few studies started to collect the ground-truth data from different social networks by asking people to label their groups [22,28]. None of these studies, however, evaluate the effect of automating group detection for grouping friends by OSNs users. Rather, they collect the ground-truth data by asking people to group their friends manually and use it for evaluating clustering algorithms.

While automating friend grouping has been discussed in previous studies, to date no academic work has explored the strengths and weaknesses of automated clustering algorithms in OSNs using an interface. In this paper, we begin by applying different automation approaches on Facebook friendship networks and evaluate the groupings qualitatively and quantitatively. In the next section, we introduce the three chosen clustering algorithms used in our study.

3   Clustering Algorithms

In choosing a subset of clustering algorithms for our study, we explored algorithms with different input information. *Network structure* is the most common input information used by clustering algorithms. This information represents people as nodes and their friendship relations as links. Algorithms with this input information are called structure-based (or structural) algorithms. Other algorithms, called feature-based algorithms, use *nodes' features and attributes* to detect groups. For example, these features can be age, gender, and education of people in OSNs. A third category combines these two inputs *network structure and nodes' features*. In this paper, we focus on structure-based clustering algorithms. One advantage is the ability to interpret why the resulting groupings emerged and to compare algorithms with a consistent evaluation metric across the same network structure [10]. Furthermore, using feature-based algorithms necessitates extracting extra data from an OSN. This extraction results in very high processing time which makes conducting studies in a limited time in the lab difficult or almost impossible.

Structured-based clustering algorithms can be further classified in to three categories based on their membership attribute: (i) '*disjoint clustering*' algorithms where each object can only belong to one group; (ii) '*overlapping clustering*' algorithms where an object can be a member of more than one group. For example, a person may belong to different groups such as 'Family', 'Main East High School', and 'Loves Red Sox'; and (iii) '*hierarchical clustering*' algorithms which categorize objects in a multi-level structure where one group can be a subset of another group [22]. For example, cousin Joe is in a group labelled 'Cousins' which is a subset of a group named 'Family'. '*Hierarchical clustering*' algorithms have been used widely in social network analysis [10]. Figure 1 shows a schematic view of these clustering algorithms based on the defined membership attributes. We chose a representative algorithm from each membership category explained above for a total of three algorithms:

- *Markov Clustering (MCL)*: This algorithm is a *disjoint clustering* algorithm that uses the concept of Markov chains to simulate stochastic flows in graphs and builds a fast and scalable unsupervised clustering algorithm. MCL has a relatively high performance and is scalable [26].
- *OSLOM*: The Order Statistics Local Optimization Method (OSLOM) is an *overlapping clustering* algorithm that is among the first to account for edge weights and overlapping groups. It has a high performance and is scalable to large networks [19].
- *Louvain*: This *hierarchical clustering* algorithm uses modularity as its objective function and maximizes it using multiple heuristics to detect the groups. While this algorithm finds groups in a hierarchical manner, the lowest level of the hierarchies, which are the subgroups, are disjoint; i.e. one person cannot be a member of more than one group in a same level. The Louvain algorithm is highly accurate and has a very low computation time which makes it appealing for our study [5].

(a) Disjoint Clustering   (b) Overlapping Clustering   (c) Hierarchical Clustering

Fig. 1. Three clustering methods with different membership attributes

4   Method

We conducted a three part mixed methods study to better understand how social media users currently create and use groups and to evaluate how an automated approach would fit into our users' intended grouping goals. Our methodology consists of: (i) a

Fig. 2. A Snapshot of the Facebook Group Detection Application

pre-interview to understand existing group usage in social media; (ii) a *lab study* using a customized Facebook grouping application to understand how users perceive and modify automated groups; and (iii) a *post-interview* to explore the advantages and disadvantages of automated group creation. We recruited 18 (11 female and 7 male) participants during two months from a large Midwest university. They were from 8 different departments and ranged in age from 18-55. The participants' Facebook friendship networks ranged in size from 139 friends to 1853 friends ($\mu = 601.7, \sigma = 367.5$). All the participants reported using Facebook daily (on average for the past 5.7 years) and the majority of them logged into Facebook several times a day (n=12).

4.1 The Pre-interview

We first asked participants about basic demographics information, the social networking sites they used, and how frequently they used their favourite social networking site. We then probed them on the perceived importance of friend grouping in social networks and asked them whether they had used any friend grouping tools and why. If the participants mentioned using Facebook lists, we asked them about the type of lists they used (regular, smart or both), their goal in using Facebook lists, and the helpfulness of Facebook smart list suggestions.

4.2 Facebook Grouping Application Use

For the second part of our study, we implemented a novel automatic Facebook grouping application [1]. We used Facebook API v1.0 to extract participants' friendship networks. Our application utilized the three structural clustering algorithms explained in

Section 3 to automatically group friends on Facebook. Figure 2 shows the groups created by each clustering algorithm in a separate tab. Each tab is named after the corresponding membership attribute of the clustering algorithm: disjoint, overlapping, and hierarchical. Each tab contains two panels: the *groups panel* (left side) and the *members panel* (right side). The groups panel shows the created groups by the corresponding algorithm. By clicking on a group in this panel, the members of that group are shown in the members panel. Users can move their friends from one group to another. They can also change the name of a group through both the groups and members panels. At the bottom of the group panel, there is a category named '*ungrouped*' which contains any friends that the algorithm did not place into existing groups. The overlapping and hierarchical tabs offer some additional features. For instance, in the overlapping tab, moving a member from one group to another group would not result in removing the member from the first group. Similarly, in the hierarchical tab, color coding distinguishes groups at different levels of the hierarchy (see [1] for a thorough explanation of the interface).

After a brief introduction to the application, we asked participants to modify each algorithm's automated groups considering the task of content selective sharing. As a first step, we asked them to look over each group and label it based on at least $\frac{2}{3}$ of the group members. If a group had no meaning for them, we asked them to delete the group. When a group was deleted, its members automatically went to the ungrouped category. After the first round, participants were asked to come back and review the members of each group individually. During the review process, they were asked to move or delete members when they did not belong to a group, create new groups, or merge the existing groups as necessary. Finally, we asked them to check the members of the ungrouped category to see whether they could find a group for any of them. The participants repeated this process for each tab. To mitigate any learning effects, order effects, and bias toward a specific algorithm, we randomized the order of tabs. Due to time constraints, the participants with large network sizes ($n > 500$) where asked to work on one or two of the algorithms only.

4.3   The Post Interview

Upon the completion of group modification in each tab and before moving to the next tab, we discussed with participants to understand how usable the interface was. We then asked them to rate the quality of the groups based on their usability before and after the modification process on a 5-point scale. We then followed up with a short semi-structured interview asking questions about each method's performance, weaknesses and strengths. We encouraged participants to discuss any interesting or challenging points they found during the modification process in that tab. After modifying the groups in all the tabs, the participants were asked to compare the performance of the algorithms by ranking the groupings before and after the modification process (see Appendix for the detailed questions).

5   Evaluation

During the study, we asked participants to compare our application with the existing recommendation-based interface of Facebook lists. The majority of the participants stated that our automated grouping interface removed the burden of verifying friends' groups individually in comparison with the Facebook interface: *"Suggesting friends by Facebook is not user friendly as I have to add each person one by one; additionally changing a list of friends is not easy because it needs many clicks! I prefer this user interface that creates groups and then I [can] modify them. It will be faster."* (P3). They declared that if Facebook had this interface, they might be more willing to manage their friendship network: *"if Facebook had this feature, I would probably use it. When Facebook came out, it didn't have the list feature and then when it had it, it was hard to do it by hand. So, this version will make it easy to manage my groups of friends."* (P9). This overwhelming preference of the proposed interface to the current recommendation-based approaches illustrates the necessity of automated friend grouping in social networks specifically when users deal with a large number of friends [8].

In the following sections, we evaluate the groups created by the algorithms to understand how well these algorithms detected users' friendship groups. We then investigate the modified groupings of users to find out whether an automated grouping technique can bias the user's ideal groups. Finally, we explore the group dynamics without ground-truth by using two group scoring metrics to see how well these metrics are able to identify human-curated groups.

5.1   Evaluating Groups and Algorithms

To assess the effectiveness of our automatic friend grouping application, we relied on both quantitative and qualitative metrics. The quantitative metrics helped us to measure the similarity of the *'predicted grouping'* (i.e. the original group structure created by our application) and the *'desired grouping'* (i.e. the final group structure modified by a participant), while the qualitative metrics were used to measure the level of the user's satisfaction with the groups created by our application.

To measure the similarity of the predicted and desired grouping, we utilized a metric named BCubed, inspired by precision and recall metrics [3,12]. For BCubed, a value of 1 represents identical groupings and 0 illustrates that none of the friends are grouped similarly in two groupings. While BCubed indicates the similarity between the predicted grouping to the desired grouping, it may not convey the user's satisfaction level with the algorithms or our interface. For example, during the study, a few participants became confused during the modification phase and they were not able to completely create their ideal grouping. Therefore, in addition to the BCubed we asked participants to state a quality rating for each of the groups prior to modification on a 5-point Likert scale (1=poor, 5=excellent).

Table 1. The Algorithms evaluation by BCubed metric and participants' rating

Algorithm	BCubed [0-1]	Participants' Rating [1-5]
MCL (Disjoint)	0.89	3.3
Louvain (Hierarchical)	0.86	3.2
OSLOM (Overlapping)	0.78	3.1

Table 1 shows the participants' Facebook friendship information and the results of each algorithm's performance using BCubed and participant ratings. As the results demonstrate, (unlike OSLOM algorithm) both MCL and Louvain final groupings are highly similar in average to the ones participants modified. This significant similarity illustrates high accuracy of these two algorithms in detecting friendship groups in Facebook networks. In contrast to the previous work [14] in which a structural clustering algorithm (SCAN) could not find the groups of friends in OSNs with such high accuracy, our results illustrate that an appropriate structural clustering algorithm such as MCL and Louvain can detect the desired friendship groups with a significantly high accuracy while preserving human satisfaction. This outcome shows that the proper selection of a structural clustering algorithm besides including some attribute-based features of social networks (such as intimacy) can lead us to an accurate automated friend grouping approach in OSNs.

5.2   Automation Bias

Kelly et al. [16] investigated how different manual grouping techniques affected the final groups created by one person. They discovered that while it was possible to have an internal 'ground-truth' as the user's desired grouping, the manual grouping strategies could bias the user in creating his/her desired groups. While automation has been suggested as a solution to mitigate the burdens of manual grouping, it can also introduce bias in the friend grouping process. To examine whether such bias exists in our automated grouping techniques, for each participant, we compared the desired groups that emerged from the MCL predicted groups with the desired groups that resulted from the Louvain predicted groups. The comparison was performed using the BCubed metric and revealed that the MCL and Louvain desired groups created by the same participant are different from each other by 14% on average. This difference suggests that automated techniques (i.e. MCL and Louvain) used for generating the predicted groups can influence the desired groups created by a participant. We did not compare the desired groups created from OSLOM with the desired groups produced by the MCL and Louvain modification since OSLOM predicted groups are not disjoint.

In order to understand the possible causes of the bias introduced by the automation techniques, for each participant we carefully examined the difference between the MCL desired groups and the Louvain desired groups. To this end, for each group from a set of desired groups, we found its corresponding group in the other set of desired groups. Then, we looked over the groups with the most difference in two sets of desired groupings. Investigating these groups, we found that this difference is caused by four main factors:

1. Following What Algorithms Create: Some participants stated that if an algorithm did not find a specific group, they would not create that extra group. For example, one of the participants mentioned that one algorithm put his 'church' friends in a separate group. If he had manually created groups, he suspected he would not have considered a 'church' group. He then admitted the group made sense, he liked it, and kept it. Such examples demonstrate that automating the friend grouping process influences users to follow what algorithms seed.

2. Existing Hierarchical Structures in Social Relationships: One potential cause of inconsistency between the two sets of desired groups created by the same participant, is the difference in the hierarchy levels of the initial predicted groups. For example, while MCL might detect a group that a participant would call "university", Louvain might divide this group into smaller groups that the same participant would label using criteria such as entrance year or closeness. Therefore, after the participant was done with the modification process, the desired groups from these two algorithms would differ (see table 2) .

Table 2. Examples of Automation Bias Causes

Automation Bias Reasons	P#	MCL Desired Groups	Louvain Desired Groups
Hierarchical Structures	P12	US High School	US HS 2010, US HS 2011
	P14	ECO	ECO, ECO close, Others
	P15	Facebook	Facebook, Facebook interns, University CS
Friends with Multiple Roles	P5	Family	Family, Brother's Friends
	P11	Industrial design	Industrial design, Roommates, Art and design
	P15	Chicago friends	Chicago friends, University Other
User's Uncertainty	P7	Not close (University)	Not Close Uni Friends, Average Uni Friends
	P9	April's Family	April's Family, Family and Family Friend
	P15	Family	Family friends, Un-Grouped

3. Having Friends with Multiple Roles: Some of our participants had a number of friends with multiple roles, but they could assign these friends to only one group due to MCL and Louvain's disjoint membership constraint. Our participants' decisions on the most appropriate group for this type of friend were affected by the available predicted groups. For example, when a friend was a member of family and also a classmate in the university, the participant assigned this friend to the predicted group which could be the 'family' group in one algorithm and the 'university' group in the other algorithm. More cases are shown at table 2.

4. User's Uncertainty: One of the main issues in the friend grouping process was the participants' uncertainty when identifying or creating groups for some friends. For example, one of the participants started to make a 'Bay Area' group and decided to make it more specific based on different organizations (Facebook, Yahoo and ...). She eventually became confused with the organization of these groups and gave up. This confusion came from the uncertainty in identifying the right group. In another case, we found some participant were unable to distinguish the intimacy levels between some friends. For example, while a participant created a group named 'closer friends (University)' after modifying the Louvain predicted groups, she divided this group to two groups of 'Average University Friends' and 'Close University Friends' in the modification process of the MCL predicted groups. Table 2 shows more examples of uncertainty.

5.3 Exploring Group Dynamics without Ground Truth

In this study, we used the BCubed metric to compare the predicted groups generated by an algorithm to the desired groups made by a participant. In most real-world cases, we cannot access the desired groups or so called ground-truth. Therefore, various group scoring metrics have been defined to evaluate groupings in the absence of ground truth. These metrics are grounded in the general definition of a group —a group has many connections between its members and few connections to the rest of the network. Recent work evaluated these metrics by applying them to social, collaboration, and information networks where the nodes had explicit group memberships. Of the thirteen evaluated metrics, we chose the two with the best consistent reported performance in identifying ground-truth communities: Conductance and Triad-Participation Ratio [20,28].

- *Conductance*: This metric measures the fraction of total links of a cluster that point outside the cluster. Since a group by definition has more connections between members than outside, a conductance of 0 represents an 'ideal' group with no connections to the rest of the network; a conductance of 1 implies no connections within that grouping [20].
- *Triad-Participation Ratio (TPR)*: TPR metric is the fraction of members in a group that belong to a triad, a set of three connected nodes, inside the group. Unlike conductance, a higher TPR represents a tighter group [28].

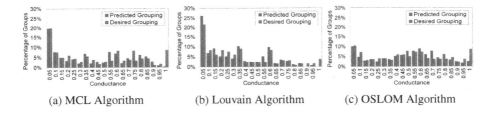

(a) MCL Algorithm   (b) Louvain Algorithm   (c) OSLOM Algorithm

Fig. 3. Histogram Percentage of Groups at various Conductance Values

We measured these metrics over predicted and desired groups to compare their values before and after users' modification. We hypothesized that the desired groups will have a lower conductance and higher TPR with respect to the predicted groups. To test our hypothesis, we calculated these metrics for predicted and desired groups produced by the three clustering algorithms. We found that the TPR metric increased significantly after the modification process as it was expected. However, the number of groups with high conductance ([0.80-1]) increased by 10% (Figure 3). That is, the number of groups with almost no inside connections between members increased after the modification process. To further explore this unexpected result, we investigated the groups which their conductance value increased after the modification process. We also coded the transcripts from our interviews where participants described their grouping process. We found out that some of our participants put some of their friends that were not linked together in one group. We found four categories of phenomena that explained this increase of conductance:

Others: This category contained the friends that participants did not care to or could not easily group. One of the participants drew Figure 4 to illustrate her grouping model.

As she explained, she saw her online friendship network as a network with three layers: (i) close friends; (ii) regular friends; and (iii) Others. She stated that he did not want to spend time to create groups for the 'others' layer.

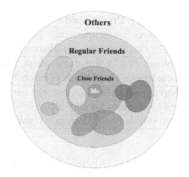

Fig. 4. A Participant's Rendering of Friend Categories

Another participant described the people in her 'others' group: *"In social media, I don't know these people very well as I meet them online and I have no more relations with them..."*. Some examples of the groups which reside in this category are shown in Table 3. The common attribute between the individuals in these groups is 'not being important to be in a labelled group'. Therefore, there is a lower chance for the members of these groups to be connected.

Functional Ties: Facebook is a social networking site, yet some people use it to maintain connections that are not reciprocally social [14]. These connections were added for professional or functional reasons. For example, one of our participants made a group labelled 'political' and said this group contained important people in policy whom he follows. However, the members of this group were not mutual friends in Facebook since they were from different political backgrounds. This resulted in a higher conductance in this group. Other examples of functional ties are shown in Table 3.

Indirect Friends: Our participants treated some of their online friends as indirect friends and consequently grouped them as friends of other friends. One of our participants made a group labelled *'friends of friends'*: *"I made a group named 'friends of friends' that contains people who friended me but are my friends' friends but they might not know each other even [if] they are in one group!"* There are similar examples in Table 3 such as 'Friend's siblings' where the members in the group may not be connected. These examples explain the high conductance in these groups containing indirect friends.

Temporal Ties: These are friendships that are bounded in time. Many of our study participants created groups such as *'People I worked with/talked to once and never again'*. One participant labelled a group ' *We win competitions and hackathons for silly ideas*' and described it as a group of people he knew during a contest. Other examples of temporal ties can be seen in Table 3. The short-term temporal tie relationships increase the probability for fewer connections in an online space such as Facebook.

Table 3. The desired groups with high conductance - () shows the conductance of each group

Category	Group Name
The Others Group	University friends who don't fit other groups (1), University other (1), Don't know (1), People I don't remember ever having talked to (0.91), Others [0.83 - 0.97], Un-Grouped [0.84 - 1]
Functional Ties	Advertising/Journalism people I met from totally different places (0.84), Political (0.87), Old Teachers (1), Bloggers and Organizations (0.83-0.94)
Indirect Friends	Friend's siblings (0.87), Stevenson close friends (0.93), Brian's friends (0.87), Sisters friends (1), Friends of Friends (1), Met via Sibs (0.95)
Temporal Ties	People I worked with/talked to once and never again (0.83), We win competitions and hackathons for silly ideas (0.84), Vineyard (0.95), Habitat for Humanity No Builds (0.89), Summer University (1), Old church (0.83)

The different characteristics of 'The Others', 'Functional Ties', 'Indirect Friends', and 'Temporal Ties' are challenging for group scoring metrics such as conductance. The conductance and other similar metrics assume intense inside group connections, however, some of the groups our participants labelled do not fit the traditional definition of a group. This suggests that for OSNs, we should explore alternate group scoring metrics compatible with the dynamic groups that exist in these networks.

6   Discussion

From the three clustering algorithms that we used in our study, MCL and Louvain performed well in terms of accuracy and human satisfaction. This result suggests that structure-based clustering algorithms such as MCL and Louvain are effective in detecting groups in OSNs. However, these algorithms do not consider some important features such as intimacy or interaction between friends. During our study, many participants said that the groups generated by the algorithms would have been more useful if they were able to separate their close friends from other friends or split some of the groups to smaller groups based on intimacy. However, since the applied algorithms in the study were structure-based, they did not have the required information to detect these types of groups. This finding which corroborates previous studies [11,14] demonstrates the necessity of adding important factors such as intimacy and interaction between friends to the current structure-based clustering algorithms.

Although the participants preferred our automated friend grouping tool to the current recommendation-based interface of Facebook smart lists, this automation introduces bias in the friend grouping process. While this bias could also exist in recommendation-based tools, creating fully populated groups from onset with our automated approach could increase it. However, we believe this bias can be reduced. For example, having both hierarchical and overlapping membership attributes for supporting subgroups and friends with multiple roles simultaneously can mitigate this bias.

In our study, participants did not care to group some of their contacts; we labeled these contacts 'others'. We believe an effective clustering algorithm should be able to find and prune this group of contacts before starting to group the friends. Pruning contacts helps increase the accuracy of the clustering algorithm when detecting the actual friendship groups. Furthermore, our results revealed that there were some other types of groups besides 'the others group' (functional ties, indirect friends and temporal ties) which current group scoring metrics cannot identify. It would be fruitful to probe alternative metrics which are compatible with such human-curated groups in OSNs.

A limitation of our study is the small sample of university students. We look forward to collecting data from additional OSN users with more diverse friendship networks. Another challenge was the time it took participants with large numbers of friends to use the three different interfaces. On average, participants completed the study in 114 minutes. This length of time could result in human fatigue and consequently, human error during the modification process. To lessen this effect, we adjusted the number of automated approaches based on the participant's number of friends. Sampling friends in a uniform way to reduce the time while still providing significant results could be a fruitful approach for future work. Finally we asked participants' perceptions of groupings rather than having them use the created groupings in a real world task. Our subjects were told to imagine groupings for selectively sharing a message/image in Facebook. Future work should observe users sending specific content using the grouping approach described in the paper.

7 Conclusion

Given the significant cost of manual grouping in OSNs, this work takes a step toward providing an automated friend grouping tool that applies three different clustering algorithms on Facebook friendship networks. Studying this tool, we found that users preferred our automated friend grouping tool to the current recommendation-based Facebook smart lists. We compared the three clustering algorithms using quantitative and qualitative evaluation methods. The evaluation results showed that the MCL and Louvain algorithms performed well in terms of accuracy and human satisfaction. While our automated friend grouping tool was well received by the participants, comparing the desired groups created by two different algorithms illustrated a significant bias in the automation approach. We believe future work should address educating users of these biases in their algorithmic interfaces. In our analysis of group composition before and after the modification process using two group scoring metrics, we found four categories of groups which do not satisfy the traditional definition of a networked group. This suggests that more exploration is needed and perhaps new metrics are necessary for understanding groupings of real world social connections. Grounded in our findings, we presented suggestions for designing future automated friend grouping tools. This work is a promising step toward designing an automated friend grouping framework for OSNs' users which can help manage their contacts efficiently.

Acknowledgements. This research is funded by the NSF (#0643502) and the Center for People and Infrastructures at the University of Illinois at Urbana-Champaign.

References

1. Mygroups,
   `http://social.cs.uiuc.edu/projects/MyGroups/CDA/index.php/`
   `frontend/intro`
2. Amershi, S., Fogarty, J., Weld, D.S.: Regroup: Interactive machine learning for on-demand group creation in social networks. In: Proceedings of CHI 2002 (2012)
3. Amigó, E., Gonzalo, J., Artiles, J., Verdejo, F.: A comparison of extrinsic clustering evaluation metrics based on formal constraints. Inf. Retr. 12(4), 461–486 (2009)
4. Bernstein, M.S., Marcus, A., Karger, D.R., Miller, R.C.: Enhancing directed content sharing on the web. In: Proceedings of the SIGCHI Conference on Human Factors in Computing Systems, CHI 2010, pp. 971–980 (2010)
5. Blondel, V.D., Guillaume, J.L., Lambiotte, R., Mech, E.L.J.S.: Fast unfolding of communities in large networks. J. Stat. Mech., P10008 (2008)
6. Brohé, S., van Helden, J.: Evaluation of clustering algorithms for protein-protein interaction networks. BMC Bioinformatics 7, 488 (2006)
7. Carr, A.: Facebook's New Groups, Dashboards, and Downloads Explained (October 2010), `http://www.fastcompany.com/1693443/`
   `facebooks-new-groups-dashboards-and-downloads-explained-video`
8. Eslami, M., Aleyasen, A., Zilouchian Moghaddam, R., Karahalios, K.: Evaluation of automated friend grouping in online social networks. In: CHI 2014, Extended Abstracts on Human Factors in Computing Systems, pp. 2119–2124. ACM (2014)
9. Estes, W.K.: Classification and Cognition. Oxford University Press (1994)
10. Fortunato, S.: Community detection in graphs. CoRR, abs/0906.0612 (2009)
11. Gilbert, E., Karahalios, K.: Predicting tie strength with social media. In: Proceedings of the SIGCHI Conference on Human Factors in Computing Systems, CHI 2009, pp. 211–220. ACM (2009)
12. Han, J., Kamber, M.: Data mining: concepts and techniques. Morgan Kaufmann Publishers Inc. (2000)
13. Johnson, M.L.: Toward usable access control for end-users: A case study of facebook privacy settings. PhD Dissertation University of Columbia US (2012)
14. Jones, S., O'Neill, E.: Feasibility of structural network clustering for group-based privacy control in social networks. In: Proceedings of the Sixth Symposium on Usable Privacy and Security, SOUPS 2010, pp. 9:1–9:13. ACM (2010)
15. Kairam, S., Brzozowski, M., Huffaker, D., Chi, E.: Talking in circles: selective sharing in google+. In: Proceedings of the SIGCHI Conference on Human Factors in Computing Systems, CHI 2012, pp. 1065–1074. ACM, New York (2012)
16. Kelley, P.G., Brewer, R., Mayer, Y., Cranor, L., Sadeh, N.: An investigation into facebook friend grouping. In: Proceedings of the 13th IFIP TC 13 International Conference on Human-Computer Interaction - Volume Part III, INTERACT 2011, pp. 216–233 (2011)
17. Kincaid, J.: Kleiner-Backed Katango Organizes Your Facebook Friends Into Groups For You (July 2011), `http://tcrn.ch/10qQ7A6`
18. Kincaid, J.: Google Acquires Katango, The Automatic Friend Sorter (November 2011), `http://tcrn.ch/1gtN9jD`
19. Lancichinetti, A., Radicchi, F., Ramasco, J.J., Fortunato, S.: Finding statistically significant communities in networks. PLoS ONE 6(4), e18961 (2011)
20. Leskovec, J., Lang, K.J., Mahoney, M.: Empirical comparison of algorithms for network community detection. In: Proc. of the 19th International Conference on World Wide Web, WWW 2010, pp. 631–640. ACM (2010)

21. MacLean, D., Hangal, S., Teh, S.K., Lam, M.S., Heer, J.: Groups without tears: mining social topologies from email. In: Proceedings of the 16th international conference on Intelligent user interfaces, IUI 2011, pp. 83–92. ACM (2011)
22. McAuley, J., Leskovec, J.: Discovering social circles in ego networks. CoRR, abs/1210.8182 (2012)
23. Ross, B.: Improved Friend Lists (September 2011), http://on.fb.me/1rbm98o
24. Slee, M.: Friend lists (December 2007), http://on.fb.me/1oHzyp2
25. Stone, B.: There's a list for that (October 2009), https://blog.twitter.com/2009/theres-list
26. Van Dongen, S.M.: Graph clustering by flow simulation. PhD Dissertation University of Utrecht, The Netherlands (2000)
27. Watson, J., Besmer, A., Lipford, H.R.: +your circles: sharing behavior on google+. In: Proceedings of the Eighth Symposium on Usable Privacy and Security, SOUPS 2012, pp. 12:1–12:9. ACM, New York (2012)
28. Yang, J., Leskovec, J.: Defining and evaluating network communities based on ground-truth. In: Proceedings of the ACM SIGKDD Workshop on Mining Data Semantics, MDS 2012, pp. 3:1–3:8. ACM (2012)

Appendix: Post-interview Questions

1. At first glance, how would you rate the quality of clusters created by this method? (1= poor, 5= excellent)
2. How well are you satisfied with the final groupings you made after the modifications? (1= Not very, 5 =Very)
3. How would you rate the groups created here by their usability? e.g. this grouping is useful for text messaging, announcing special events or ... (1= unusable, 5= usable)
4. How comfortable were you with the interface of this method? e.g. working with groups, moving friends, ... (1= Not very, 5= Very)
5. What worked well about this method? Can you give specific scenarios?
6. In what circumstances did this method not work well? Can you give specific scenarios?
7. If you decide to continue working on this grouping, is there any group you want to work on to make it better?
8. How cautious and accurate do you think you made your groups?

The Influence of Indirect Ties
on Social Network Dynamics

Xiang Zuo[1], Jeremy Blackburn[2], Nicolas Kourtellis[3], John Skvoretz[4],
and Adriana Iamnitchi[1]

[1] Computer Science and Engineering, University of South Florida, FL, USA
[2] Telefonica Research, Barcelona, Spain
[3] Yahoo Labs, Barcelona, Spain
[4] Department of Sociology, University of South Florida, FL, USA
xiangzuo@mail.usf.edu, jeremyb@tid.es, kourtell@yahoo-inc.com,
jskvoretz@usf.edu, anda@cse.usf.edu

Abstract. While direct social ties have been intensely studied in the
context of computer-mediated social networks, indirect ties (e.g., friends
of friends) have seen less attention. Yet in real life, we often rely on friends
of our friends for recommendations (of doctors, schools, or babysitters),
for introduction to a new job opportunity, and for many other occasional
needs. In this work we empirically study the predictive power of indirect
ties in two dynamic processes in social networks: new link formation and
information diffusion. We not only verify the predictive power of indirect
ties in new link formation but also show that this power is effective over
longer social distance. Moreover, we show that the strength of an indirect
tie positively correlates to the speed of forming a new link between the
two end users of the indirect tie. Finally, we show that the strength of
indirect ties can serve as a predictor for diffusion paths in social networks.

Keywords: indirect ties, social network dynamics, information diffusion.

1 Introduction

Mining the huge corpus of social data now available in digital format has led
to significant advances of our understanding of social relationships and confirmed long standing results from sociology on large datasets. In addition, social
information (mainly relating people via declared relationships on online social
networks or via computer-mediated interactions) has been successfully used for
a variety of applications, from spam filtering [1] to recommendations [2] and
peer-to-peer backup systems [3].

All these efforts, however, focused mainly on direct ties. Direct social ties (that
is, who is directly connected to whom in the social graph) are natural to observe
and reasonably easy to classify as strong or weak [4,5]. However, indirect social
ties, defined as relationships between two individuals who have no direct relation
but are connected through a third party, carry a significantly larger potential [6].

This paper analyzes the quantifiable effects that indirect ties have on network
dynamics. Its contributions are summarized as follows:

L.M. Aiello and D. McFarland (Eds.): SocInfo 2014, LNCS 8851, pp. 50–65, 2014.
© Springer International Publishing Switzerland 2014

- We quantitatively confirm on real datasets several well-established sociological phenomena: triadic closure, the timing of tie formation, and the effect of triadic closure on information diffusion.
- We extend the study of the indirect ties' impact on network dynamics to a distance longer than 2 hops.
- We show that indirect ties accurately predict information diffusion paths.

The rest of paper is organized as follows. Section 2 provides the context for this work. Section 3 introduces the datasets used in this study. Section 4 shows that the strength of *indirect* ties can be used to predict the formation of direct links at longer social distance. Section 5 refines this quantification to classify an indirect tie as weak or strong, showing that the classification meets theoretical expectations of a positive correlation between the strength of a tie and the speed at which a link forms. We also show in Section 6 that pairs with a strong indirect tie end up having more interactions after link formation when compared to pairs with a weaker indirect tie. In Section 7, we examine indirect tie strength as a predictor for diffusion paths in a network. Finally, Section 8 concludes with a discussion of lessons and future work.

2 Related Work

In sociology, two theories are closely related to the properties of indirect ties. First, the theory of *homophily* [7] postulates that people tend to form ties with others who have similar characteristics. Moreover, a stronger relationship implies greater similarity [8]. Second, the principle of *triadic closure* [10] states that two users with a common friend are likely to become friends in the near future. The triadic closure has been demonstrated as a fundamental principle for social network dynamics. For example, Kossinets and Watts [12] showed how it amplifies homophily patterns by studying the triadic closure in e-mail relations among college student. Kleinbaum [13] found that persons with atypical careers in a large firm tend to lack triadic closure in their email communication network and so have their brokerage opportunities enhanced.

Lately, large online social networks provided unprecedented opportunities to study dynamics of networks. Thus, many studies examined the evolution of groups or analyzed membership and relationship dynamics in these networks. For example, Backstrom et al. analyze how communities or groups evolve over time and how a community dies or falls apart [14]. Patil et al. use models to predict a group's stability and shrinkage over a period of time [15]. Yang and Counts examine the diffusion of information and innovations and the spread of epidemics and behaviors [16].

Compared to previous studies, we quantitatively investigate the effects of indirect ties on network dynamics, specifically on tie formation, the speed of tie formation, and information diffusion. More importantly, we study the impact of longer indirect ties on network dynamics: while previous work focused on 2-hop indirect ties, we also show the impact of 3-hop indirect ties.

3   Datasets

In this paper we use several datasets from different domains. Our datasets are varied, from fast non-profound dynamics to slow professional networks and more traditional social networks augmented with heavy interactions.

Team Fortress 2 (TF2) is an objective-oriented first person shooter game released in 2007. We collected more than 10 months of gameplay interactions (from April 1, 2011 to February 3, 2012) on a TF2 server [17]. The dataset includes game-based interactions among players, timestamp information of each interaction, declared relationship in the associated gaming OSN, Steam Community [18], and the time when the declared friendship was recorded. The resulting TF2 network is thus composed of edges between players who had at least one in-game interaction while playing together on this particular server, and also have a declared friendship in Steam Community. This dataset has three advantages. First, it provides the number of in-game interactions that can be used to quantify the strength of a social tie. Second, each interaction and friendship formation is annotated with a timestamp, which is helpful for examining the dynamics of links under formation. Third, over a pure in-game interaction network, it has the advantage of selecting the most representative social ties, as shown in [17].

Table 1. Characteristics of the social networks used in the following experiments. APL: average path length, CC: clustering coefficient, A: assortativity, D: diameter, EW: range of edge weights, OT: observation time.

Networks	Nodes	Edges	APL	Density	CC	A	D	EW	OT
TF2	2,406	9,720	4.2	0.0034	0.21	0.028	12	[1–21,767]	300 days
IE	410	2,765	3.6	0.0330	0.45	0.225	9	[1–191]	90 days
CA-I	348	595	6.1	0.0098	0.28	0.173	14	[1–52]	N/A
CA-II	1,127	6,690	3.4	0.0100	0.33	0.211	11	[1–127]	N/A

Infectious Exhibition (IE) held at the Science Gallery in Dublin, Ireland, from April 17th to July 17th in 2009 was an event where participants explored the mechanisms behind contagion and its containment. Data were collected via Radio-Frequency Identification (RFID) devices that recorded face-to-face proximity relations of individuals wearing badges [19]. Each interaction was annotated with a timestamp. We translated the number of interactions into edge weights.

Co-authorship networks (CA-I and CA-II) are the two largest connected components of the co-authorship graph of Computer Science researchers extracted by Tang et al. [20] from ArnetMiner[1]. Nodes in these graphs represent authors, edges are weighted with the number of papers co-authored. Because the dataset does not include time publication information, the observation window is unspecified in Table 1.

Note that IE is a smaller but much denser network than TF2, while TF2's interactions frequency is higher than IE's, as shown by the range of edge weights.

---

[1] http://arnetminer.org/

We use the TF2 and IE networks to study link formation and delay as they contain timestamps of the links formed and interactions between users. We use the TF2 and CA networks to study diffusion as they are larger, sparser and based on longer lasting relationships compared to IE's ad-hoc interactions.

4 Predicting Link Formation

According to Granovetter's idea of the *forbidden triad* [8], a triad between users u, v and w in which there are strong ties between u and v and between v and w, but no tie between u and w is unlikely to exist. When it does, according to the theory of *triadic closure*, it is typically quickly closed with the formation of a tie between u and w.

In this section, we not only empirically verify the theory of triadic closure by using multiple measures of the strength of indirect ties, but we also examine this theory over paths of length 3.

4.1 Methodology

The link prediction problem asks whether two unconnected nodes will form a tie in the near future [21]. Link prediction models that use an estimation of the tie strength from graph structure [22] or interaction frequency and users' declared profiles similarities [23] have been proposed in the past.

We use a group of tie strength metrics and classifiers to quantitatively demonstrate how indirect ties can be used for inferring new links formation. Specifically, given a snapshot of a social network, we use the strength of indirect ties to infer which relationships or interactions among users are likely to occur in the near future. Because people can be aware of others' behaviors within 2 hops [24] and be influenced by indirect ties up to 3 hops [25], we focus this task for pairs of users at social distance 2 and 3.

To investigate how such indirect ties materialize into actual links between users, we compare the performance of three different metrics of indirect tie strength: 1) Jaccard Index (J) [21], 2) Adamic-Adar (AA) [26], and 3) Social Strength (SS), a recently proposed metric [27,28] that quantifies the strength of indirect ties. We note that Jaccard Index and Adamic-Adar consider only the number of shared friends between users, while Social Strength also takes into account interaction intensity.

Social Strength. For completeness, we briefly describe next the Social Strength metric. For measuring the Social Strength of an indirect social tie between users i and m, we consider relationships at n ($n = 2$ or $n = 3$) social hops, where n is the shortest path between i and m. A weighted interaction graph model that connects users with edges weighted based on the intensity of their direct social interactions is assumed. Assuming that $\mathcal{P}_{i,m}^n$ is the set of different shortest paths of length n joining two indirectly connected users i and m and $\mathcal{N}(p)$ is the set of nodes on the shortest path $p, p \in \mathcal{P}_{i,m}^n$, we define the social strength between i and m from i's perspective over an n-hop shortest path as:

$$SS_n(i,m) = 1 - \prod_{p \in \mathcal{P}^n_{i,m}} (1 - \frac{\min\limits_{j,...,k \in \mathcal{N}(p)} [NW(i,j), ..., NW(k,m)]}{n})$$ (1)

This definition uses the normalized direct social weight $NW(i,j)$ between two directly connected users i and j, defined as follows:

$$NW(i,j) = \frac{\sum_{\forall \lambda \in \Lambda_{i,j}} \omega(i,j,\lambda)}{\sum_{\forall k \in N_i} \sum_{\forall \lambda \in \Lambda_{i,k}} \omega(i,k,\lambda)}$$ (2)

Equation 2 calculates the strength of a direct relationship by considering all types of interactions $\lambda \in \Lambda$ between the users i and j such as, phone calls, interactions in online games, and number of co-authored papers. These interactions are normalized to the total amount of interactions of type λ that i has with other individuals. This approach ensures the asymmetry of social strength in two ways: first, it captures the cases where $\omega(i,j,\lambda) \neq \omega(j,i,\lambda)$ (such as in a phone call graph). Second, by normalizing to the number of interactions within one's own social circle (e.g., node i's neighborhood N_i), even in undirected social graphs, the relative weight of the mutual tie will be different from the perspective of each user.

Prediction Task. The link prediction task decides whether the edge (u,v) will form during the observation time. We studied this task on the TF2 and IE datasets. The TF2 network has a timestamp of when a declared relationship was created in Steam Community. However, since for the IE network we do not have formally declared relationships, we use the timestamp of the first recorded face-to-face interaction between two individuals as a proxy for relationship creation.

In TF2, there are $5,984$ 2-hop ($2,475$ 3-hop) pairs that had a relationship formed within the observation time (OT) and $161,561$ 2-hop ($676,863$ 3-hop) pairs who didn't. In IE, there are $1,886$ 2-hop (484 3-hop) pairs that had a relationship formed within OT, and $4,111$ 2-hop ($24,631$ 3-hop) pairs who didn't. This means our datasets are imbalanced with respect to pairs who closed the 2-hop or 3-hop distance or not. There are two common approaches for dealing with unbalanced data classifications: under-sampling [29] and over-sampling [30]. We chose to under-sample pairs of users with no relationships materializing within OT, thus in our experiment they appear at the same empirical frequency as the pairs who formed relationships within OT.

In this prediction task, we used two classic machine learning classifiers: *Random Forest* (RF) and *Decision Tree* (J48). They are tested using tie strength values calculated from the three metrics (Jaccard Index, Adamic-Adar and Social Strength) as features. We used standard prediction evaluation metrics: Precision, Recall, F-Measure and Area Under Curve (AUC) to evaluate the performance of prediction of each classifier and tie strength metric.

4.2   Experimental Results

Table 2 shows the link prediction results of nodes 2 and 3 hops away. Clearly, all three indirect tie metrics demonstrate their power in predicting the formation of links between pairs of non-connected 2-hop users. We note that the AUC reaches 0.77 for the TF2 network using social strength as the metric and J48 as the classifier, and reaches 0.88 for the IE network when using social strength as the metric with random forests as the classifier, greatly outperforming the other two tie strength predictor metrics.

Given that the Jaccard Index and Adamic-Adar metrics are restricted to predictions within 2 hops, we test only the social strength metric for the 3-hop distant link predictions. The results in Table 2 show that while the social strength's effectiveness to predict link formation is reduced, it still manages to properly discriminate between links formed or not by up to about 70% of the time in TF2 and 68% of the time in IE. Overall, while it is expected to see a decrease in performance when we cross the horizon of observability of 2 hops [24], our results show that indirect ties are able to predict the formation of links.

Table 2. Results of link prediction between pairs of n-hop distant users. Only SS is applicable to $n = 3$.

Network	n	Classifier	Metric	Precision	Recall	F-Measure	AUC
TF2	2	RF	SS	**0.71±0.005**	**0.71±0.005**	**0.71±0.006**	**0.76±0.006**
			AA	0.68±0.003	0.67±0.003	0.67±0.003	0.70±0.005
			J	0.67±0.004	0.66±0.003	0.66±0.003	0.70±0.003
		J48	SS	**0.75±0.012**	**0.74±0.008**	**0.74±0.006**	**0.77±0.009**
			AA	0.71±0.004	0.71±0.004	0.71±0.004	0.71±0.006
			J	0.51±0.007	0.51±0.006	0.50±0.008	0.51±0.008
IE	2	RF	SS	**0.81±0.005**	**0.81±0.002**	**0.81±0.003**	**0.88±0.005**
			AA	0.67±0.004	0.66±0.0114	0.66±0.011	0.71±0.002
			J	0.67±0.001	0.66±0.0172	0.66±0.005	0.72±0.002
		J48	SS	**0.84±0.013**	**0.84±0.002**	**0.84±0.002**	**0.87±0.001**
			AA	0.69±0.002	0.69±0.002	0.68±0.003	0.70±0.003
			J	0.69±0.007	0.68±0.005	0.68±0.001	0.68±0.004
TF2	3	RF	SS	**0.653±0.01**	**0.651±0.01**	**0.651±0.01**	**0.709±0.02**
		J48	SS	0.630±0.02	0.627±0.01	0.624±0.01	0.644±0.03
IE	3	RF	SS	**0.659±0.01**	**0.650±0.004**	**0.646±0.004**	**0.682±0.01**
		J48	SS	0.636±0.01	0.633±0.01	0.631±0.01	0.664±0.01

5   Timing of Link Formation

Network dynamics can also be examined from the perspective of *link delays* [31]. If we consider that a link between two nodes is *possible* when all the enabling conditions are met, then the link delay is the time lag between the conditions being met and the link forming. In this section, we investigate if there is a connection between the strength of a tie of indirectly connected users and the delay the link experiences before it is formed.

5.1 Methodology

Let us consider the toy networks in Figure 1. We define the link formation delay for 2-hop indirect ties (Figure 1a) as:

$$\Delta_{(b,c)} = t_{(b,c)} - max\{t_{(a,b)}, t_{(a,c)}\},$$

where $t_{(a,b)}$ is the time when the direct link between two nodes is established. This formulation can also be thought of as the triadic closure delay [31]. Δ thus is a proxy of the "speed" at which two indirectly connected nodes become directly connected: small Δ indicates that the triangle closes quickly, and vice versa.

Similarly, the link formation delay for 3-hop indirect ties (Figure 1b) is:

$$\Delta_{(c,d)} = t_{(c,d)} - max\{t_{(a,b)}, t_{(a,d)}, t_{(b,c)}\}.$$

Although no direct analogue for the 3-hop link formation delay was explored in [31], an n-hop link delay can be considered a form of the general link delay scenario with the restriction that an n-hop path must exist between the two nodes under consideration.

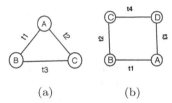

(a)   (b)

Fig. 1. (a) B and C have a 2-hop relationship before t_3, since $t_1, t_2 < t_3$, and a 1-hop relationship thereafter. (b) C and D have a 3-hop relationship before t_4, since $t_1, t_2, t_3 < t_4$, and a 1-hop relationship thereafter.

To measure the strength of indirect ties, we employ the social strength metric to quantify the strength of a social connection between indirectly connected nodes. We are primarily interested in whether the latent tie strength between indirectly connected nodes corresponds to different delays in a direct connection forming. Intuitively, if the strength of a user's indirect tie is stronger than any of the user's strong direct ties, we consider it a strong indirect tie. Because we have no information regarding the strength of a direct tie (other than the edge weight), we consider an indirect tie of a's as strong if its strength is larger than the minimum/average/maximum weight of all of a's direct edges. These alternative criteria are formally presented below. (We note that the social strength metric is asymmetric, i.e., $SS(a, b) \neq SS(b, a)$):

[C-min]: $SS(a,b) \geq \min\limits_{i \in Neigh(a)} [NW(a,i)]$ or $SS(b,a) \geq \min\limits_{a \in Neigh(i)} [NW(i,a)]$

[C-mean]: $SS(a,b) \geq \dfrac{\sum\limits_{i \in Neigh(a)} [NW(a,i)]}{size(Neigh(a))}$ or $SS(b,a) \geq \dfrac{\sum\limits_{a \in Neigh(i)} [NW(i,a)]}{size(Neigh(i))}$

[C-max]: $SS(a,b) \geq \max\limits_{i \in Neigh(a)} [NW(a,i)]$ or $SS(b,a) \geq \max\limits_{a \in Neigh(i)} [NW(i,a)]$

In each criterion, $NW(a, b)$ is the normalized weight of the edge between nodes a and b, and the normalization is conducted by the total weight of node a's edges. If an indirect tie (a, b) satisfies the conditions for a given criterion, it is marked as a strong indirect tie; otherwise it is marked as a weak indirect tie. Table 3 summarizes the tie classification results when these criteria are applied to the networks TF2 and IE.

Table 3. The statistics of 2- and 3-hop indirect ties in TF2 and IE networks where ties are divided into strong and weak ties under three criteria

Dist.	Network	Tie classification criterion	# strong ties	# weak ties
2	TF2	C-min	6,868	164
2	TF2	C-mean	5,470	1,562
2	TF2	C-max	2,780	4,252
3	TF2	C-min	2,351	90
3	TF2	C-mean	297	2,144
3	TF2	C-max	12	2,429
2	IE	C-min	1,555	42
2	IE	C-mean	1,235	344
2	IE	C-max	715	882
3	IE	C-min	193	258
3	IE	C-mean	11	440
3	IE	C-max	0	451

5.2 Experimental Results

We use the TF2 and IE networks described in Section 3 to analyze link delays when examining 2- and 3-hop indirect ties. We compare the link delay of weak and strong ties classified by the previously defined criteria. For TF2, we use *days* as the time unit, but for IE we use *minutes* due to the ephemeral nature of its face-to-face interactions.

The link delay distributions are plotted in Figure 2, where we see that pairs with strong indirect ties formed direct links with shorter delay than those with weak indirect ties. We note that strong ties formed their link with less delay than weak ties throughout all scenarios and when the tie is stronger, its link formed even quicker. For example, when using 3-hop indirect ties in TF2 and criterion *C-max* for classifying strong vs. weak, 33% of strong indirect ties formed a direct link within a day, compared to only 7% for weak indirect ties. In contrast, over 40% of weak indirect ties formed direct links with a large delay (over 60 days). Overall, these results indicate several things. First, when indirect ties are stronger, there is an increased chance for them to establish a link quicker. Second, even quantifying the strength of the tie from 3-hops away, strong indirect ties led to faster link creation.

6 Interaction Intensity along Newly Formed Links

A key characteristic of social interactions is their continuous change, and this change is likely to affect user behavior related to network dynamics. E.g.,

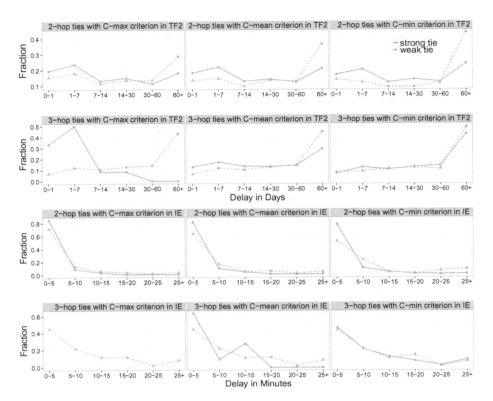

Fig. 2. Link delay comparison between 2- and 3-hop strong vs. weak ties in TF2 and IE. Note that for the IE network, when 3-hop ties are divided by criterion *C-max*, no strong ties exist.

frequent interactions lead to the formation of new links, and by interacting with each other, information can be disseminated in the network. Thus, we believe the changes in the interactions between nodes previously connected by indirect ties also can predict the dynamic status of the network.

Note that among all four datasets introduced in Section 3, only the online game social network (TF2) supplies a timestamp for each friendship formation and interaction. More importantly, because gamers can play with each others without being declared friends in Steam Community OSN, we can measure interaction intensity in the absence of a declared relationship. Thus, in the following our analysis is based on TF2 network.

We analyze the intensity of user interactions before and after a pair of users, who are 2- and 3-hop away, form a new edge. Figure 3 shows that in both scenarios (2 and 3 hops), more pairs of users have interactions after their link formation than before the link formation. For example, 54% pair of users have no interactions before they establish an edge with each other, while this number decreases to 17% after a 2-hop indirect tie is closed with a direct tie. This result shows

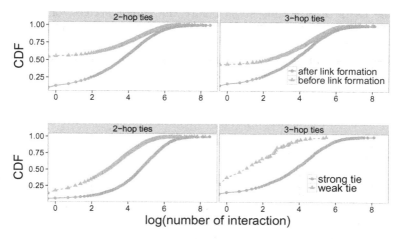

Fig. 3. Interaction intensity before vs. after 2- and 3-hop link formation in TF2, and strong vs. weak ties' (identified by the C-min criterion) interaction intensity after link formation of TF2

that after users form a direct link, their interactions not only continue but also increase, implying that users actively maintain their newly formed relationship.

As a further step, we investigate the difference of interaction intensity between nodes previously connected via strong vs. weak indirect ties after forming their direct links. We use the *C-min* criterion introduced in Section 5 to classify indirect ties into strong and weak. Figure 3 plots interaction intensity after link formation. The figure shows that strong indirect ties lead to direct ties with more interactions than weak indirect ties do.

7   Predicting Information Diffusion Paths

Information diffusion is a fundamental process in social networks and has been extensively studied in the past (e.g., [32,33,34,35,36]). In fact, some studies have shown that the evolution of a network is affected by the diffusion of information in the network [35] and vice versa [34]. Our results from the previous sections show that indirect ties affect the process of network evolution. In this section, we go a step further and investigate if the strength of indirect ties can predict diffusion paths between distant nodes in the graph. That is, departing from the step-wise diffusion processes examined in the past, and given that a user received a piece of information at time step t, can we predict which other users will receive this information at time step $t + n$ ($n \geq 2$)? I.e., if we know someone who received the information at t_0, then can we directly predict the infected users at t_n ($n \geq 2$) instead of step-wise (e.g., at t_1)?

Predictions over such longer intervals could help OSN providers customize strategies for preventing or accelerating information spreading. For example, to contain rumors, OSN providers could block related messages sent to the susceptible users several time steps before the rumor arrives, or disseminate official anti-rumor messages in advance. Similarly, marketers could accelerate their advertisements spreading in the network by discovering who will be the next susceptible to infection. This n-hop path prediction can supply more time for decision makers to contain harmful disseminations, and to choose users who are pivotal in information spreading for targeted advertisements.

This section describes our experiments of applying several indirect-tie metrics to predict information diffusion paths.

7.1   Experimental Setup

The strength of an indirect tie decreases with the length of the shortest path between the two individuals. This has been quantitatively observed by Friedkin [24], who concluded that people's awareness of others' performance decreases beyond 2 hops. Three degrees of influence theory, proposed by Christakis et al. [25], states that social influence does not end with people who are directly connected but also continues to 2- and 3-hop relationships, albeit with diminishing returns. This theory has held true in a variety of social networks examined [37,38] and in accordance with these observations, we set our experiments up to 3 hops. A single node is chosen as the original source of information at t_0. We then predict the nodes that will accept the information at t_n with the knowledge from t_0.

Diffusion Simulation. As ground truth, we applied the basic and widely studied *Linear Threshold (LT)* diffusion model [39] to simulate a diffusion process, i.e., which nodes are affected during each time step.

The LT model is a threshold-based diffusion model where nodes can be in one of two states: active or inactive. We say a node has accepted the information if it is active, and once active, it can never return to the inactive state. In the LT model, a node v is influenced by each of its neighbors $Neigh_v$ according to an edge weight $b_{v,w}$. Each node v chooses a *threshold* θ_v that is randomly generated from the interval [0,1]. The diffusion process is simulated as follows: first, an initial set of active nodes A_0 is chosen at random and these are the seed nodes. Then, at each step t, all nodes that were active in step $t-1$ remain active, and we activate any node v for which the total weight of its active neighbors is at least θ_v, that is $\sum_{w \in Neigh_v} b_{v,w} \geq \theta_v$. Thus, the threshold θ_v intuitively represents the different latent tendencies of nodes to adopt the behavior exhibited by neighbors, and a node's tendency to become active increases as more of its neighbors become active. The input to the simulation is a weighted graph where edge weights represent the intensity of interactions between nodes (n.b., the LT model considers only the status of a node's directly connected neighbors).

We controlled the effectiveness of the diffusion by gradually changing the upper bound of the thresholds applied on the nodes to simulate different diffusion processes; from almost no diffusion to fully dissemination to all nodes in the graph. To do so, we set a threshold $\theta_v = random(0,1)/w$ where w is empirically selected based on the range of edge weights in each of the tested networks, i.e., w in the range of [1-10] for the CA-I, [1-30] for the CA-II and [1-60] for the TF2.

Predicting Diffusion Paths via Indirect Ties. Once we generate the ground truth from the LM model, we then use the strength of indirect ties to predict the path of diffusion. To measure the strength of indirect ties, we also employ social strength, Adamic-Adar and Jaccard metrics introduced in Section 4.1 where the social strength metric considers the edge weight while Adamic-Adar and Jaccard only consider the neighborhood overlap. We calculate the strength of indirect tie values between the seed and its n-hop nodes, then convert the values to a social rank. Each user has a rank list for all her n-hop nodes according to the strength of the indirect tie value between the user and the node.

After obtaining social ranks, we need a cut-off threshold to decide whether or not a node's n-hop nodes will be active at t_{0+n}(n=2 or 3). Our strategy requires that the social ranks from information recipient's perspective must be high, e.g., $socialrank_n(A, B)$ ranks among the top 10% of user A's contacts. Then, the cut-off threshold can classify a node's n-hop nodes into two categories: active or inactive at t_{0+n}(n=2 or 3). The intuition of this cut-off is that users will likely believe the information from their "closest" social ties. The cut-off threshold can be calculated as $\theta_{pred} = |Neigh_{nhops}|/q$ where q is empirically selected to have an inversely proportional relationship to w, which decides the diffusion process from almost no diffusion to full dissemination to all nodes. In other words, when no diffusion happens the θ_{pred} should be small enough to select the strongest indirect ties while in a fully diffused scenario a larger θ_{pred} is needed to cover a large portion of indirect ties.

7.2   Results and Evaluation

In literature, co-authorship networks capture many general features of social networks [40] and have been studied in information cascades [39], and diffusion dynamics have been observed in online game social networks [41,42]. Therefore, in our experiments, we use the three datasets—CA-I, CA-II and TF2—as described in the previous section. To better demonstrate indirect ties' effective power on inferring diffusion processes, we compare indirect tie metrics with a baseline method, which randomly selects a information recipient's 2 and 3-hop friends to accept the information.

We compare the prediction results with the ground truth obtained from the diffusion simulation to verify the effectiveness of indirect ties in predicting diffusion paths. We evaluate our method using accuracy, sensitivity and specificity [43]. Figures 4 and 5 depict the prediction results in a 2- and 3-hop social distance, respectively. We see that for both 2- and 3-hop path predictions, overall the accuracies of indirect tie metrics are higher than the baseline's, reaching a maximum of 0.90 with social strength metric in 2-hop path predictions. Also, the accuracies of the three indirect tie metrics in all cases are always higher than 0.56, and social strength outperforms the other two metrics in most of the scenarios. Although 3-hop predictions (generated by the Social Strength metric) show decreased sensitivity, specificity and accuracy compared to 2-hop results, they remain above 0.64. It is important to note that these three networks have very different network structure (from sparse to dense), yet the performance of

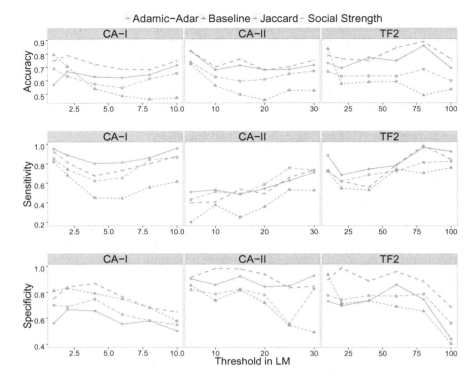

Fig. 4. Performance of different measures of strength for 2-hop indirect ties, in the prediction of information diffusion paths in the networks CA-I, CA-II and TF2

indirect tie metrics are consistently higher than the baseline in all three networks and for different diffusion thresholds. From these results, we conclude that indirect ties can be used in the prediction of information diffusion, i.e., along which paths information will propagate and which users will be activated, at least 2-3 steps before a susceptible node is even in contact with an infected node.

8   Summary and Discussions

In this paper, we empirically examine the predictive power of indirect ties in network dynamics. By using four real-world social network datasets and three indirect measurements, we empirically show that indirect ties can be used for predicting the newly formed edges and the stronger an indirect tie is, the quicker the tie will form a link. In addition, strong indirect ties correlate to more interactions, and the interaction has the tendency to be continued after the link formed. Finally, we show that indirect ties can also be used for predicting information diffusion paths in social networks.

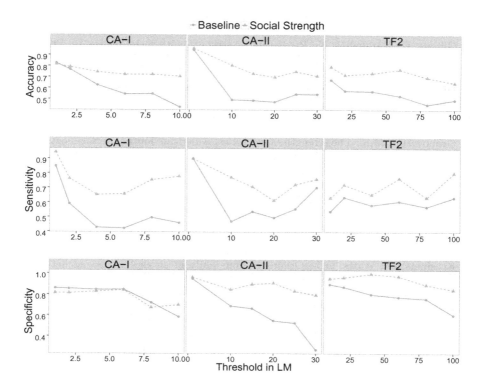

Fig. 5. Performance of different measures of strength for 3-hop indirect ties, in the prediction of information diffusion paths in the networks CA-I, CA-II and TF2

This is our first step to investigate the influences of indirect ties on network dynamics. In the future, we will further study the effects of indirect ties on information diffusion paths with various diffusion models and real-world cascades.

Acknowledgment. This research was supported by the U.S. National Science Foundation under Grant No. CNS 0952420 and by the MULTISENSOR project partially funded by the European Commission under contract number FP7-610411.

References

1. Li, Z., Shen, H.: Soap: A social network aided personalized and effective spam filter to clean your e-mail box. In: Proceedings IEEE INFOCOM (2011)
2. Basu, C., Hirsh, H., Cohen, W.: Recommendation as classification: Using social and content-based information in recommendation. In: AAAI/IAAI, pp. 714–720 (1998)
3. Li, J., Dabek, F.: F2F: reliable storage in open networks. In: Proceedings of the 4th International Workshop on Peer-to-Peer Systems, IPTPS (2006)

4. Kahanda, I., Neville, J.: Using transactional information to predict link strength in online social networks. In: ICWSM (2009)
5. Gilbert, E., Karahalios, K.: Predicting tie strength with social media. In: Proceedings of the SIGCHI Conference on Human Factors in Computing Systems, pp. 211–220. ACM (2009)
6. Granovetter, M.S.: A study of contacts and careers. Cambridge, Mass. Harvard (1974)
7. McPherson, M., Smith-Lovin, L., Cook, J.: Birds of a feather: Homophily in social networks. Annual review of sociology, 415–444 (2001)
8. Granovetter, M.S.: The strength of weak ties. American Journal of Sociology 78(6) (1973)
9. Coleman, J.: Social capital in the creation of human capital. American Journal of Sociology, S95–S120 (1988)
10. Rapoport, A.: Spread of information through a population with socio-structural bias: I. assumption of transitivity. The bulletin of mathematical biophysics 15(4), 523–533 (1953)
11. Szell, M., Thurner, S.: Measuring social dynamics in a massive multiplayer online game. Social Networks 32(4), 313–329 (2010)
12. Kossinets, G., Watts, D.J.: Origins of homophily in an evolving social network1. American Journal of Sociology 115(2), 405–450 (2009)
13. Kleinbaum, A.M.: Organizational misfits and the origins of brokerage in intrafirm networks. Administrative Science Quarterly 57(3), 407–452 (2012)
14. Backstrom, L., Huttenlocher, D., Kleinberg, J., Lan, X.: Group formation in large social networks: memberships, growth, and evolution. In: The International Conference on Knowledge Discovery and Data Mining, KDD (2006)
15. Patil, A., Liu, J., Gao, J.: Predicting group stability in online social networks. In: Proceedings of the 22nd International Conference on World Wide Web, pp. 1021–1030 (2013)
16. Yang, J., Counts, S.: Predicting the speed, scale, and range of information diffusion in twitter. In: ICWSM, vol. 10, pp. 355–358 (2010)
17. Blackburn, J., Iamnitchi, A.: Relationships under the microscope with interaction-backed social networks. In: 1st International Conference on Internet Science, Brussels, Belgium (2013)
18. Blackburn, J., Kourtellis, N., Skvoretz, J., Ripeanu, M., Iamnitchi, A.: Cheating in online games: A social network perspective. ACM Transactions on Internet Technology (TOIT) 13(3), 9 (2014)
19. Isella, L., Stehlé, J., Barrat, A., Cattuto, C., Pinton, J.-F., den Broeck, W.V.: What's in a crowd? Analysis of face-to-face behavioral networks. Journal of Theoretical Biology 271(1), 166–180 (2011)
20. Tang, J., Sun, J., Wang, C., Yang, Z.: Social influence analysis in large-scale networks. In: International Conference on Knowledge Discovery and Data Mining, KDD (2009)
21. Liben-Nowell, D., Kleinberg, J.: The link-prediction problem for social networks. Journal of the American society for information science and technology 58(7), 1019–1031 (2007)
22. Lü, L., Zhou, T.: Physica A: Statistical Mechanics and its Applications. Physica A: Statistical Mechanics and its Applications 390(6), 1150–1170 (2011)
23. Xiang, R., Neville, J., Rogati, M.: Modeling relationship strength in online social networks. In: 19th International Conference on World Wide Web, Raleigh, NC, USA, pp. 981–990 (2010)
24. Friedkin, N.E.: Horizons of observability and limits of informal control in organizations. Social Forces 62(6), 54–77 (1983)

25. Christakis, N.A., Fowler, J.H.: Connected: The surprising power of our social networks and how they shape our lives. Hachette Digital, Inc. (2009)
26. Adamic, L., Adar, E.: Friends and neighbors on the web. Social Networks 25(3), 211–230 (2003)
27. Kourtellis, N.: On the design of socially-aware distributed systems. Ph.D. dissertation, University of South Florida (2012)
28. Zuo, X., Blackburn, J., Kourtellis, N., Skvoretz, J., Iamnitchi, A.: The power of indirect ties in friend-to-friend storage systems. In: 14th IEEE International Conference on Peer-to-Peer Computing (September 2014)
29. Kubat, M., Matwin, S., et al.: Addressing the curse of imbalanced training sets: one-sided selection. ICML 97, 179–186 (1997)
30. Chawla, N.V., Bowyer, K.W., Hall, H.O., Kegelmeyer, W.P.: Smote: synthetic minority over-sampling technique. Journal of Artificial Intelligence Research 16, 321–357 (2002)
31. Zignani, M., Gaito, S., Rossi, G.P., Zhao, X., Zheng, H., Zhao, B.Y.: Link and triadic closure delay: Temporal metrics for social network dynamics. In: ICWSM 2014 (2014)
32. Yildiz, M.E., Scaglione, A., Ozdaglar, A.: Asymmetric information diffusion via gossiping on static and dynamic networks. In: 49th IEEE Conference on Decision and Control (CDC), pp. 7467–7472 (December 2010)
33. Guille, A., Hacid, H.: A predictive model for the temporal dynamics of information diffusion in online social networks. In: Proceedings of the 21st International Conference Companion on World Wide Web, pp. 1145–1152 (2012)
34. Bakshy, E., Rosenn, I., Marlow, C., Adamic, L.: The role of social networks in information diffusion. In: Proceedings of the 21st International Conference on World Wide Web, WWW 2012, pp. 519–528 (2012)
35. Weng, L., Ratkiewicz, J., Perra, N., Gonçalves, B., Castillo, C., Bonchi, F., Schifanella, R., Menczer, F., Flammini, A.: The role of information diffusion in the evolution of social networks. In: Proceedings of the 19th ACM International Conference on Knowledge Discovery and Data Mining, KDD 2013, pp. 356–364 (2013)
36. Guille, A., Hacid, H., Favre, C., Zighed, D.A.: Information diffusion in online social networks: A survey. SIGMOD Rec. 42(2), 17–28 (2013)
37. Fowler, J.H., Christakis, N.A., Roux, D.: Dynamic spread of happiness in a large social network: longitudinal analysis of the framingham heart study social network. BMJ: British Medical Journal, 23–27 (2009)
38. Christakis, N.A., Fowler, J.H.: The spread of obesity in a large social network over 32 years. New England Journal of Medicine 357(4), 370–379 (2007)
39. Kempe, D., Kleinberg, J., Tardos, É.: Maximizing the spread of influence through a social network. In: Proceedings of the Ninth ACM International Conference on Knowledge Discovery and Data Mining, KDD 2003, pp. 137–146. ACM, New York (2003)
40. Newman, M.E.: The structure of scientific collaboration networks. Proceedings of the National Academy of Sciences 98(2), 404–409 (2001)
41. Wei, X., Yang, J., Adamic, L.A., Araújo, R.M.d., Rekhi, M.: Diffusion dynamics of games on online social networks. In: Proceedings of the 3rd conference on Online Social Networks, p. 2. USENIX Association (2010)
42. Blackburn, J., Simha, R., Kourtellis, N., Zuo, X., Ripeanu, M., Skvoretz, J., Iamnitchi, A.: Branded with a scarlet "c": cheaters in a gaming social network. In: Proceedings of the 21st International Conference on World Wide Web (2012)
43. Fawcett, T.: An introduction to ROC analysis. Pattern Recognition Letters 27(8), 861–874 (2006)

Predicting Online Community Churners Using Gaussian Sequences

Matthew Rowe

School of Computing and Communications, Lancaster University, Lancaster, UK
m.rowe@lancaster.ac.uk

Abstract. Knowing which users are likely to churn (i.e. leave) a service enables service providers to offer retention incentives for users to remain. To date, the prediction of churners has been largely performed through the examination of users' social network features; in order to see how churners and non-churners differ. In this paper we examine the social and lexical development of churners and non-churners and find that they exhibit visibly different signals over time. We present a prediction model that mines such development signals using Gaussian Sequences in the form of a joint probability model; under the assumption that the values of churners' and non-churners' social and lexical signals are normally distributed at a given time point. The evaluation of our approach, and its different permutations, demonstrates that we achieve significantly better performance than state of the art baselines for two of the datasets that we tested the approach on.

1 Introduction

The churn (leaving) of a user from a service represents a loss to the service owner, be it: a telecommunications operator, where a customer leaving represents a loss of financial income; a question-answering platform, where an expert leaving could lead to a reduction of *know-how* in the community, or: a discussion forum, in which a user leaving could result in the forum's social capital, and perhaps *vibrancy*, being reduced. Therefore, predicting which users will leave a given service is important to a range of service providers; and an effective means of doing so enables retention strategies to be applied to those potential churners.

The majority of work within the area of churn prediction has focussed on building a prediction model using information about a user's social network position [2], and thus the extent to which he is interacting with other users, and/or the activity of a given user up until a given point in time. Our prior work [8] proposed an approach based upon the *lifecycle* of a user (i.e. the period of time between a user joining a service to either churning or remaining) in which social and lexical dynamics of the user were mined and a model fitted to the development curve of the user; properties of those models were then used as features for prediction models to differentiate between churners and non-churners. However, this approach was limited by only concentrating on a fixed number of lifecycle stages (e.g. 20) and did not examine how churners

L.M. Aiello and D. McFarland (Eds.): SocInfo 2014, LNCS 8851, pp. 66–79, 2014.

and non-churners developed differently. In this paper we attempt to fill these gaps by exploring the following two questions: (i) *How do churners and non-churner develop?* And; (ii) *How can we exploit development information to detect churners?*

In exploring these questions we found that churners and non-churner do indeed differ in how they develop, both socially and lexically, over their lifetimes, and that by assuming a Gaussian distribution at each lifecycle stage then we can chain together *Gaussian Sequences* for use in prediction. This paper makes the following contributions:

- Examination of different lifecycle patterns for both churners and non-churners across three online community datasets; with different lifecycle fidelity settings (5, 10 and 20 stages).
- Prediction models based on Gaussian sequences with slack variables for tuning, and a new model learning approach called *Dual-Stochastic Gradient Descent*.
- Evaluation of the proposed models against state of the art baselines showing significantly better performance for two of the tested datasets.

We begin this paper with a review of existing churn prediction approaches before then moving on to detail the datasets used for our work and how we label both churners and non-churners within them.

2   Related Work

Churner prediction has been studied across a number of domains, for instance Zhang et al. [9] predicted churners in a Chinese telecommunications network by inducing a decision tree classifier from user activity features (e.g. call duration) and network properties (e.g. 2nd order ego-network clustering coefficient). Mc-Gowan et al. [6] also predicted churners from a telecommunications provider by experimenting with different dimensionality reduction and boosting methods. Lewis et al. [5] examined Facebook networks of college students over a 4 year period and found an association between friendship maintenance and geographical proximity and shared tastes. Quercia et al. [7] analysed Facebook friendships and users' personality traits, finding that churn was likely to happen if the ages of connected users differed and if one of the users was neurotic or introverted. Kwak et al. have examined churners from Twitter networks in [3] and [4]: in the former the authors analysed the differences between social network snapshots, separated by 6 weeks, of Korean Twitter users finding that users unfollowed other users when the users talked about uninteresting topics; while in the latter work [4], the authors induced a logistic regression model to predict churners based on pairwise features (formed between the user and each of his subscribers).

Similar to our work, the work by Karnstedt et al. in [1] and [2] examined the prediction of churners from the Irish online community platform Boards.ie, finding that the probability of a user churning was related to the number of prior users with whom the individual has communicated having churned before.

The authors examined the social network properties of churners against non-churners (i.e. in-degree, out-degree, clustering coefficient, closeness centrality, etc.), inducing a J48 decision tree to differentiate between churners and non-churners when using social network properties formed from the reply-to graph of the online communities. In this paper, we implement this model as our baseline by engineering the same features using the same experimental setup. Our approach differs from existing work by assessing churners' and non-churners' development signals, and inducing a joint-probability function from such information.

Table 1. Splits of users within the datasets and the churn window duration

Platform	#Churners	#Non-churners	Churn Window
Facebook	1,033	1,199	[04-11-2011, 28-08-2012]
SAP	10,421	7,255	[29-11-2009,07-09-2010]
Server Fault	12,314	11,144	[13-06-2010,24-12-2010]
Boards.ie	65,528	6,120,008	[01-01-2005,13-02-2008]

3 Datasets

To provide a broad examination of user lifecycles across different online community platforms we used data collected from four independent platforms:

1. *Facebook:* Data was obtained from Facebook groups related to Open University degree course discussions. Although Facebook provides the ability to collect social network data for users, we did not collect such data in this instance and instead used the reply-to graph within the groups to build social networks for individual users.
2. *SAP Community Network (SAP):* The SAP Community Network is a community question answering system related to SAP technology products and information technologies. Users sign up to the platform and post questions related to technical issues, other users then provide answers to those questions and should any answers satisfy the original query, and therefore solve the issue, the answerer is awarded points.
3. *Server Fault.* Similar to SAP, Server Fault is a platform that is part of the Stack Overflow question answering site collection.[1] The platform functions in a similar vein to SAP by providing users with the means to post questions pertaining to a variety of server-related issues, and allowing other community members to reply with potential answers.
4. *Boards.ie* This platform is a community message board that provides a range of dedicated forums, where each forum is used to discuss a given topics (e.g. Rugby Union, Xbox360 games, etc.). We were provided with data covering the period 1998-2008 and, like SAP and ServerFault, we also had access to the reply-to graph in each forum.

---

[1] http://stackoverflow.com/

Unlike on subscription-based services where a *churner* is identifiable by the cancellation of the service (e.g. cancelling a contract), on online community platforms we do not have such information from which to label churners and non-churners. Instead, we examined users' activity and then decided on a suitable *inactivity* threshold where, should a user remain inactive for more than that period (i.e. x days), then we can say he has *churned*. To derive this threshold, we first defined Δ as the maximum number of days between posts across the platforms' datasets for each user, and then plotted the relative frequency distribution of Δ across the platforms in Figure 1. We then selected each distribution's mean as the *churn control window* size: 149 days, 141 days, 97 days and 198 days for Facebook, SAP, ServerFault and Boards.ie respectively.

To derive churners and non-churners we took the final post date in a given dataset and went back n (size of the churn control window) days, this date gives the *churn window end point* (t''). We then went back a further $2n$ days to give the *churn window start point* (t'); thus the *churn window* is defined as a closed date interval $[t', t'']$. Users who posted for the final time in $[t', t'']$ were defined as *churners* and users who posted after $[t', t'']$ were labeled as a *non-churners*. Table 1 shows the number of churners and non-churners derived using this approach. We split each platform's users up into a *training* and *test* set using an 80:20% split respectively - using the former set to inspect how users develop and evolve and the latter set (test) to detect churners. All analysis that follows and the features engineered for our experiments use data from *before* the churn window, thereby not biasing our prediction experiments and reflecting a real-world churn prediction setting where we only have information up until a given time point.

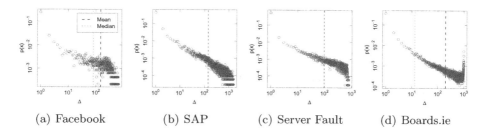

(a) Facebook   (b) SAP   (c) Server Fault   (d) Boards.ie

Fig. 1. Gap distributions across users of the different platforms. The mean and median of the distributions are shown using blue dashed and red dotted lines respectively.

4 User Lifecycles

In this section we briefly describe our approach for representing the lifecycles of users on the online community platforms - for a more comprehensive description we refer the reader to our prior work [8]. We begin by segmenting a user's lifecycle into k stages, where each stage contains the same number of posts. The setting of k controls the *fidelity* of a user's lifecycle and in this paper we experiment with various settings where $k = \{5, 10, 20\}$. For each lifecycle stage (i.e. $s \in S = \{1, 2, \ldots, k\}$) we wish to inspect the social and lexical dynamics of the user, as follows:

4.1 Social Dynamics

For examining the *social dynamics* of each user we looked at the distribution of his *in-degree* and *out-degree* - i.e. the number of edges that connect to a given user and the the number of edges from the user. As we are dealing with conversation-based platforms for our experiments we can use the *reply-to* graph to construct these edges, where we define an edge connecting to a given user u if another user v has replied to him. Given our use of lifecycle periods we use the discrete time intervals that constitute $s \in S$ to derive the set of users who replied to u, defining this set as $\Gamma_s^{IN} = \{author(q) : p \in P_u, q \in P, time(q) \in s, q \rightarrow p\}$.[2] We also define the set of users that u has replied to within a given time interval using Γ_s^{OUT} with the reply direction reversed. From these definitions we can then form a discrete probability distribution that captures the distribution of repliers to user u, using Γ_s^{IN}, and user u responding to community users using Γ_s^{OUT}. For an arbitrary user $(v \in \Gamma_s^{IN})$ who has contacted user u within lifecycle stage s we define this probability of interaction as follows:

$$Pr(v \mid \Gamma_s^{IN}) = \frac{|\{q : p \in P_u, q \in P_v, time(q) \in s, q \rightarrow p\}|}{\sum_{x \in \Gamma_s^{IN}} |\{q : p \in P_u, q \in P_x, time(q) \in s, q \rightarrow p\}|}$$

Given this formulation we now have time-dependent discrete probability distributions for a given user's in-degree and out-degree distributions, thereby allowing the *social* changes of users to be analysed in terms of the users communicating with a given user over time.

4.2 Lexical Dynamics

We modelled the *lexical dynamics* of users based on their term usage over time. We first retrieved all posts made by a given user within a lifecycle period and then removed stop words and filtered out any punctuation. We defined a multiset of the set of terms used by u in a given time period: $t \in C_s$ and a mapping function $g : C_s \rightarrow \mathbb{N}$ that returns the multiplicity of a given term's usage by the user at a given time period. We then defined the discrete conditional probability distribution for a given user u and lifecyle stage s based on the relative frequency distribution of terms used by u in that lifecycle period.

4.3 Modelling User Evolution

Given the use of discrete probability distributions derived for each dynamic (e.g. in-degree) and lifecycle stage (s) we can gauge the changes that each user goes

---

[2] We use $p \rightarrow q$ to denote message q replying to message p, P_u to denote posts authored by u, P to denote all posts.

through by assessing for changes in this distribution. To this end we assess: (i) period variation, using entropy; (ii) historical contrasts to assess how the user is diverging from prior dynamics, using cross-entropy measured between a given stage's distribution and all prior stage distributions and then taking the minimum, and; (iii) community contrasts to assess how the user is diverging from the general behaviour of the community, using cross-entropy also.

Period Variation. For each platform (Facebook, SAP, and ServerFault) we derived the entropy of each user in each of his individual lifecycle periods based on the in-degree, out-degree and term distributions; we then recorded the mean of these entropy values over each lifecycle period for churners and non-churners. Figure 2 shows the differences in the development signals between churners and non-churners for ServerFault users,[3] where, although the development signals remain relatively level across the lifecycle stages, there are clear differences in the magnitude of the entropy values - in particular for lower fidelity settings the 95% confidence intervals of the signals do not overlap. Such distinct signals between churners and non-churners resonate with the theory of social exchange: users who remain in the community share more connections (higher in-degree and out-degree entropy) and thus invest more and get more out of the community, churners meanwhile are the converse.

Historical Comparisons. Figure 3 shows the in-degree, out-degree and lexical period cross-entopies for Server Fault, deriving the values as above for the period variation measures for both the *churners* and *non-churners*. We note that across all of the plots churner signals are lower in magnitude than non-churners signals, indicating that the properties of the non-churners tend to have a greater divergence with respect to earlier properties than the churners. This suggests that churners' behaviour is more *formulaic* than non-churners, that is they exhibit less divergence from what has occurred beforehand. In general, the curve of churners and non-churners diminishes towards the end of their lifecycles but with different gradients.

Community Comparisons. For examining how users diverged from the community in which they were interacting we used users' in-degree, out-degree and lexical term distributions and compared them with the same distributions derived globally over the same time periods. For the global probability distributions we used the same means as for forming user-specific distributions, but rather than using the set of posts that a given user had authored (P_u) to derive the probability distribution, we instead used all posts to return Q.[4] We then calculated the the cross-entropy as above between the distributions. ($H(P_u, Q)$)

---

[3] We use this platform throughout as an example due to brevity. The remaining platforms exhibit similar curves.

[4] For instance, for the global in-degree distribution we used the frequencies of received messages for all users.

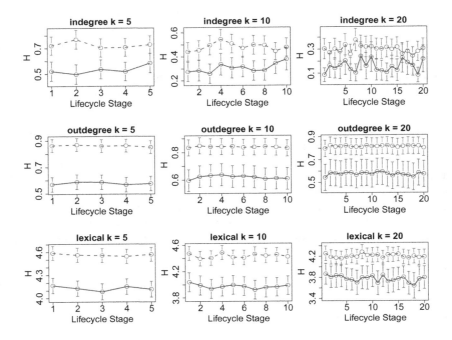

Fig. 2. Period entropy distribution on ServerFault for different fidelity settings (k) for users' lifecycles and different measures of social (indegree and out degree) and lexical dynamics. The green dashed line shows the non-churners, while the red solid line shows the churners.

over the different lifecycle stages. Again, as with period cross-entropies, we find churners' signals to have a lower magnitude than non-churners suggesting that non-churners' properties tend to diverge from the community as they progress throughout their lifetime within the online community platforms.

5   Churn Prediction from Gaussian Sequences

Above we plotted the 95% confidence intervals of a given measurement m (e.g. the period entropy of users' in-degree at lifecycle stage 1) for both churners and non-churners. If we assume that the distribution of a given measurement (m) at a particular lifecycle stage (s) is normally distributed, then for each measurement we have two signals (one for churners and one for non-churners) that each correspond to a *sequence* of Gaussians measured over the k lifecycle stages:

Definition 1 (Gaussian Sequence). *Let m be a given measurement, s be a given lifecycle stage drawn from the set of lifecycle stages $s \in S$, then m is said to be normally distributed on s and defined by $\mathcal{N}\big(\hat{\mu}_{m,s}, (\hat{\sigma}_{m,s})^2\big)$ where $\hat{\mu}_{m,s}$*

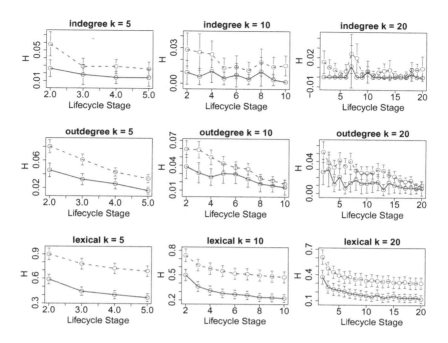

Fig. 3. Period cross-entropy distribution on ServerFault for different fidelity settings (k) for users' lifecycles and different measures of social (indegree and out degree) and lexical dynamics

and $\hat{\sigma}_{m,s}$ denote the maximum likelihood estimates of the mean and standard deviation respectively. Then the Gaussian Sequence of m is defined as follows:
$$G_m = \left(\mathcal{N}\!\left(\hat{\mu}_{m,1}, (\hat{\sigma}_{m,1})^2\right), \mathcal{N}\!\left(\hat{\mu}_{m,2}, (\hat{\sigma}_{m,2})^2\right), \ldots, \mathcal{N}\!\left(\hat{\mu}_{m,|S|}, (\hat{\sigma}_{m,|S|})^2\right) \right).$$

5.1 Single-Gaussian Sequence Model

Under the assumption that a given measurement has a Gaussian distribution at s then for an arbitrary user (u) we may measure the likelihood that the user belongs within a given distribution given his measurement at that stage. Using the convenience function $f(u, m, s)$ we can compute the probability that the user belongs to the *churn* gaussian, at that time step (s), using:

$$P(u|\beta_{m,s}) \propto \beta_{m,s} \mathcal{N}\!\left(f(u, m, s)|\hat{\mu}_{m,s}^c, (\hat{\sigma}_{m,s}^c)^2\right)$$

In the above equation, $\mathcal{N}\!\left(f(.)|\hat{\mu}, \hat{\sigma}^2\right)$ defines the conditional probability of the observed measurement $f(.)$ being drawn from the given gaussian of measure m in lifecycle stage s. We have also included a slack variable $\beta_{m,s}$ to control for *influence* on the churn probability; its inclusion is necessary because we may have an outlier measure for u and should limit over fitting as a consequence

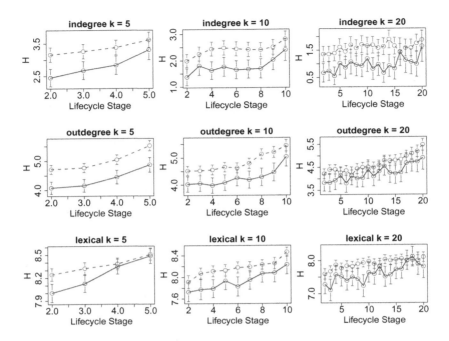

Fig. 4. Community cross-entropy distribution for different fidelity settings (k) for users' lifecycles and different measures of social (indegree and out degree) and lexical dynamics

- note that this variable is indexed by both m and s as it is specific to both the lifecycle stage, and the measure under inspection. Given our formulation of the churn probability in a particular lifecycle stage s and based on measure m, we can therefore derive the joint probability of u churning over the observed sequence of measures $(m \in M)$ and his lifecycle stages $(s \in S)$ as follows - we term this the *Single-Gaussian Sequence Model*:

$$Q(u|\mathbf{b}) = \prod_{m \in M} \prod_{s \in S} \rho\Big(\beta_{m,s} \mathcal{N}\big(f(u,m,s)|\hat{\mu}^c_{m,s}, (\hat{\sigma}^c_{m,s})^2\big)\Big)$$

The parameter ρ *smooths* zero probability values given our joint calculation. Assuming we have $|S|$ lifecycle stages, and $|M|$ measures, then the slack variables are stored within a parameter vector: $\mathbf{b}$ where - where $\beta_{m,s} \in \mathbf{b}$.

5.2   Dual-Gaussian Sequence Model

The above formulation can be extended further to include two *competing* Gaussian distributions at a particular lifecycle stage: the *churn gaussian*, formed from measurements of the known churner users, and; the *non-churn gaussian*, formed from measurements of known non-churners. We can therefore adapt the probability of the user belonging to the churn gaussian to be as follows:

$$P(u|\beta_{m,s}) \propto \Big[\beta_{m,s}\mathcal{N}\big(f(u,m,s)|\hat{\mu}^c_{m,s},(\hat{\sigma}^c_{m,s})^2\big)$$

$$-\,(1-\beta_{m,s})\mathcal{N}\big(f(u,m,s)|\hat{\mu}^n_{m,s},(\hat{\sigma}^n_{m,s})^2\big)\Big]_+$$

In this instance we wrap the subtraction of the churn-distribution membership probability by the β-scaled non-churn-distribution membership probability within the positive value operand $[]_+$ in order to return a non-negative value. As above, we can then calculate the joint churn probability over observed measures and lifecycle stages as follows - we term this the *Dual-Gaussian Sequence Model*:

$$Q(u|\mathbf{b}) = \prod_{m\in M}\prod_{s\in S}\rho\Big[\beta_{m,s}\mathcal{N}\big(f(u,m,s)|\hat{\mu}^c_{m,s},(\hat{\sigma}^c_{m,s})^2\big)$$

$$-\,(1-\beta_{m,s})\mathcal{N}\big(f(u,m,s)|\hat{\mu}^n_{m,s},(\hat{\sigma}^n_{m,s})^2\big)\Big]_+$$

5.3   Model Learning: Dual-Stochastic Gradient Descent

For both the single and dual-gaussian models, our objective is to minimise the squared-loss between a user's forecasted churn probability and the observed churn label - given that the former is in the closed interval $[0,1]$ and the latter is from the set $\{0,1\}$ - our parameters are L2-regularised to control for over-fitting:

$$\underset{\mathbf{b}^*}{\arg\min} \sum_{(\mathbf{x}_i,y_i)\in\mathcal{D}} \big(y_i - Q(u|\mathbf{b})\big)^2 + \lambda||\mathbf{b}||_2^2 \qquad (1)$$

Using this objective, we can then use gradient descent to calculate the setting of each $\beta \in \mathbf{b}$ by minimising the loss between a single user's forecasted churn probability and his actual churn label (i.e. either 0 - did not churn - or 1 - did churn). We experimented with two learning procedures: stochastic gradient descent (SGD), and dual-stochastic gradient descent (D-SGD) - the latter being a novel contribution of this paper. This latter procedure learns $\mathbf{b}$ with the approach in Algorithm 1, which takes as input a given regularisation weight λ, learning rate η, smoothing variable ρ, the dataset to use for parameter tuning D, the dimensionality of the feature space m, and the convergence threshold ϵ. The algorithm runs two loops: the *outer* loop (lines 4-12) shuffles the order of the dataset's instances and iterates through them, the inner loop (lines 7-11) then shuffles the order of the features. For each feature, the error of predicting the churn label of the user is derived (line 9) and this is used to update the parameter for feature j in the model; this process is repeated until the model's parameters have converged to a degree less than ϵ. We used D-SGD here to avoid sequential updating of parameters, and to examine its effects.

Algorithm 1. Learning the model's parameters using dual-stochastic gradient descent. **Input:** λ, η, ρ, D, m, ϵ. **Output: b**

1. $\mathbf{b} = \mathbf{0}^m$; $\mathbf{b}_{OLD} = random(m, [0, 1])$; $J = \{0, 1, \ldots m\}$
2. **while** $|\beta - \beta_{OLD}| < \epsilon$ **do**
3. $\beta_{OLD} = \beta$
4.   Shuffle D
5.   **for** $(\mathbf{x}_i, y_i) \in D$ **do**
6.   Shuffle J
7.   **for** $j \in J$ **do**
8. $e = y_i - Q(i|\mathbf{b})$
9. $\beta_j \leftarrow \beta_j + \eta * (e - \lambda\beta_j)$
10.   **end for**
11.   **end for**
12. **end while**
13. **return b**

6 Evaluation

We now turn to the evaluation of the above models, there are two stages to this: we begin by first tuning the various models' hyperparameters, before then applying the best performing model hyperparameters to the held-out test split of users. All proposed models use a fixed smoothing setting of $\rho = 0.1$ and the convergence threshold to be $\epsilon = 10^{-7}$.

6.1 Model Tuning: Setup

For the above proposed gaussian sequence models we have two hyperparameters that are to be tuned: (i) λ the regularisation weight, and; (ii) η the learning weight. For each model and learning routine (i.e. stochastic or dual-stochastic gradient descent) we set the possible settings for the hyperparameters of each be from $\{10^{-8}, 10^{-7}, \ldots, 10^{-1}\}$. To tune the hyperparameters we used 10-fold cross-validation over the training split with 9 segments for training using a given hyperparameter vector ($\theta = \{\lambda, \eta\}$) to derive the parameter vector $\mathbf{b}$, we then applied this to the 1 segment held-out and recorded the Area Under the ROC Curve (ROC). We repeated this 10 times for the 10 different segments and recorded the mean of these as the 10-fold CV average ROC. Appendix A presents the tuned hyperparameters for the proposed models and learning procedures.

6.2 Baselines

In order to judge how well our approach, and its variant models, performs against existing work we included two baselines. The first baseline we denote as B1-J48: for this we induced a J48 decision tree classifier using the above mentioned features (e.g. in-degree entropy of a user in lifecycle stage 1) using the training split users and applied this to the test split. For the second baseline, that we denote by B2-NB, we implemented the approach from [2] using features derived from the social network of users: *in-degree, out-degree, closeness-centrality, betweenness-centrality, reciprocity, average number of posts in initiated threads, average number of posts within participated threads, popularity (% of user authored posts that receive replies), initialisation (% of threads authored by the user), and polarity.* We first tested the J48 classifier, as used in [2], but found this to be poor performing[5] therefore we used the Naive Bayes classifier instead.

Table 2. Area under the Receiver Operator Characteristic (ROC) Curve results for the different Gaussian Sequence Models and Learning Procedures

Platform	Lifecycle Fidelity	Baselines		SGD		D-SGD	
		B1-J48	B2-NB	Single-$\mathcal{N}$	Dual-$\mathcal{N}$	Single-$\mathcal{N}$	Dual-$\mathcal{N}$
Facebook	5	0.559	0.461	**0.570**	0.472	0.548	0.478
	10	0.531	0.491	0.569	0.554	**0.593**	0.545
	20	0.478	0.444	**0.664**	0.500	0.528	0.583
SAP	5	**0.594**	0.497	0.573	0.527	0.545	0.533
	10	0.533	0.494	0.553	0.503	0.584	**0.590**
	20	0.478	**0.582**	0.500	0.500	0.540	0.525
ServerFault	5	0.583	0.530	0.522	0.556	**0.583**	0.577
	10	0.534	0.546	0.500	0.557	0.569	**0.589**
	20	0.463	0.530	0.500	**0.634**	0.486	0.484
Boards.ie	5	0.504	**0.611**	0.524	0.547	0.526	0.518
	10	0.512	**0.593**	0.500	0.539	0.501	0.496
	20	**0.560**	0.553	0.500	0.501	0.500	0.502

6.3 Results: Churn Prediction Performance

For the model testing phase of the experiments, we took the best performing hyperparameters for each model and learning procedure, trained the model using this setting using with entire training split, and then applied it to the test split; we did this twenty-times for each model (as each induction of the parameter vector is affected by the stochastic nature of the learning procedure) and took the average ROC value. These ROC values for the different models and baselines are shown in Table 2. The results show that for certain proposed models we significantly outperformed the baselines for two of the datasets.[6] Surpassing B1-J48 indicates that our proposed Gaussian models beat a widely-used classification model when detecting churners - given that this baseline makes use of the same features as our proposed model.

The results indicate variance across the prediction model as to which model performs best and under what conditions. For instance, the single-gaussian model performs better overall than the dual-gaussian model: this is largely due to the latter model smoothing zero-probability values through the setting of ρ.

---

[5] We also tested support vector machines and the perceptron classifier.

[6] Testing for significance using the Student T-test for independent samples.

Future work will experiment with ρ, either by indexing prediction models by this value or by tuning it as a hyperparameter. There appears to be no discernible *winner* in terms of the learning procedure to adopt, thus with dual-stochastic gradient descent being more computationally expensive we would lean towards using stochastic gradient descent in its place.

7   Conclusions and Future Work

In this paper we have presented a means to predict churners based on Gaussian Sequences. Our approach assumes that measures of user development are normally distributed, at each discrete lifecycle stage, and from which the probability of a user belonging to a churner or non-churner class can be gauged. We proposed two models to detect churners: the first using a single Gaussian Sequence formed from known churners' development information, and a second approach using dual-Gaussian Sequences from both churners' and non-churners' development information. Evaluation demonstrated that our detection models outperformed the two baselines - including the popular J48 decision tree classifier - for two of the tested online community datasets.

To the best of our knowledge this is the first work to *directly* compare the development signals of churners and non-churners and use that information to inform predictions. Our own prior work [8] focused on inducing regression models that capture the development trajectory of the above-mentioned measures. We implemented this same approach across the reduced lifecycle fidelity settings (i.e. $k = \{5, 10\}$) and found that: (i) its fit was poor for lower lifecycle fidelities; and (ii) prediction experiments resulted in low ROC values, in many cases zero. The presented approach in this paper therefore surpasses our own prior work in terms of its applicability to lower settings of lifecycle fidelities, and thus more users.

The first area of further work will explore the use of different objective functions that are to be optimised: above we used a reduction in the squared-error, yet an objective that accounts for rankings of users, based on their churn probability, would be better suited given the use ROC as our evaluation measure. The second area of future work will explore the task of *churn point prediction*: forecasting the day at which the user posts for the last time, our approach is amenable to such a setting via changing predictive function's codomain.

References

1. Karnstedt, M., Hennessy, T., Chan, J., Basuchowdhuri, P., Hayes, C., Strufe, T.: Churn in social networks. In: Handbook of Social Network Technologies and Applications, pp. 185–220. Springer (2010)
2. Karnstedt, M., Rowe, M., Chan, J., Alani, H., Hayes, C.: The effect of user features on churn in social networks. Proceedings of the ACM WebSci. 11, 14–17 (2011)
3. Kwak, H., Chun, H., Moon, S.: Fragile online relationship: a first look at unfollow dynamics in twitter. In: Proceedings of the SIGCHI Conference on Human Factors in Computing Systems, pp. 1091–1100. ACM (2011)

4. Haewoon Kwak, Sue B Moon, and Wonjae Lee. More of a receiver than a giver: Why do people unfollow in twitter? In: ICWSM (2012)
5. Lewis, K., Gonzalez, M., Kaufman, J.: Social selection and peer influence in an online social network. Proceedings of the National Academy of Sciences 109(1), 68–72 (2012)
6. McGowan, D., Brew, A., Casey, B., Hurley, N.J.: Churn prediction in mobile telecommunications. In: Proceedings of the 22nd Irish Conference on Artificial Intelligence and Cognitive Science (2011)
7. Quercia, D., Bodaghi, M., Crowcroft, J.: Loosing friends on facebook. In: Proceedings of the 3rd Annual ACM Web Science Conference, pp. 251–254 (2012)
8. Rowe, M.: Mining user lifecycles from online community platforms and their application to churn prediction. In: 2013 IEEE 13th International Conference on Data Mining (ICDM), pp. 637–646. IEEE (2013)
9. Zhang, X., Liu, Z., Yang, X., Shi, W., Wang, Q.: Predicting customer churn by integrating the effect of the customer contact network. In: 2010 IEEE International Conference on Service Operations and Logistics and Informatics (SOLI), pp. 392–397. IEEE (2010)

A   Appendix: Model Tuning Results

Table 3. Tuned hyperparameters for the various proposed models as λ, η pairs

Platform	Fidelity	SGD Single-$\mathcal{N}$	SGD Dual-$\mathcal{N}$	D-SGD Single-$\mathcal{N}$	D-SGD Dual-$\mathcal{N}$
Facebook	5	$0.1, 0.01$	$0.1, 0.01$	$10^{-5}, 0.1$	$0.1, 0.01$
	10	$0.1, 0.01$	$10^{-5}, 0.01$	$0.01, 0.1$	$10^{-5}, 0.01$
	20	$0.1, 0.01$	$0.001, 0.1$	$0.1, 0.1$	$0.01, 0.1$
Sap	5	$0.1, 0.001$	$10^{-5}, 0.01$	$10^{-6}, 0.01$	$0.01, 0.01$
	10	$0.1, 0.01$	$0.1, 0.01$	$0.001, 0.1$	$0.01, 0.1$
	20	$0.1, 0.01$	$0.1, 0.01$	$0.001, 0.1$	$0.001, 0.1$
ServerFault	5	$10^{-5}, 0.01$	$0.1, 0.01$	$0.01, 0.1$	$0.1, 0.01$
	10	$0.1, 0.1$	$0.01, 0.01$	$0.001, 0.1$	$0.01, 0.1$
	20	$10^{-6}, 0.1$	$0.01, 0.01$	$10^{-5}, 0.1$	$0.01, 0.1$
Boards.ie	5	$10^{-6}, 0.001$	$0.1, 0.001$	$0.001, 0.1$	$0.1, 0.001$
	10	$0.1, 0.1$	$0.01, 0.001$	$0.001, 0.1$	$0.001, 0.001$
	20	$0.1, 0.1$	$0.1, 0.01$	$0.1, 10^{-6}$	$10^{-5}, 0.1$

GitHub Projects.
Quality Analysis of Open-Source Software

Oskar Jarczyk[1], Błażej Gruszka[1], Szymon Jaroszewicz[2,3], Leszek Bukowski[1], and Adam Wierzbicki[1]

[1] Polish-Japanese Institute of Information Technology
Department of Social Informatics
Warsaw, Poland
{oskar.jarczyk,blazej.gruszka,bqpro,adamw}@pjwstk.edu.pl
[2] National Institute of Telecommunications
Warsaw, Poland
s.jaroszewicz@itl.waw.pl
[3] Institute of Computer Science
Polish Academy of Sciences
Warsaw, Poland

Abstract. Nowadays Open-Source Software is developed mostly by decentralized teams of developers cooperating on-line. GitHub portal is an online social network that supports development of software by virtual teams of programmers. Since there is no central mechanism that governs the process of team formation, it is interesting to investigate if there are any significant correlations between project quality and the characteristics of the team members. However, for such analysis to be possible, we need good metrics of a project quality. This paper develops two such metrics, first one reflecting project's popularity, and the second one - the quality of support offered by team members to users. The first metric is based on the number of 'stars' a project is given by other GitHub members, the second is obtained using survival analysis techniques applied to issues reported on the project by its users. After developing the metrics we have gathered characteristics of several GitHub projects and analyzed their influence on the project quality using statistical regression techniques.

Keywords: OSS, online collaboration, performance metrics, survival analysis.

1 Introduction

Very often *Open-Source Software* (*OSS*) is developed by decentralized teams of programmers, who cooperate globally using web-based source code repositories. There are several features typically associated with such *Collaborative Innovation Networks* (*COINs*): (a) voluntary work; (b) low organizational costs; (c) meritocracy. In recent years COINs have proved their ability to produce high quality software. Leading example of such network is the *GitHub website*, an

L.M. Aiello and D. McFarland (Eds.): SocInfo 2014, LNCS 8851, pp. 80–94, 2014.

online social network that supports development of software by virtual teams of programmers. Every *GitHub user* can create his/her own repository and work on it with other registered users. They may also join projects created by somebody else and make their own contributions there. *GitHub* has no recommendation system for developers, which would support their decisions on how to contribute effectively from a scratch. Every GitHub user makes his/her own decision on how to manage their personal time and professional skills - that is the reason why the process of *team formation* on GitHub is decentralized and might be considered as self-organizing. In other words, there is no central mechanism governing the formation of OSS teams.

It sheds light on the puzzling fact that even though open source software (OSS) constitutes public good, it is being developed for free by highly qualified, young, motivated individuals, and evolves at a rapid pace. We show that when OSS development is understood as the private provision of public good, these features emerge quite naturally. We adapt a dynamic private provision of public goods model to reflect key aspects of the OSS phenomenon, such as play value or homo ludens payoff, user-programmers' and gift culture benefits. Such intrinsic motives feature extensively in the wider OSS literature and contribute new insights to the economic analysis

1.1   Problem Definition

Foregoing facts awaken our interest in investigating if there are any significant correlations between *project quality* and characteristics of the *team members*. We consider project quality as consisting of two aspects: one is the *number of stars* that any project might receive from GitHub users, and the second one is the response of project team to issues reported for a given repository. In case of the first indicator, we believe that community reaction to the project is a proper measure of its quality. Any GitHub user is able to gratitude a chosen projects with a *star* – it shows his admiration and positive attitude towards chosen repository.

It is very important for every kind of software to have a good support - that is a team of people, who are able to respond, in case when the community of users reports some *bugs* and *feature requests*. Considering any piece of software, bugs constitutes an almost inevitable part of its lifetime – even alpha and beta tests are not able to rule out all possible problems with the software. Moreover, community of users is the best source of information about the performance of solutions that have been implemented and about the lack of some features, which might significantly improve usability of the created system. Open-Source Software developed on GitHub by COINs is no exception here – it also needs maintenance of issues reported by community of users.

Github platform has a distinct functionality for reporting issues on a project: it allows GitHub users to report such things as bugs, feature requests or enhancements to the team of developers. The categories of all possible issues might be defined by the owner of the repository. For each repository from GitHub we have a record of its issues survival - data about moment when a particular issue had been opened, and eventually when it was closed. The analysis of survival of those issues gives us insights about typical life duration of issues in different kinds of GitHub projects.

We believe that *issue survival* is one of the indicators of GitHub *team quality*. Our assumption is, that well organized and motivated teams tend to maintain and swiftly close issues associated with their repositories, and measuring the time of issue closure, together with other *predictor variables* describing GitHub repositories, is a good metric of quality for the team maintaining a given repository.

We have also gathered several characteristics of GitHub projects and analyzed their influence on project popularity and the quality of support offered to users. Several interesting conclusions were made, for example, it is better to attract focused active developers to the project than to attract popular members.

2   Related Work

Questions concerning the problem of *quality* in *Open-Source Software* (OSS) and *COIN*s has been investigated by several researchers. A generic review of the empirical research on Free/Libre and Open-Source Software (*FLOSS*) development and assessment of the state of the literature might be found in ACM article by Crowston et.al. (2008) *"Free/Libre Open-source Software Development: What We Know and What We Do Not Know"*.[2]. In publication *"Software Product Quality Models"* by Ferenc et.al. (2014) authors provide a brief overview about the history of software product quality measurement, focusing on software maintainability, and the existing approaches and high-level models for characterizing software product quality. Based on objective aspects, the implementations of the most popular software maintainability models are compared and evaluated. This paper also presents the result of comparing the features and stability of the tools and the different models on a large number of open-source Java projects.[5] However, we have not found many papers that directly investigate relations between projects issues survival and its quality.

A solid understanding of online collaboration is provided by research on wikiteams. Wikipedia is a laboratory for open, virtual teamwork.[17] It is also a community similar to GitHub, because collaboration manifests through a swarm creativity which is a part of COIN model. Scholars Hupa et.al. (2010) enhance expert matchmaking and recommender systems with multidimensional social networks (MDSN).[8] According to them, dimensions of: trust, acquaintance and knowledge store information about the social context of an individual, as well as team's social capital, intra-group trust and skill difference. Social network is based on the entire Wikipedia edit history, and therefore is a summary of all recorded author interactions.[18] Using information from these dimensions they define a criteria that predict team performance.[8] A dimension of distrust is added to model because of its beneficial behaviour to teams quality.[19]

Rahmani, Khazanchi (2010) published *"A Study on Defect Density of Open Source Software"*, where they present an empirical study of the relationship between defect density and download number, software size and developer number as three popular repository metrics. Contrary to theoretical expectations, their regression analysis discovered no significant relationship between defect density and number of downloads in OSS projects. Yet, researcher find that the number of developers and software project size together present some promise in explaining defect density of OSS projects. They plan to explore other potential predictors for defect density in OSS projects, together with the use of non-linear regression to explain the trends in defect density associated with OSS project.[16]

Michlmayr, Senyard (2006) in their paper *"A Statistical Analysis of Defects in Debian and Strategies for Improving Quality in Free Software Projects"* analyse 7000 tickets from the Debian issue tracking system. This data accumulated during over 2.5 years allowed them to make conclusions regarding a high-maturity project through analysing their issues closure. Scholars found that the number of issues is increasing together with the decrease in a defect removal rate. Scientist found that frequent releases lead to shorter defect removal times and possibly to more user feedback. Secondly, they argued that a close interaction with the upstream authors of free software is beneficial, and upstream authors gain from wide testing and more user feedback. Finally, working in groups increases the reliability of volunteer maintainers and leads to shorter defect removal times.[13]

Related to our approach is work by Fischer et.al. (2003) *"Analyzing and relating bug report data for feature tracking""*, where bug reports tracks were used to investigate software evolution. Authors method has been validated using the large open source software project of *Mozilla* and its bug reporting system *Bugzilla*. Their approach uncovers hidden relationships between features via problem report analysis and presents them in easy to evaluate visual form.[6]

As one can observe, there is a lot of research concerning quality in Open-Source Software, and their number happens to grow after the success of the *SourceForge* and *GitHub* portal. Internet databanks, which aggregate data from different web-based online source repositories, make for the analysis of Open-Source Software easier and wider. Researchers Farah et.al. (2014) published work titled *"Open-Hub: A scalable architecture for the analysis of software quality attributes"* where they analyze 140, 000 *Python* repositories under quality attributes - performance, testability, usability, maintainability. Scientists merged information on Python repositories collected from GitHub with metrics generated by OpenHub (formerly Ohloh) - an internet aggregator for OSS repositories.[4]

Interesting analysis of activity fade-out in OSS projects are presented in *"Is it all lost? A study of inactive open source projects"* by Khondhu et.al. (2013). Researchers quote an informal rule, according to which "when developers lose interest in their project, their last duty is to hand it off to a competent successor". However, mechanism of such hand-off is not widely known among OSS users. Paper goal is to differentiate projects that had maintainability issues from those that were inactive for other reasons.[11]

A discussion about central management vs. OSS is covered in book by O'Reilly Media *"Making Software: What Really Works, and Why We Believe It"*.[15] Mahony, Ferraro (2007) prove that successful communities structure their work and that good communities and teams are self-governing.[14] There also have been attempts to support programmers work, by recommendation engines. In paper from Zimmermann et.al. (2014) *"Mining version histories to guide software changes"* data mining has been applied to version histories, in order to guide programmers along the related changes. Their system prototype called ROSE was able to correctly predict 26% of further files to be changed—and 15% of the precise functions or variables.[20]

In work of Kalliamvakou et.al. (2014) researchers indicate that, although GitHub is a rich source of data, there are also potential perils that should be taken into consideration. Among other things they show that the majority of the projects on GitHub are personal and inactive. According to their research two thirds of projects (71.6% of repositories) are personal – the number of committers per project is very skewed: 67% of projects have only one committer, 87% have two or less, and 93% three or less. This findings shows that most of the users do not use GitHub for collaboration on projects. However, in our studies we have been focused on most popular repositories, which eliminates one-person projects[10].

Finally, different case studies of chosen GitHub repositories reveal even more interesting conclusions. In paper *"Social coding in GitHub: transparency and collaboration in an open software repository"* by Dabbish et.al. (2012) a series of in-depth interviews with central and peripheral GitHub users was performed. Authors claim that people make a surprisingly rich set of social inferences from the networked activity information in GitHub, such as inferring someone else's technical goals and vision when they edit code, or guessing which of several similar projects has the best chance of thriving in the long term. It is suggested that users combine these inferences into effective strategies for coordinating work, advancing technical skills and managing their reputation.[3] Another series of interviews with GitHub developers reader might be found in *"Performance and participation in open source software on GitHub"* McDonald et.al. (2013). Authors conducted qualitative, research with lead and core developers on three successful projects on GitHub. They aim was to understand how OSS communities on GitHub measure success. Two main findings were reported: first, lead and core members of the projects display a nuanced understanding of community participation in their assessment of success; second, they attribute increased participation on their projects to the features and usability provided by GitHub.[12]

3   Dataset

3.1   Predictor Variables

We now describe the dataset containing representative GitHub projects used in this paper. We used *Google BigQuery* online tool to create a list of GitHub repositories (or 'repos' for short), sorted descending by their highest peak in trend (received attention from Internet users) during a month. We define trend by the biggest increase in popularity during a month. Popularity of a repository is measured by its number of stars (number of 'stargazers'). We analysed mature repositories existing for at least two years. There are together 164418 GitHub repositories on the list. This list simply consists of repositories, but lacks information on their team members (contributors and collaborators). From this big set of repositories we selected 2000 of them with the highest increase in popularity during a month. In this way we avoid taking into consideration projects which are personal or inactive – inactive repositories were one of the main perils of GitHub data described in Kalliamvakou et.al. (2014)[10].

Each record in our dataset has 12 columns, which are: 'repository owner', 'repository url', 'repository name', 'biggest increase in popularity', 'repository description', 'is a fork', 'wiki enabled', 'when pushed at', 'master branch', 'issues enabled', 'downloads enabled', 'repository creation date'. Additionally, we also have information on below values on any moment of time during the repository existence: 'number of stars', 'number of forks', 'number of pushes', 'how many issues open/closed etc.

Next step is to receive information on developers in those entry teams, which we call x-axis attributes for a repository. For this purpose, we use *GitHub API* to parse missing information. For the mentioned 2000 repositories we downloaded through a script (available freely here: https://github.com/wikiteams/supra-repos-x) below additional information on a developer: 'developer username (login)', 'developer name', 'developer followers count', 'developer following count', 'developer company', 'number of repos developer contributed to', 'number of repos he owns', 'date when developer registered', 'developer location', 'is developer hireable', 'is developer working during business hours', 'developer typical working period', 'gists count', 'private repos count'. Also, more properties for a repository (*y-axis*) are downloaded: 'repository default branch', 'opened issues count', 'repository organization', 'repository language (main technology)'.

Good source of general developer activity on GitHub is a data source called *OSRC* report card, from where we download aggregated data regarding the *user activity time*. We calculate two additional attributes. Firstly, we want to check whether the developer contributes mostly during working hours (between 9 and 17 o'clock) in his local time, or he is an active GitHub user but committing beyond this period of time. Secondly, we calculate a working period (in hours) for this developer. We define a working period as a sum of hours in the biggest rectangle drawn on the daily activity histogram.

3.2   Repositories Issues

Any change in an *Issue* is recorded in a databank called *GitHub Archive* (in short - 'GHA'). It is a third party project to record the public *GitHub timeline* and make it easily accessible for further analysis. In GHA, every time when some issue is opened, closed or reopened, 'IssuesEvent' is stored to a database. Firstly, we downloaded all data collected in year 2013 from the *GitHub Archive*. Secondly, we selected IssuesEvents to create a history of issue creation on all *GitHub repositories* during that year. IssuesEvent is triggered whenever an issue is created, closed or reopened, and the GHA collects those events.

Data was merged into full information record on each single issue. It contained opening and closing date of the issue and a calculated difference: the time span. Once created, the issue in a GitHub repository cannot be deleted, it can be only closed. Finally, we used the *GitHub API* to query for issue labels, which GHA didn't provide.

We managed to create a dataset of issues with following attributes: repository owner (a person or organization who manages the code repository as a privileged user), repository url (an 1-1 identifying repository key, a web address which allows to view the repository in a browser), repository name, issue number, issue status (opened or closed), 'opened at' (when was the issue created), 'closed at' (when was the issue last time resolved), difference in minutes (also hours and days, difference between fields 'opened at' and 'closed at').

3.3   Data Preprocessing

Since we are drawing conclusions on the whole projects, not individual developers, characteristics of project members had to be aggregated into single attributes. Here we simply computed means of each attribute over all project members. Such attributes are prefixed by 'average.' (alias 'avg.').

Many attributes exhibited highly skewed, power-law like distributions, which are difficult to model with statistical methods. Logarithmic transformation $x' = \log_{10}(x + 10)$ has been applied to the following attributes to decrease the skew:

'forks_count', 'network_count', 'average.developer_followers',
'average.developer_following', 'average.developer_contributions',
'average.developer_total_public_repos', 'average.developers_works_period',
'average.public_gists', 'commits_count', 'branches_count', 'releases_count',
'contributors_count'.

4   Measures of Project's Quality

In order to discover factors influencing project quality we need to be able to precisely measure project quality. Unfortunately, the task is not easy, as there are many possible criteria, which are not always easy neither to measure nor to evaluate. In this chapter, we are going to introduce and describe two GitHub project *quality metrics*, based on project popularity and the quality of user support offered by team members.

4.1   Attractiveness and Popularity – Stargazers

The first metric we analyze is the number of *stars* the project has, i.e. how many times it has been endorsed by members of the GitHub community. For each project, we gathered the number of stargazers - users who starred a given project.

Since the stargazers count follows a *power-law distribution* (it means there are lots of projects with few stars and a few projects with a very large number of stars), it is not suitable for e.g. regression analysis. We applied logarithmic transformation $x' = \log_{10}(x + 10)$ before using it as a metric of project quality. The resulting quantity has a well behaved distribution as can be seen on its histogram shown in Figure no. 1. The offset 10 is provided to avoid taking logarithms of zero and to reduce skew for small values.

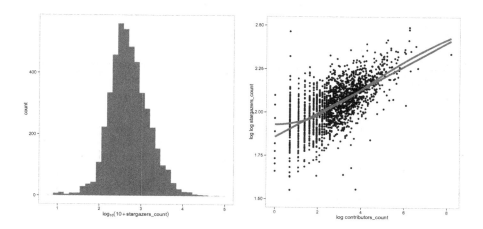

Fig. 1. Left - histogram of the number of stargazers for different project. Right - dependence of the number of stargazers on the number of contributors to a project.

The metric may than be used to analyze the factors influencing project quality. As an example of the type of analysis for which it can be used we show the dependence of the metric on the logarithm of the number of contributors to the project, see the right part of Figure no. 1.

The red line shows the linear regression fit and the blue line a nonlinear LOESS regression fit [1]. Clearly, the larger the number of contributors is, the larger the number of stars. This by itself is not surprising; however, what is interesting is that the number of stars grows exponentially with the number of contributors. Indeed, the nonlinear fit is almost identical to the linear one (recall that we use a double logarithm of the number of stars). It is not yet clear to us whether it is the project's popularity that attracts contributors or, vice-versa, the large number of contributors results in good and, consequently, popular projects.

4.2   Quality of Support – Survival of Issues

We now describe the second metric of a project quality introduced in this paper. It is based on the time it takes the project team members to close issues related to the project. From now on, we will use the tools of *survival analysis*.

Survival analysis focuses typically on times to a given event – it might be loss of some user, customer migration, or death in case of biological research. One of the typical questions, which survival analysis attempts to answer, is: what is the proportion of a population which survived a certain amount of time? Of course in case of *GitHub issues* our question is - what proportion of issues was not closed before some particular point in time.

A key aspect of survival analysis [9] is censoring - if an issue was opened just a month ago, at the current time point we do not know, whether it will be closed within a year or not. In order to handle censoring in a statistically proper way we use the *Kaplan-Meier estimates* of survival time for issues of a given project. The left part of Figure no. 2 shows the survival curve for issues of an example project. It can be seen that a certain percentage of issues is closed very rapidly, indicating active support of users by the project's team. However, older issues are often not closed at all. In total, about 50% of issues is not being addressed, suggesting that there is a high chance, that user problems will not be resolved. The '+' marks on the curve indicate the age of issues opened recently, which have not yet been closed and have not reached the maximum time displayed on the x axis. The right part of the figure shows the combined survival curve for issues of all analyzed GitHub projects. It can be seen that the response times are usually fast, but many issues have not been addressed at all.

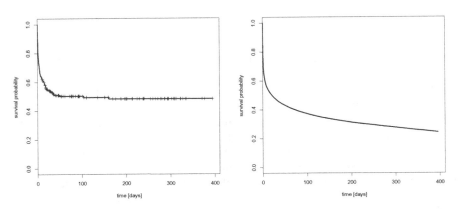

Fig. 2. Survival curves for issues of the GateOne project and for all projects combined

In order to facilitate the assessment of project quality, we devised summary metric for survival curves. To this end, we computed survival probabilities for issues after 1, 2, 3, 7, 30, 100, and 365 days. Performing the PCA *(Principal Components Analysis* on those probabilities revealed that just two components are enough to explain 96% of the variance of the seven probabilities. Further, the first component describes (roughly) the average of the percentages of bugs closed after different amounts of time, and the second one differentiates the probabilities of issues being closed rapidly, in a matter of days.

Consequently, we decided to summarize the survival curve for each project with just two numbers: the percentage of issues closed after 3 and 365 days. It turned out that those two numbers explain about 94% of the total variability of the seven issue survival probabilities—almost the same as the first two principal components—while being much easier to understand.

To summarize, we have provided the concise metric of GitHub project quality based on the time needed to respond to the issues. The metric is based on two numbers measuring how quickly the user may expect a response and what are the chances that his or her issue will be eventually resolved.

5 What Makes a Good GitHub Project?

In this section we are going to analyze how different characteristics of a project and its developers influence the two aspects of its project quality.

5.1 What Affects the Project Popularity?

We have conducted a regression analysis to discover factors influencing project popularity. Since the programming language specified for the project is a categorical attribute with many (55) values, we decided to exclude it from the initial analysis and analyze it separately in the next chapter (no. 5.2). Only the information whether the project's language was specified is included.

Application of linear regression resulted in a model with a very high multiple R^2 coefficient of 0.779. In other words, almost 80% of the variability of projects popularity (after the logarithmic transform) is explained by project features. The most significant variable in the model was `forks_count` - the number of times the fork of this project was created. The number of forks reflects general amount of activity in the project, so it is a logical conclusion, that active projects are more popular. Few other attributes turned out to be highly correlated with the number of forks and they also reflect general amount of project activity and a project size. Those attributes are commit_count, contributor_count, releases_count, branches_count and network_count. Another pair of mutually correlated attributes, repo.updated_at and repo.pushed_at, also reflects the amount of activity in the project.

Unfortunately, those correlations are of little practical use, since project activity is likely an *effect* of its popularity (developers are more likely to fork a project, if it is well known and attractive), not necessarily its cause, at least not the primary one. Those attributes have thus been removed from the dataset before further analysis.

A new regression model was built on the reduced dataset. Variables with statistically significance (p-value below 0.05) coefficients are shown in Table no. 1. The first column gives us the attribute's name, the second its coefficient in the regression model, and the third one is the p-value indicating statistical significance of the coefficient. Higher values of coefficients mean that a given quantity has a positive influence on project's popularity. Since the number of stars has been logarithmized, the coefficients should be interpreted multiplicatively. For example, the coefficient for repository creation time, -0.144 means that an increase in creation time by one year corresponds to the number of stars increasing $10^{-0.144} = 0.72$ times (in fact decreasing).

It can be seen, that the most significant attribute is project creation time. The coefficient is negative, so projects created later have, in general, less stars. This is obvious since, the stars accumulate over time giving older projects a natural advantage. A similar attribute is the average time at which developer accounts were created. Surprisingly this time the relation is opposite, having newer developers in the project is positively correlated with its popularity. This phenomenon can probably be explained by the fact that programmers may join Github in order to contribute to attractive projects.

Another observation is, that forks of other projects are in general much less popular. This is plausible since forks can be created very easily on Github and are often used as a part of the development process, not necessarily constituting separate projects. The language of the project being specified is correlated with popularity (see a dedicated section below for a discussion).

Another significant attribute is whether the project is owned by an organization. Projects owned by companies and other organizations are in general more popular.

Another two significant attributes: the average number of followers of developers in the project and the average number of developers/projects followed by them can be seen as an approximation of social relations of project members. Actually, it turns out that project whose developers follow many others are in general more popular. Surprisingly the effect for developers being followed by many others (i.e. having popular developers in the project) is much weaker. This discovery results in a practical advice for projects: finding developers engaged in the community is good for the project popularity and can be measured by a simple proxy quantity.

Another two related attributes are the average number of repositories owned by project members and the total amount of their contributions (including other projects). The coefficients here are negative offering another practical advice to project managers: try getting people who will be able to concentrate on your project without spreading attention on too many other projects.

5.2   Programming Language

We will now analyse how the project's programming language is related to its popularity. To discover this, we have built a regression model based on just one attribute, *repository language*. As previously mentioned, there are 54 programming languages used in the analyzed projects, plus an extra value for no language specified.

It turns out that the programming language has little effect on the projects popularity, with a few exceptions - significant influence was observed for only 4 cases. The most significant effect was that projects written in *Common Lisp* are

Table 1. Regression model

attribute	coefficient	p-value
repo.created_at	-0.144	$p < 2.0 \cdot 10^{-16}$
is_fork	-0.272	$p = 2.2 \cdot 10^{-6}$
language.specified	0.290	$p = 8.8 \cdot 10^{-7}$
organization.specified	0.163	$p = 5.0 \cdot 10^{-12}$
average.developer_followers	0.057	$p = 0.029$
average.developer_following	0.174	$p = 0.002$
average.developer_contributions	-0.094	$p = 0.009$
average.developer_created_at	0.120	$p = 2.8 \cdot 10^{-11}$
average.developer_total_public_repos	-0.163	$p = 0.042$
average.developers_works_period	0.108	$p = 5.3 \cdot 10^{-6}$
Observations	1755	
R^2	0.203	

less popular. The coefficient was -0.702, significant with p-value of $2.3 \cdot 10^{-3}$. After taking into account the log-transformation of the number of stars, this roughly translates to those projects having 5 times less stars than similar projects written in other languages. The probable explanation is that Common Lisp is an old technology, nowadays used only by a small fraction of developers for very specialized purposes.

Another significant fact was that projects that did not specify the programming language were also significantly less popular. An inspection revealed that many of those projects include color themes, documentation etc. which may not be very popular among users. Moreover, GitHub assigns project language automatically, based on file contents, so projects with no particular files are assigned to this category.

On the contrary, projects based on CSS styles were more likely to be popular, probably due to the growing popularity of web-based technologies.

5.3   What Affects the Quality of a Support?

We now move on to the analysis of the quality of projects from a short and long term support. Since the survival probabilities are not normally distributed, we have used *binomial regression* (a variant of logistic regression) to model it. Each data record has a number of trials n and a number of successes n_1 assigned. A generalized linear model is then built, which predicts the probability of success p, assuming that n_1 follows, in each record, the binomial distribution with parameters n and p (for more details, see *Hosmer, Lemeshow* book [7]). In our case n corresponds to the total number of project's issues, p to the estimated probability of bug survival. We set the n_1 to

$$n_1 = n \cdot p_t, t \in \{3, 365\}$$

where p_t is the *Kaplan-Meier* estimate of the fraction of surviving issues defined in the previous section. Hence, that n_1 needs to have non-integer values - however, this is not a problem for the implementation of logistic regression available in the R statistical package.

We begin by analyzing the influence of attributes related to project size and general activity. For both - short term (3 days bug survival) and long term (365 days bug survival) support - the most important attribute is the number of branches, which is negatively correlated with bug survival. This means that a large number of branches has a positive influence on the project. Since a typical *git workflow* for fixing a bug involves creating a branch, making the changes and merging the branch back, this correlation is logical.

Unfortunately, the number of branches is correlated with general project activity and thus, as was the case with project popularity, we removed this and correlated attributes before further analysis. Those attributes are commit_count, contributor_count, releases_count, branches_count and network_count. Another pair of mutually correlated attributes, repo.updated_at and repo.pushed_at.

The model was then rebuilt. Table no. 2 shows the significant regression coefficients for both short and long term bug survival. Note that negative values of the coefficients are desired here as they translate to lower numbers of surviving bugs.

Table 2. Regression coefficients for short and long term bug survival

3 day bug survival		
attribute	coefficient	p-value
repo.created_at	−0.014	$p = 0.049$
is_fork	−0.926	$p = 0.017$
has_downloads	−0.055	$p = 0.021$
average.developer_following	2.158	$p = 0.002$
average.developer_contributions	1.787	$p < 2.0 \cdot 10^{-16}$
average.developer_hireable	−1.598	$p = 5.7 \cdot 10^{-7}$
average.developer_total_public_repos	0.235	$p = 0.009$
average.developers_works_period	0.314	$p = 0.004$
365 day bug survival		
attribute	coefficient	p-value
organization.specified	0.079	$p = 0.005$
average.dev_name_given	−0.182	$p = 0.031$
average.developer_following	−1.194	$p = 0.001$
average.developer_contributions	1.337	$p < 2.0 \cdot 10^{-16}$
average.developer_hireable	2.317	$p = 0.002$
average.developer_works_during_bd	0.329	$p = 0.029$
average.public_gists	0.391	$p = 0.022$

Let us now comment on the significant attributes. First of all, it can be seen that having developers making many contributions (including other projects) and owning many repositories negatively influences the number of bugs fixed.

The same advice can be offered as in the case of project popularity - try getting into the project focused developers who will concentrate all their efforts on it.

The number of projects/developers followed by team members is again an important factor. However, surprisingly, it has a positive effect only on fixing bugs in long term, not on 'rapid response' to user issues. This issue needs to be investigated further.

The effect of projects being run by employed developers (attributes 'organization.specified' and 'average.developer_works_during_bd') is significantly negative towards addressing longstanding bugs. The probable reason is that organizations are unwilling to commit resources to fixing user issues and prefer to concentrate on aspects of the project which are important to them.

6   Conclusions and Future Research

Our paper presented *two measures of quality* for GitHub Open-Source Software projects. One is based on a project popularity, the other one is based on how fast the project's team solves issues reported by users. We have also collected several attributes describing projects and their developers and analyzed their influence on those quality measures. Together, it resulted in making several interesting discoveries. For example, it is better for a software project to have focused developers involved in the community rather than having in the team popular, often followed developers. Future work will focus on detailed studies of what aspects of a team collaboration affect a project quality.

Acknowledgements. This work is supported by Polish National Science Centre grant 2012/05/B/ST6/03364

References

1. Cleveland, W.S., Devlin, S.J.: Journal of the American Statistical Association 83(403), 596–610 (1988)
2. Crowston, K., Wei, K., Howison, J., Wiggins, A.: Free/libre open-source software development: What we know and what we do not know. ACM Comput. Surv. 44(2), 7:1–7:35 (2008)
3. Dabbish, L., Stuart, C., Tsay, J., Herbsleb, J.: Social coding in github: Transparency and collaboration in an open software repository. In: Proceedings of the ACM 2012 Conference on Computer Supported Cooperative Work, CSCW 2012, pp. 1277–1286. ACM, New York (2012)
4. Farah, G., Tejada, J.S., Correal, D.: Openhub: a scalable architecture for the analysis of software quality attributes. In: Proceedings of the 11th Working Conference on Mining Software Repositories, pp. 420–423. ACM (2014)
5. Ferenc, R., Hegedus, P., Gyimothy, T.: Software product quality models. In: Mens, T., Serebrenik, A., Cleve, A. (eds.) Evolving Software Systems, pp. 65–100. Springer, Heidelberg (2014)
6. Fischer, M., Pinzger, M., Gall, H.: Analyzing and relating bug report data for feature tracking. In: 2013 20th Working Conference on Reverse Engineering (WCRE), p. 90. IEEE Computer Society (2003)

7. Hosmer, D.W., Lemeshow, S.: Applied logistic regression. Wiley-Interscience Publication (2000)
8. Hupa, A., Rzadca, K., Wierzbicki, A., Datta, A.: Interdisciplinary matchmaking: Choosing collaborators by skill, acquaintance and trust. In: Abraham, A., Hassanien, A., Snášel, V. (eds.) Computational Social Network Analysis. Computer Communications and Networks, pp. 319–347. Springer, London (2010)
9. Kalbfleisch, J.D., Prentice, R.L.: The statistical analysis of failure time data. John Wiley & Sons (2002)
10. Kalliamvakou, E., Gousios, G., Blincoe, K., Singer, L., German, D.M., Damian, D.: The promises and perils of mining github. In: Proceedings of the 11th Working Conference on Mining Software Repositories, MSR 2014, pp. 92–101. ACM, New York (2014)
11. Khondhu, J., Capiluppi, A., Stol, K.-J.: Is it all lost? a study of inactive open source projects. In: Open Source Software: Quality Verification, pp. 61–79. Springer (2013)
12. McDonald, N., Goggins, S.: Performance and participation in open source software on github. In: CHI 2013 Extended Abstracts on Human Factors in Computing Systems, CHI EA 2013, pp. 139–144. ACM, New York (2013)
13. Michlmayr, M., Senyard, A.: A statistical analysis of defects in debian and strategies for improving quality in free software projects. The Economics of Open Source Software Development, 131–148 (2006)
14. O'Mahony, S., Ferraro, F.: The emergence of governance in an open source community. Academy of Management Journal 50(5), 1079–1106 (2007)
15. Oram, A., Wilson, G.: Making Software: What Really Works, and Why We Believe It. O'Reilly Media (2010)
16. Rahmani, C., Khazanchi, D.: A study on defect density of open source software. In: 2010 IEEE/ACIS 9th International Conference on Computer and Information Science (ICIS), pp. 679–683. IEEE (2010)
17. Turek, P.: Wikiteams: How do they achieve success? IEEE Potentials 30(5), 15–20 (September 2011)
18. Turek, P., Wierzbicki, A., Nielek, R., Hupa, A., Datta, A.: Learning about the quality of teamwork from wikiteams. In: 2010 IEEE Second International Conference on Social Computing (SocialCom), pp. 17–24 (August 2010)
19. Wierzbicki, A., Turek, P., Nielek, R.: Learning about team collaboration from wikipedia edit history. In: Proceedings of the 6th International Symposium on Wikis and Open Collaboration, WikiSym 2010, pp. 27:1–27:2. ACM, New York (2010)
20. Zimmermann, T., Weissgerber, P.: Mining version histories to guide software changes. In: 26th International Conference on Software Engineering (ICSE 2004), pp. 563–572 (2004)

Improving on Popularity as a Proxy for Generality When Building Tag Hierarchies from Folksonomies

Fahad Almoqhim, David E. Millard, and Nigel Shadbolt

Electronics and Computer Science, University of Southampton, Southampton, United Kingdom
{fibm1e09,dem,nrs}@ecs.soton.ac.uk

Abstract. Building taxonomies for Web content manually is costly and time-consuming. An alternative is to allow users to create folksonomies: collective social classifications. However, folksonomies have inconsistent structures and their use for searching and browsing is limited. Approaches have been proposed for acquiring implicit hierarchical structures from folksonomies, but these approaches suffer from the "generality-popularity" problem, in that they assume that popularity is a proxy for generality (that high level taxonomic terms will occur more often than low level ones). In this paper we test this assumption, and propose an improved approach (based on the Heymann-Benz algorithm) for tackling this problem by direction checking relations against a corpus of text. Our results show that popularity works as a proxy for generality in at most 77% of cases, but that this can be improved to 81% using our approach. This improvement will translate to higher quality tag hierarchy structures.

Keywords: Folksonomies, Taxonomies, Collective Intelligence, Social Information Processing, Social Metadata, Tag similarities.

1 Introduction

The transition from the Document Web, where content is produced mainly by the owners of websites, to the Social Web where users are not only information consumers but also content contributors, means that web content today is huge and constantly growing. Building and maintaining taxonomies for organizing such content manually by experts is costly and time-consuming. Consequently, an alternative approach is to allow users to contribute by tagging, this is a process that allows individuals to freely assign tags, descriptive metadata, to a web object or resource, producing a folksonomy (a set of user, tag, resource triples) as a result of that process [1].
Collaborative tagging is one of the most successful examples of the power of Collective Intelligence (CI) [2] for constructing and organizing knowledge in the Web. It has become a key part on most online portals, such as Delicious, Blogger, Flickr, Twitter and Facebook.

In recent years, folksonomies have emerged as an alternative to traditional classifications of organizing information [3,4]. They benefit from the power of collective intelligence to offer an easier (in terms of time, effort and cognitive costs) approach to organizing web resources [5]. However, they share the inconsistent structure problem

L.M. Aiello and D. McFarland (Eds.): SocInfo 2014, LNCS 8851, pp. 95–111, 2014.
© Springer International Publishing Switzerland 2014

that is inherited from uncontrolled vocabularies, which causes many problems such as ambiguity, homonymy, synonymy, and basic level variation [6,7]. Consequently, many researchers have been working on approaches for acquiring latent hierarchical structures from folksonomies and constructing tag hierarchies [8,9,10]. Constructing tag hierarchies from folksonomies can be useful in different tasks, for example:

- **Improving Content Retrieval:** Although folksonomies have become a very popular method to describe web contents due to their simplicity of use [5], the lack of structure in folksonomies makes content retrieval tasks, like searching, subscription and exploration, limited [11,12]; they tend to have low recall performance and do not support efficient query refinement [13]. Tag hierarchies, therefore, can improve content retrieval tasks by making the relations between tags explicit [14,15]. In addition, Morrison found that searches conducted with tag hierarchies achieved better results than those conducted with search engines [16].
- **Building Lightweight Ontologies:** Ontology is the backbone of the semantic web [17], and an important knowledge structure for improving the organization, retrieval and management of heterogeneous content and widespread understanding of a specific domain. However, building and maintaining ontologies is so costly and time-consuming that it obstructs the progress of the Semantic Web development [18]. The large number of folksonomies offers a promising way to build tag hierarchies and then to construct lightweight ontologies. For instance, Mika provides a model of semantic and social networks for building lightweight ontologies from Delicious [19]. Also, Schmitz proposes subsumption-based model for constructing ontology from Flickr [13].
- **Enriching Knowledge Bases:** Since users constantly and freely tag new web contents, the tag hierarchies are up-to-date and hence can be used to update existing knowledge bases or enlarge their scope. For example, Kiu and Tsui present TaxoFolk, an algorithm that uses tag hierarchies for enriching existing taxonomies by unsupervised data mining techniques and augmented heuristics [20]. Furthermore, Zheng et al. propose an approach for enriching WordNet with tag hierarchies that extracted from Delicious [21]. Also, Van Damme et al. offer a comprehensive method for building and maintaining ontologies from tag hierarchies alongside some online resources [22].

However, current approaches to automatic tag hierarchy construction come with limitations [12] and [23], one of the most significant of which is the "generality-popularity" problem. This arises from the tendency of hierarchy construction algorithms to use popularity as a proxy for generality (this is explained further in Section 2.4). For example, if users tend to tag a picture of London attractions with "London" much more than "UK", then "London" will have higher popularity and thus be placed in a more general position than "UK" despite the fact that the relation makes more sense semantically if "UK" is the more general term. In this research, we present an experiment to quantify the extent of the "generality-popularity" problem, and combine and extend prior research in tag hierarchy building and lexico-syntactic patterns to propose an improved approach to building tag hierarchy that tackles this problem. Our approach works by correcting the taxonomic direction between popular and more general tags by using Hearst's lexico-syntactic patterns [24] that are commonly used for acquiring taxonomic relations from large text corpora [25].

2 Related Work

2.1 Learning Concept Hierarchy from Text

The origins of automatic acquisition of latent hierarchical structures from unstructured content can be found in approaches to learning lexical relations from free text. These approaches can be seen in two directions: approaches that exploit clustering techniques based on Harris' distributional hypothesis [26], e.g. [25] and [27]; or approaches that use lexico-syntactic patterns to acquire a certain semantic relation in texts, e.g. "is-a" or "such-as" relationship, e.g. [24] and [28]. Many of the latter direction of the approaches have focused on a key insight first expressed by Hearst in [24], that certain lexico-syntactic patterns (Table 1) can acquire a particular semantic relationship (hyponym/hypernym relationship) between terms in large text corpora [29].

Table 1. Hearst's lexico-syntactic patterns for detecting hyponym/hypernym relations

No	Pattern	Example
1	P such as $\{C_1, C_2 ..., (and \mid or)\}$ C_n	**European countries** such as *England* and *Spain*.
2	Such P as $\{C_1,\}$ * $\{(or \mid and)\}$ C_n	... works by such **authors** as *Herrick*, *Goldsmith*, and *Shakespeare*.
3	$C_1 \{, C_n\}$ * $\{,\}$ $\{(or \mid and)\}$ other P	... *apple*, *orange*, *banana* or other **fruits**.
4	$P \{,\}$ including $\{C_1,\}$ * $\{or \mid and\}$ C_n	... all **common-law countries**, including *Canada* and *England*.
5	$P \{,\}$ especially $\{C_1,\}$ * $\{or \mid and\}$ C_n	... most **European countries**, especially *England*, *Spain*, and *France*.

Lexico-syntactic patterns can capture different semantic relations, though hyponym/hypernym relationship seems to produce the most accurate results, even with no pre-encoded knowledge. Additionally, they occur frequently in texts and across their genre boundaries [24] and [30].

2.2 Learning Tag Hierarchy from Folksonomies

Recently there have been several promising approaches proposed for learning tag hierarchies from folksonomies. These approaches can be seen in three directions based on using: clustering techniques, relevant knowledge resources or a hybrid of both to infer semantics from folksonomies.

Clustering Techniques Based Approaches. Clustering techniques are mostly based on agglomerative, bottom-up, approaches. First pair-wise tag similarities are computed and then divided into groups based on these similarities. After that, pair-wise group similarities are computed and then merged as one until all tags are in the same group [31].

Heymann and Garcia-Molinay [8] introduce an extensible greedy algorithm that automatically constructs tag hierarchies from folksonomies, extracted from Delicious and CiteULike. They use graph centrality [32] in the tag-tag co-occurrence network to identify the generality order of the tags. Their algorithm hypothesis is that the tag with the highest centrality is the most general tag thus it should be added to the tag hierarchy before others. Benz et al. [10] present an extension of Heymann's algorithm by applying tag co-occurrence as the similarity measure and the degree centrality as the generality measure. They tested their algorithm with the dataset gathered from Delicious and succeed to produce clearer and more balanced tag hierarchies compared to the original algorithm.

C. Schmitz et al.[33] and P. Schmitz [13] used statistical models of tag subsumption for constructing tag hierarchies. C. Schmitz et al used the theory of association rule mining to analyze and structure folksonomies from Delicious. P. Schmitz adapted the work of [34] to introduce a subsumption-based model for building tag hierarchy from Flickr. Schwarzkopf et al. [35] extend the two algorithms in [8] and [33] by taking into account the tag context.

Mika[19] presents a graph-based model for constructing two tag hierarchies from folksonomies, extracted from Delicious, using statistical techniques. The first tag hierarchy is based on the overlapping set of user-tag networks, whereas the second is based on the overlapping set of object-tag networks. Hamasaki et al. [36] extended the work of Mika while considering the user-user relationship. In particular, the first tag hierarchy is modified by considering tagging information of the user's neighbors.

Solskinnsbakk and Gulla [9] constructed tag hierarchies from folksonomies extracted from Delicious by using morpho-syntactic and semantic similarity measures. Morpho-syntactic similarities are found by the Levenshtein distance, whereas the Cosine similarity has been used to find the semantic similarity between tags. Plangprasopchok et al. [37] adapted affinity propagation proposed by Frey & Dueck [38] to build deeper and denser tag hierarchies from folksonomies. However, Strohmaier et al. [4] have proved that generality-based approaches to learning tag hierarchy, with degree centrality as generality measure and co-occurrence as similarity measure, e.g. [10] have a superior performance compared to probabilistic models, e.g. [37].

Knowledge Resources Based Approaches. Several existing knowledge resources, such as Wikipedia, WordNet and online ontologies, can be used to discover the meaning of tags and their relationships.

Laniado et al. [15] use WordNet to disambiguate and structure tags from Delicious. Angeletou et al. [39] present FLOR, an automatic approach for enriching folksonomies, extracted from Flickr, by linking them with related concepts in WordNet and online ontologies, using the Watson semantic search engine. Cantador et al. [40] introduce an approach that automatically maps tags, extracted from Delicious and

Flickr, with Wikipedia concepts, and then associates those tags with domain ontologies. Similarly, Tesconi et al. [41] use Wikipedia as an intermediate representation between tags, extracted from Delicious, and some semantic resources, namely: YAGO and WordNet. Garcia et al. [42] propose an approach to automatically disambiguate polysemous, multiple related meanings, tags through linking them to DBpedia entries.

Hybrid Approaches. Some approaches to learning tag hierarchies are based on the combination of both previous directions, clustering techniques and knowledge resources.

Specia and Motta [43] present a semi-automatic approach rely on clustering techniques and using WordNet and Google to structure tags, extracted from Delicious and Flickr. Giannakidou et al. [44] introduce a co-clustering approach for identifying the tag semantics by clustering tags, from Flickr, and relevant concepts from a semantic resource, WordNet. Lin et al. [45] propose an approach based on data mining techniques and WordNet concepts to discover the semantics in the tags and build tag hierarchies.

2.3 Limitations of Current Approaches

Although the **approaches that based on lexico-syntactic patterns** provide reasonable precision, their recall is low [46]. In addition, they are not appropriate to use them for acquiring semantic relations in tag collections since these collections tend to be much more inconsistent than text collections [47]. Moreover, Strohmaier et al., in their study of tag hierarchy building algorithms, show that the approaches tailored towards collaborative tagging systems outperform the approaches based on traditional hierarchical clustering techniques [4].

While several **approaches based on clustering techniques** have been offered solutions to structure folksonomies, they come with limitations [12] and [23]. These include the suffering from the "generality-popularity" problem. In practice a tag could be used more frequently not because it is more general, but because it is more popular among users. For instance, Plangprasopchok and Lerman [48] found, on Flickr, that the number of photos tagged with "car" are ten times as many as that tagged with "automobile". By applying clustering techniques, the tag "car" is likely to have higher centrality, and thus it will be perceived as more general than "automobile".

Knowledge resources based approaches are developed to partially solve the limitations of clustering techniques approaches. However, such resources are limited and they can only deal with standard terms [12]. This limitation is due to the tags nature in which they may contain spelling errors, abbreviations, idiosyncratic terms etc. Furthermore, tags can be multi-lingual, which make these sources even harder to handle [23].

In this paper, we combine these approaches in order to benefit from the accuracy of lexico-syntactic patterns, while maintaining the flexibility and scalability of clustering techniques. We do this by using hyponym/hypernym patterns to check and correct the *direction* of taxonomic tag pairs in a tag hierarchy generated via clustering, thus addressing the "generality-popularity" problem.

3   Our Approach to Building High-Quality Tag Hierarchies

In previous work [49], we have shown that applying generality-based approaches to folksonomies constructed of user provided *tag pairs* results in a better quality hierarchy than those constructed of user provided *tags*. However, asking users to provide tag pairs rather than tags results in a poorer set of terms, and a less expressive hierarchy. This leads us to the insight of our new approach that if we could improve the *accuracy of directions* in relations constructed between tags by a generality-based approach, we would be able to improve the quality of the resulting tag hierarchy structure and semantics without sacrificing richness.

It has been shown that generality-based approaches of tag hierarchy construction show a superior performance compared to other approaches [4]. However, they suffer from the "generality-popularity" problem. To tackle this problem, our proposed approach

Table 2. Pseudo-code for the proposed algorithm

Input: *user-generated terms (tags)*
Output: *tag hierarchy*

1. Filter the tags by an occurrence threshold *occ*.
2. Order the tags in descending order by generality (measured by degree centrality in the tag–tag co-occurrence network).
3. Starting from the most general tag, as the root node, add all tags ti subsequently to an evolving tag hierarchy:
   (a) Calculate the similarities (using the co-occurrence weights as similarity measure) between the current tag ti and each tag currently present in the hierarchy, and append the current tag ti underneath its most similar tag *tag_sim*.
   (b) If ti is very general (determined by a generality threshold *min_gen*) or no sufficiently similar tag exists (determined by a similarity threshold *min_sim*), append ti underneath the root node of the hierarchy.
   (c) Check the taxonomic direction (ti → its suggested hypernym; i.e. *tag_sim* or the root) by using the proposed lexico-syntactic patterns, and calculate p_occ_1; i.e. in total, how many (ti → its suggested hypernym), with using the proposed patterns, found in Wikipedia.
   (d) Check the taxonomic direction (ti ← its suggested hypernym; i.e. *tag_sim* or the root) by using the proposed lexico-syntactic patterns, and calculate p_occ_2; i.e. in total, how many (ti ← its suggested hypernym), with using the proposed patterns, found in Wikipedia.
   (e) Correct the taxonomic direction if needed based on p_occ_1 and p_occ_2.
4. Apply a post-processing to the resulting hierarchy by re-inserting orphaned tags underneath the root node in order to create a balanced representation. The re-insertion is done based on step 3.

extended a promising generality-based algorithm, based on [4], by using lexico-syntactic patterns applied to a large text corpus specifically the text of English Wikipedia. The patterns that our approach used are a combination of the well-known Hearst's lexico-syntactic patterns (Table 1) and other two other direct patterns:

- " *C* is a **P** "
- " *C* is an **P** "

While lexico-syntactic patterns suffer from low recall [46], our approach leverages their reasonable precision to correct the taxonomic direction between popular and more general tags before using them to build the tag hierarchy. The algorithm we have used in our approach is an extension of Benz's algorithm [10], which itself is an extension of Heymann's algorithm [8]. Table 2 demonstrates the pseudo-code for the proposed algorithm.

The algorithm is affected by several parameters, including: occurrence threshold *occ* (the number of tag occurrences); similarity threshold *min_sim* (the number of tag co-occurrences with another tag); generality threshold *min_gen* (the number of tag co-occurrences with other tags); and patterns matching occurrences p_occ_1 and p_occ_2. Empirical experiments were performed to optimize these parameters.

4   Experimental Setup

To test the performance of our approach, we applied the original algorithm and our proposed algorithm, using five common tag similarity measures and with different similarity thresholds, to a large-scale folksonomy dataset collected from Delicious (see Section 4.2), yielding 20 different tag hierarchies. The five common similarity measures between *Tag 1* and *Tag 2* can be mathematically defined as follows:

$$Matching = |A \cap B| \tag{1}$$

$$Dice = \frac{2|A \cap B|}{|A| + |B|} \tag{2}$$

$$Jaccard = \frac{|A \cap B|}{|A \cup B|} \tag{3}$$

$$Overlap = \frac{|A \cap B|}{\min(|A|, |B|)} \tag{4}$$

$$cosine = \frac{|A \cap B|}{\sqrt{|A| \times |B|}} \tag{5}$$

Where "A" is the set of the folksonomies that contains *Tag 1*, and "B" is the set of the folksonomies that contains *Tag 2*.

In this paper, we are focusing on checking and correcting the taxonomic tag pairs that we get from our proposed algorithm. Therefore, we evaluate all the taxonomic tag pairs from all the resulting 20 tag hierarchies against a gold-standard dataset, namely: WordNet. The detailed experimental setup is presented next.

4.1   Experimental Design

Fig. 1 Summarized the process of the experimental design that we have used for performing our experiments detailed in this paper.

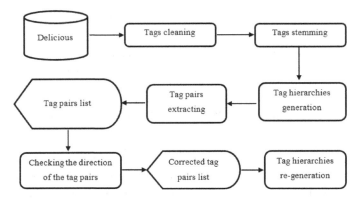

Fig. 1. The Process diagram of our experimental design

The above process consists of four main components, as follows:

- **Tags Normalising**: Before running the Tag Hierarchy Generation component, the tags are passed to the normalisation process that applies two steps: 1) **Tags Cleaning**, including: *Letters lower-case, symbol deleting* and *non-English letters deleting*. 2) **Tags stemming**, by using the will-known Porter Stemmer [50].
- **Tag Hierarchy Generation**: This component uses our proposed algorithm, except the steps (3.c – 3.e), to construct tag hierarchies from the tags.
- **Tag Pairs Direction Checking**: This is the most important component of our approach. It uses the steps (3.c – 3.e) of our proposed algorithm to check, and to correct if needed, the direction of the tag pairs that generated from the previous component. Note that since the produced tag pairs are stemmed, the Wikipedia and WordNet datasets are stemmed as well.
- **Tag Hierarchy Re-Generation:** It uses the Tag Hierarchy Generation to re-generate the tag hierarchy after correcting the direction of the taxonomic tag pairs.

4.2   Datasets

In our experiments, we have used two large datasets, as detailed follows:

- **Delicious Dataset:** To compare the performance of our proposed algorithms of building tag hierarchy compared to the original algorithm, we have used a large-scale folksonomy dataset from the PINTS experimental dataset[1] containing a systematic

---

[1] http://www.uni-koblenz-landau.de/koblenz/fb4/AGStaab/
Research/DataSets/PINTSExperimentsDataSets/index_html

crawl of Delicious during 2006 and 2007. Table 3 summarized the statistics of the dataset.

Table 3. Statistics of the Delicious dataset

Dataset	Users	Tags	Resources	Tag assignments
Delicious	532,924	2,481,698	17,262,480	140,126,586

- **Wikipedia Dataset:** To solve the "generality-popularity" tags problem by using the proposed lexico-syntactic patterns, we have chosen Wikipedia dataset. We selected to use Wikipedia since it is currently the largest knowledge repository available on the Web. The dataset that we have used contains 4,487,682 English Wikipedia articles[2].

4.3   Evaluation Methodology

To evaluate our proposed approach to building tag hierarchy against the original approach, we have chosen WordNet [51] dataset for two reasons:
  - It is considered to be a gold-standard dataset for testing hyponym/hypernym relations building algorithms [29].
  - And we avoided any dataset that was constructed automatically or based on Wikipedia since we have used it in our approach.

WordNet is a structured lexical database of the English language that build manually by experts. It contains 206,941 terms grouped into 117,659 synsets[3]. The synsets are connected by several lexical relations. The most important and frequently of these relations is the hyponym/hypernym relation. For our purpose we have extracted the taxonomic terms among synsets in WordNet.

5   Results and Analysis

In the first round of our experiment, we have applied our proposed algorithm and the original algorithm, using the five selected tag similarity measures, to the Delicious dataset, yielding 10 tag hierarchies. Then, we have rerun the experiment again but with a tag similarity threshold equal 0 to examine the effectiveness of using similarity threshold that suggested by the original algorithm. Finally, we have evaluated the direction correctness of all the taxonomic tag pairs from all the produced 20 tag hierarchies against WordNet. To give an impression of the results, Table 4 shows a few examples of the produced taxonomic tag pairs, using the five similarity measures under study.

---

[2]   As collected in March 2014.
[3]   http://wordnet.princeton.edu/wordnet/man/wnstats.7WN.html, as visited on June 2014.

Table 4. Examples of produced tag pairs for each of the selected similarity measures

Measure	Rank	Tag A	Tag B	Rank	Tag A	Tag B
Matching			design			technology
Dice			design			lcd
Jaccard	1	blog	design	1000	display	lcd
Overlap			blogger-beast			tft
Cosine			daily			lcd
Matching			blog			php
Dice			news			willow
Jaccard	100	daily	news	5000	maple	willow
Overlap			blog			willow
Cosine			news			willow
Matching			news			dress
Dice			forecast			bridal
Jaccard	500	weather	forecast	10000	bridesmaid	bridal
Overlap			noaa			dress
Cosine			forecast			bridal

And to get an overall view of how different each of the selected similarity measures is to others in terms of generating taxonomic tag pairs, Table 5 displays the overlap between the produced tag hierarchies based on these similarity measures.

Table 5. Overlap between tag hierarchies generated using selected similarity measures

	Matching	Cosine	Overlap	Jaccard
Dice	0.15	0.71	0.16	0.57
Jaccard	0.09	0.40	0.10	
Overlap	0.71	0.24		
Cosine	0.22			

To give a comprehensive view of the evaluation against WordNet, we investigated the WordNet coverage of the investigated delicious dataset. Table 6 shows the Word-Net coverage of the top delicious tags, whereas Table 7 illustrates the WordNet coverage of all the tags appeared in the produced tag hierarchies.

Table 6. WordNet coverage of tags in delicious dataset

	Top 10	Top 100	Top 500	Top 1000
WordNet coverage	80.00%	77.00%	74.20%	71.10%

Table 7. WordNet coverage of tags in produced hierarchies

	With using similarity threshold				
	Matching	*Dice*	*Jaccard*	*Overlap*	*Cosine*
WordNet coverage	39.50%	41.18%	41.91%	37.85%	40.61%

	Without using similarity threshold				
	Matching	*Dice*	*Jaccard*	*Overlap*	*Cosine*
WordNet coverage	38.89%	39.93%	31.30%	37.52%	39.90%

A number of factors limit the WordNet coverage of the tags and the taxonomic tag pairs. First, WordNet is a static knowledge resource, while the delicious dataset is an open-ended collection. Also, WordNet only covers the English language, whereas the delicious dataset contains multi-language tags. However, WordNet can be a reasonable reference for our purpose, i.e. tackling the "generality-popularity" problem, since a significant fraction of the popular tags in delicious is covered by WordNet; as shown in Table 6. Having established this the next step is to compare the tag pair directions produced by the original algorithm and our variation of the algorithm against the directions as defined in WordNet. This will give us a measure of how many times generality was a successful proxy for popularity in the original algorithm, and also the extent to which our approach improves on this.

Table 8 shows the results. For further improvement, we added a *min_p_occ* threshold in our proposed algorithm; to correct the generated taxonomic tag pairs, the occurrences number found in Wikipedia, by using the proposed lexico-syntactic patterns, need to be more than the *min_p_occ* threshold. The last column of Table 8 shows the improvement of using the *min_p_occ* threshold, which was more effective with the Matching, Dice and Jaccard similarity measures.

The first observation that can be drawn is that the original algorithm is moderately successful (as much as 76.96%), even though it blindly accepts popularity as a measure of generality. So while "generality-popularity" has been identified as a weakness of clustering approaches, using this assumption over three quarters of the generated relationships are in the right direction.

Table 8. Taxonomic tag pairs evaluation, using selected similarity measures and a similarity threshold for each measure, agains WordNet

	No of Tag Pairs found in WordNet	*% Agreement with WordNet*		
		Original Algorithm	**Our Algorithm**	**Our strict Algorithm**
Matching	**305**	75.74%	77.38%	79.34%
Dice	130	47.22%	55.56%	61.11%
Jaccard	114	47.37%	64.91%	64.04%
Overlap	217	**76.96%**	**81.11%**	**81.11%**
Cosine	161	54.90%	64.71%	64.71%

The second observation that can be drawn is that there is a modest improvement achieved by our proposed algorithm compared to the original algorithm among all the selected tag similarity measures. This means, regardless of the similarity measure, our approach has succeeded in correcting the direction of taxonomic tag pairs that were generated in the wrong direction by the original algorithm. In the best case (Overlap) this leads to an accuracy of over 81%.

Table 9. Examples of wrong direction taxonomic tag pairs generated by original algorithm

Similarity Measure	Tag A	Tag B	Similarity Measure	Tag A	Tag B
Matching	Faith	christian	Dice	Meat	Beef
	Footwear	shoes		primates	Monkey
	Society	culture		Road	Highway
	Wealth	money		Search	Google
	Poultry	chicken		Sweet	Candy

Similarity Measure	Tag A	Tag B	Similarity Measure	Tag A	Tag B
Jaccard	Coffee	espresso	Overlap	broadcast	Video
	Drink	alcohol		Canine	Dog
	Ireland	dublin		Footwear	shoes
	Pastry	tart		Poultry	chicken
	Puzzle	sudoku		Ride	Bike

Similarity Measure	Tag A	Tag B
Cosine	bag	purses
	sweet	candy
	meat	beef
	search	google
	broadcast	radio

Table 9 shows examples of these taxonomic tag pairs, which the original algorithm has generated them in the form of (*Tag A* is-a *Tag B*), where they have been found in WordNet as (*Tag B* is-a *Tag A*).

Given the large numbers of pairs generated by the algorithm and the moderate inter-section of tags with WordNet (around 40%, as shown in Table 7) the low number of matched pairs is surprising. It may reflect the relatively small size of WordNet as compared to the delicious dataset, but it also may reflect the fact that our algorithm looks for *direct matches* in WordNet. One approach to increase the number of matches would be to use the transitivity of the generality relationship, this would match (and possibly correct the direction of) a tag pair, even if those tags were not directly linked in WordNet, but instead were part of a chain of generality relationships.

Another observation from Table 8 is that, among all the selected tag similarity measures, the Overlap measure yields the best performance of generating taxonomic tag pairs against WordNet, whereas Matching measure yields the biggest amount of generated tag pairs that found in WordNet regardless of the taxonomic direction.

Table 10. Taxonomic tag pairs evaluation, using selected similarity measures and without using a similarity threshold, agains WordNet

	No of matched tag pairs	*Original Algorithm*	*Our Algorithm*	*Our strict Algorithm*
Matching	**329**	76.90%	77.81%	80.55%
Dice	150	51.33%	66.00%	66.67%
Jaccard	246	47.56%	66.26%	62.60%
Overlap	230	**77.39%**	**80.00%**	**81.30%**
Cosine	178	59.55%	67.42%	67.98%

Table 10 shows the results of rerunning the experiment but with a tag similarity threshold = 0. In addition to the previous observations on Table 8, Table 10 demonstrates that without using a similarity threshold, as suggested by the original algorithm, both the original algorithm and our variations can generate more taxonomic tag pairs that can be found in WordNet. Also, by using all selected tag similarity measures, both algorithms yield better taxonomic tag pairs structure and semantics.

6   Conclusion

Building and maintaining taxonomies for organizing Web content manually by experts is costly and time-consuming. Therefore, folksonomy has emerged as an alternative approach for organizing online resources. Yet, folksonomies are beset by many problems, due to the lack of consistent structure, such as ambiguity, homonyms, and synonymy. Thus many approaches have been proposed to resolve these problems by proposing mechanisms for acquiring latent hierarchical structures from folksonomies and constructing tag hierarchies. Among these approaches, it has been revealed that generality-based approaches show a superior performance compared to other approaches. However, it has been argued that generality-based automatic tag hierarchy algorithms suffer from a "generality-popularity" problem, where they (sometimes inaccurately) assume that because a tag occurs more frequently it must be more general and thus appear higher in the hierarchy. Therefore, we have presented an experiment to measure this effect, and proposed an approach to reduce its impact. Our proposed approach extends a promising generality-based algorithm by using lexico-syntactic patterns for discovering hyponym/hypernym relations in order to distinguish between popular and general tags. For this purpose we have used Wikipedia as the text corpus, and for evaluation we have used WordNet as a gold-standard reference.

Our experiment reveals that generality acts as a successful proxy for popularity in 47% to 76% of cases (depending on the similarity measure used), and that the performance of our proposed algorithm outperforms the original algorithm, among all the selected tag similarity measures (correct in between 56% and 81% of cases). This means, regardless of the similarity measure, our approach has succeeded in correcting the direction of taxonomic tag pairs that were wrongly generated by the original algorithm. This improvement will result in building higher quality tag hierarchy structure and semantics.

In term of the comparison between the selected tag similarity measures, the Overlap measure yields the best performance of generating taxonomic tag pairs against WordNet. Finally, we have shown that removing the similarity threshold (in both the original algorithm and our variations) results in better taxonomic tag pairs, in terms of quantity and quality, irrespective of tag similarity measures.

For future work, we plan to investigate which lexico-syntactic patterns are most successful in correcting errors, and whether any introduce significant errors. This should give us a clear explanation of which patterns are more reliable in correcting the wrong direction of taxonomic tag pairs. Secondly, based on the results we achieved, we are planning to use a dynamic knowledge repository, such as a search engine, instead of a static knowledge resource, like Wikipedia. This should increase the coverage and occurrences of the tags in any tag collection. Finally, we intend to evaluate the tag hierarchies produced using our approach against more than one large reference taxonomies, this should give a measure of how the improvements in tag pair directions presented here translate into improved tag hierarchies.

Tagging has become an established method of crowd-sourcing structure on the Web, but folksonomies based on tags have serious weaknesses for both search and browsing, which is a primary use of structure on websites. Our hope is that our work will contribute towards the growing understanding of how more sophisticated hierarchical structure can be successfully derived from folksonomies, and that this will ultimately improve our interaction with the Social Web.

References

1. Vander Wal, T.: Folksonomy Coinage and Definition (2007),
   `http://vanderwal.net/folksonomy.html`
2. O'Reilly, T.: What is web 2.0: design patterns and business models for the next generation of software (2005), `http://oreilly.com/web2/archive/what-is-web-20.html` (June 20, 2013)
3. Gupta, M., Li, R., Yin, Z., Han, J.: An Overview of Social Tagging and Applications. In: Aggarwal, C. (ed.) Social Network Data Analytics, pp. 447–497. Springer, New York (2011)
4. Strohmaier, M., Helic, D., Benz, D., Körner, C., Kern, R.: Evaluation of Folksonomy Induction Algorithms. ACM Transactions on Intelligent Systems and Technology 3(4), Article 74 (2012)
5. Mathes, A.: Folksonomies-cooperative classification and communication through shared metadata. Computer Mediated Communication 47(10) (2004),
   `http://adammathes.com/academic/computer-mediated-communication/folksonomies.pdf`

6. Golder, S., Huberman, B.: Usage patterns of collaborative tagging systems. Journal of Information Science 32(2), 198–208 (2006)

7. Guy, M., Tonkin, E.: Tidying up tags. D-Lib Magazine 12(1) (January 2006), ISSN 1082-9873

8. Heymann, P., Garcia-Molinay, H.: Collaborative Creation of Communal Hierarchical Taxonomies in Social Tagging Systems. InfoLab Technical Report, Stanford (2006)

9. Solskinnsbakk, G., Gulla, J.: A Hybrid Approach to Constructing Tag Hierarchies. In: International conference on: On the move to meaningful internet systems: Part II, Hersonissos, Crete, Greece, pp. 975–982 (2010)

10. Benz, D., Hotho, A., Stutzer, S.: Semantics made by you and me: Self-emerging ontologies can-capture the diversity of shared knowledge. In: 2nd Web Science Conference (WebSci 2010), Raleigh, NC, USA (2010)

11. Begelman, G., Keller, P., Smadja, F.: Automated tag clustering: Improving search and exploration in the tag space. In: Collaborative Web Tagging Workshop at WWW 2006, Edinburgh, Scotland, pp.15-33 (2006)

12. Lin, H., Davis, J.: Computational and crowdsourcing methods for extracting ontological structure from folksonomy. In: 7th Extended Semantic Web Conference (ESWC 2010), Heraklion, Greece, pp.472-477 (2010)

13. Schmitz, P.: Inducing ontology from flickr tags. In: Collaborative Web Tagging Workshop at WWW 2006, Edinburgh, Scotland (2006)

14. Angeletou, S., Sabou, M., Specia, L., Motta, E.: Bridging the gap between folksonomies and the semantic web: An experience report. In : 4th European Semantic Web Conference (ESWC 2007), Innsbruck, Austria, pp.30-43 (2007)

15. Laniado, D., Eynard, D., Colombetti, M.: Using WordNet to turn a folksonomy into a hierarchy of concepts. In: 4th italian semantic web workshop: Semantic web application and perspectives, Bari, Italy, pp. 192–201 (2007)

16. Morrison, P.: Tagging and searching: Search retrieval effectiveness of folksonomies on the world wide web. Information Processing and Management 44(4), 1562–1579 (2008)

17. Berners-Lee, T., Hendler, J., Lassila, O.: The Semantic Web. Scientific American 284(5), 28–37 (2001)

18. Park, Y., Byrd, R., Boguraev, B.: Towards Ontologies On Demand. In: Workshop on Semantic Web Technologies for Searching and Retrieving Scientific Data (ISWC-03), Florida, USA (2003)

19. Mika, P.: Ontologies are us: A unified model of social networks and semantics. Web Semantics: Science, Services and Agents on the World Wide Web 5(1), 5–15 (2007)

20. Kiu, C.-C., Tsui, E.: TaxoFolk: a hybrid taxonomy–folksonomy classification for enhanced knowledge navigation. Knowledge Management Research & Practice 8(1), 24–32 (2010)

21. Zheng, H., Wu, X., Yu, Y.: Enriching WordNet with Folksonomies. In: Washio, T., Suzuki, E., Ting, K.M., Inokuchi, A. (eds.) PAKDD 2008. LNCS (LNAI), vol. 5012, pp. 1075–1080. Springer, Heidelberg (2008)

22. Van Damme, C., Hepp, M., Siorpaes, K.: Folksontology: An integrated approach for turning folksonomies into ontologies. In: ESWC Workshop Bridging the Gap between Semantic Web and Web 2, pp. 57–70. Innsbruck, Austria (2007)

23. Solskinnsbakk, G., Gulla, J.: Mining tag similarity in folksonomies. In: 3rd international workshop on Search and mining user-generated contents (SMUC 2011), Glasgow, Scotland, pp. 53–60 (2011)

24. Hearst, M.: Automatic acquisition of hyponyms from large text corpora. In: 14th Conference on Computational Linguistics, Morristown, NJ, USA, pp. 539–545 (1992)

25. Cimiano, P., Hotho, A., Staab, S.: Learning concept hierarchies from text corpora using formal concept analysis. Journal of Artificial Intelligence Research 24(1), 305–339 (2005)
26. Harris, Z.: Mathematical structures of language. John Wiley and Son (1968)
27. Faure, D., Nedellec, C.: A corpus-based conceptual clustering method for verb frames and ontology. In: The LREC Workshop on Adapting Lexical and Corpus Resources to Sublanguages and Applications, pp. 5–12 (1998)
28. Berland, M., Charniak, E.: Finding parts in very large corpora. In: the 37th Annual Meeting of the Association for Computational Linguistics (ACL), Stroudsburg, PA, USA, pp. 57–64 (1999)
29. Snow, R., Jurafsky, D., Ng., A.: Learning syntactic patterns for automatic hypernym discovery. In: The Eighteenth Annual Conference on Neural Information Processing Systems (NIPS 2004), Vancouver, Canada, vol. 17 (2004)
30. Hearst, M.: Automated discovery of wordnet relations. In: Fellbaum, C. (ed.) WordNet: An Electronic Lexical Database, MIT Press, Cambridge (1998)
31. Wu, H., Zubair, M., Maly, K.: Harvesting social knowledge from folksonomies. In: 17th Conference on Hypertext and Hypermedia, Odense, Denmark, pp. 111–114 (2006)
32. Hoser, B., Hotho, A., Jäschke, R., Schmitz, C., Stumme, G.: Semantic network analysis of ontologies. In: Sure, Y., Domingue, J. (eds.) ESWC 2006. LNCS, vol. 4011, pp. 514–529. Springer, Heidelberg (2006)
33. Schmitz, C., Hotho, A., Jäschke, R., Stumme, G.: Mining association rules in folksonomies. In: 10th IFCS Conference: Studies in Classification, Data Analysis and Knowledge Organization, Ljubljana, Slovenia, pp. 261–270 (2006)
34. Sanderson, M., Croft, B.: Deriving concept hierarchies from text. In: 22nd ACM Conference of the Special Interest Group in Information Retrieval, Berkeley, California, USA, pp. 206–213 (1999)
35. Schwarzkopf, E., Heckmann, D., Dengler, D., Kröner, A.: Mining the structure of tag spaces for user modeling. In: Workshop on Data Mining for User Modeling at the 11th International Conference on User Modeling, Corfu, Griechenland, pp. 63–75 (2007)
36. Hamasaki, M., Matsuo, Y., Nishimura, T., Takeda, H.: Ontology extraction using social network. In: International Workshop on the Semantic Web for Collaborative Knowledge Acquisition, Hyderabad, India (2007)
37. Plangprasopchok, A., Lerman, K., Getoor, L.: From saplings to a tree: Integrating structured metadata via relational affinity propagation. In: Proceedings of the AAAI Workshop on Statistical Relational AI, Menlo Park, CA, USA (2010)
38. Frey, B., Dueck, D.: Clustering by passing messages between data points. Science 315(5814), 972–976 (2007)
39. Angeletou, S., Sabou, M., Motta, E.: Semantically Enriching Folksonomies with FLOR. In: 1st International Workshop on Collective Semantics: Collective Intelligence & the Semantic Web (CISWeb 2008), Tenerife, Spain (2008)
40. Cantador, I., Szomszor, M., Alani, H., Fernández, M., Castells, P.: Enriching ontological user profiles with tagging history for multi-domain recommendations. In: 1st International Workshop on Collective Semantics: Collective Intelligence & the Semantic Web (CISWeb 2008), Tenerife, Spain (2008)
41. Tesconi, M., Ronzano, F., Marchetti, A., Minutoli, S.: Semantify del.icio.us: Automatically Turn your Tags into Senses. In: Social Data on the Web Workshop at the 7th International Semantic Web Conference, Karlsruhe, Germany (2008)
42. Garcia, A., Szomszor, M., Alani, H., Corcho, O.: Preliminary results in tag disambiguation using dbpedia. In: 1st International Workshop in Collective Knowledge Capturing and Representation, California, USA (2009)

43. Specia, L., Motta, E.: Integrating Folksonomies with the Semantic Web. In: 4th European Conference on The Semantic Web: Research and Applications, Innsbruck, Austria, pp. 624–639 (2007)
44. Giannakidou, E., Koutsonikola, V., Vakali, A., Kompatsiaris, Y.: Co-clustering tags and social data sources. In: 9th International Conference on Web-Age Information Management, Zhangjiajie, China, pp. 317–324 (2008)
45. Lin, H., Davis, J., Zhou, Y.: An integrated approach to extracting ontological structures from folksonomies. In: 6th European Semantic Web Conference, Heraklion, Greece, pp. 654–668 (2009)
46. Cimiano, P.: Ontology Learning and Population from Text: Algorithms, Evaluation and Applications, vol. 27. Springer (2006)
47. Plangprasopchok, A., Lerman, K., Getoor, L.: Growing a Tree in the Forest: Constructing Folksonomies by Integrating Structured Metadata. In: 16th ACM SIGKDD International Conference on Knowledge Discovery and Data Mining, Washington, DC, USA, pp. 949–958 (2010)
48. Plangprasopchok, A., Lerman, K.: Constructing Folksonomies from User-Specified Relations on Flickr. In: 18th International World Wide Web Conference, Madrid, Spain, pp. 781–790 (2009)
49. Almoqhim, F., Millard, D.E., Shadbolt, N.: An approach to building high-quality tag hierarchies from crowdsourced taxonomic tag pairs. In: Jatowt, A., Lim, E.-P., Ding, Y., Miura, A., Tezuka, T., Dias, G., Tanaka, K., Flanagin, A., Dai, B.T. (eds.) SocInfo 2013. LNCS, vol. 8238, pp. 129–138. Springer, Heidelberg (2013)
50. Porter, M.: An algorithm for suffix stripping. Program 14(3), 130–137 (1980)
51. Miller, G.: WordNet: a lexical database for English. Communications of the ACM 38(11), 39–41 (1995)

Evolution of Cooperation in SNS-norms Game on Complex Networks and Real Social Networks

Yuki Hirahara[1], Fujio Toriumi[2], and Toshiharu Sugawara[1]

[1] Department of Computer Science and Engineering,
Waseda University, Tokyo 169-8555, Japan
`y.hirahara@toki.waseda.jp`, `sugawara@waseda.jp`
[2] Department of Systems Innovation,
The University of Tokyo, Tokyo 113-8656, Japan
`tori@sys.t.u-tokyo.ac.jp`

Abstract. Social networking services (SNSs) such as Facebook and Google+ are indispensable social media for a variety of social communications, but we do not yet fully understand whether these currently popular social media will remain in the future. A number of studies have attempted to understand the mechanisms that keep social media thriving by using a meta-rewards game that is the dual form of a public goods game. However, the meta-rewards game does not take into account the unique characteristics of current SNSs. Hence, in this work we propose an SNS-norms game that is an extension of Axelrod's metanorms game, similar to meta-rewards games, but that considers the cost of commenting on an article and who is most likely to respond to it. We then experimentally investigated the conditions for a cooperation-dominant situation in which many users continuing to post articles. Our results indicate that relatively large rewards compared to the cost of posting articles and comments are required, but optional responses with lower cost, such as "Like!" buttons, play an important role in cooperation dominance. This phenomenon is of interest because it is quite different from those shown in previous studies using meta-rewards games.

Keywords: SNS, Agent-based simulation, Facebook, Metanorms game.

1 Introduction

Social media are now an almost indispensable infrastructure for a variety of social activities such as personal information and opinion exchange, advertising, marketing, and political participation/campaigns [9]. Providers of social media merely set up the platforms for information exchange on the Internet and the actual content is mostly created and published by individual users. Because users expend personal effort and time in writing articles and comments, incentives or psychological rewards for doing so should be provided for users to keep social media active. These incentives can be achieved, for example, by providing comments on posted articles and responses to these comments; such interactions can provide users with feelings of connection to other people. Thus, the incentives themselves are also provided by SNS users, which means that users incur some cost for giving the incentives. Obviously, there is a trade-off between cost and incentive, but the conditions between them that enable networks to thrive are poorly

L.M. Aiello and D. McFarland (Eds.): SocInfo 2014, LNCS 8851, pp. 112–120, 2014.

understood. Clarifying the conditions or mechanisms that keep SNSs thriving is a key challenge in the design of social media, which are clearly already essential tools in human society.

A number of studies aimed at analyzing social media. For example, Myers et al. [7] investigated the dynamics of information diffusion in Twitter networks while Gonzalez et al. [4] analyzed the growth of Google+ in its first year. Ghanem et al. [3] studied the different patterns of interaction in social media. Toriumi et al. [10] proposed a meta-rewards game, which was a part of a general metanorms game, to identify evolved behaviors of users in social media. They analyzed the conditions under which cooperation is dominant — that is, in which many users continue to post articles and comments — and found that meta-rewards corresponding to responses to comments on articles, such as "comments on comments" and "Like!" buttons for comments, play an important role in social media. Because [10] assumed that the agent networks are complete graphs, Hirahara et al. [5] conducted the same analysis using WS and BA model networks [2,12], which are more similar to real-world social networks than complete graphs. However, the meta-rewards game does not take into account a number of the key characteristics of current SNSs. For example, a user that responds to comments on a certain article tends to be the user who posted the original article. The structure of this type of interaction on SNSs may restrict who receives rewards and thus may lead to different behaviors and consequently different conditions for cooperation-dominant situations.

Therefore, we propose an SNS-norms game in which we have modified the basic meta-rewards game to reflect the interaction structures in current SNSs and investigate the conditions for cooperation dominance. Our proposed SNS-norms game is based on a generalized Axelrod's metanorms game [10], but the interaction between agents that correspond to SNS users on the social networks is restricted by considering (1) who is likely to respond to articles and comments and (2) the cost and rewards associated with various response methods, such as posting a comment, clicking a "Like!" button, and showing a "read" mark automatically. The instance of social networks we use in our experiments is that observed on Facebook [8]. We experimentally demonstrate that relatively high rewards compared to the cost values are required for cooperation dominance. This is a quite strict condition in actual SNSs. We also show that introducing an optional response that is a low-cost but low-reward feedback mechanism is significant in terms of promoting thriving. We believe our results can provide helpful guidelines for designing social media that will continue to flourish.

2 SNS-norms Game

2.1 SNS as Public Goods Game

We can observe the following three characteristics of social media:

1. Social media only become meaningful if many participants post articles and mutually comment on the posted articles.
2. Some cost in terms of personal time and effort is incurred to post/comment, but participants can receive responses that can be considered rewards (i.e., psychological rewards as incentives).
3. There are free riders who only read content and do not produce anything.

From the above, we conclude that social media and thus SNSs have the properties of being public goods that are produced cooperatively and shared by a community [10].

The mechanism driving the contribution to public goods has been analyzed using a public goods game whose basic structure is the n-person prisoners' dilemma (PD) game. For example, Axelrod proposed norms and metanorms games that were evolutionary games based on the PD game in order to analyze public goods problems [1]. The metanorms games were based on punishments imposed by promoters on non-cooperative people. However, because social media have no mechanisms to deliver punishments, Toriumi et al. [10] proposed a meta-rewards game that is a dual part of the metanorms game. In this game, rewards are delivered to cooperative participants as dual structures of punishments, which makes the game suitable for modeling SNSs.

2.2 SNS and SNS-norms Game

We propose an *SNS-norms game* to express behaviors in SNSs as a public goods game. This game is designed to model only SNSs, although many types of social media exist. An overview picture of the proposed SNS-norms game is provided in Fig. 1. It is a modified version of the meta-reward game and features two key improvements. First, the SNS-norms game considers the tendency that a comment on a comment on an article is posted by the person who posted the original article. Thus, from now, a comment on a comment is called a *response* to a comment. Second, it takes into account a number of reaction types and reward delivery types with different costs. For example, conventional comments or responses by posting sentences/words can give relatively higher rewards to receivers with high cost, while simplified responses such as clicking the "Like!" button on Facebook and showing "read" marks automatically (response by marking mechanism is called *read mark delivery*, after this)[1] provides low rewards to receivers but with low or zero cost. Note that these simplified responses are usually implemented as an optional response method in actual SNS systems and agents can thus make optional responses regardless of commenting on an article or not.

An agent network is denoted by graph $G = (A, E)$, where A is the set of n agents representing users and E is the set of edges representing the friend relationship between agents. Let N_i be the set of i's neighbors, i.e., the set of i's friends. Agents in an SNS-norms game select a strategy of either *cooperation* or *defeat*. Cooperation corresponds to posting articles/comments and defeat corresponds to doing nothing (just reading them; free-riding). Agent i has two learning parameters: the probability of cooperation (i.e., posting a new article) B_i and the probability of giving rewards (e.g., posting a comment on the article) L_i. Assuming the gene expression in the genetic algorithm described below, these learning parameters are expressed with three bits, meaning that they have discrete values $0/7, 1/7, \cdots, 7/7$, the same as the metanorms game in [1]. Parameter S ($0 \leq S \leq 1$), which is defined randomly each time an article is posted, expresses "awareness", i.e., the probability of discovering the posted article and the comments on it. The meanings of the other parameters in Fig. 1 are listed in Table 1. We assume that the parameters in this table have the same value in all agents.

---

[1] This type of read mark delivery is implemented in a number of SNS systems and includes the "Who's Viewed Your Updates?" function in LinkedIn, "read label" in LINE [6]. The access counter in a blog system can also be considered an example of this type of optional response.

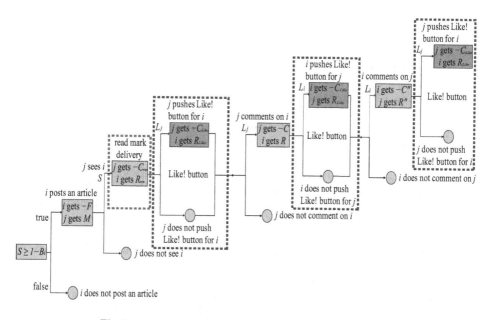

Fig. 1. Models of SNS-norms game with optional responses

Table 1. Parameters in SNS-norms game

Description	Parameter	Description	Parameter
Cost of article post	F	Reward by comment return	R''
Reward by post	M	Cost of Like! button	C_{Like}
Cost of comment	C	Reward by Like! button	R_{Like}
Reward by comment	R	Cost of "read" mark	C_{rm}
Cost of comment return	C''	Reward by "read" mark	R_{rm}

2.3 Chain in SNS-norms Game

The overall chain in the SNS-norms game is as follows (see also Fig. 1). First, the value of parameter S is randomly selected. Agent i decides whether to post an article or not according to parameter B_i: if $S < 1 - B_i$, i does not post the article (i.e., defeat), and the game chain ends, and if $S \geq 1 - B_i$, i posts the article (i.e., cooperation) with cost F. Another agent $j \in N_i$, a friend of i, gains reward M by reading the posted article. If j does not comment on i's article with probability $1 - L_j$, the game for j ends here. Otherwise, j posts a comment on the article with probability L_j and pays cost C. Then, i gains reward R through j's comment. The game chain so far is a rewards step, and this part of the SNS-norms game is referred to as the *SNS-reward game*. After the SNS-rewards game, i (that is, the agent who posted the original article) reads j's comment and posts a response to the comment with probability L_i. If i posts it, i pays cost C'' and j gains benefit R''. This subsequent step is different from that in the original meta-rewards game. The game explained so far is the basic part of the SNS-norms game (so this part is often referred as the basic SNS-norms game) and corresponds to the chain diagram excluding the regions boxed by green and blue dashed lines in Fig. 1.

A low-cost optional response such as a "Like!" button or "read" mark is added to the SNS-norms game. We assume that after i posts an article, the responses from $j \in N_i$ using the "Like!" button will occur with probability L_j, which is also the probability of comments by j. The probability of optional response is identical for a comment on a comment by i. This means that, for example, after agent i posts the article, $j \in N_i$ will comment on i's article with probability L_j and j will also give a response using the "Like!" button with probability L_j. Thus, j both gives a comment and clicks the "Like!" button with probability $(L_j)^2$. Then, after j clicks the "Like!" button, it pays cost C_{Like} and i receives gain R_{Like}. The SNS-norms game with the "Like!" button is represented by the regions unboxed and boxed with blue dashed lines in Fig. 1.

The SNS-norms game with the read mark delivery mechanism corresponds to the regions unboxed and boxed with green dashed lines in Fig. 1. The responses by the "read" mark are done automatically when a friend j reads the article posted by i. Note that no "read" mark is delivered to j when i reads j's comment on i's article. Then, i, who received the "read" mark, gains R_{rm}, and j, who read it, provides the "read" mark and pays cost C_{rm}. However, in both optional responses, we can assume that their costs C_{Like} and C_{rm} are almost zero. Note that the value of S used in determining whether or not agents post an article is also used in determining whether agents read the posted articles, as in the metanorms game. Thus, if S is low, it is difficult for other agents to read/notice the posted article and its comments, and only agents having relatively high B_i post articles.

2.4   Evolution by Genetic Algorithm

SNS-norms and SNS-rewards games are evolutionary games, as is Axelrod's metanorms game. One generation of the game is defined as the term in which each agent has four chances to post articles. The agent selects two agents as parents from the set of itself and its neighboring agents using roulette wheel selection based on *fitness values* and generates a child agent for the same node of the network in the next generation. The fitness values are defined as the cumulative rewards received minus cumulative costs incurred during the current generation. This process is continued up to the 10,000th generation. As stated in Section 2.2, i has two learning parameters, B_i and L_i. Each of these parameters is represented in three bits, so agents have six-bit genes. This encoding is also based on that in Axelrod [1]. The initial values of the six bits are set randomly at the beginning of each experimental trial. Child agent for the next generation is then created as follows.

Selection of Parents: Agent i selects two parents from its adjacent agents and itself according to the probability distribution $\{\Pi_h\}_{h \in N_i^+}$, where

$$\Pi_h = \frac{(v_h - v_{h,min})^2}{\sum_{k \in N_h^+} (v_k - v_{h,min})^2}, \tag{1}$$

for $\forall i \in A$, $N_i^+ (= N_i \cup \{i\})$ is the set of i and its adjacent agents, v_i is i's fitness value, and $v_{i,min} = \min_{h \in N_i^+} v_h$. If all agents in N_i^+ have the same fitness value, we define $\Pi_i = 1/|N_i^+|$.

Crossover: Two new genes are generated using uniform crossover from the genes in the selected parent agents. Then, one agent is created as a child by randomly selecting one of the two genes.

Mutation: Each bit of the gene of the child agent is inverted with a probability of 0.01. This means that if there are 20 agents in the network, 1.2 bits will mutate on average.

3 Experiments

3.1 Experimental Settings

We conducted two experiments; the first one is to investigate the conditions for cooperation dominance using basic SNS-norms games. The agent network used in our experiments is a network extracted from a real social network on Facebook. We then examine the effect of optional response methods in SNSs in the second experiment. We also conducted the same experiments using CNN networks [11]. However, we omit the results due to the page-length limit, although we have obtained the results that are quite similar to those on Facebook network.

We set the values of the parameters in Table 1 to $F = 3.0$, $M = 1.0$, $C = C'' = 2.0$, and $R = R'' = 9$ and also varied the values of $R = R''$ to examine the effect of the rewards on evolved agent behaviors. The values of the parameters for optional responses were set as $C_{Like} = C_{rm} = 0.0$ and $R_{Like} = R_{rm} = 1.0$ so R_{Like} and R_{rm} are considerably smaller than R and R''. These parameter values, except the parameters for optional responses, were determined by referring to those that were used in previous norms and metanorms games [1]. The initial values for B_i and L_i were defined randomly, as stated in Section 2.4. All results in our experiments were the average values of 100 independent trials with different random seeds.

3.2 Evolved Behaviors in SNS-norms Games

We explored the effect of cost-to-reward ratios, that is, how rewards R and R'' affected the evolved behaviors of agents if costs C and C'' were fixed in SNS-norms games (with no optional response). We show the average values of $B = \sum_{i \in A} B_i/|A|$ and $L = \sum_{i \in A} L_i/|A|$ when rewards by comments, R and R'', were changed in Fig. 2. This graph plotted their average values according to data obtained between the 1,001st and 10,000th generations.

Figure 2 indicates realistic curves: with increasing R and R'' rewards, the value of B gradually increased ahead of the value of L. However, this suggests strict requirements for cooperation dominance: specifically, a considerably high reward is necessary for all users to continue posting articles and comments. This means that it is quite difficult to keep SNSs thriving merely by posting articles, making comments, and commenting on comments.

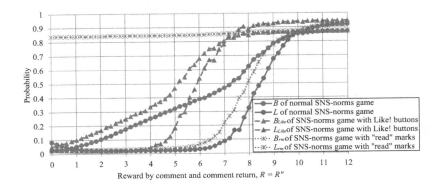

Fig. 2. Convergent values of B and L in SNS-norms game with optional responses (Facebook network)

3.3 Effect of Optional Responses on Evolved States

Next, we investigated how optional responses affect the evolved behaviors in the second experiment. Figure 2 plots the values of B and L when the rewards $R = R''$ are varied in the SNS-norms games with "Like!" buttons or with the read mark delivery. We denote the average probability of posting articles in an SNS-norms game with "Like!" buttons (with the read mark delivery) as B_{Like} (B_{rm}). Similarly, the average probabilities of posting comments with one of the optional responses are denoted by L_{Like} and L_{rm}.

We first focus on the results in an SNS-norms game with the read mark delivery. As shown in Fig. 2, B_{rm} converged around 0.9, meaning that agents in Facebook network always continue to post articles actively even if the reward is quite small. However, the average probability of posting comments was almost identical to that in the basic SNS-norms game. This is because (1) the read marks are automatically delivered when articles are read, and (2) the read marks were not delivered only when agents read comments, therefore posting a comment is not so important. However, this result shows that the SNS converges to a cooperation-dominant situation and that the read mark delivery is an effective mechanism to keep SNS thriving because the marks are delivered even when free-riders read the posted articles. Problematically, this mechanism often results in privacy invasion, an actual example of which we discuss in Section 3.4.

Figure 2 indicates that the optional response of clicking "Like!" buttons was also effective, as the probability of posting articles B_{Like} gradually increased according to the increase of rewards and was always larger than B in the basic SNS-norms game. The value of L_{Like} also increased with rewards smaller than those in the basic SNS-norms game and the SNS-norms game with the read mark delivery. This means that even with small rewards, a certain ratio of users continue to post articles and thus a certain degree of flourishing on the part of SNSs can be assured on the basis of the value of the reward.

3.4 Remarks

Our experimental results indicate that optional responses are important for keeping SNSs thriving. However, there are potential privacy issues with some of the optional responses, especially read mark delivery. Although the read mark delivery and other low-cost response methods are quite powerful functions, some users are nervous about automatic notification of SNS activities to other users, which highlights the major difference between the read mark delivery and "Like!" button mechanisms. Note that, aside from rewards (comments) from friends, we can consider a number of other important factors that may affect the attitudes to SNSs, such as user's characters and motivations. These factors are external to the SNSs, and thus are not included in SNS-norms games. We believe that rewards from friends are the most important factor for contributing in SNS.

4 Conclusion

We proposed an SNS-norms game to model SNSs and investigated the conditions required for cooperation-dominant situations using it on a social network extracted from Facebook. To keep SNSs flourishing, many articles and comments on articles must be posted continuously, a situation that corresponds to that where cooperation is dominant in an SNS-norms game. Our experimental results indicate that very large psychological rewards compared with the cost of writing and posting comments are necessary for cooperation dominance in the SNS-norms game, which corresponds to bare SNSs that have only article posting and comment functions. The results also show that optional responses that could provide a small reward with nearly zero cost were the most effective means of keeping SNSs thriving.

References

1. Axelrod, R.: An Evolutionary Approach to Norms. The American Political Science Review 80(4), 1095–1111 (1986)
2. Barabasi, A.L., Albert, R.: Emergence of Scaling in Random Networks. Science 286(5439), 509–512 (1999)
3. Ghanem, A.G., Vedanarayanan, S., Minai, A.A.: Agents of Influence in Social Networks. In: Proceedings of the 11th International Conference on Autonomous Agents and Multiagent Systems, AAMAS 2012, vol. 1, pp. 551–558 (2012)
4. Gonzalez, R., Cuevas, R., Motamedi, R., Rejaie, R., Cuevas, A.: Google+ or Google-?: Dissecting the Evolution of the New OSN in its First Year. In: Proceedings of the 22nd International Conference on World Wide Web, WWW 2013, pp. 483–494 (2013)
5. Hirahara, Y., Toriumi, F., Sugawara, T.: Evolution of Cooperation in Meta-Rewards Games on Networks of WS and BA Models. In: Proceedings of the 2013 IEEE/WIC/ACM International Joint Conferences on Web Intelligence (WI) and Intelligent Agent Technologies (IAT), WI-IAT 2013 - Volume 03, vol. 3, pp. 126–130. IEEE Computer Society (2013)
6. LINE, http://line.me/
7. Myers, S.A., Leskovec, J.: The Bursty Dynamics of the Twitter Information Network. In: Proceedings of the 23rd International Conference on World Wide Web, WWW 2014, pp. 913–924 (2014)

8. Stanford Large Network Dataset Collection, `http://snap.stanford.edu/data/`
9. Stieglitz, S., Dang-Xuan, L.: Social Media and Political Communication: A Social Media Analytics Framework. Social Network Analysis and Mining 3(4), 1277–1291 (2013)
10. Toriumi, F., Yamamoto, H., Okada, I.: Why Do People Use Social Media? Agent-Based Simulation and Population Dynamics Analysis of the Evolution of Cooperation in Social Media. In: Proceedings of the 2012 IEEE/WIC/ACM International Joint Conferences on Web Intelligence and Intelligent Agent Technology, WI-IAT 2012, vol. 2, pp. 43–50. IEEE Computer Society (2012)
11. Vázquez, A.: Growing network with local rules: Preferential Attachment, Clustering Hierarchy, and Degree Correlations. Phys. Rev. E 67, 056104 (2003)
12. Watts, D.J.: Collective Dynamics of 'Small-World' Networks. Nature 393, 440–442 (1998)

International Gender Differences and Gaps in Online Social Networks*

Gabriel Magno[1] and Ingmar Weber[2]

[1] Universidade Federal de Minas Gerais, Belo Horizonte, Brazil
magno@dcc.ufmg.br
[2] Qatar Computing Research Institute, Doha, Qatar
iweber@qf.org.qa

Abstract. Article 1 of the United Nations Charter claims "human rights" and "fundamental freedoms" "without distinction as to [...] sex". Yet in 1995 the Human Development Report came to the sobering conclusion that "in no society do women enjoy the same opportunities as men"[1]. Today, gender disparities remain a global issue and addressing them is a top priority for organizations such as the United Nations Population Fund. To track progress in this matter and to observe the effect of new policies, the World Economic Forum annually publishes its Global Gender Gap Report. This report is based on a number of offline variables such as the ratio of female-to-male earned income or the percentage of women in executive office over the last 50 years.

In this paper, we use large amounts of network data from Google+ to study gender differences in 73 countries and to link online indicators of inequality to established offline indicators. We observe consistent global gender differences such as women having a higher fraction of reciprocated social links. Concerning the link to offline variables, we find that online inequality is strongly correlated to offline inequality, but that the directionality can be counter-intuitive. In particular, we observe women to have a higher online status, as defined by a variety of measures, compared to men in countries such as Pakistan or Egypt, which have one of the highest measured gender inequalities. Also surprisingly we find that countries with a larger fraction of within-gender social links, rather than across-gender, are countries with *less* gender inequality offline, going against an expectation of online gender segregation. On the other hand, looking at "differential assortativity", we find that in countries with more offline gender inequality women have a stronger tendency for withing-gender linkage than men.

We believe our findings contribute to ongoing research on using online data for development and prove the feasibility of developing an automated system to keep track of changing gender inequality around the globe. Having access to the social network information also opens up possibilities of studying the connection between online gender segregation and quantified offline gender inequality.

---

* This work was done while the first author was at Qatar Computing Research Institute.

[1] http://hdr.undp.org/en/content/human-development-report-1995

L.M. Aiello and D. McFarland (Eds.): SocInfo 2014, LNCS 8851, pp. 121–138, 2014.
© Springer International Publishing Switzerland 2014

1   Introduction

Gender equality and full empowerment of women remains elusive in most countries around the world. Women are often at a significant disadvantage in fields such as economic opportunities, educational attainment, political empowerment and in terms of health. Reducing and ultimately erasing the "Gender Gap" in these fields is both an intrinsic, moral obligation but also a crucial ingredient for economic development. By limiting women's access to education and economic opportunities an immeasurable amount of human resource is lost and huge parts of the population are not able to develop their full potential.

To quantify gender inequality around the globe and to track changes over time, for example in response to policies put in place, the World Economic Forum annually publishes "The Global Gender Gap Report" in collaboration with the Center for International Development at Harvard University and the Haas School of Business at the University of California, Berkeley. This report ranks countries according to a numerical gender gap score. These scores can be interpreted as the percentage of the inequality between women and men that has been closed and so a large gap score is desirable. In 2013 the leading country Iceland had an aggregate score of 0.87, whereas Yemen scored lowest with 0.51. Scores are based on publicly available "hard data", rather than cultural perceptions, and variables contributing include the ratio of female-to-male earned income and the ratio of women to men in terms of years in executive office (prime minister or president) for the last 50 years. The emphasis of the report is on the relative gender difference for the variables considered rather than the absolute level achieved by women.

This paper contributes to this line of work by quantifying gender differences around the globe using existing methodology and applying it to *online* data, concretely data derived from Google+ for tens of millions of users. We start our analysis by describing the absolute differences along dimensions such as the number of male vs. female users or their virtual, social ranking in terms of number of followers. Our main emphasis is on studying correlations between online indicators of inequality and existing offline indicators. We do this both for the purpose of validation, to be sure that what we measure is linked to phenomena in "the real world", and for the purpose of devising new indicators, where a seemingly important online measure does not seem to be in good agreement with existing indicators.

Our current study is deliberately done *without* doing analysis of the content shared by men and women in different countries, and we are only relying on network structure data. One reason for this choice was one of global coverage: doing any type of content analysis for languages spanning all continents and having results comparable across languages and countries remains a fundamental challenge. Doing something only for English would have beaten the purpose of measuring gender inequality online in virtually all developing countries. A second reason for our choice was the fact that current indices are based on "hard data". Whereas the number of followers is well-defined, things such as the sentiment or mood of a user are hard to measure in an objective manner and are difficult to compare across cultures.

Analyzing gender differences for 73 countries we find both expected and surprising trends. Our main findings are:

- Countries with more men than women online are countries with more pronounced gender inequality.
- Women are more tightly cliqued and their links are more reciprocated.
- In countries with higher offline inequality women are, suprisingly, followed more than men. This result holds both using the mean and the median, and it holds for other "status" metrics such as PageRank.
- Countries with a larger fraction of within-gender social links, rather than across-gender, are countries with *smaller* offline gender inequalities.
- Countries with larger offline gender inequalities have a larger "differential assortativity" where women have a stronger preference for within-gender links than men.
- Applying existing gap-based methodology to online data yields a strong negative correlation, up to $r = -0.76$, with existing offline measures.

Generally our analysis is more quantitative and descriptive rather than qualitative and diagnostic. Though we describe the gender differences we find and comment on whether they agree with (at least our) expectations, we do not attempt to give explanations. We hope that experts in domains such as gender studies or social psychology will find our analysis useful and that it can save as a starting point for more in-depth studies focused at the root causes of what we observe.

As more and more economic activity becomes digital and moves online, as more and more education happens online through MOOCs and other initiatives, and as more and more of political engagement happens online we are convinced that, ultimately, quantifying gender inequality also has to crucially take into account online activity.

2   Related Work

As far as we are aware, this is the first study that links online gender differences in dozens of countries to existing quantitative offline indicators. However, lots of valuable research has been done looking at gender differences and gender inequality offline and online separately and such work has considered various psychological, sociological and economical differences. It is not within this paper's scope to serve as a complete review of literature in gender studies but, rather, it should give the reader a good overview of aspects than have been investigated.

2.1   Offline

Feingold conducted a meta-analysis to investigate differences in personal traits between genders as reported in literature [13]. For some traits such as extroversion, anxiety and tender-mindedness, women were higher, while for others such as assertiveness and self-esteem, men had higher scores. And, as one might hope, there are also traits with no observed gender differences such as social anxiety and impulsiveness.

Pratto et al. studied gender differences in political attitudes [30]. By analyzing a sample of US college students, they found that men tend to support more conservative ideology, military programs, and punitive policies, while women tend to support more equal rights and social programmes. They also show that males were in general more social dominance oriented than females.

Costa et al. [10] aggregated results of psychological tests from different countries for the so-called "Big Five" basic factors of personality: Neuroticism, Extroversion, Openness to Experience, Agreeableness, and Conscientiousness [29]. They observed that, contrary to predictions from the social role model, gender differences concerning personality were most pronounced in western cultures, in which traditional sex roles are comparatively weak compared to more traditional cultures. In a similar line of work, Schmitt et al. [34] conceived the General Sex Difference Index and observed that sex differences appear to diminish as one moves from Western to non-Western cultures.

Hyde performed a meta-analysis on psychological gender differences to show that, according to the gender similarities hypothesis, males and females are alike on most psychological variables, contrasting the differences model that states that men and women are vastly different psychologically [19].

2.2   Online

Gender Gap. Bimber analyzed data from surveys in the United States, in which people were asked about Internet access and frequency of utilization [4]. His analysis showed that there is a gap in access regarding the gender, but that this gap is not related to the gender itself, but rather to socioeconomic factors, such as education and income. Collier and Bear investigated the low participation of women in terms of contributions to Wikipedia [9]. They found strong support that the gender gap is due to the high levels of conflict in discussions, and also due to a lack of self-confidence in editing others' work. Iosub et al. investigated the communication between editors in Wikipedia and observed that female editors communicate in a way that develop social affiliation [20]. In terms of online social network usage in the US in 2013, women had higher rates of users for Facebook, Pinterest or Instagram, whereas usage was similar for both genders for Twitter and Tumblr [8]. In our data for the US, we have more male users. A possible explanation for this is an increased concern for privacy with a corresponding choice to reveal less information about themselves. See related work further down on this subject.

Privacy and Interests. Researchers investigated whether there is a difference between genders regarding the kind and amount of information shared online. Thelwall conducted a demographic study of MySpace members, and observed that male users are more interested in dating, while female users are more interested in friendship, and also tend to have more friends [36]. When analyzing the privacy behavior, women were found to be more likely to have a private profile. Joinson analyzed reports on motivation to utilize Facebook [21]. He found that female users are more likely to use Facebook for social connections, status updates and photographs than male users. Also, female users are more prone to make an effort to make their profile private. Bond conducted a survey among undergraduate students regarding their utilization in OSNs and found that female participants disclose more images and information on OSN profiles than male participants [7]. They also observed that the kind of content shared between genders are different. For instance, female users tend to share more content about friends, family, significant others, and holidays, while male users are more likely to post content related to sports. Other works also investigated the vocabulary used by users in OSNs, and found that there is differences regarding the semantic category of words between women and men [28,12]. Quercia et al. studied the relationship between information disclosure and personality by using information from personality tests done by Facebook users, and found out that women are less likely than men to publicly share privacy-sensitive fields [32].

Network. Szell and Thurner analyzed the interactions between players of a massive multiplayer online game [35]. They constructed the interactions graphs and observed that there are differences between male players and female players for all kinds of connections. For instance, females have higher degrees, clustering coefficient and reciprocity values, while males tend to connect to players with higher degree values. Ottoni et al. also investigated the friendship connections of the users in Pinterest and observed that females are more reciprocal than males [28]. In our analysis, we also found women to have a higher clustering coefficient and a larger fraction of reciprocated friendship links on Google+. Heil et al. analyzed Twitter data from 300 thousand users, and found that males have 15% more followers than women. When looking at homophily, they found that on average men are almost twice as likely to follow other men than women, and, surprisingly, women are also more likely to follow men [18,26]. In our analysis, we observed homophily for both genders in Google+, i.e. females tend to follow more females and males to follow more males. Recent work has also looked at generalizing concepts from the "Bechdel Test"[2] to Twitter [14]. The authors look at tweets from the US for users sharing movie trailers, which are then linked to Bechdel Test scores, and they find larger gender independence for urban users in comparison to rural ones, as well as other relations with socio-economic indicators.

Socio-Economic Indicators from Online Data. Putting aside the concrete issue of gender inequality, we are essentially interested in using online data as a socio-economic indicator. This idea in itself is not new and previous research has attempted to estimate things such as unemployment rates [1], consumer confidence [27], migration rates [37,17], values of stock market and asset values [6,5,38] and measures of social deprivation [33]. Work in [31] is also related as it looked at search behavior, in this case "forward looking searches" and links such queries to estimates of economic productivity around the globe.

---

[2] http://en.wikipedia.org/wiki/Bechdel_test

3   Data Set

Our dataset was created by collecting public information available in user profiles in the Google+ network. We inspected the *robots.txt* file and followed the sitemap to retrieve the URLs of Google+ profiles. Since we retrieved the complete list of profiles provided by Google+, we believe our data set covers almost all users with public profiles in Google+ by the time of the data collection. The data collection ran from March 23rd of 2012 until June 1st of 2012. When inspecting the sitemap we found 193,661,503 user IDs. In total we were able to retrieve information from 160,304,954 profiles. Some IDs were deleted or we were not able to parse their information. With the social links of the users, we have constructed a directed graph that has 61,165,224 user nodes and 1,074,088,940 directed friendship edges.

Country Identification. To identify a user's country in Google+, we extracted the geographic coordinates of the last location present on the *Places lived* field and identified the corresponding country. We were able to identify the country of 22,578,898 users.

Gender. Google+ provides a self-declared gender field where the user can choose between three categories: *female, male* and *other*. As any other profile field in Google+ (except for the name), it is possible to put this information as private, so we do not have this information for all users. Of the 160 millions users, 78.9% provided the gender field publicly, from which 34.4% are female, 63.8% are male and 1.8% selected "other".

Details of the Google+ platform and a data characterization of an early version of the dataset are discussed in a previous work [25]. A summary of the number of users for each country can be found in Table A.1 (appendix). We only selected countries with at least 5,000 users for each gender.

3.1   Online Variables

As doing any type of content analysis for dozens of languages and cultures is extremely challenging, we decided to study how *network* metrics could be indicators for gender gaps. At the country-level, we looked at the following metric which we hypothesized could be an indication of online gender segregation.

– The *assortativity*[3] is the fraction of links to the same gender rather than across genders. A large value can be indicative of either strong same-gender linkage preference, or simply a highly imbalanced gender distribution of the users, which trivially makes cross-gender links less likely.

We also computed the following metrics for each user from one of the 73 countries in our data set.

---

[3] We use "assortativity" rather than "homophily" to emphasize the correlation rather than necessary a causal link.

- The *in-degree*, also referred to as the number of followers, counts the number of "circles" a user is in. A large in-degree can be seen as an indicator of popularity or status.
- The *out-degree*, also referred to as the number of followees or friends, counts the number of users a user has in their circles.
- The *reciprocity* is the fraction of reciprocal links in relation to the out-degree, i.e. the fraction of times where the act of following is reciprocated by the receiving user.
- The *clustering coefficient* for a particular node is the probability of any two of its neighbors being neighbors themselves. It is calculated by the fraction of the number of triangles that contain the node divided by the maximum number of triangles possible (when all the neighbors are connected), which for a directed graph is equal to $n(n-1)$, where n is the number of neighbors that reciprocate the connection. A large value typically indicates a large degree of "cliqueness" and more tightly connected social groups.
- The *PageRank* measures the relative importance of a user in the network and, unlike the mere in-degree, is influenced by the "global" social graph structure. A damping factor $d = 0.85$ was used for the iterations of the algorithm. A large PageRank value is often thought of as an indicator of "centrality" or "importance" in the social graph.
- The *differential assortativity* is the "lift" of the fraction of users of the same gender followed by a particular user. It is calculated by dividing the fraction of links to the same gender by the share of that gender for the country of the user. A large value means that users are more likely than by random chance to follow other users of their same gender. The comparison against random chance corrects for the fact that in an online population of, say, 80% males are trivially more likely to follow other males even without any same-gender homophily.

These per-user metrics are then aggregated into a per-country score as described in the next paragraph. Though we group the results by country, connections across countries are included in our analysis. So a reciprocal link between two users in Brazil and Qatar would contribute to the statistics of both countries.

Gender Gap. One of the goals of our study was to devise an "Online Gender Gap" score and to see how this relates to the existing offline Gender Gap scores. We therefore followed the same methodology of computing a "gap" score: First, we group the users by country and gender, and calculate the average of the variable for each country-gender group. After having the aggregated value for each country-gender group, we calculate the gender ratio by dividing the female value by the male value, for each country. Differently from the Global Gender Gap score methodology, we do not truncate the ratio at 1, since we want to analyze the trend even when the value is higher for female users, especially as some of our variables, such as the number of followers, exhibited a counter-intuitive trend. Furthermore, for some of our variables such as the Differential Assortativity, it is also not intuitively obvious if a high or a low gender-specific value is desirable and, correspondingly, it is unclear if high or low values should be truncated.

Note that, in line with the Global Gender Gap report, a large "gap value" is actually *desirable* in the sense that it typically indicates gender equality or female dominance for the variable considered, whereas a very low gap value is undesirable as it indicates that the variable considered is lower for women than for men.

3.2   Offline Variables

The Global Gender Gap Index [4] is a benchmark score that captures the gender disparities in each country. It takes into account social variables from four categories (economy, politics, education and health), such as life expectancy, estimated income, literacy rate and number of seats in political roles. The index is built by (1) calculating the female by male ratio of the variables, (2) truncating the ratios at a certain level (1.0 for most variables), (3) calculating subindexes for each one of the four categories (weighted average in relation to the standard deviation) and (4) calculating the un-weighted average of the four subindexes to create the overall index. The scores range from 0 (total inequality) to 1.0 (total equality). For this study we use the 2013 Global Gender Gap report [16].

We also use additional economic variables and demographic information to see if these are linked to online gender gaps. For population and internet penetration information we use information from the Internet World Stats website[5] on internet usage for 2012. The GDP per capita information was collected from the World Bank website[6] and is for 2011. Information for more recent years was missing for some countries which is why we selected data from 2011.

4   Gender Differences Online

Before we link online variables to offline indicators of gender gaps, we first describe how men and women in 73 countries differ in their usage of Google+. Figure 1 shows the gender ratio of the variables for each country. We observe that for some variables there is a female predominance (such as for "Reciprocity" and "Clustering Coefficient"), while for others there's a male predominance (such as "Number of followees"). In most cases, the gender predominance is the same across countries, but for some variables ("Number of followers") there are divergences.

5   Online and Offline Gender Gaps

To test the significance of the difference between female and male values of the variables we conducted a permutation test that does not make assumptions

---

[4] http://www.weforum.org/issues/global-gender-gap

[5] http://www.internetworldstats.com

[6] http://www.worldbank.org

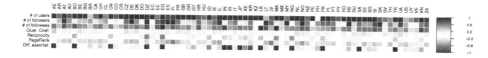

Fig. 1. A color plot of the logarithm, base 2, of the (female value)/(male value) gender ratio (GR), i.e. $\log_2(\mathrm{GR})$, for the variables in each country. The scale is truncated at -1.0 and 1.0. A value higher than 0 (blue) indicates male predominance, and lower than 0 (red) means female predominance.

about the distribution of the variables.[7] First, for each country we compute the average of a variable across all female users and compare the value with the one obtained for the male users. Let δ be the observed difference. Then we use the same set of users, but now randomly permute the gender label. The basic idea is to see if the observed difference could have arisen due to random variance or whether it is more systematically linked to the gender of the users. We now calculate the average of the two groups derived from the permutation, and calculate the difference δ_p. We repeat this process 1,000 times to estimate the level of variability of δ_p. Finally, we mark the δ as significant if it was in the bottom/top 0.5% (or 2.5%) of the percentiles of the δ_p. In Table 1 we present the significance test result for some variables for a fraction of the countries. In Table B.2 (appendix) we present the values for all the countries. For most countries and most variables the difference between female and male is significant.

6   Linking Online and Offline Gender Gaps

Whereas the previous section looked exclusively at online gender differences, here we focus on linking online and offline gender gaps across 73 countries.

Figure 2 shows the linear regression between online variables and the Global Gender Gap scores. GR stands for Gender Ratio (female divided by male value). We observe that the gap score for the number of users is positively correlated with the gender gap score. Countries with a roughly equal number of male and female users online tend to score better (= higher) for the offline gap scores. Surprisingly, at least to us, we also find that the number of followers and other measures of "status" are negatively correlated for both networks. For example, Pakistan has an offline Gender Gap score of 0.546 (with 1.0 indicating equality) but, at the same time, women who are online in Pakistan have on average (and in median) more followers than their male counterparts. . We discuss potential reasons later in the paper.

The two plots in the right column of Figure 2 show the linear regression plots of the assortativity variables in Google+. When we analyze the Differential

---

[7] See http://en.wikipedia.org/wiki/Resampling_%28statistics%29# Permutation_tests for background information on permutation tests in statistics.

Table 1. Significance test results for variables in Google+ for a subset of our 73 countries, ranked in descending order of the number of users. The value on the left is the average female value and the value on the right is the average male value, followed by the significance result ('*' is 95% significant, '**' is 99% significant). The full list of results can be found in Table B.2 (appendix).

Country	In-degree ♀/♂	Out-degree ♀/♂	Recipr. ♀/♂	Clust. Coeff. ♀/♂	PageRank ♀/♂
United States	34.8/47.1**	20.6/30.3**	0.49/0.50**	0.31/0.28**	2.0e-08/2.6e-08**
Russian Federation	17.7/20.8**	31.0/36.1**	0.45/0.41**	0.38/0.32**	1.5e-08/1.8e-08**
Italy	34.7/22.0	22.7/33.3**	0.51/0.48**	0.33/0.29**	1.8e-08/2.0e-08**
Viet Nam	36.9/57.4**	41.7/78.3**	0.41/0.34**	0.29/0.29	1.8e-08/2.0e-08**
Philippines	11.6/16.6**	28.8/38.5**	0.42/0.41	0.40/0.36**	1.4e-08/1.6e-08**
Pakistan	25.4/15.8**	35.3/49.1**	0.40/0.31**	0.32/0.29**	1.6e-08/1.3e-08**
Saudi Arabia	39.3/24.6**	30.2/47.4**	0.37/0.33**	0.29/0.26**	1.7e-08/1.6e-08
Bangladesh	17.4/15.2	30.4/54.1**	0.41/0.30**	0.32/0.30**	1.4e-08/1.3e-08
United Arab Emirates	19.6/18.4	21.4/33.6**	0.46/0.42**	0.28/0.22**	1.7e-08/1.7e-08
Greece	19.0/22.1	26.5/40.3**	0.47/0.44**	0.34/0.30**	1.5e-08/1.8e-08**
Norway	16.8/40.3**	17.6/30.8**	0.57/0.56**	0.35/0.31**	1.7e-08/2.5e-08**
Sri Lanka	20.9/21.1	23.7/50.7**	0.47/0.36**	0.31/0.30*	1.6e-08/1.6e-08
El Salvador	12.8/11.5	31.7/28.7	0.38/0.39	0.21/0.24**	1.4e-08/1.5e-08*
Guatemala	10.1/12.1	21.2/26.2**	0.46/0.40**	0.27/0.29*	1.5e-08/1.5e-08
Slovenia	10.0/18.2**	16.8/30.2**	0.56/0.53**	0.27/0.28	1.6e-08/2.1e-08**

assortativity we observe that most countries, clustered together on the dashed line, have similar values for female and male, meaning that the level of gender assortativity is the same for women and men. On the other hand, in countries with a low Gender Gap score there's a female predominance, meaning that women in these countries connect much more among themselves than expected when compared to men. This could be seen as an indication of women "shying away" from cross-gender linkage in such countries. When we analyze not the gap but the actual assortativity of a country we observe a positive correlation with the gap score, meaning that in countries with higher Gender Gap score (= little inequality), there is higher assortativity (= more within-gender linkage). We discuss potential hypotheses explaining this arguably surprising finding in Section 7.

Figure 3 presents the matrix of correlation between the online and offline variables, essentially summarizing the linear regression fits from Figure 2 and adding more variables. As in Figure 2, the Gender Gap Score is positively correlated with the gender gap of the number of users in Google+, and, surprisingly, negatively correlated with the gap of the number of followers, reciprocity and PageRank. In terms of assortativity, there is a negative correlation for differential assortativity, meaning that female users connect more among themselves in countries with a low Gap score, while the actual assortativity of the network is positively correlated, implying more segregation in countries with high Gender Gap score.

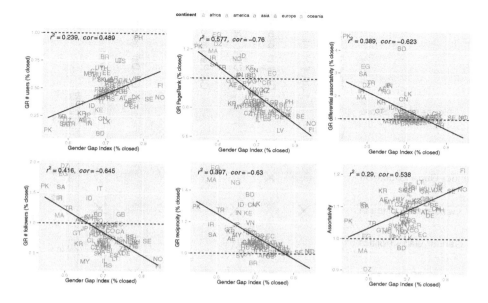

Fig. 2. Linear regression and correlation between online social network metrics and the Global Gender Gap score. GR stands for Gender Ratio (female by male value). See Table A.1 (appendix) for a list of 2-letter country codes. The p-values for the correlation were all lower than 0.01.

7 Discussion

One of our main motivation for this work was to see if online data could be used to derive global indicators of gender inequality and whether these indicators were in some sense "grounded" in that they are linked to existing indicators. Our findings indicate that this indeed the case.

Surprisingly, the directionality of important indicators was *opposite* from what we had expected. Concretely, we found that all indicators of gaps in online social status such as the average number of followers, or the Pagerank on Google+ all had noticeable *negative* correlations (.65 and -.76 correspondingly) with the aggregated offline gender gap score. For example in Pakistan, with a gender gap score of 0.55, indicating a large inequality, we found that women have on average 50% more followers on Google+ than men. Note that the number of followers is typically heavy-tailed [22] and for such distributions it is known that the observed average will increase as the sample size increases[8]. As we have fewer women and men for countries where we observe these effects, the actual effect might hence be even stronger. We also mention that we observed the same effect by looking at medians, rather than averages, indicating a robust result not caused by outliers.

---

[8] See, e.g., http://en.wikipedia.org/wiki/Pareto_distribution which has an infinite mean when $\alpha \leq 1$.

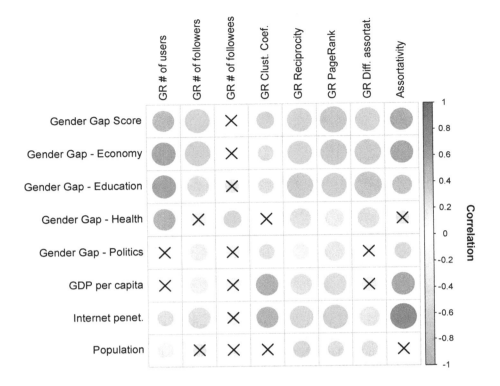

Fig. 3. Correlation between offline variables and the ratio of online variables of the countries. GR stands for Gender Ratio (female by male value). The relation is marked with an X when the p-value of the correlation is lowen than 0.05.

Our current hypothesis is that this unexpected result might be due to the so-called "Jackie Robinson Effect"[9]. Jackie Robinson was a baseball player who who became the first African-American to play in Major League Baseball in the modern era. If he had been only good, rather than great, it is unlikely that he would have been given a chance to play rather than a slightly less talented white alternative. Similarly, one might imagine that women that are online in countries where women have more limited online access compared to men must be extraordinary to begin with. In a similar vein it was found that female politicians perform better than their male counter-parts as doing just as well would not suffice to "make it" [2].

The effect above might also be linked to our observation of more within-gender linkage for countries such as Finnland or Norway, compared to Egypt or Pakistan. Other potential explanations for this observation could be acts of online "stalking" or "staring" where women attract follow links from men, causing more cross-gender linkage. This latter hypothesis is also consistent with

---

[9] http://en.wikipedia.org/wiki/Jackie_Robinson

our observation that in countries with more offline gender inequality women have a stronger tendency for withing-gender linkage than men, potentially indicative of shying away from cross-gender linkage.

Of course, our current data set and methodology are by no means perfect. Clearly, our user set is by no means representative of the overall population. Generally, we expect people over a higher social status to be overrepresented in our data. But even the fact that for Pakistan we find about 8 times as many male Google+ users as female ones is in itself a signal. Also note that for certain applications the selection bias might be irrelevant. If, for example, the main purpose of using online data is to have a low-cost and real-time alternative to compute the offline gender gap index then as long as it works, despite the selection bias, the selection bias itself becomes irrelevant. As a comparison, if it is possible to accurately predict current levels of flu activity from social media data then there is no reason to question this approach, assuming that the prediction remains valid as the online population continues to change [3,23,11].

The example of monitoring flu activity also points to another limitation of our study: the use of only one data source. For flu monitoring using online data, Google Flu Trends [15] is the de-facto standard and baseline to beat. Recently, its use as a figurehead has however been questioned [24]. Still, it seems promising to look at, say, the relative search volume of topics associated with gender roles to see if their search volume could be indicative of gender gaps. Additionally, gender differences on comments on national, political sites could be indicators for political engagement.

Another big limitation is our decision to ignore the content/topics that are discussed. The main reasons for this are (i) technical difficulties when dealing with content analysis for dozens of different languages and character sets, in particular if the results need to be comparable across countries, and (ii) the emphasis of existing offline indices on "hard data" rather than sentiments or more qualitative analysis. Still, it seems valuable to look at the topics discussed by, say, men and women in Mali to get better insights into their lived online experiences. In future work we plan to focus on a limited set of countries and languages and study topical differences in depth. Integrating content could also lead to an improvement of the already decent fit between a combination of online indicators and the offline gender gap scores. Finally, it could provide hypotheses for the root causes of the differences we observe.

Our current analysis is based on a static snapshot of time. However, our declared goal is to design a system that frequently calculates the latest online indicators of gender gaps and makes these publicly available. This is done with initiatives such as the United Nations Global Pulse in mind. "The Global Pulse initiative is exploring how new, digital data sources and real-time analytics technologies can help policymakers understand human well-being and emerging vulnerabilities in real-time."[10] Similarly, the United Nations Population Fund supports use of Data for Development and "women's roles and status, spatial mobility of populations and differentials in morbidity and mortality within

---

[10] http://www.unglobalpulse.org/

population subgroups were singled out as pressing concerns"[11]. At a broader level, more and more non-profit organizations are advocating the use of data mining "for good" and, as an example, the US Center for Disease Control and Prevention is organizing a competition to encourage the use of social media to predict flu activity[12].

Ultimately, of course, the goal is not just to describe and quantify gender gaps but to close these gaps. Here, a large amount of responsibility undoubtedly lies with politicians and people in positions of power. As good policy making needs to be linked to quantifying the progress made, and there is a necessity to observe the impact of new policies, measurement efforts are a valid objective in their own right. However, it is well worthwhile thinking about how social media and online social networks could in itself be used as a tool to facilitate the process of closing the gap, rather than as a mere data source. It might for example be possible to automatically strengthen the social capital of underprivileged women or, if nothing else, it could be used as communication channel to support the cause of gender equality.

8   Conclusion

We presented a large-scale study of gender differences and gender gaps around the world in Google+. Our analysis is based on 17,831,006 users from 73 countries with an identified gender and, to the best of our knowledge, is the first study that links online indicators of gender inequality to existing offline indicators.

Our main contribution is two-fold. First, we describe gender differences along a number of dimensions. Such insights are valuable both as a starting point for in-depth studies on identifying the root causes of these differences, but also when it comes to designing gender-aware systems. Second, we show how applying existing offline methodology for quantifying gender gaps can be applied to online data and that there is a respectable match in form of a 0.8 correlation across 73 countries.

Looking at individual variables we also find surprising patterns such as a tendency for women in less developed countries with larger gender differences to have a *higher* social status online as measured in terms of number of followers or Pagerank. We hypothesize the existence of an underlying "Jackie Robinson Effect" where women who decided to go online in a country such a Pakistan are likely to be more self-confident and tech-savvy than random male counterparts. Such an effect might also be linked to the fact that we observe a *higher* within-gender link assortativity for countries with *less* offline gender inequality, though alternative explanations include men "stalking" women online.

As more and more economic activity, education, and political engagement happens online we are convinced that, ultimately, quantifying gender inequality has to crucially take into account online activity.

---

[11] http://www.unfpa.org/public/datafordevelopment
[12] http://www.cdc.gov/flu/news/predict-flu-challenge.htm

Acknowledgments. We thank Ricardo Hausmann at the Harvard Center for International Development and Martina Viarengo at the Graduate Institute of International and Development Studies of Geneva for their valuable input.

References

1. Antenucci, D., Cafarella, M., Levenstein, M., Ré, C., Shapiro, M.D.: Using social media to measure labor market flows. Tech. Rep. 20010, National Bureau of Economic Research (March 2014)
2. Anzia, S.F., Berry, C.R.: The jackie (and jill) robinson effect: Why do congresswomen outperform congressmen? American Journal of Political Science 55, 478–493 (2011)
3. Aramaki, E., Maskawa, S., Morita, M.: Twitter catches the flu: Detecting influenza epidemics using twitter. In: Proceedings of the Conference on Empirical Methods in Natural Language Processing, EMNLP 2011, pp. 1568–1576. Association for Computational Linguistics, Stroudsburg (2011)
4. Bimber: Measuring the Gender Gap on the Internet. Social Science Quarterly 81(3) (Sep 2000)
5. Bollen, J., Mao, H., Pepe, A.: Modeling public mood and emotion: Twitter sentiment and socio-economic phenomena. In: ICWSM (2011)
6. Bollen, J., Mao, H., Zeng, X.J.: Twitter mood predicts the stock market. J. Comput. Science 2(1), 1–8 (2011)
7. Bond, B.J.: He posted, she posted: Gender differences in self-disclosure on social network sites. Rocky Mountain Communication Review 6(2), 29–37 (2009)
8. Women's Media Center: The status of women in the u.s. media 2014 (2014), http://www.womensmediacenter.com/page/-/statusreport/WMC-2014-status-women-with-research.pdf
9. Collier, B., Bear, J.: Conflict, criticism, or confidence: an empirical examination of the gender gap in wikipedia contributions. In: CSCW, pp. 383–392 (2012)
10. Costa, P.T., Terracciano, A., McCrae, R.R.: Gender differences in personality traits across cultures: robust and surprising findings. Journal of Personality and Social Psychology 81(2), 322–331 (2001), http://view.ncbi.nlm.nih.gov/pubmed/11519935
11. Culotta, A.: Lightweight methods to estimate influenza rates and alcohol sales volume from twitter messages. Language Resources and Evaluation 47(1), 217–238 (2013)
12. Cunha, E., Magno, G., Gonçalves, M.A., Cambraia, C., Almeida, V.: How you post is who you are: Characterizing google+ status updates across social groups. In: Proceedings of the 25th ACM Conference on Hypertext and Social Media, HT 2014, pp. 212–217. ACM (2014)
13. Feingold, A.: Gender differences in personality: a meta-analysis. Psychological bulletin 116(3), 429–456 (1994), http://www.ncbi.nlm.nih.gov/pubmed/7809307
14. Garcia, D., Weber, I., Garimella, V.R.K.: Gender asymmetries in reality and fiction: The bechdel test of social media. In: ICWSM (2014)
15. Ginsberg, J., Mohebbi, M., Patel, R., Brammer, L., Smolinski, M., Brilliant, L.: Detecting influenza epidemics using search engine query data. Nature 457, 1012–1014 (2009), http://www.nature.com/nature/journal/v457/n7232/full/nature07634.html doi:10.1038/nature07634

16. Hausmann, R., Tyson, L.D., Zahidi, S. (eds.): The global gender gap report 2013 (2013), http://www3.weforum.org/docs/WEF_GenderGap_Report_2013.pdf
17. Hawelka, B., Sitko, I., Beinat, E., Sobolevsky, S., Kazakopoulos, P., Ratti, C.: Geolocated twitter as proxy for global mobility patterns. Cartography and Geographic Information Science 41, 260–271 (2014)
18. Heil, B., Piskorski, M.: New twitter research: Men follow men and nobody tweets (June 2009),
http://blogs.hbr.org/2009/06/new-twitter-research-men-follo/
19. Hyde, J.S.: The Gender Similarities Hypothesis. American Psychologist 60(6), 581–592 (2005), http://bama.ua.edu/~sprentic/672%20Hyde%202005.pdf
20. Iosub, D., Laniado, D., Castillo, C., Fuster Morell, M., Kaltenbrunner, A.: Emotions under discussion: Gender, status and communication in online collaboration. PLoS ONE 9(8), e104880 (2014)
21. Joinson, A.N.: Looking at, Looking Up or Keeping Up with People?: Motives and Use of Facebook. In: Proceedings of the SIGCHI Conference on Human Factors in Computing Systems, CHI 2008, pp. 1027–1036. ACM (2008), http://doi.acm.org/10.1145/1357054.1357213
22. Kwak, H., Lee, C., Park, H., Moon, S.B.: What is twitter, a social network or a news media? In: WWW, pp. 591–600 (2010)
23. Lampos, V., Cristianini, N.: Nowcasting events from the social web with statistical learning. ACM Trans. Intell. Syst. Technol. 3(4), 1–72 (2012)
24. Lazer, D., Kennedy, R., King, G., Vespignani, A.: The parable of google flu: Traps in big data analysis. Science 343, 1203–1205 (2014)
25. Magno, G., Comarela, G., Saez-Trumper, D., Cha, M., Almeida, V.: New kid on the block: exploring the google+ social graph. In: Proceedings of the 2012 ACM Conference on Internet Measurement Conference, IMC 2012, pp. 159–170. ACM, New York (2012)
26. Miritello, G., Lara, R., Cebrian, M., Moro, E.: Limited communication capacity unveils strategies for human interaction. Sci. Rep. 3 (June 2013), http://dx.doi.org/10.1038/srep01950
27. O'Connor, B., Balasubramanyan, R., Routledge, B.R., Smith, N.A.: From tweets to polls: Linking text sentiment to public opinion time series. In: ICWSM (2010)
28. Ottoni, R., Pesce, J.P., Las Casas, D., Franciscani Jr, G., Meira Jr, W., Kumaraguru, P., Almeida, V.: Ladies first: Analyzing gender roles and behaviors in pinterest. In: Proceedings of the Seventh International Conference on Weblogs and Social Media, ICWSM 2013 (2013)
29. Costa, P.T., McCrae, R.R.: The Revised NEO Personality Inventory (NEO-PI-R), pp. 179–199. SAGE Publications Ltd (2008)
30. Pratto, F., Stallworth, L.M., Sidanius, J.: The gender gap: Differences in political attitudes and social dominance orientation. The British Journal of Social Psychology 36(1), 49–68 (1997), http://www.ncbi.nlm.nih.gov/pubmed/9114484
31. Preis, T., Moat, H.S., Stanley, H.E., Bishop, S.R.: Quantifying the advantage of looking forward. Nature Scientific Reports 2, 350 (2012)
32. Quercia, D., Casas, D.B.L., Pesce, J.P., Stillwell, D., Kosinski, M., Almeida, V., Crowcroft, J.: Facebook and privacy: The balancing act of personality, gender, and relationship currency. In: ICWSM (2012)
33. Quercia, D., Sáez-Trumper, D.: Mining urban deprivation from foursquare: Implicit crowdsourcing of city land use. IEEE Pervasive Computing 13(2), 30–36 (2014)

34. Schmitt, D.P., Realo, A., Voracek, M., Allik, J.: Why can't a man be more like a woman? Sex differences in Big Five personality traits across 55 cultures. Journal of Personality and Social Psychology 94(1), 168–182 (2008), http://psycnet.apa.org/index.cfm?fa=search.displayRecord&uid=2007-19165-013
35. Szell, M., Thurner, S.: How women organize social networks different from men. Scientific Reports 3 (July 2013), http://dx.doi.org/10.1038/srep01214
36. Thelwall, M.: Social networks, gender, and friending: An analysis of myspace member profiles. JASIST 59(8), 1321–1330 (2008)
37. Zagheni, E., Garimella, V.R.K., Weber, I., State, B.: Inferring international and internal migration patterns from twitter data. In: WWW (Companion Volume), pp. 439–444 (2014)
38. Zhang, X., Gloor, H.F.P.A.: Predicting asset value through twitter buzz. Advances in Intelligent and Soft Computing 113, 23–34 (2012)

Appendix

A List of Countries

Table A.1. List of countries with their respective 2-letter country codes and the total number of female and male users. We select only countries with at least 5,000 females and males.

Code	Name	Female	Male	Total	Code	Name	Female	Male	Total
US	United States	2,186,509	2,910,470	5,096,979	KR	South Korea	16,570	60,696	77,266
IN	India	363,956	1,964,070	2,328,026	SE	Sweden	22,342	54,815	77,157
BR	Brazil	563,173	716,455	1,279,628	BE	Belgium	21,755	55,223	76,978
GB	United Kingdom	210,801	445,343	656,144	AE	United Arab Emirates	12,250	57,399	69,649
ID	Indonesia	136,013	396,028	532,041	DK	Denmark	20,219	47,470	67,689
RU	Russian Federation	140,024	326,464	466,488	CZ	Czech Republic	19,409	46,548	65,957
CA	Canada	147,247	255,750	402,997	SG	Singapore	20,798	43,515	64,313
MX	Mexico	129,566	261,958	391,524	FI	Finland	21,831	41,072	62,903
DE	Germany	98,500	275,813	374,313	GR	Greece	17,578	41,393	58,971
ES	Spain	116,997	221,343	338,340	IE	Ireland	21,277	35,959	57,236
IT	Italy	87,028	226,777	313,805	RS	Serbia	16,458	40,241	56,699
FR	France	98,628	211,602	310,230	CH	Switzerland	14,255	42,085	56,340
JP	Japan	57,234	221,049	278,283	AT	Austria	15,487	37,185	52,672
CN	China	45,551	199,300	244,851	NO	Norway	15,246	35,795	51,041
AU	Australia	87,605	156,493	244,098	IL	Israel	15,101	33,752	48,853
VN	Viet Nam	64,539	152,459	216,998	EC	Ecuador	15,611	31,654	47,265
TH	Thailand	80,655	117,904	198,559	NZ	New Zealand	17,462	29,547	47,009
AR	Argentina	68,877	116,617	185,494	SK	Slovakia	16,061	27,749	43,810
TR	Turkey	25,974	147,023	172,997	LK	Sri Lanka	7,186	35,540	42,726
CO	Colombia	62,590	110,004	172,594	BG	Bulgaria	13,136	25,260	38,396
PH	Philippines	78,760	81,601	160,361	HR	Croatia	13,612	23,944	37,556
MY	Malaysia	60,607	95,842	156,449	MA	Morocco	7,170	29,434	36,604
UA	Ukraine	46,132	105,582	151,714	DO	Dominican Republic	10,750	23,303	34,053
PL	Poland	48,381	102,802	151,183	SV	El Salvador	11,891	19,049	30,940
NL	Netherlands	40,074	104,336	144,410	DZ	Algeria	5,176	24,887	30,063
PK	Pakistan	15,420	128,150	143,570	CR	Costa Rica	9,632	20,186	29,818
IR	Iran	27,153	112,444	139,597	KE	Kenya	6,868	22,522	29,390
CL	Chile	53,286	81,165	134,451	NG	Nigeria	5,050	23,523	28,573
EG	Egypt	19,414	113,495	132,909	GT	Guatemala	7,342	20,189	27,531
ZA	South Africa	34,153	66,871	101,024	UY	Uruguay	9,966	14,552	24,518
SA	Saudi Arabia	15,173	85,416	100,589	LT	Lithuania	10,416	13,801	24,217
PE	Peru	32,296	66,141	98,437	KZ	Kazakhstan	5,727	12,555	18,282
RO	Romania	28,907	63,982	92,889	PY	Paraguay	6,273	10,731	17,003
PT	Portugal	32,218	59,238	91,456	SI	Slovenia	5,644	11,269	16,913
VE	Venezuela	32,623	56,556	89,179	LV	Latvia	5,722	9,979	15,701
BD	Bangladesh	7,029	74,221	81,250	EE	Estonia	5,337	8,337	13,674
HU	Hungary	30,525	48,858	79,383					

B   Significance Test Results

Table B.2. Significance test results for variables in Google+ for our 73 countries, ranked in descending order of the number of users. The value on the left is the average female value and the value on the right is the average male value, followed by the significance result ('*' is 95% significant, '**' is 99% significant).

Country	In-degree ♀/♂	Out-degree ♀/♂	Recipr. ♀/♂	Clust. Coeff. ♀/♂	PageRank ♀/♂
United States	34.8/47.1**	20.6/30.3**	0.49/0.50**	0.31/0.28**	2.0e-08/2.6e-08**
India	25.5/23.2	20.3/38.2**	0.52/0.41**	0.25/0.23**	2.0e-08/2.0e-08
Brazil	20.4/28.7**	38.0/48.0**	0.37/0.39**	0.16/0.17**	1.7e-08/2.2e-08**
United Kingdom	30.9/26.8	20.5/28.9**	0.47/0.46**	0.33/0.29**	1.8e-08/2.1e-08**
Indonesia	25.0/17.7**	39.5/53.4**	0.43/0.33**	0.36/0.34**	1.9e-08/1.6e-08**
Russian Federation	17.7/20.8**	31.0/36.1**	0.45/0.41**	0.38/0.32**	1.5e-08/1.8e-08**
Canada	33.9/38.9	19.6/29.1**	0.48/0.48	0.31/0.28**	1.8e-08/2.2e-08**
Mexico	10.5/12.6**	22.8/28.0**	0.45/0.41**	0.28/0.27*	1.5e-08/1.6e-08**
Germany	21.5/42.2**	21.9/31.6**	0.49/0.47**	0.35/0.31**	1.6e-08/2.1e-08**
Spain	13.7/29.2**	20.4/29.1**	0.50/0.47**	0.32/0.29**	1.6e-08/2.2e-08**
Italy	34.7/22.0	22.7/33.3**	0.51/0.48**	0.33/0.29**	1.8e-08/2.0e-08**
France	15.6/24.7**	19.8/30.5**	0.49/0.46**	0.33/0.29**	1.6e-08/2.1e-08**
Japan	32.0/35.0	30.8/49.1**	0.44/0.37**	0.34/0.32**	1.9e-08/1.9e-08
China	45.1/46.3	48.0/76.5**	0.41/0.31**	0.27/0.25**	1.9e-08/1.8e-08
Australia	14.8/21.5**	18.5/27.2**	0.48/0.48	0.33/0.29**	1.5e-08/2.0e-08**
Viet Nam	36.9/57.4**	41.7/78.3**	0.41/0.34**	0.29/0.29	1.8e-08/2.0e-08**
Thailand	19.4/29.1**	34.0/48.2**	0.41/0.39**	0.34/0.31**	1.6e-08/2.2e-08**
Argentina	13.4/17.8**	22.7/29.7**	0.43/0.43*	0.29/0.27**	1.6e-08/1.9e-08**
Turkey	18.8/15.1**	29.0/45.7**	0.46/0.36**	0.32/0.28**	1.5e-08/1.4e-08
Colombia	9.6/10.9**	24.8/31.0**	0.44/0.40**	0.28/0.27**	1.4e-08/1.6e-08**
Philippines	11.6/16.6**	28.8/38.5**	0.42/0.41	0.40/0.36**	1.4e-08/1.6e-08**
Malaysia	11.8/32.7**	26.5/38.1**	0.45/0.40**	0.33/0.30**	1.4e-08/1.8e-08**
Ukraine	20.1/37.9**	31.8/43.0**	0.48/0.45**	0.37/0.31**	1.6e-08/1.9e-08**
Poland	8.1/13.6**	17.0/23.9**	0.53/0.50**	0.37/0.32**	1.5e-08/1.8e-08**
Netherlands	15.7/22.3**	18.6/27.5**	0.51/0.50**	0.33/0.28**	1.6e-08/2.1e-08**
Pakistan	25.4/15.8**	35.3/49.1**	0.40/0.31**	0.32/0.29**	1.6e-08/1.3e-08**
Iran	50.2/35.6	34.9/49.0**	0.46/0.39**	0.30/0.29**	1.9e-08/1.7e-08
Chile	9.7/13.5**	17.7/23.4**	0.50/0.50*	0.27/0.26**	1.6e-08/2.0e-08**
Egypt	34.2/18.9**	30.3/62.4**	0.38/0.25**	0.31/0.28**	1.7e-08/1.3e-08**
South Africa	10.5/17.9**	19.4/31.0**	0.45/0.42**	0.29/0.26**	1.4e-08/1.8e-08**
Saudi Arabia	39.3/24.6**	30.2/47.4**	0.37/0.33**	0.29/0.26**	1.7e-08/1.6e-08
Peru	12.2/11.3	27.7/34.9**	0.41/0.36**	0.28/0.28	1.5e-08/1.5e-08
Romania	22.8/24.0	34.4/52.7**	0.43/0.38**	0.35/0.31**	1.5e-08/1.7e-08**
Portugal	13.3/20.4**	22.6/35.9**	0.47/0.46**	0.27/0.26**	1.5e-08/1.9e-08**
Venezuela	13.5/14.4	28.6/34.9**	0.42/0.39**	0.28/0.26**	1.5e-08/1.7e-08**
Bangladesh	17.4/15.2	30.4/54.1**	0.41/0.30**	0.32/0.30**	1.4e-08/1.3e-08
Hungary	10.0/12.4**	17.9/22.5**	0.55/0.53**	0.34/0.31**	1.5e-08/1.8e-08**
South Korea	17.7/26.8**	26.8/42.1**	0.48/0.42**	0.33/0.31**	1.6e-08/2.0e-08**
Sweden	16.8/23.6**	17.6/28.2**	0.58/0.57*	0.37/0.31**	1.7e-08/2.3e-08**
Belgium	13.8/17.6*	17.9/26.4**	0.50/0.49**	0.34/0.29**	1.6e-08/1.9e-08**
United Arab Emirates	19.6/18.4	21.4/33.6**	0.46/0.42**	0.28/0.22**	1.7e-08/1.7e-08
Denmark	12.7/18.4**	14.8/23.5**	0.57/0.57	0.34/0.29**	1.7e-08/2.2e-08**
Czech Republic	12.2/20.2**	17.0/27.1**	0.56/0.52**	0.38/0.31**	1.6e-08/2.1e-08**
Singapore	14.8/20.6**	19.5/30.0**	0.51/0.49**	0.27/0.24**	1.7e-08/2.1e-08**
Finland	13.4/47.0**	13.7/23.5**	0.60/0.59*	0.37/0.35**	1.6e-08/2.5e-08**
Greece	19.0/22.1	26.5/40.3**	0.47/0.44**	0.34/0.30**	1.5e-08/1.8e-08**
Ireland	13.9/22.2**	17.3/27.4**	0.49/0.48	0.35/0.31**	1.6e-08/2.1e-08**
Serbia	13.9/46.9*	19.8/31.8**	0.53/0.47**	0.31/0.30	1.5e-08/2.0e-08**
Switzerland	22.4/29.2	20.6/33.3**	0.50/0.48**	0.31/0.28**	1.7e-08/2.2e-08**
Austria	14.2/27.9**	17.9/31.4**	0.52/0.49**	0.37/0.33**	1.5e-08/1.9e-08**
Norway	16.8/40.3**	17.6/30.8**	0.57/0.56**	0.35/0.31**	1.7e-08/2.5e-08**
Israel	23.2/61.5	24.5/37.4**	0.50/0.49	0.26/0.23**	1.8e-08/2.5e-08**
Ecuador	8.5/8.5	27.6/31.4**	0.40/0.36**	0.32/0.31**	1.4e-08/1.3e-08
New Zealand	14.3/22.4**	16.7/27.8**	0.51/0.50**	0.33/0.29**	1.6e-08/2.0e-08**
Slovakia	6.4/12.8**	13.1/21.1**	0.61/0.58**	0.32/0.30**	1.6e-08/2.0e-08**
Sri Lanka	20.9/21.1	23.7/50.7**	0.47/0.36**	0.31/0.30*	1.6e-08/1.6e-08
Bulgaria	14.9/19.1**	25.2/36.2**	0.48/0.46**	0.34/0.31**	1.5e-08/1.8e-08**
Croatia	8.9/14.5**	15.0/26.4**	0.54/0.50**	0.32/0.30**	1.4e-08/1.7e-08**
Morocco	20.7/18.3	27.1/57.9**	0.44/0.30**	0.27/0.25	1.7e-08/1.4e-08**
Dominican Republic	16.7/16.0	27.5/39.3**	0.43/0.38**	0.27/0.27	1.6e-08/1.7e-08
El Salvador	12.8/11.5	31.7/28.7	0.38/0.39	0.21/0.24**	1.4e-08/1.5e-08*
Algeria	20.7/10.6**	27.6/51.4**	0.34/0.22**	0.25/0.27	1.3e-08/1.0e-08**
Costa Rica	14.6/15.1	20.3/27.6**	0.50/0.46**	0.27/0.27	1.7e-08/1.8e-08
Kenya	13.1/14.8	28.6/42.0**	0.42/0.34**	0.27/0.26	1.6e-08/1.5e-08
Nigeria	8.7/8.4	31.9/47.7**	0.31/0.21**	0.26/0.27	1.2e-08/1.1e-08*
Guatemala	10.1/12.1	21.2/26.2**	0.46/0.40**	0.27/0.29*	1.5e-08/1.5e-08
Uruguay	13.2/13.9	23.9/29.6*	0.46/0.46	0.27/0.27	1.5e-08/1.7e-08**
Lithuania	7.9/19.3**	19.3/34.5**	0.51/0.49**	0.30/0.28**	1.5e-08/2.0e-08**
Kazakhstan	16.5/16.8	33.6/35.6	0.38/0.37	0.33/0.32	1.4e-08/1.5e-08
Paraguay	16.8/18.2	28.1/34.0**	0.45/0.42**	0.23/0.23	1.8e-08/1.8e-08
Slovenia	10.0/18.2**	16.8/30.2**	0.56/0.53**	0.27/0.28	1.6e-08/2.1e-08**
Latvia	11.8/19.7**	26.2/35.3*	0.51/0.48**	0.34/0.31**	1.5e-08/2.3e-08**
Estonia	8.9/15.0**	15.0/25.7**	0.54/0.51**	0.26/0.25	1.6e-08/1.9e-08**

Gender Patterns in a Large Online Social Network

Yana Volkovich[1], David Laniado[1], Karolin E. Kappler[2],
and Andreas Kaltenbrunner[1]

[1] Barcelona Media, Barcelona, Spain
[2] Fernuniversität in Hagen, Hagen, Germany

Abstract. Gender differences in human social and communication behavior have long been observed in various contexts. This study investigates such differences in the case of online social networking. We find a general tendency towards gender homophily, more marked for women, however users having a large circle of friends tend to have more connections with users of the opposite gender. We also inspect the temporal sequences of adding new friends and find that females are much more likely to connect with other females as their initial friends. Through studying triangle motifs broken down by gender we detect a marked tendency of users to gender segregation, i.e. to form single gender groups; this phenomenon is more accentuated for male users.

Keywords: Gender, Homophily, Social Network, Data Mining.

1 Introduction

It is a common believe that men are more frequently early adopters of new technologies. However, in the case of many social media websites and services women are in the vanguard. Thus, women outnumbered men by a considerable amount for most social networking sites [6,16] with Pinterest having the largest gender inequality [22] and LinkedIn being the only exception [15]. With technology entering the mass market, women lean in and overtake males not only in spending time on social networking platforms, but also in owning gadgets or playing casual social games [3].

Differences in styles of social interactions for males and females have been documented for centuries [4]. A seminal work [20] on quantitative analysis of gender differences introduces a network terminology to describe social relations between children and evolution of these relations over time. Many of the successive studies rely on questionnaires, surveys or direct observations by adults. We refer to [18,26,30] for further reading on this subject.

The technological advances led to the emergence of new ways to investigate human behavioral patterns. Examples of such new tools can be the analysis of data obtained from wearable sensors (see again [26] and references therein) or the exploration of mobile [23] and online social traces. Among the first works focused on gender differences in online friendship preferences were Lewis et al. [14]

L.M. Aiello and D. McFarland (Eds.): SocInfo 2014, LNCS 8851, pp. 139–150, 2014.

for Facebook and Thelwall [28] for MySpace. A recent study [27] analyzed online social interactions in the setting of a massive multi-player online game. Gender homophily, the tendency of individuals to bond with similar others, was also reported for interactions in Wikipedia, a community with strong female minority; a higher presence of women was found in discussions with a more positive tone [13]. Finally, in [10,12,25] authors studied how gender influences linguistic style of messages in Twitter, Facebook and Wikipedia.

Nevertheless there is still a lack of understanding of gender roles in online social communications. As most of the studies rely on analysis of US-based users [1] some of these findings can be less relevant in non-US contexts. Gender influence on access to information and communication technologies often varies according to local and cultural practices [5,17]. In this work we use a complete dump of a large Spanish social networking service to present an extensive analysis of *online gender homophily*, i.e. gender preferences emerging online. Spain is among the most "social media addicted" countries in the European Union [7] with almost 75% of the Spaniards using Internet as an instrument for communication and interaction with others.

In this study we explore dissimilarities between men and women in the way they sign up to a social network platform and they make friends online. We further discuss how gender homophily observed in the offline world is translated into the case of online social communications.

2   Paper Roadmap and Main Results

To detect the fundamental differences between male and female usage of the SNS (social networking service) under analysis, we first compare the process of building their *ego networks*, i.e. online personal networks. Of particular interest is to inspect the gender of the first friend of each user to estimate the influence of gender on the adoption of a new technology. So, our first research question is:

(RQ1). *How does gender homophily affect SNS-adoption? Do men show a preference to accept invitations from men and women from women? Do online ego-networks grow in a gender-biased way?*

In our invitation-only Spanish SNS, we find that female users in most cases join the new social platform by following invitation by another female, and they add women as their initial friends, while for male users we don't observe any strong preference.

Next, we study gender homophily in more detail by answering the the following questions:

(RQ2). *Do females and males have similar friendship networks, both in size and composition? Is there a preference for connection among same gender users?* We find that males and females are almost indistinguishable with respect to their network size. We observe a relation between user popularity and the gender of a user's friends: users having an around average number of friends exhibit gender homophily (more marked for females), while users with few friends tend to have more female friends, and users with a large number of friends have more opposite gender friends.

Finally, we inspect the effect of gender on the network structure with our third research question:

(RQ3). *How does gender affect the network structure and the formation of transitive relationships (triangles)?*

We find evidence for gender segregation, as we observe a much larger proportion of single gender triangles than expected. This result is particularly marked for male only triangles. So, while we find in general a higher homophily for women, men exhibit a higher tendency to form gender homogeneous groups.

3   Dataset Description

In contrast to many recent studies on gender difference based on large-scale online data, our dataset is complete in the sense that it contains the entire friendship network. Another advantage is that the SNS under analysis is gender-balanced, i.e. the number of male and female subscribers is practically the same. This is different from many other online platforms. Finally, it is also worth mentioning that we focus on a non US-located community, a category that is underrepresented in the literature.

The dataset (see detailed descriptions in [11,31]) is a fully anonymized snapshot of friendship connections from the invitation-only (at the time this dataset was collected) Spanish social networking service Tuenti (`www.tuenti.com`). Similar to many other popular social networking platforms Tuenti allows users to set up their profiles, connect with friends and share links and media items. Users can interact by writing messages on each other's walls. The dataset includes about 9.8 million registered users (25% of Spain's population), their bidirectional friendship links (with the temporal order of link formation), and the directed interactions (an interaction is an exchange of a wall message) generated by the users during a three months period. There are small differences between the numbers of male and female participants (see Appendix A for the exact numbers) similar to those reported in surveys [9].

4   Building Social Environment

Gender has been observed to play a crucial role in defining people's decisions about adopting and using new technologies. Thus, men are more driven by instrumental factors (i.e. perceived usefulness) while women are more motivated by process and social factors [29]. We examine differences in how males and females start their online social experience, i.e. how they organize their online social environment, by comparing the order in which they are making friends.

The dataset under analysis comes from an invitation-only online platform, therefore we assume that the first friend of a user is the one who invited her or him. Although some data limitations (we only have successful, i.e. accepted, invitations, and no information about unfriending) we believe in the importance of this analysis for better understanding of social media involvement mechanisms.

The First Friend: We schematically draw the difference in gender for the first and second friends. In Figure 1 (left) we look at the gender of the users who successfully invited a male user to join the SNS. We observe that males sign up through the invitation sent by another male in 55% of the cases and only in 45% of the cases after the invitation by a female. The gender bias however is much more significant for female users (Figure 1 (right)): in 72% of the cases women accept an invitation to join the online platform from another woman, and just in 28% of the cases from a man. We observe a similar trend for the second friend of a female user in the case that the first friend was already a female. However if, on the contrary, the first friend of a female was a male, the probability of being the second friend as well a male rises to 42%. For male users the dependency of the genders of the first two friends is even stronger: the second friend has in almost 6 of 10 cases the same gender as the first friend.

Friendship Order: We go beyond the first two friends and look at the average number of same gender friends added by users given their gender and degree. In Figure 2 we plot the average fraction of same gender friends for the kth friend of male and female users form $k = 1$ to $1\,000$ (the Tuenti friendship limit). In the same plot we also show the average fraction of female friends for all users. We find than most women, as they join the new social platform, connect primarily to their female friends, creating female dominated ego networks. Women prefer to add other female users until their degrees grow larger than 150. When they have over 150 friends they tend to connect more with males. In Section 5.1 we confirm that females with many friends have a smaller fraction of same gender friends. For men we do not observe pronounced preferences. The only observation is that at the very beginning of their online social experience, and also when they have between 50 and 200 friends approximately, males have a slight tendency to connect preferentially with other males.

To sum up, women do organize their online social environment different from men especially in the initial steps, which suggests that they are more likely to add other women as their initial friends and to try a new service or enter a new social environment following an invitation by another woman. As there are many

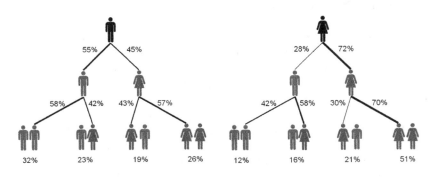

Fig. 1. Gender differences in making the first friends for males (left) and females (right)

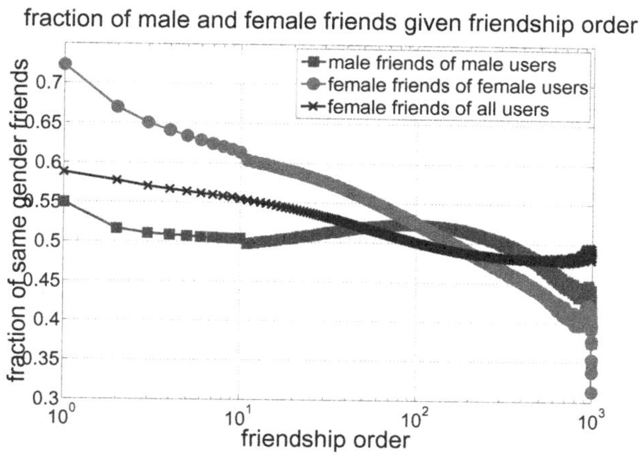

Fig. 2. Gender of the *k*th friend: fraction of same gender friends for male (blue squares) and female (red circles) users, and fraction of female friends of all users (black crosses) given friendship order

different ways for users to find new friends (e.g. by using search or recommendation tools provided by SNS, through direct invitations, or by exploring friends of users' friends) further investigation is needed to explore this result.

5 Gender Homophily

Exploring online friendship homophily we first find that users have just a small preference to make friends of the same gender (see detailed statistics in Appendix A). This preference is larger for females: on average male users have 82 male and 78 female friends, while females have 85 females and 76 males. The corresponding percentages are smaller in comparison to the offline world, where men are reported to have 65% and women 70% of same gender friends [24].

5.1 Gender Homophily by Degree

Previous work on Facebook [14] reported that males and females are almost indistinguishable with respect to their network size. In our case we also do not find any differences for degree distributions for male and female users (data not shown). However, by looking at gender ratios of users having a given degree we find that users with low (< 100) or high (> 300) numbers of friends are slightly more often females (Figure 3(a)).

In Figure 3(b) we plot the ratio of female friends given the degree of a user. That is, for all users with exactly k friends, the figure shows what fraction of their friends are females, on average. We find that users with few friends tend to have more female friends; their proportion decreases with increasing degree, and

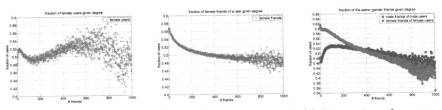

(a) Proportion of female users with a given degree.

(b) Proportion of female friends of a user given her/his degree.

(c) Proportion of same gender friends of a user given his/her degree.

Fig. 3. Gender differences given the number of friends (degree) of a user

falls below 50% for users with more than 350 friends: users having many friends have more male friends.

To understand more deeply gender preference in friendship relationships, we also consider the fraction of same gender friends, given the degree, for male and female users separately (see Figure 3(c)). The figure shows, for women with few friends, a marked preference for connection with other females: around 60% for women having less than 50 friends. This preference tends to decrease with increasing degree, until women with more than 450 friends, who tend to have more male friends. For male users we observe a more balanced pattern, while we still find that users with many friends prefer to friend opposite gender users. Interestingly, males with a low number of friends also have a higher proportion of female friends. This finding is in contrast with the slight tendency of men to add other men as their initial friends, observed in Figure 2, suggesting that a preference for female friends applies only to male users having a small circle of friends (less than 25) in the SNS.

5.2   Triangle Motifs

To investigate the interplay between gender and the structure of the network we next inspect gender composition of friendship triangles, i.e. triples of nodes in which each node is connected to the other two. A high presence of triangles (or a high *clustering coefficient*) is one of the key elements that distinguish social networks from other kinds of networks, such as biological or technological networks [21]. In other words, the presence of transitive relationships can be seen as a sign of a community structure, which is typical of social networks. Therefore it is particularly relevant to assess how gender affects the formation of this distinguishing pattern.

For this analysis, beyond the friendship network we consider the *interaction network*: the friendship network filtered by reciprocal interactions (i.e. keeping only connections between users who have exchanged messages on each other walls). More details about the methodology used for this analysis can be found in Appendix B.

Table 1. Proportion of triangle motifs with different gender composition (blue=male, red=female) in the friendship and interaction networks. The differences between observed (obser.) and expected proportions (shuff., calculated via reshuffling the gender of users having the same degree) are highly significant (stdv. of reshuffling < 0.03%).

Type of triangle		friendship		interaction	
		obser.	shuff.	obser.	shuff.
males only	▷	16.0%	11.6%	9.9%	6.2%
1 female, 2 males	▷	32.5%	36.6%	24.4%	28.4%
2 females, 1 male	▷	34.5%	38.4%	37.3%	43.3%
females only	▷	17.0%	13.4%	28.4%	22.1%
total		3.64×10^{10}		1.24×10^{8}	

Explicit Friendship Triangles: In total we find more than 3.64×10^{10} triangles in the friendship network. The second and third column of Table 1 list the proportion of triangles of different composition together with the expected values based on the networks with randomly reshuffled genders. We clearly observe a much larger proportion of single gender friendship triangles than expected. In particular, although the number of female only triangles is higher, if we compare the results with the ones obtained in the reshuffled networks we find a stronger deviation for male only triangles (+38%, versus +27% for female only triangles). This indicates that the trend to form gender homogeneous groups is more accentuated for males.

Interaction Triangles: When analyzing only the connections which mutually exchange messages, i.e. the interaction network, we find a striking difference between males and females, as can be observed in the two rightmost columns in Table 1. The number of female only triangles is about 3 times larger than the number of male only triangles. This difference seems high, however reshuffling shows that again we would actually have to expect an even larger disproportionality between male- and female only triangles, given that females are much more active in sending (and receiving) messages. So the tendency to form gender homogeneous groups is more marked for male users also in the interaction network. In this case the proportion of male only triangles exceeds by 60% the expected value, while the proportion of female only triangles is only 28.5% higher than expected. This indicates that male users are in general less active in the SNS, but when they interact they tend to do it in gender homogeneous groups in a much more marked measure than females.

The above results show that users do not only tend to connect preferentially with others of the same gender, but they also tend to group more by gender, and to create gender-homogeneous groups of friends. As demonstrated in [19], gender segregation is a widespread characteristic of offline social behavior. Our findings show that, in this sense, online social behavior reproduces this offline phenomenon, and that this happens more markedly for male users.

6   Conclusions

Recent studies on digital inequalities treat gender in very different ways. Some only concentrate on the influence of gender on human behavior [5], others such as Zillien [32] consider gender only as one of many variables in the emergence of digital inequalities, and yet others like boyd [2] completely ignore the gender dimension. This lack of consensus in considering gender and its influence on digital experience indicates that there are still many open questions that need to be addressed. This study is one of the first intents to shed light on emerging gender patterns in the growth of users' online personal networks.

There is growing evidence that men and women use online social platforms differently [8,16,27]. These differences are generally neglected when all users are treated *de-gendered* and *equally*. The analysis we present here reveals fundamental differences in how male and female users organize their online friendship networks. One of our most important findings is that females show in general a higher homophily than male users, and that this phenomenon is particularly prominent in the first steps they take in the new social environment. Women join the SNS following significantly more often an invitation from a female, and they add much more frequently other females as their initial friends.

Our findings also suggest a popularity effect, with *heterophily* characterizing users having many connections. At the same time, users having smaller circles of friends exhibit a preference for female friends irrespectively of their gender. For males, in the case when their personal network is still growing, this does not correspond to the general behavior: men tend to add slightly more frequently other men as their initial friends. For females instead we find clear evidence for homophily among women having a small or average sized personal network, as well as for women in general at their early stages in the social network (until having about 150 friends). Further research could explain whether also women who get to have large personal circles of friends (and have more male friends) still tended to exhibit homophily in their first stages.

Finally, we found evidence of homophily also in the formation of groups: the proportion of single-gender triangles is much higher than expected, reproducing the offline phenomenon of gender segregation in social behavior [19]. In contrast with the results about homophily in one-to-one friendship connections and interactions, this tendency to gender segregation is stronger for male users. Further research would be needed to investigate the gender composition of richer motifs, such as cliques and dense clusters.

Our findings show how gender affects the growth of a user's personal network and the composition and structure of friendship circles. They also unveil the importance of gender when entering a new digital social environment, and can help to understand the gender gap observed in some online communities: when females are a minority, it is less likely that other females will join, as the perceived presence of other females appears to be fundamental in the first stages.

References

1. Ahn, J.: Teenagers and social network sites: Do off-line inequalities predict their online social networks? First Monday, 17(1) (2011)
2. Boyd, D.: It's complicated: the social lives of networked teens (2014)
3. Cook, S.G.: Women lead in adopting new technologies. Women in Higher Education 21(2), 24–25 (2012)
4. Darwin, C.: The Descent of Man and Selection in Relation to Sex. John Murray (1871)
5. Drabowicz, T.: Gender and digital usage inequality among adolescents: A comparative study of 39 countries. Computers & Education 74, 98–111 (2014)
6. Duggan, M., Brenner, J.: The demographics of social media users – 2012. Pew Research Center (2013)
7. Eurobarometer. E-communications household survey summar. Technical report, Public Opinion Analysis, European Comission (2010)
8. Hoffman, A.: The Social Media Gender Gap. Bloomberg Businessweek (May 2008)
9. Instituto Nacional de Estadistica. Mujeres y hombres en españa. Technical report (2013)
10. Iosub, D., Laniado, D., Castillo, C., Morell, M.F., Kaltenbrunner, A.: Emotions under discussion: Gender, status and communication in online collaboration. PloS one 9(8), 104880 (2014)
11. Kaltenbrunner, A., Scellato, S., Volkovich, Y., Laniado, D., Currie, D., Jutemar, E.J., Mascolo, C.: Far from the eyes, close on the Web: impact of geographic distance on online social interactions. In: Proceedings of ACM SIGCOMM Workshop on Online Social Networks (WOSN 2012). ACM (2012)
12. Kivran-Swaine, F., Brody, S., Naaman, M.: Effects of gender and tie strength on twitter interactions. First Monday 18(9) (2013)
13. Laniado, D., Kaltenbrunner, A., Castillo, C., Morell, M.F.: Emotions and dialogue in a peer-production community: the case of Wikipedia. In: Proc. WikiSym (2012)
14. Lewis, K., Kaufman, J., Gonzalez, M., Wimmer, A., Christakis, N.: Tastes, ties, and time: A new social network dataset using Facebook.com. Social Networks 30(4), 330–342 (2008)
15. MacManus, R.: Study: Women outnumber men on most social networks. Report by Rapleaf (2008), http://readwrite.com/2008/07/29/social_networks_women_outnumber_men
16. Madden, M., Lenhart, A., Cortesi, S., Gasser, U., Duggan, M., Smith, A., Beaton, M.: Teens, social media, and privacy. Technical report, Pew Internet Research (2013)
17. Magno, G., Weber, I.: International gender differences and gaps in online social networks. In: Social Informatics - Third International Conference, SocInfo 2014 (2014)
18. McPherson, M., Smith-Lovin, L., Cook, J.M.: Birds of a feather: Homophily in social networks. Annual Review of Sociology, 415–444 (2001)
19. Mehta, C.M., Strough, J.: Sex segregation in friendships and normative contexts across the life span. Developmental Review 29(3), 201–220 (2009)
20. Moreno, J.L.: Who shall survive? Foundations of sociometry, group psychotherapy and socio-drama. Beacon House (1953)
21. Newman, M.E., Park, J.: Why social networks are different from other types of networks. Physical Review E 68(3), 36122 (2003)

22. Ottoni, R., Pesce, J.P., Las Casas, D., Franciscani Jr, G., Meira Jr, W., Kumaraguru, P., Almeida, V.: Ladies First: Analyzing Gender Roles and Behaviors in Pinterest. In: Proc. ICWSM (2013)
23. Palchykov, V., Kaski, K., Kertész, J., Barabási, A.-L., Dunbar, R.I.: Sex differences in intimate relationships. Scientific Reports, 2 (2012)
24. Reeder, H.M.: The effect of gender role orientation on same-and cross-sex friendship formation. Sex Roles 49(3-4), 143–152 (2003)
25. Schwartz, H.A., Eichstaedt, J.C., Kern, M.L., Dziurzynski, L., Ramones, S.M., Agrawal, M., Shah, A., Kosinski, M., Stillwell, D., Seligman, M.E., et al.: Personality, Gender, and Age in the Language of Social Media: The Open-Vocabulary Approach. PloS One 8(9), e73791 (2013)
26. Stehlé, J., Charbonnier, F., Picard, T., Cattuto, C., Barrat, A.: Gender homophily from spatial behavior in a primary school: a sociometric study. Social Networks 35(4), 604–613 (2013)
27. Szell, M., Thurner, S.: How women organize social networks different from men. Scientific reports 3 (2013)
28. Thelwall, M.: Social networks, gender, and friending: An analysis of MySpace member profiles. Journal of the American Society for Information Science and Technology 59(8), 1321–1330 (2008)
29. Venkatesh, V., Morris, M.G.: Why don't men ever stop to ask for directions? Gender, social influence, and their role in technology acceptance and usage behavior. MIS Quarterly, 115–139 (2000)
30. Vigil, J.M.: Asymmetries in the friendship preferences and social styles of men and women. Human Nature 18(2), 143–161 (2007)
31. Volkovich, Y., Scellato, S., Laniado, D., Mascolo, C., Kaltenbrunner, A.: The Length of Bridge Ties: Structural and Geographic Properties of Online Social Interactions.. In: Proc. ICWSM (2012)
32. Zillien, N.: Digitale Ungleichheit. Springer (2008)

A   Detailed Statistics for Gender Homophily

Table A1 reports the number of male and female users in the Tuenti SNS. Quantities are shown for both the whole dataset and the filtered dataset (i.e. considering only users having more than 10 friends).

Table A2 shows the average number of friends for male and female users, broken down by gender. In Figure A1 we plot the complete distribution of the percentage of same gender friends for users with more than 10 friends. We observe that the red bars are more shifted to the right, indicating greater homophily for females.

Table A1. Number of users in the dataset broken down by gender. The second column shows these numbers for users with more than 10 friends.

# users	total	> 10 friends
male	4 899 659	3 269 611
female	4 784 975	3 350 189

Table A2. Basic friendship statistics by gender (all averages are taken over users with more than 10 friends) together with 25% and 75% quantiles.

friends	avg # male	avg # female	avg % same gender
male	82[20, 116]	78[19, 106]	51.48%
female	76[15, 104]	85[23, 122]	56.46%

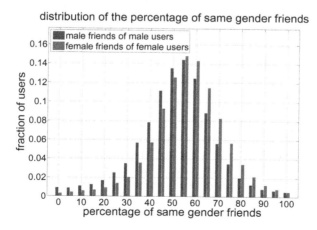

Fig. A1. Distribution of the percentage of the number of same gender friends for users with more than 10 friends

B   Methodology for Assessing Gender Homophily in Transitive Relationships (Triangle Motifs)

To explore the gender composition of friendship triangles we first focus on the entire friendship network and then restrict our analysis only to friends for which we observe reciprocal interactions. In the latter case we only consider a connection between two friends if they have sent to each other at least one wall message. We call this filtered network the interaction network. To construct it we use the information of all wall message exchanges over a period of 3 months. The resulting network is composed of 2 247 992 male and 2 521 200 female users. The number of connections for both networks, broken down by gender, is reported in Table B1. Note that for this analysis we did not filter out users having less than 10 connections. The higher number of connections involving females in the interaction network indicates that women are much more active than men in sending (and receiving) wall messages in the SNS.

Table B1. Number of connections in the friendship network and in the network of reciprocal interactions, broken down by gender.

# connections	male-male	female-female	mixed
# friendship	135 064 946	143 740 462	256 894 050
# interactions	12 236 165	22 698 114	27 346 769

There are four possibilities for the gender composition of the triangles: 3 females, 3 males, 1 male and 2 females, or 2 males and 1 female. In case of a perfectly gender balanced network, one could expect, using the binomial distribution, to have exactly 12.5% male-only triangles, 12.5% female-only triangles, and 37.5% of the triangles in each of the two mixed triangle possibilities. However, the numbers of males and females in the networks are not equal, and more importantly, the degree distributions are not equal. Females have more connections, especially in the interaction network, and this leads to a higher number of triangles involving females.

To compensate for the bias we assess how the results we observe differ from the results one should expect given the user composition of the networks. We produce randomized equivalents of our networks by re-shuffling user genders. To maintain the same gender proportions, and the same degree distribution for each gender, we randomly re-shuffle the gender of all users having the same degree. The resulting networks have the same structure and the same number of connections involving males and females as the original network. Comparing the proportion of triangles observed in the real networks with the average proportion obtained in 10 of these reshuffled networks, we are able to assess how gender influences the formation of transitive relationships. The results presented in Section 5.2 are highly significant: the standard deviation of the values observed for the reshuffled networks is smaller than 0.03%.

User Profiling via Affinity-Aware Friendship Network

Zhuohua Chen[1], Feida Zhu[2], Guangming Guo[3], and Hongyan Liu[1]

[1] Tsinghua University, Beijing, China
chenzhh3.12@sem.tsinghua.edu.cn, hyliu@tsinghua.edu.cn
[2] Singapore Management University, Singapore, Singapore
fdzhu@smu.edu.sg
[3] University of Science and Technology of China, Hefei, China
guogg@mail.ustc.edu.cn

Abstract. The boom of online social platforms of all kinds has triggered tremendous research interest in using social network data for user profiling, which refers to deriving labels for users that characterize their various aspects. Among different kinds of user profiling approaches, one line of work has taken advantage of the high level of label similarity that is often observed among users in one's friendship network. In this work, we identify one critical point that has been so far neglected — different users in one's friendship network play different roles in user profiling. In particular, we categorize all users in one's friendship network into (I) close friends whom the user knows in real life and (II) online friends with whom the user forms connection through online interaction. We propose an algorithm that is affinity-aware in inferring users' labels through network propagation. Our divide-and-conquer framework makes the proposed method scalable to large social network data. The experiment results in three real-world datasets demonstrate the superiority of our algorithm over baselines and support our argument for affinity-awareness in label profiling.

1 Introduction

The recent blossom of social network services has provided everyone with an unprecedented level of ease and fun in sharing information of all sorts. These public social data therefore reveal a surprisingly large amount of information about an individual which is otherwise unavailable. A central task in leveraging this big social data for business, consumer and social insights is user profiling, which is to derive labels (also called attributes) that characterize various aspects of a user. These labels range from simple demographic ones such as gender, age and education, to more sophisticated ones including income levels, personal interests and expenditure propensities. Accurate user profiling is a crucial foundation to support a wide range of business intelligence tasks including targeted marketing and customer relation management.

The most straightforward way of user profiling is to derive labels based solely on a user's own information [13,2]. Such methods, however, suffer from the data

L.M. Aiello and D. McFarland (Eds.): SocInfo 2014, LNCS 8851, pp. 151–165, 2014.

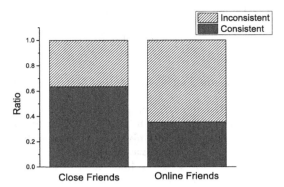

Fig. 1. Consistency on education background for close friends vs online friends in Renren

sparsity problem, i.e., labels might not be obtainable for some users, which could result from a number of reasons including privacy concern and missing information. Fortunately, researchers have identified the phenomenon that, for many labels, a user would share the same value with a group of other users in his online social network, as a result of the context in which they become socially connected in the first place, e.g., education background, work place, location, etc. Algorithms such as [11,7] have therefore been proposed to propagate labels across users' online social network to recover the missing labels.

However, a critical point neglected so far in this line of research is that, *different users in one's online social network play different roles in terms of label propagation.* Indeed, on platforms like Twitter, users do connect with a variety of different users apart from those in their real-life social circles. In particular, here we classify nodes in a user's online social network into two kinds — (I) those friends whom the user knows in offline real life, which we call "close friends", and (II) those whom the user connects with through online interaction only, which we call "online friends". It is obvious that, to derive labels such as education background (e.g., alumni from the same alma mater), close friends are more likely to contribute, as the underlying assumption that these labels come from the shared context does not hold for online friends such as celebrities in the user's online network. We use an example to further illustrate this point. In Figure 1, we show the fraction of close friends and online friends whose college is consistent with the target user, based on a result computed on our real Renren dataset (details of the dataset are given in Section 4.1). As shown in Figure 1, on average nearly 65% of a user's close friends share the same college label with the target user while only less than 35% of the online friends do. This drives home the importance of differentiating nodes in a user's online social network for label propagation, which is a key contribution of this work.

We propose in this paper an optimization-based label profiling algorithm to assign labels for all the label-missing nodes in a given partially-labeled social network. To handle large social networks in real-life applications, we adopt a

Table 1. Average value of AFP and AFN

Dataset	Avg. AFP	Avg. AFN
Sina	34.66%	4.48%
Renren	37.24%	0.19%
Pokec	43.19%	0.65%

divide-and-conquer framework to deal with the scalability issue. Our main contribution can be summarized as follows.

First, to the best of our knowledge, this is the first work to consider the difference between a user's close and online friends in the task of user profiling.

Second, we observe the homophily among users in social network that only a^*-labeled users connect to many a^*-labeled friends (see Section 2), based on which we propose an optimization-based label profiling algorithm (see Section 3).

Third, we give a divide-and-conquer framework to scale up to large social networks (see Section 3). The experiments show that, we can accurately profile the labels for around 83% of the label-unknown users even if we only know the labels of 20% of all the users (see Section 4).

2 Problem Analysis and Formulation

2.1 General Homophily

Homophily [10] is a well-known principle in social network describing the phenomenon that friends tend to be similar. In other words, a user would share the same labels with many of her friends. For example, consider the user label of "location": a user is likely to have many friends in online social network who live in the same city, simply due to their offline interaction in their daily life. Indeed, both Li et al. [8] and Backstrom et al. [1] have found that the likelihood of friendship is inversely proportional to the distance in online social network. Take "education" as another example. Being alumni of the same institution, a user would have many friends with the same label of their alma mater. In the work of profiling user's interest, Yang et al. [16] discovered that people with similar interests tend to connect to one another. In general, for a large class of labels including "location" and "education", comparing between users with any given label value a^* and users with label values other than a^*, the fraction of their respective friends with the same label value a^* is much higher for the former than the latter.

The observation has also been verified by our real-world datasets. For each label value a_i, we calculate the average fraction of friends with label a_i of a user whose label is a_i, which is named as AFP, and the average fraction of friends with label a_i of a user whose label is not a_i, which is named as AFN. Table 1 shows the average value of AFP and AFN for all different label values that appear in our three datasets. In all datasets the average value of AFP is notably higher than the average value of AFN.

2.2   Affinity-Aware Homophily

Even though homophily has been observed for both labels of user location and user interest, the underlying reasons could be very different. For "location", homophily arises mostly as a result of offline social interaction of geographical proximity — those friends sharing the same label value largely know the target user in offline world. For "interest", however, it follows from the nature of social network platforms that users would often seek information and interaction from others with the same interest online — those friends sharing the same label value might not know the target user in offline world. The message is that, when inferring different types of labels based on homophily, one should distinguish different types of friends. In particular, here we classify nodes in a user's online social network into (I) those whom the user knows in offline real life, which we call "close friends", and (II) those whom the user forms connection with through purely online interaction, which we call "online friends". It is evident that, to derive labels such as education background and geographic location, the close friends would be more reliable sources, while for labels such as user interest, the online friends could give more informative clues.

Yet, it is a technical challenge how to distinguish a user's close friends and online friends from online social network. We have adopted Xie's method [15] which achieves great performance in identifyng users' close friends based on the friendship network structure.

2.3   Problem Formulation

Given a network $G(V, E)$, where V is the set of users, $E \subseteq V \times V$ is the set of edges each representing an undirected friendship among the users. Let $E_c \subseteq E$ be the set of all undirected close friendships, and $E_o = E \setminus E_c$ be the set of all undirected online friendships. We also define dc_i as the number of close friends of user $v_i \in V$ and do_i as the number of online friends of user $v_i \in V$.

As we mentioned in Section 2.2, close friendship and online friendship should be treated differently when profiling users for a specific label. On the other hand, it is imprudent to completely disregard one or the other. We therefore give different weights to close friendship and online friendship when profiling users. Specifically, the weight of close friendship is always 1, the weight of online friendship is a parameter w. We define the *propagating importance* $PI_{i \leftarrow j}$ from user v_j to user v_i as follows, which is the normalized value of the friendship's weight with respect to user v_i.

$$PI_{i \leftarrow j} = \begin{cases} \frac{1}{dc_i + w \times do_i} & \text{, if } e_{i,j} \in E_c \\ \frac{w}{dc_i + w \times do_i} & \text{, if } e_{i,j} \in E_o \\ 0 & \text{, otherwise} \end{cases} \quad (1)$$

One thing to note is that $PI_{i \leftarrow j}$ does not equal $PI_{j \leftarrow i}$ in general. Given a specific label value a^*, let $V_{a^*} \subseteq V$ denote the set of users whose label is a^*. We define $f(V_{a^*})$ in (2).

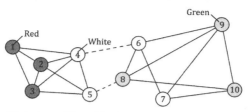

Fig. 2. An example of label profiling. Each solid edge represents a close friendship, and each dashed edge represents an online friendship. The red nodes represent the users whose label is known as a^*, the green nodes represent the users whose label is known as not a^*, and the white nodes represent the users whom we want to profile.

$$f(V_{a^*}) = \frac{\sum_{v_i \in V_{a^*}} \sum_{v_j \in V_{a^*}} PI_{i \leftarrow j}}{|V_{a^*}|} - \frac{\sum_{v_i \notin V_{a^*}} \sum_{v_j \in V_{a^*}} PI_{i \leftarrow j}}{|V \setminus V_{a^*}|} \qquad (2)$$

Equation (2) is actually the difference between average total propagating importance from a^*-labeled friends of a a^*-labeled user and average propagating importance from a^*-labeled friends of a user whose label is not a^*. To capture the affinity-awareness in user profiling, our goal is to maximize the value of $f(V_{a^*})$ as in Equation (2). We take Figure 2 as an example to illustrate (2). Let parameter w equal 0.5. If all label-unknown users (i.e., the white nodes) are assigned a^*, then $V_{a^*} = \{1, 2, 3, 4, 5, 6, 7\}$ and $V \setminus V_{a^*} = \{8, 9, 10\}$. The average total propagating importance from a^*-labeled friends of a user whose label is a^* is $(1 + 1 + 1 + 1 + \frac{6}{7} + \frac{3}{7} + \frac{1}{3})/7 \approx 0.80$, and the average total propagating importance from a^*-labeled friends of a user whose label is not a^* is $(\frac{3}{7} + \frac{1}{2} + \frac{1}{3})/3 \approx 0.42$. So $f(V_{a^*}) \approx 0.38$ for this assignment. If a^* is only assigned to user 4 and 5, then $V_{a^*} = \{1, 2, 3, 4, 5\}$ and $V \setminus V_{a^*} = \{6, 7, 8, 9, 10\}$. The average total propagating importance from a^*-labeled friends of a user whose label is a^* is $(1 + 1 + 1 + \frac{8}{9} + \frac{6}{7})/5 \approx 0.95$, and the average total propagating importance from a^*-labeled friends of a user whose label is not a^* is $(\frac{1}{7} + 0 + \frac{1}{7} + 0 + 0)/5 = 0.06$. So $f(V_{a^*}) \approx 0.89$ for this assignment, which is the optimal solution for the example shown in Figure 2. We now give our label profiling problem statement.

Definition 1. [Label Profiling] *Given a network $G(V, E)$ and a label value a^*, let $K_{a^*} \subseteq V$ be the set of users whose label is already known as a^*, $K_{\neg a^*} \subseteq V$ be the set of users whose label is already known as not a^*, and $U = V \setminus (K_{a^*} \bigcup K_{\neg a^*})$ be the set of users whose label is unknown. The problem of* Label Profiling *is to find a subset $U_{a^*} \subseteq U$ to assign the label value of a^* such that $f(K_{a^*} \bigcup U_{a^*})$ as defined in (2) is maximized.*

3   User Profiling Algorithm

3.1   Label Profiling Algorithm

Given a network $G(V, E)$, a specific label value a^* and a set V_{a^*} which is the set of a^*-labeled users in V. Define $h(V_{a^*})$ in (3) as the fraction of the number of a^*-labeled users over the number of users whose label is not a^* under the assignment V_{a^*}. We define $g(V_{a^*})$ in (4).

$$h(V_{a^*}) = \frac{|V_{a^*}|}{|V \setminus V_{a^*}|} \tag{3}$$

$$\begin{aligned} g(V_{a^*}) &= \frac{\sum_{v_i \in V_{a^*}, v_j \notin V_{a^*}} PI_{i \leftarrow j}}{|V_{a^*}|} + \frac{\sum_{v_i \notin V_{a^*}, v_j \in V_{a^*}} PI_{i \leftarrow j}}{|V \setminus V_{a^*}|} \\ &= \frac{\sum_{v_i \in V_{a^*}, v_j \notin V_{a^*}} (PI_{i \leftarrow j} + h(V_{a^*}) \times PI_{j \leftarrow i})}{|V_{a^*}|} \end{aligned} \tag{4}$$

It is easy to prove (5).

$$f(V_{a^*}) = 1 - g(V_{a^*}) \tag{5}$$

According to (5), the V_{a^*} that minimizes $g(V_{a^*})$ in (4) would maximize $f(V_{a^*})$ in (2). However, the term $h(V_{a^*})$ in (4) makes it difficult to solve the optimization problem. Therefore, we estimate $h(V_{a^*})$ by an estimator $\hat{h}_{a^*}$ as defined in (6).

$$\hat{h}_{a^*} = \frac{|K_{a^*}|}{|K_{\neg a^*}|} \tag{6}$$

where K_{a^*} is the set of users whose label is already known as a^* and $K_{\neg a^*}$ is the set of users whose label is already known as not a^*.

Replacing $h(V_{a^*})$ in (4) with $\hat{h}_{a^*}$, one gets $\hat{g}(V_{a^*})$ as shown in (7), which is an approximation to $g(V_{a^*})$. Accordingly, the U_{a^*} that minimizes $\hat{g}(K_{a^*} \bigcup U_{a^*})$ is an approximate solution to our label profiling problem defined in Section 2.3.

$$\hat{g}(V_{a^*}) = \frac{\sum_{v_i \in V_{a^*}, v_j \notin V_{a^*}} (PI_{i \leftarrow j} + \hat{h}_{a^*} \times PI_{j \leftarrow i})}{|V_{a^*}|} \tag{7}$$

Finding the optimal U_{a^*} that minimizes $\hat{g}(K_{a^*} \bigcup U_{a^*})$ is a fractional programming problem. Dinkelbach et al. [4] proposed an approach to solve one such kind of fractional programming problem by transforming it to a parametric programming problem.

Let $N(x)$ and $D(x)$ be two continuous functions of $x \in S$ where $S \subseteq R^n$ is the domain of x, and $D(x)$ is non-negative when $x \in S$. Considering two equations:

$$F(x) = N(x)/D(x) \tag{8}$$

$$G(q) = min\{N(x) - qD(x)|x \in S\} \tag{9}$$

Three lemmas can be derived according to [4].

Lemma 1. [Monotonicity] $G(q)$ *is strictly monotonic decreasing, i.e., if* $q_1 < q_2$, $G(q_1) > G(q_2)$.

Lemma 2. [Dinkelbach Property] *Let q^* be the minimal value of $F(x)$ defined in (8). Then,*

$$\begin{cases} G(q) = 0 \Leftrightarrow q = q^* \\ G(q) < 0 \Leftrightarrow q > q^* \\ G(q) > 0 \Leftrightarrow q < q^* \end{cases}$$

Lemma 3. [Optimality] *Let q^* be the minimal value of $F(x)$. x^* that minimizes $N(x) - q^* D(x)$ also minimizes $F(x)$.*

Let $V_{a^*} = K_{a^*} \bigcup U_{a^*}$ be the final a^*-labeled user set. In our problem setting, $N(x) = \sum_{v_i \in V_{a^*}, v_j \notin V_{a^*}} (PI_{i \leftarrow j} + \hat{h}_{a^*} \times PI_{j \leftarrow i})$ and $D(x) = |V_{a^*}|$. Define $l_q(V_{a^*})$ for a constant q as follows.

$$l_q(V_{a^*}) = \sum_{v_i \in V_{a^*}, v_j \notin V_{a^*}} (PI_{i \leftarrow j} + \hat{h}_{a^*} \times PI_{j \leftarrow i}) - q|V_{a^*}| \tag{10}$$

Accordingly, $G(q) = min\{l_q(V_{a^*}) | K_{a^*} \subseteq V_{a^*} \subseteq V\}$. Our key problem therefore is to calculate $G(q)$ for a given q and find corresponding V_{a^*} that satisfies $K_{a^*} \subseteq V_{a^*} \subseteq V$ and minimizes $l_q(V_{a^*})$. To solve this problem, a flow-network-based algorithm can be applied, which is omitted due to space limit.

To summarize, our label profiling algorithm is straightforward. Based on Lemma 1 and Lemma 2, we use a binary search method to find the q^* that satisfies $G(q^*) = 0$. Based on Lemma 3, the U_{a^*} that minimizes $l_{q^*}(K_{a^*} \cup U_{a^*})$ is the set of a^*-labeled users according to our problem formulation.

3.2   Speed-Up for Large Networks

The bottleneck of our algorithm is finding the minimal value of $l_q(V_{a^*})$ for a given q, to which we applied a maximum flow algorithm. However, even the most efficient maximum flow algorithm can only handle networks with around ten thousand nodes, hardly scalable enough to solve our label profiling problem on large social networks in reality.

Our solution is to adopt a divide-and-conquer framework in which we first divide the entire network into many sub-networks, efficiently solve the label profiling problem on these sub-networks, and integrate the profiling results to eventually determine users' labels. The sub-networks should be large enough to capture the affinity-awareness, yet small enough to admit efficient computation. A natural choice is users' ego network, in which dense connections among users sharing the same label with the central user can usually be observed. In each ego network, we infer the set of users whose label is the same as the central user.

Label Profiling in Ego Networks. To infer the set of users whose label is consistent with the central user's label in an ego network, one can directly apply the label profiling algorithm by treating the ego network simply as the entire network. However, the challenge here is that the information of the neighbors outside the ego network is missing, which may lead to incorrect profiling. For example, if a user v only links to the central user whose label is A in a ego network, v must be labeled as A when profiling users in this ego network, since it optimizes our objective function. But user v actually connects to many users outside the ego network whose label is B, which strongly indicates that v's real label is B but mistakenly profiled as A. This example demonstrates that the outside information is supposed to be considered to improve the prediction performance.

Given a partially labeled ego network $G^e(V^e, E^e)$ and central user's label a^*, let $K^e_{a^*} \subseteq V^e$ be the set of users whose label is known as a^* in the ego network, $K^e_{\neg a^*} \subseteq V^e$ be the set of users whose label is known as not a^*, $U^e = V^e \setminus (K^e_{a^*} \bigcup K^e_{\neg a^*})$ be the set of label-unknown users in the ego network.

Suppose all the labels of users outside the ego network are already known, let $G(V, E)$ denote the whole network, $V^o_i = \{v_j | e_{i,j} \in E\} \setminus V^e$ be the set of neighbors of each user $v_i \in V^e$ external to V^e, and $V^o_{i,a^*} \subseteq V^o_i$ be the users in V^o_i whose real label is a^*, and $V^o_{i,\neg a^*} \subseteq V^o_i$ be the set of users in V^o_i whose real label is not a^*. Considering a set $V^e_{a^*} \subseteq V^e$ which is the set of a^*-labeled users in V^e, we define $f^e(V^e_{a^*})$ in (11), which is the difference between average total propagating importance from a^*-labeled friends of a a^*-labeled user inside the ego network and average propagating importance from a^*-labeled friends of a user whose label is not a^* inside the ego network. To capture the affinity-awareness, our goal is to maximize the value of $f^e(V^e_{a^*})$ in (11).

$$f^e(V^e_{a^*}) = \frac{\sum_{v_i \in V^e_{a^*}} \sum_{v_j \in V^e_{a^*} \cup V^o_{i,a^*}} PI_{i \leftarrow j}}{|V^e_{a^*}|} - \frac{\sum_{v_i \in V^e \setminus V^e_{a^*}} \sum_{v_j \in V^e_{a^*} \cup V^o_{i,a^*}} PI_{i \leftarrow j}}{|V^e \setminus V^e_{a^*}|}$$

(11)

Analogous to the definition of $g(V^e_{a^*})$, we define $h^e(V^e_{a^*})$ in (12) and $g^e(V^e_{a^*})$ in (13) satisfying (14).

$$h^e(V^e_{a^*}) = \frac{V^e_{a^*}}{|V^e \setminus V^e_{a^*}|}$$

(12)

$$g^e(V^e_{a^*}) = \frac{\sum_{v_i \in V^e_{a^*}} \sum_{v_j \in V^e \setminus V^e_{a^*}} PI_{i \leftarrow j} + h^e(V^e_{a^*}) \times PI_{i \leftarrow j}}{|V^e_{a^*}|}$$
$$+ \frac{\sum_{v_i \in V^e_{a^*}} \sum_{v_j \in V^o_{i,\neg a^*}} PI_{i \leftarrow j}}{|V^e_{a^*}|}$$
$$+ \frac{h^e(V^e_{a^*}) \times \sum_{v_i \in V^e \setminus V^e_{a^*}} \sum_{v_j \in V^o_{i,a^*}} PI_{i \leftarrow j}}{|V^e_{a^*}|}$$

(13)

$$f^e(V^e_{a^*}) = 1 - g^e(V^e_{a^*})$$

(14)

In reality, only partial labels of the users in V^o_i are known for each $v_i \in V^e$, so the exact value of term $\sum_{v_j \in V^o_{i,\neg a^*}} PI_{i \leftarrow j}$ and $\sum_{v_j \in V^o_{i,a^*}} PI_{i \leftarrow j}$ in (13) cannot be calculated if the unknown labels are not given. However, we assume that the label-known users in V^o_i are uniformly sampled from V^o_i. Let $K^o_{i,a^*} \subseteq V^o_i$

be the users in V_i^o whose label is known to be a^*, and $K_{i,\neg a^*}^o \subseteq V_i^o$ the users in V_i^o whose label is known not to be a^*. We approximate $\sum_{v_j \in V_{i,\neg a^*}^o} PI_{i \leftarrow j}$ by $\frac{|V_i^o|}{|K_{i,a^*}^o \cup K_{i,\neg a^*}^o|} \times \sum_{v_j \in K_{i,\neg a^*}^o} PI_{i \leftarrow j}$ and $\sum_{v_j \in V_{i,a^*}^o} PI_{i \leftarrow j}$ by $\frac{|V_i^o|}{|K_{i,a^*}^o \cup K_{i,\neg a^*}^o|} \times \sum_{v_j \in K_{i,a^*}^o} PI_{i \leftarrow j}$. After that, we can approximate $g^e(V_{a^*}^e)$ by $\hat{g}^e(V_{a^*}^e)$, as defined in (15).

$$\hat{g}^e(V_{a^*}^e) = \frac{\sum_{v_i \in V_{a^*}^e} \sum_{v_j \in V^e \backslash V_{a^*}^e} PI_{i \leftarrow j} + h^e(V_{a^*}^e) \times PI_{i \leftarrow j}}{|V_{a^*}^e|}$$
$$+ \frac{\sum_{v_i \in V_{a^*}^e} \frac{|V_i^o|}{|K_{i,a^*}^o \cup K_{i,\neg a^*}^o|} \sum_{v_j \in K_{i,\neg a^*}^o} PI_{i \leftarrow j}}{|V_{a^*}^e|} \qquad (15)$$
$$+ \frac{h^e(V_{a^*}^e) \times \sum_{v_i \in V^e \backslash V_{a^*}^e} \frac{|V_i^o|}{|K_{i,a^*}^o \cup K_{i,\neg a^*}^o|} \sum_{v_j \in K_{i,a^*}^o} PI_{i \leftarrow j}}{|V_{a^*}^e|}$$

To solve the label profiling problem in an ego network, we want to find the optimal $U_{a^*}^e \subseteq U^e$ to assign label a^* such that $\hat{g}^e(K_{a^*}^e + U_{a^*}^e)$ is minimized. Notice that label assignment in each ego network is independent, the result of which would be integrated eventually.

Results Integration across Ego Networks. After solving the label assignment problem for each and every ego network, we use majority-voting to determine the label for each user with at least one assignment in any ego network. To summarize, our label profiling framework is an iteration of the following two steps until (I) all users' labels are known, or (II) no user is assigned a label value at the current iteration. Note that other early termination criteria can be implemented as accuracy decreases with increasing iterations.

1. **Step I:** For each user v_i whose label value is known, take v_i's ego network with v_i as the central user, and idenfity the set of users whose label value is the same as v_i's label value.
2. **Step II:** Use majority-voting to decide on the labels of users with at least one label assignment in Step I.

4   Experiments

4.1   Dataset

Sina Weibo. Sina Weibo[1] is a Twitter-like Chinese online social network with over 500 million users. To crawl the data, we initially chose 1000 seed users, and crawled both their followees and followers. At this step, about 240 thousand first-level neighbors were crawled. In order to compute their close friendships,

---

[1] http://weibo.com/

we also crawled the followees and followers of the first-level neighbors. At this step, about 3.5 million second-level neighbors were crawled. At each step, we crawled the users whose followers did not exceed 2000, since users with more than 2000 followers are more of an information hub than a normal user in Sina Weibo. For each user, we also crawled her location in terms of province which we profile in this dataset. About 20% of users did not provide their province location when we crawled the data.

Renren. Renren[2] is a Facebook-like Chinese online social network with over 280 million users. Most users in Renren are undergraduate and graduate students, and most of them provide their education information. However, due to privacy protection mechanism, only about 15% of the users' college information are publicly available. The friend relationship in Renren is undirected. One can befriend another user only after the friend invitation request is accepted. We have collected over 1.9 million users' profile information and the friendship links among them. We profile the label of user's college in this dataset.

Pokec. Pokec[3] is the most popular online social network in Slovakia. Pokec has been launched for more than 10 years and connects more than 1.6 million people. 99.9% of the users provide their location information in Pokec. The friend relationship on Pokec is directed. Other information of the dataset can be found in [14]. We profile user's location label on this dataset.

In this paper, we consider both followers and followees as friends of a user for the directed network. We adopted Xie's method [15] to identify close friendships. The average ratio of close friends among users' online friendships is 59% in Sina Weibo, 87% in Renren and 69% in Pokec.

4.2 Algorithm Comparison

We compare our affinity-awareness label profiling algorithm (denoted as AA algorithm) with three state-of-the-art propagation-based algorithms, all of which are designed to identify the set of users with the same label in a central user's ego network.

- CP algorithm: Liu et at. [7] proposed the co-profiling algorithm that models the latent correlation between labels and social connections. CP infers the labels of the label-unknown central user as well as her label-unknown neighbors in an ego network originally. In our experiment, we use almost the same model in [7], except that we reveal the label of the central user.
- GSSL algorithm: The label profiling problem can be viewed as a graph-based semi-supervised learning problem (GSSL) [18]. We use the most widely used GSSL method [17] as one of our baselines.
- MRF algorithm: Markov random field [9] can be used to model the interaction between nodes in a network and predict the labels of the nodes. We use a basic MRF model that has been used in [3] to label the users in ego network.

---

[2] http://www.renren.com/
[3] http://snap.stanford.edu/data/soc-pokec.html

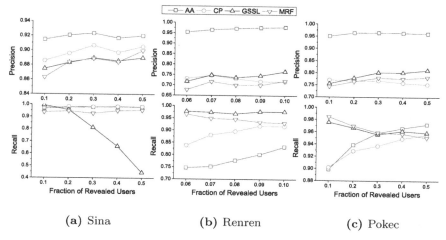

Fig. 3. Evaluation of label profiling in ego networks

Label Profiling in Ego Networks. We first compare our AA algorithm with 3 baseline algorithms for the task of label profiling in ego networks.

In our experiments, we randomly hide some of the labels in the whole network, and then infer the labels for label-unknown users in ego networks where the central user's label is revealed. We change the fraction of revealed users to see how this factor affects the results. To make sure the results are statistically sound, we repeat this experiment 5 times with different concealed nodes. In our experiments, all the algorithm parameters are empirically set as the optimal values.

Figure 3 shows the precision and recall of the 4 algorithms in three datasets. The precision of our algorithm outperforms other algorithms in all datasets. In Renren and Pokec dataset, the precision of our algorithm achieves 0.95, while the precision of other algorithms is less than 0.8. The precision of our algorithm is stable when the fraction of revealed users varies. The recall of our algorithm is lower than the recall of other algorithms, when the fraction of revealed users is lower than 0.1. Compared with other algorithms, our algorithm is relatively conservative in propagating labels when information in the network is highly insufficient. However, the recall of our algorithm grows as the fraction of revealed users increases. When information in the network is sufficient, the recall of our algorithm achieves the best level among all the algorithms.

Label Profiling in Whole Network. In our global user profiling framework, we can use either of the 4 algorithms to profile users in ego networks. We compare the performance of these 4 algorithms under our user profiling framework.

We only experiment in Pokec dataset, since it is the only dataset where all the users and their connections are collected. In our experiments, we randomly reveal 20% of the labels on the whole network, and then iteratively use the

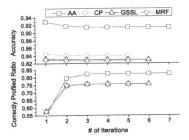
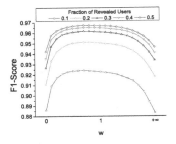

Fig. 4. Evaluation of label profiling in whole network

Fig. 5. Parameter sensitivity analysis

profiling framework to profile users. We repeat this experiment 5 times with different concealed nodes to make sure the results are statistically sound. In our experiments, all algorithm parameters are empirically set at the optimal values.

The experiment results are shown in Figure 4. The number of iterations is 7 for AA algorithm, 6 for GSSL algorithm and 5 for CP and MRF algorithm. After each iteration, we calculate the accuracy and the ratio of correctly profiled users over all the label-concealed nodes (i.e., correctly profiled ratio). For all algorithms, the accuracy gradually decreases as we mentioned in Section 3.2. Both the accuracy and correctly profiled ratio of our algorithm are better than other algorithms at every iteration, except for the correctly profiled ratio after first iteration. One thing to note is that the global accuracy after the first iteration is higher than the precision of profiling users in ego networks for the three baseline algorithms, which reflects the effectiveness of our global user profiling framework in integrating the propagation results to determine users' labels. However, for our algorithm, global accuracy after the first iteration is lower than the precision of profiling users in ego networks, since some users are correctly profiled in different ego networks, which are counted multiple times when calculating the local precision but counted only once when calculating the global accuracy.

4.3   Parameter Sensitivity Analysis

In our algorithm, there is a parameter w which distinguishes the different propagating importance between close friendships and online friendships. We investigate how parameter w affects the prediction results. In our experiments, we randomly hide some of the labels in the whole network, and then infer the labels for label-unknown users in ego network of the label-revealed users using different ws, which are supposed to vary from 0 to $+\infty$. We sample w on $w = 0.0, 0.1, ..., 1.0$ and $w = \frac{1}{0.9}, \frac{1}{0.8}, ..., \frac{1}{0.0}$. Also, we vary the fraction of revealed users to see whether w is related to this factor.

In Figure 5, we plot the f_1-score of the experiment results in Pokec dataset, from which we reach the following conclusion. For a fixed fraction of revealed users, the shape of the function of f_1-score on w is concave. And the optimal w that maximizes the f_1-score is consistent in networks with different fractions

Table 2. Optimal w in three datasets

Dataset	Optimal w
Sina	0.4 - 0.5
Renren	0.2 - 0.3
Pokec	0.7 - 0.8

of revealed users, which indicates that w is a latent variable dependent on the social network in question.

In Tabel 2, we give the range of optimal w in the three datasets based on our experiments. In all datasets , the optimal w is less than 1. It indicates that close friends contribute more when profiling labels like geographic location and education background, supporting our argument for affinity-awareness in label profiling.

5   Related Work

User profiling in social network has been studied for long. [11,7,16] adopted propagation-based approach to profile users. Mislove et al. [11] used community discovery method [5] to find communities on the entire network, and assigned the same label to users in the same community. Li et al. [7] proposed an optimization model to simultaneously profile different labels and determine relationship types for a target user in her ego network, which is called co-profiling. Yang et al. [16] proposed a probabilistic model in a heterogeneous network where the connections consist of both edges between users and edges between users and service items to propagate interest implied by the service items among users. [13,2] purely used users' own information such as user profile data and tweets to infer their labels. Rao et al. [13] used an SVM model whose features were identified from tweets to classify user's gender, age, regional origin and political orientation. Cheng et al. [2] proposed a probabilistic framework for estimating a Twitter user's city location based on the content of the user's tweets. [8,12,1] used both users' own information and network structure to profile users. Li et al. [8] proposed a unified discriminative influence model incorporating both tweets and network structure to profile user's home location. Pennacchiotti et al. [12] used gradient boosted decision trees framework [6] to classify user's labels like political orientation based on the profile features, linguistic content features and social network features. Backstrom et al. [1] used IP location and network structure to profile user's location information. However, all the previous works did not distinguish the difference between close and online friends when profiling users.

The single label profiling problem can be viewed as a graph-based semi-supervised learning problem (GSSL) [18]. Zhou et al. [17] proposed a widely used GSSL method where a weighted graph was constructed to capture the pairwise relationships between data points. In contrast, we directly use the social network as the input graph when profiling users' labels.

Label inference based on network structure has also been studied in other fields like biology. Deng et al. [3] inferred the functions for proteins in proteins interaction network. A markov random field [9] was used to model the relation between the proteins. Deng et al. used logistic regression to learn the parameter of the markov random field and used MCMC method to solve the inference problem.

Real-life friendship discovery in online social network is a relatively novel problem. [15] was the first work that identified users' real-life friends in Twitter network only based on network structure. Three principles helpful in identifying real-life friendship were proposed from ground-truth data. Based on these principles, an algorithm was proposed to iteratively identify target user's real-life friends.

6   Conclusion

In this paper, we proposed a propagation-based method to profile users' labels with an awareness of the context in which the friendship connections were established. We observed homophily among users in social network that only a^*-labeled users connect to many a^*-labeled friends. Also, we argued that we should differentiate role of close friendships and online friendships when profiling users, which has been neglected so far. Based on above analysis, we proposed an optimization-based algorithm to profile users where a parameter w is used to distinguish the different propagating importance between close and online friendships. To scale up to large social networks, we adopted a divide-and-conquer user profiling framework. The experiment results in three real-world datasets demonstrated the superiority of our algorithm over baselines and supported our argument for affinity-awareness in label profiling.

Acknowledgement. This research is supported in part by the National Natural Science Foundation of China under Grant No.71272029 and the Singapore National Research Foundation under its International Research Centre @ Singapore Funding Initiative and administered by the IDM Programme Office, Media Development Authority (MDA).

References

1. Backstrom, L., Sun, E., Marlow, C.: Find me if you can: improving geographical prediction with social and spatial proximity. In: Proceedings of the 19th International Conference on World Wide Web, pp. 61–70. ACM (2010)
2. Cheng, Z., Caverlee, J., Lee, K.: You are where you tweet: a content-based approach to geo-locating twitter users. In: Proceedings of the 19th ACM International Conference on Information and Knowledge Management, pp. 759–768. ACM (2010)
3. Deng, M., Zhang, K., Mehta, S., Chen, T., Sun, F.: Prediction of protein function using protein-protein interaction data. Journal of Computational Biology 10(6), 947–960 (2003)

4. Dinkelbach, W.: On nonlinear fractional programming. Management Science 13(7), 492–498 (1967)

5. Fortunato, S.: Community detection in graphs. Physics Reports 486(3), 75–174 (2010)

6. Friedman, J.H.: Greedy function approximation: a gradient boosting machine. Annals of Statistics, 1189–1232 (2001)

7. Li, R., Wang, C., Chang, K.C.-C.: User profiling in an ego network: co-profiling attributes and relationships. In: Proceedings of the 23rd International Conference on World Wide Web. International World Wide Web Conferences Steering Committee, pp. 819–830 (2014)

8. Li, R., Wang, S., Deng, H., Wang, R., Chang, K.C.-C.: Towards social user profiling: unified and discriminative influence model for inferring home locations. In: Proceedings of the 18th ACM SIGKDD International Conference on Knowledge Discovery and Data Mining, pp. 1023–1031. ACM (2012)

9. Li, S.Z.: Markov random field modeling in computer vision. Springer-Verlag New York, Inc. (1995)

10. McPherson, M., Smith-Lovin, L., Cook, J.M.: Birds of a feather: Homophily in social networks. Annual Review of Sociology, 415–444 (2001)

11. Mislove, A., Viswanath, B., Gummadi, K.P., Druschel, P.: You are who you know: inferring user profiles in online social networks. In: Proceedings of the third ACM International Conference on Web Search and Data Mining, pp. 251–260. ACM (2010)

12. Pennacchiotti, M., Popescu, A.-M.: Democrats, republicans and starbucks afficionados: user classification in twitter. In: Proceedings of the 17th ACM SIGKDD International Conference on Knowledge Discovery and Data Mining, pp. 430–438. ACM (2011)

13. Rao, D., Yarowsky, D., Shreevats, A., Gupta, M.: Classifying latent user attributes in twitter. In: Proceedings of the 2nd International Workshop on Search and Mining User-Generated Contents, pp. 37–44. ACM (2010)

14. Takac, L., Zabovsky, M.: Data analysis in public social networks. In: Intl. Scientific Conf. & Intl. Workshop Present Day Trends of Innovations (2012)

15. Xie, W., Li, C., Zhu, F., Lim, E.-P., Gong, X.: When a friend in twitter is a friend in life. In: Proceedings of the 3rd Annual ACM Web Science Conference, pp. 344–347. ACM (2012)

16. Yang, S.-H., Long, B., Smola, A., Sadagopan, N., Zheng, Z., Zha, H.: Like like alike: joint friendship and interest propagation in social networks. In: Proceedings of the 20th International Conference on World Wide Web, pp. 537–546. ACM (2011)

17. Zhou, D., Bousquet, O., Lal, T.N., Weston, J., Schölkopf, B.: Learning with local and global consistency. In: NIPS, vol. 16, pp. 321–328 (2003)

18. Zhu, X., Ghahramani, Z., Lafferty, J., et al.: Semi-supervised learning using gaussian fields and harmonic functions. In: ICML, vol. 3, pp. 912–919 (2003)

Disenchanting the World:
The Impact of Technology on Relationships

Paolo Parigi and Bogdan State

Stanford University

Abstract. We explore the impact of technology on the strength of friendship ties. Data come from about two millions ties that members of CouchSurfing—an international hospitality organization whose goal is to promote travelling and friendship between its members—developed between 2003 and 2011 as well as original and secondary ethnographic data. The community, and the data available about its members, grew exponentially during our period of analysis, yet friendships between users tended to be stronger in the early years of CouchSurfing, when the online reputation system was still developing and the whole network was enmeshed in considerable uncertainty. We argue that this case illustrates a process of disenchantment created by technology, where technology increases the ease with which we form friendships around common cultural interests and, at the same time, diminishes the bonding power of these experiences.

Keywords: On-line Communities, Uncertainty, Networks.

Technology has greatly facilitated the establishment and growth of communities centered on unique interests. What is the impact of this development on social networks? The technological revolution of the last decades years has greatly increased opportunities for interaction and the speed with which new relationships develop (DiMaggio et al. 2001). Indeed, finding like-minded people with whom to share a passion, no matter how obscure, has never been easier than it is today (Kairam, Wang, and Leskovec 2012). The sharing of a common interest is a powerful factor influencing the emergence of ties between individuals (Lizardo 2006), and more arcane, more specific interests produce stronger ties. Online communities have ermeged as powerful foci (Feld, 1981) through which social ties are formed. Here we investigate how risk and uncertainty influence the strenght of these ties.

A key aspect of these online communities is their reliance on reputation systems, which accumulate information about the members of the community in the form of reviews, ratings, comments, etc. The communities and institutions in which individuals are embedded have always stored information about members' past interactions in the form of reputations and collective memories. Yet the processes by which this information was transmitted were less formalized and more contested than those made possible by current technology. Here we consider how the nature of online reputation systems impacts the strength of ties they facilitate.

Data for our analysis come from CouchSurfing (CS hereafter), an international hospitality organization established in 2003. The goal of CS is to promote cultural

L.M. Aiello and D. McFarland (Eds.): SocInfo 2014, LNCS 8851, pp. 166–182, 2014.

understanding between strangers. "CouchSurfers" (CSers) engage in hospitality interactions with other members of the organization with no exchange of money. Host and guest often are previously unknown to each other, save for information provided through the organization's website. The interaction between the two is not strictly instrumental, thus giving it many characteristics of an altruistic exchange (Bialski 2009). In the decade since the CS's founding, the organization has developed an elaborate system for reporting reputation. In addition to these hospitality exchanges, local CS chapters organize events where members can meet and interact. While these events are not directly related to the <u>experience</u> of travelling and meeting strangers, they help instead bring together people that have a general interest in travel.

We use the entire friendship network of CSers over time, from 2003, when the website had just a few members, to 2011, when the website had about four million registered users. Of these users, about 650,000 were active members in the sense that they had participated in at least one hospitality exchange (visiting or hosting a stranger) or event organized by a local chapter of CS (a potluck at another member's house, for instance). We focus the analysis on the set of active users and on the more than two million friendships that formed as a result of their participation in CS. In the great majority of the cases, CS members meet online first and then meet in person as either guest or host. The resulting relationships are therefore a mix of the online and offline worlds—i.e., of the unified social reality most people live in. We think that the data we analyzed is uniquely suited for studying the impact of technology on relationships. Further, we complement the quantitative data with ethnographic observations from a secondary source and with interviews we conducted.

We present evidence that the friendship ties that formed between strangers as the result of guest-host interactions were stronger than interactions formed as result of participation in a local event. In line with prior research (DiMaggio 1987; Lizardo 2006), the sharing of a more unique cultural product—travelling and meeting with strangers in this case—produced stronger ties than just the sharing of a broader interest in travelling. At the same time, we found that the greater the amount of information available about potential hospitality partners, the weaker the friendships that emerged from the experience. These findings highlight a complex processes by which technology is impacting relationships. On one hand, the ways technology facilitates the aggregation of people into cultural communities makes ties easier to develop. On the other hand, the growing amount of information circulating in a community about potential others makes the friendships that are the byproduct of their interactions weaker. The larger point this paper underscores is that technology may be making people more connected then ever before, but these connections have less binding force to meaningfully structure our lives.

1   Research Hypotheses

Cultural sociologists interested in social networks have for a long time highlighted a direct connection between the emergence of relationships and the consumption of cultural goods. Long (2003) found that women who belonged to reading groups in

Houston, Texas, also shared multiple ties based on neighborhood and affiliations with local religious and educational groups. This study and similar others (Erickson 1996) illustrate how social structure, in the form of pre-existing social ties, impacts cultural tastes and the consumption of cultural goods. At the same time, the consumption of cultural goods such as books also has an effect on social networks in that it facilitates the emergence of new ties. Lizardo has gone farthest in extending the idea that cultural tastes and social networks are mutually constitutive. Using Bourdieu's theory about the fungibility of various forms of capital (1986), Lizardo shows how cultural capital, in the form of individual tastes, generates social capital, in the form of new ties. In investigating this conversion, Lizardo's analysis provides evidence that consumption of more specialized, highbrow cultural content produces stronger ties than consumption of popular culture goods.

While Lizardo's model uses the dichotomy popular and highbrow to explain the mix of weak and strong ties in a community, others have related the strength of ties to the consumption of exclusive versus popular cultural goods. According to Collins (1988), for instance, particularized cultural capital sustains rituals and produces solidarity among community members, making it capable of generating stronger bonds than generalized cultural capital (1988). That is, two strangers are more likely to become closer friends by discovering a common love for a little-known sports team than by discovering a common passion for a world-famous one. While almost any two strangers can talk about a famous team, sustaining a conversation about a lesser-known object activates a more precise symbolic identification of the two partners. Synthetizing this body of work, we expect that:

H1: The more exclusive the experience related to the consumption of a cultural good, the stronger the emerging relationship between the individuals in a given community (everything else equal).

With respect to CS, H1 suggests that stronger hospitality interactions would generate friendship ties than ties that emerge from participation in a common local event. The interaction between host and guest, we argue, generates greater unique and memorable experiences that could become the basis for stronger bonds. Further, H1 suggests that as a novel product spreads and becomes more popular it loses its power to create meaningful bonds. One might compare, for instance, backpackers who meet in a foreign city before and after mass tourism discovers it, or fans who meet at a band's concert before and after the band becomes popular. For early adopters, the circumstances of the meeting are unique and are likely to create strong bonds, but for late adopters the circumstances of meeting are banal and result in weaker bonds.

Social exchange scholars have also investigated the factors shaping the emergence of strong ties. Coleman (1975) showed that negotiated exchanges, or exchanges where the terms of the interactions are known beforehand (e.g., an exchange based on a contract), generate lower levels of trust than reciprocated exchanges. Building on this finding, social exchange scholars have investigated the conditions that create stronger, more trusting bonds between exchange partners. The majority of exchange scholars have focused on the development of trust in market transactions (Kollock 1994; Molm, Takahashi, and Peterson 2000), but others have analyzed trust and strength of ties with respect to the institutions of a given society. We use findings from this latter group to build our theoretical argument as to how online rating systems are impacting the relationships that develop out of the consumption of cultural goods.

Yamagishi and Yamagishi (1994) note that while the Japanese are less likely to attribute trust to individuals than are Americans, their society has a higher level of trust than U.S. society. They explain this puzzle by distinguishing between trust—an inference about the interaction based on the partner's personal traits—and assurance, which is based on the knowledge of the incentive structure surrounding the relationship. Thus, Yamagishi et al. explain, the Japanese have more relationships backed up by assurance structures than do Americans (1998). Cook, Hardin and Levi (2007) extend this approach to the role of institutions, at least partially departing from a strict rational actor perspective based on incentives and punishments. They argue that institutions facilitate a type of cooperation that does not require trust. Institutions substitute for trust because they provide contexts for creating expectations about the future behavior of interacting partners.

Cheshire (2011) uses the distinction between assurance and trust to draw attention to the impact of online assurance structures on the relationships we form. In environments characterized by high uncertainty—where little is known about the potential partners—strong relationships are more likely to develop. On the contrary, when relationships are assured by a third party—a network administrator or the scores of a rating system—ties that emerge between partners tend to be weaker (Cheshire 2011; Fiore and Cheshire 2010). Reframed with respect to Coleman's argument, greater information about individuals reduces the uncertainty in dealing with strangers and on average makes interactions closer to a negotiated exchange. We argue that a rating system in a given community operates similarly to an assurance structure, in that it decreases the interpersonal trust necessary between partners:

H2: The more information circulating in the community about potential partners, the weaker the emerging friendship between two strangers after the interaction.

H2 suggests that a potential impact of the ubiquitous online ratings system is a reduction of the binding force emerging from the shared experience of consuming cultural goods. The argument here is that, independent of the exclusivity of the good, the experience of travelling generates weaker bonds between CSers in the presence of a ratings system than without one to act as an assurance structure. Considered together, our two hypotheses suggest a technology-driven process of progressive disenchantment of the world: relationships may be easier to form now than ever, but each of these new relationships has a lower binding force and ability to fill our lives with meaning.

2   The Case Study

CouchSurfers engage in hospitality exchanges with other members of the organization. Host and guest often are previously unknown to each other except for information provided through the organization's website. Visiting a stranger's house—or hosting a stranger in one's house—poses the risk of some particularly devastating events, as well as some minor inconveniences. Despite such risks and the fact that no money changes hands to compensate for them, there are now millions of CSers worldwide engaging in thousands of hospitality exchanges every day. Friendship ties are a byproduct of these exchanges. After the hospitality interaction, the host and the guest have the option of voluntarily and independently reporting to CS the formation of a new friendship binding them. Furthermore, each partner is also asked to rate the strength of that new tie. While this reporting is completely voluntary, a large majority of CSers in our data fulfilled the request.

We purposefully use the concept of exchange to characterize the interaction between a host and a guest. Scholarly work on the CS community suggests that while exchanges do not necessarily recur with the same partners (i.e., many pairs do not share more than one exchange with each other), CSers interpret hospitality interactions through the lens of reciprocity (Molz and Gibson 2007). CSers alternate between roles, sometimes serving as guest and sometimes as host, thereby generating a "pay-forward reciprocity" that informs their behavior and expectations (Bialski 2009). The word "exchange" captures the idea that, despite the fact that a given pair often does not interact more than once, most users in the CS community feel the binding that comes from belonging to a community.

As previously detailed, hospitality exchanges are not the only mechanism through which CSers form friendships. Members organize many informal events for locals and travelers, and CSers may also meet each other casually for, say, a meal or conversation. As with hospitality exchanges, participants in CS events and have the option of reporting new friendships with people that they meet in these ways. Because the CS reporting mechanism asks users to specify how they met, it is possible to distinguish between friendship ties formed from hospitality exchanges and those formed from other interactions. We term the latter "non-hospitality exchanges." There are two key differences between hospitality and non-hospitality exchanges. First, there is higher perceived risk in a hospitality exchange. Second, non-hospitality interactions are less in line with the concept of exchange as described above. However, for clarity we maintain the term *exchange* with the caveat that we use the non-hospitality interactions mainly as a benchmark against which to compare hospitality exchanges. It is useful to apply Simmel's (1950) analysis of interactions occurring within a dyad and interactions occurring in groups of three or more when considering hospitality versus non-hospitality exchanges. In a dyad, ties are personal because a tie has to bind both actors for it to exist. In larger groups, ties are social because the group can continue even if the ties produced do not bind all actors. Thus, hospitality ties are personal while non-hospitality ties are social.

Since its founding, CS has enjoyed growing popularity and media attention. Unsurprisingly, the fact that hundreds of thousands of people around the world are brave enough to open their houses to strangers strikes many members of both the public and the social scientific community as remarkable. So focused on potential negative outcomes is the public discourse surrounding CS that the word "risk" itself appears in about one out of every six online mentions of the organization.[1] Yet members of CS engage in these seemingly risky interactions almost as a matter of routine. CS thus offers the opportunity to study a rare kind of data—well-documented, real-life, risky interactions between strangers and the evolution of the ties they form through those interactions.

Perhaps because of its combination of more and less dangerous behaviors or because of an appetite for adventure among the key demographic of its members, CS has become for many a sort of idealistic lifestyle community (Marx 2012). However, the community aspect of CS did not exist at the beginning. It emerged over time. Key to its development was the implementation of a reputation system for collecting information about others, facilitating the calculation of risk. The cornerstone of CS's reputation system is the personal reference. Members may write testimonials about others (usually a few sentences but sometimes several paragraphs), describing their experiences with their interaction partners. References, as well as other reputation signals, may be submitted unilaterally, but are often reciprocated.[2]

References may be exchanged between any pair of users, but the reference form solicits information regarding the circumstances in which two individuals met and, importantly for our analysis, whether they met through the organization or knew each other beforehand. Data gathered also includes whether an individual hosted another and, if so, for how many nights.

The reputation system makes it possible for a member to anticipate the type of interaction she will have by hosting or by being hosted by a particular other user. The reputation system thus represents a capital asset of the organization and the key element that facilitates the millions of CS interactions worldwide between strangers. However, when CS began, members could gain very little information about one another from the website, other than self-completed profiles whose credibility could always be cast into doubt.

The reputation system developed over the years. Its expansion favored the circulation of information about members and facilitated the rise of interactions among travelers—interactions which, in turn, were folded into the reputation system, helping to further its development. During this period the organization grew steadily by adding new members, but hospitality exchanges increasingly took place between repeat users—i.e., individuals who had already had the experience of offering or receiving hospitality through the organization.

3   Data and Methods

As previously mentioned, CS has gathered social network information from its members through its online platform from its very beginning as an organization. Users are

---

[1] The statistic is based on a Google search performed February 16, 2012. At that time there were 2.2 million mentions of either Couchsurfing.com or Couchsurfing.org, about 375,000 of which included the word "risk" or its derivatives.

[2] Lauterbach et al. (2009) report that about three quarters of CS "vouches" were reciprocated in 2009.

encouraged to record their ties with other members of the organization—both with friends they know from elsewhere who also happen to be users of the platform and with friends they met through the organization (through a hospitality exchange or a non-hospitality exchange). These different circumstances of meeting are captured in the CS data, as well as the date of tie formation and an explicit, self-reported measure of tie strength.

Our quantitative analysis is based on a set of 2.2m observations of social ties mediated by CS, with valid and non-missing data in all the relevant variables, recorded on the CS platform from 2003 to 2011. The ties under scrutiny are only those between people who did not know each other before they met on CS; interactions for which no time stamp was reported were excluded from the analysis. Further, we distinguished those social ties that developed as a result of a hospitality exchange from those that developed from a non-hospitality exchange. There are 645,411 unique users represented in the dataset. Because our unit of analysis is a tie (i.e., a pair of two users), repeated experiences at the individual level do not imply repeated experiences for the pair. Indeed, only a tiny fraction of the interactions in our dataset occurred more than once. We operationalize our key concepts below:

Tie strength: In the CS reputation system, users may characterize their relationships with other users as "Acquaintances," "CS Friends," "Friends," "Good Friends," "Close Friends," and "Best Friends." Tie strength is measured on this scale for 98% of all ties reported between CSers who did not know each other before joining CS. Figure 1 plots the cumulative log-count of friendship ties generated by hospitality exchanges, separated by strength.[3] An interesting pattern appears. Early in the life of CS, strong friendships ("Best Friends" and "Close Friends") were significantly more prevalent than weak ties ("Acquaintances," "Friends" and "CS Friends"), but the reverse was true from about the 40th month onward. The category "Good Friends" remained much in the middle, before and after the 40th month.

It appears that hospitality exchanges produced stronger relationships on average when CS was new than later in its existence. In our analysis, we collapse the above six categories into three: ties rated by users as "Acquaintances" and "CS Friends," are coded as Acquaintances; ties rated as "Friends" and "Good Friends" are coded as Friends; ties coded as "Close Friend" and "Best Friend" are coded as Best Friends. This classification preserves the underlying ordered nature of the recorded variable while at the same time protecting us from the ambiguity of distinctions such as "Acquaintances" versus "CS Friends" or "Close Friend" versus "Best Friend."

Ties: As previously mentioned, we distinguish between two kinds of interactions facilitated by the organization. Hospitality exchanges represent the raison d'être of CS. There is arguably a great deal at stake for both host and guest in this kind of interaction.

---

[3] When the two users' reports of tie strength do not coincide, we randomly assign the tie strength to one of the categories reported by the users. Because we have no longer access to the CS dataset, this decision cannot be reversed. While we agree that studying discrepancies of ratings could be very interesting, we think that such a study is outside the scope of this paper. We are here interested on the average strength of ties at the systemic level, not on the dyadic perception of these experiences.

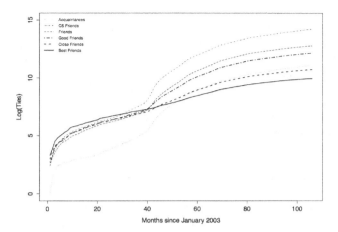

Fig. 1. Distribution of ties over time

In the worst-case scenario, both participants ultimately expose themselves to risks of theft, abuse or injury. We contrast hospitality exchanges with ties formed through less risky, non-hospitality exchanges, such as informal CS events (parties, collectives, organizational meetings, common meals, etc.) or any other occasion that does not involve one party hosting the other (one-on-one dinners, conversations, etc.).

Information: We measure the amount of information circulating in CS about potential partners by counting the number of ties (hospitality and non-hospitality) a member had prior to the focal exchange. That is, for every dyad in our set, we counted and summed the number of prior ties each user had up to the point of the latest interaction. For instance, if user A had one prior tie before interacting with user C, and user C had no prior ties, the variable cumulative ties for the dyad A-C would take the value of 1. Furthermore, the relationship between A and C would increase the number of prior ties for the two users so that the next time A established a tie with another user, say D, the total number of prior ties she contributed to the new interaction would be 2; for C, the new number would be 1.

Given that the overwhelming majority of CS interactions left a trace in the form of a review, we think that number of prior ties before the focal interaction is a good proxy for the amount of information circulating in the system about the exchanging partners. Because we wish to contrast the presence or the absence of information, rather than quantifying the impact of one extra review on the strength of the emerging relationship, we further segmented this variable in three categories: No prior information; Information about one partner; and Information about both partners.

Friends in common: To disentangle any potential confounding influences of triadic closure on our analysis, we include the number of prior friends in common as a control in our model by counting the number of triads a newly created tie closes. Furthermore, we distinguish embedding triads according to the strength of the ties they contain. We collapse the tie strengths the two users reported into a global measure of open triad strength. As in the case of contradictory tie strength reports, whenever the two ties forming an open triad are not the same strength, we randomly assign the strength of the triad to either of the strength scores the open triad received.

We include a number of other control variables in our analysis. Appendix B contains summary statistics for all the variables included in our sample. The majority of ties in our sample were between opposite-sex pairs (57%) who were on average in their mid-twenties (24 years old) and had been members of CS for a bit more than one year (13 months). Reflecting the growth of the website itself, most of the ties in our sample occurred during CS's seventh year (81 months after January 2003). Additionally, the average dyad was composed of members whose tenures in the organization were 13.88 months apart.

Our analysis also takes into account the organizational tenure of both members participating in the interaction. The average pair of CSers involved in an interaction had been members of the organization for 13 months at the time when the interaction occurred. However, the average difference in organizational tenure between the two members is just under 14 months, suggesting that a large number of interactions were between established and novice CSers.

Because of the nature of our dependent variable, *Tie strength*, we employed an ordinal logistic model with three categories. Technical details of this model are provided in the Appendix A. We also support our quantitative analysis with interview data from two sources – a series of 2005-2006 interviews reported by Bialsky (2009),[4] as well as our own ethnographic interviews conducted in 2010.[5]

4   A Community of Like-Minded People

Often stated reasons for joining and using CS included an idealistic desire to create a better world through travel and a search for opportunities for personal. This type of idealism still runs high among the CSers we interviewed and is in full display even to a cursory look at the current version of the website. Indeed, creating a better world through travel is still the motto of CS: "To make the world a smaller and friendlier place, one life-changing experience at a time."

Perhaps because of a common mindset or because of a shared ease in forming relationships, the friendship network of CSers grew rapidly from its 2003 origins. The opportunity

---

[4]   Bialsky's study contains interviews and observations capturing the early CS (from February 2005 to the summer of 2006), when the community consisted of 200,000 users worldwide (a tiny fraction of what it is now). The scope of this study was rather different from ours. Bialsky's goal was to study the impact of CS on tourism and to suggest technology's potential disruption of how people travel. Bialsky's interviews are useful to us because they reflect a time when CSers had fewer references and, as a consequence, interactions with strangers were enmeshed in greater uncertainty than present-day interactions.

[5]   We conducted a series of 18 interviews we conducted during the summer of 2010 in Amherst, MA; Santa Fe, NM; and Reykjavik, London and Milan. These interviews took place when one of the authors surfed as a guest on the couches of the interviewees. The sample includes eight females and ten males ranging in age from their early 20s to late 50s. All interviewees knew that we were conducting a research study on CS.

to make new friends was a main reason for joining CS among many early members.[6] Friendships developed from the uncertainty about how to properly interpret the roles of host and guest. This uncertainty created opportunities for a process of friendship discovery, albeit in an accelerated format. Long conversations about life with the (more) unknown alter were central to this process, and quite common at the time.[7]

If a common sentiment of bettering the world through travelling and meeting strangers has been part of CS since its very beginning, a crucial difference between the early years of CS and the more recent period is the amount of information available about potential others before an interaction takes place. Figure 2 plots for each month in the organization's life the average number of prior hospitality exchange experiences

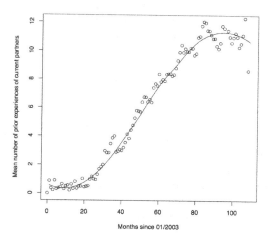

Fig. 2. Average amount of information about potential partners over time (LOESS regression)

---

[6] Johan, a 27-year-old Dutch CSer, stated: "I look back on my friends, I've been in Holland and I think I got into more meaningful relationships with CSers than people whom I've known for years.... I just see them being so static, as they are, they didn't get out. Even though we spent all these years together as friends, drinking together, or whatever, they still stay static, they're still in the same place." (PB, 33). Likweise, Ulla, a 26-year-old Finnish woman remarked, "all the Finnish culture and the Helsinki culture is just so closed down somehow. It's tough to break into circles and meet people for the first time.... There are people who do not understand this side of me... CSers all have the same needs to see." (PB, 33).

[7] This is, for instance, what Paula Bialski wrote about her very first hosting experience in 2004: "He [the guest] would speak, and I would often listen. It was the first time I ever invited a stranger into my home, and the first time I ended up speaking to a stranger until the late hours of the night" (PB, 9). Not knowing what to expect or how to behave when playing the role of host or guest also represented a challenge. Yet despite the perils of uncertainty, the psychological and emotional rewards of a successful interaction were substantial. Karen, a 27-year-old Australian traveling in Ireland, was "extremely nervous" (PB, 46) before meeting her first host in Dublin. Nevertheless, they ended up talking until two in the morning (PB, 46).

that any two partners had prior to a given interaction.[8] The figure shows an upward trend in the number of prior hospitality exchanges starting about 20 months into the life of the organization, when interactions were on average based on less than one prior experience, and leveling off around month 85, when hospitality exchanges reached an equilibrium level of about eleven prior experiences.

A key difference between early CSers and the ones we interviewed is the importance that the latter group attributes to reading references before selecting with whom to spend the night either as a host or a guest.[9] In contrast to early CSers' openness to the uncertain, the people we interviewed are more calculating of the type of "transformative experience" they are looking for when choosing other CSers. It is not the case that later CSers are less idealistic about the importance of the website for transforming the world. Rather, information has rationalized the process of selection and reduced the uncertainty associated with meeting strangers. Next we explore how greater information has impacted the strength of ties.

5   Results

Table 1 below presents the results of a partial proportional odd logistic model. We decided to use this model instead of a more standard ordered logistic model because the proportional odds assumption built in the ordered logistic model breaks down for three covariates—*Report tie*, *Exchange tie* and *Information*. A way to think about the partial proportional odd model is to see it as divided into two parts—one part with all the variables whose estimated coefficients do not change with the levels of the dependent variable (*proportional odds variables*, in Table 1) and one part made of variables whose coefficient estimates change across the levels of the dependent variables (see Appendix A for details). In Table 1, all the coefficients are standardized so that their magnitudes can be compared. Also, given the statistical power of our test all effects are significant at .0001 unless otherwise noted.

The table has some potentially-surprising results. Gender emerged as significant in affecting tie strength: opposite-sex dyads formed stronger ties than same-sex dyads.

---

[8]   For example, if A, who has participated in three prior hospitality exchanges, hosts B, who has engaged in one prior hospitality exchange, then A and B's interaction is assumed to be informed by four prior experiences. We averaged these figures across the dyads in our data.

[9]   Lisa, a woman in her 20s in London, told us she was concerned about safety when she joined CS. She had since become confident because "...the first experience was great and because, I suppose, the community's existing for many years now so the reference system is also increasing." For Lisa, experiences with other CSers appear to be mediated by the organization's reputation system. When we asked if safety ever became a concern for him, Peter, a new CSer in his 20s from Reykjavik, told us, "I will check my references. That's the only [...] thing that I learned—just check people's references." Now that information about others is available on CS, it plays an important role in maintaining a sense of safety within the community even when meeting with strangers. Roberto, a man from Milan in his 30s, told us, "Every time you write a message, there is a message that it is recorded for safety reasons. This is guaranteed and it's important because it's true that strangers are friends that you haven't met yet, but at the same time, strangers are always strangers."

Table 1. Estimates of partial proportional odds logistic model

Variable	Log odds	std. error
Proportional Odds Variables		
Female to female tie	-0.077	(0.007)
Female to male tie	0.039	(0.0048)
Mean age	0.003	(0.0018) ˏ
Age difference	-0.059	(0.0018)
Difference join CS	0.009	(0.0018)
Mean join CS	0.176	(0.0027)
Month tie creation	-0.281	(0.0027)
Friend in common	0.039	(0.0015)
Pr(Friend+ vs. Acquaintances):		
(Intercept)	-0.83	(0.007)
Report ties within one month	-0.449	(0.003)
Hospitality Exchange Tie	0.091	(0.0033)
Information about one partner	-0.009	(0.0086) ˏ
Information about both partners	0.115	(0.0065)
Pr(Best Friend vs. Friend, Acquaintances):		
(Intercept)	-2.773	(0.0134)
Report ties within one month	-0.872	(0.008)
Hospitality Exchange Tie	0.474	(0.0081)
Information about one partner	-0.285	(0.0018)
Information about both partners	-0.501	(0.013)
N = 2,171,966		
Residual deviance: 2940528 on 4343914 d.f.		
Log-likelihood: -1470264		
N. of iterations: 4		

Note: All coefficients are significant at .001 level unless marked (ˏ)

This finding runs counter to the expectations that similarities reinforce ties and that, with a given level of information, females would perceive higher risk in staying overnight with males than in staying overnight with females. Indeed, the model shows a significant positive effect for ties that are heterogeneous across gender and a significant negative effect for "Female to Female" ties. The decrease in tie strength for ties between two members of the same sex was greater than the increase for heterogeneous ties, suggesting that meeting people of the opposite sex was a sought-after experience (perhaps for individuals using CS for intimate encounters; see Zigos, 2013) among this community of travelers. The results also show that CSers place importance on their partner's tenure with the organization (*Difference join CS*)—the greater the difference between the two members of the pair, the stronger the resulting friendship. Further, the greater age difference between the two partners the lower the odds ratio of a strong tie.

We also considered the possibility that changes in tie strength over the life of the organization were influenced by the number of friends the two members of the pair had in common, with the assumption that the stronger the relationships between a CSer's friends' friends, the more binding the new tie would be. Table 1 shows that triadic closure operated in the expected direction and that the proportion of friends in common with whom the two members of the dyad had strong ties greatly reinforced the likelihood of a strong tie. Finally, we considered when the two members of the dyad joined CS (*Mean join CS*). On average, Table 1 shows a positive and large effect, suggesting that a tie between two early members was weaker than a tie between two later members or of a tie between an early member and a later member.

Against this background, we tested our two hypotheses. First, we establish that sharing the experience of travelling produced greater bonds than just participating in the local activities of CS. Focusing on the effect of *Exchange tie*, the log-odds of describing the relationship as *Friends or Best Friends vs. Acquaintances* increases by .091, while the same effect for answering *Best Friends vs. Friends* or *Acquaintances* is (.091 + .474) = 0.565, or an odds-ratio of 1.76. That is, the effect of common experience on strength of ties is stronger for higher categories of the dependent variable. Both results support H1.

The effect of information on the strength of ties is more complex. In broad terms, we can say that while information increases the likelihood of moving a new friendship from the lowest category to the middle, it decreases the likelihood that the friendship moves further. Broadly speaking this provides evidence in support of H2. A closer look reveals however that the effect for low- to mid-level strength occurs only when information is available on both exchanging partners. When information exists on just one individual in the exchange, the effect is negative. However, this effect is not statistically significant. The narrative for the formation of strong ties is more straightforward: as predicted in H2, greater information decreases the chances that a strong tie will emerge.

6   Conclusion

We used a unique dataset and ethnographic fieldwork to capture how technology influences the strength of friendship ties. Our data span several years, thousands of users worldwide and millions of interactions. We parsed all of this to discover that the relatively exclusive cultural experience of travelling and discovering oneself through overnight stays with a stranger, created stronger relationships than just a common interest in travelling. At the same time, the rating system's accumulation of information about users took something away from the experience of travelling and meeting strangers. As a consequence, the binding force of the ties that developed later among CSers was on average lower than the ties that developed in a regime of greater uncertainty.

We see this process as one of progressive disenchantment. While finding a community suited to arcane cultural tastes is easier now than ever, the relationships that develop out of the shared experiences are becoming weaker. Rating systems are the key aspect of this process because they reduce the overall uncertainty present in the environment. To the extent to which online rating systems provide assurance structures for relationships they will supplant the need for interpersonal trust between partners and thus result in the formation in fewer deep ties.

These findings may apply to other Internet platforms, especially those of companies in the emerging "sharing economy." Despite the fact that many such platforms are commercial in nature, personalized interaction is arguably a touchstone value of the sharing economy. Our results suggest that inasmuch as personalization is concerned, sharing economy platforms may become victims of their own success. As these platforms mature they acquire more information about more users. But this very process makes the platforms' use less distinctive and more automatic resulting in more impersonal interactions. This paradoxical process is the result of the sectors' growing institutionalization. That we can observe its effects at the interpersonal level speaks to the enormous potential the online world has to change our understanding of society.

References

Bialski, P.: Intimate Tourism. Solilang, London (2009)

Bourdieu, P.: The Forms of Capital. In: Swedberg, R., Granovetter, M. (eds.) The Sociology of Economic Life, pp. 96–111. Westview Press, Boulder (1986)

Cheshire, C.: Online Trust, Trustworthiness, Ot Assurance? Daedalus 50, 49–58 (2011)

Coleman, J.S.: Social Structure and a Theory of Action. In: Blau, P. (ed.) Approaches to the Study of Social Structure, pp. 76–93. Free Press, New York (1975)

Cook, C., Hardin, R., Levi, M.: Cooperation Without Trust. Russell Sage Foundation, New York (2007)

DiMaggio, P.: Classification in Art. American Sociological Review 52, 440–455 (1987)

DiMaggio, P., Hargittai, E., Neuman, W.R., Robinson, J.P.: Social Implications of the Internet. Annual Review of Sociology 27, 307–336 (2001)

Erickson, B.: Culture, Class, and Connections. American Journal of Sociology 102, 217–252 (1996)

Feld, S.L.: The Focused Organization of Social Ties. American Journal of Sociology 86, 1015–1035 (1981)

Fiore, A., Cheshire, C.: The Role of Trust in Online Relationship Formation. In: Latusek, D., Gerbasi, A. (eds.) Trust and Technology in a Ubiquitous Modern Environment: Theoretical and Methodological Perspectives, pp. 55–70. IGI Global, PA (2010)

Kairam, S., Daniel, W., Jure, L.: The Life and Death of Online Groups. In: ADM International Conference on Web Search and Data Mining, WSDM (2012)

Kollock, P.: The Emergence of Exchange Structures. American Journal of Sociology 100(2), 313–345 (1994)

Lauterbach, D., Truong, H., Shah, T., Adamic, L.: Surfing a Web of Trust: Reputation and Reciprocity on CouchSurfing.com. In: International Conference on Computational Science and Engineering (CSE 2009), vol. 4, pp. 346–353 (2009)

Lizardo, O.: How Cultural Tastes Shape Personal Networks. American Sociological Review 71, 778–807 (2006)

Long, E.: Book Clubs: Women and the Uses of Reading in Everyday Life. Chicago University Press, Chicago (2003)

Marx, P.: You're Welcome. Couch-Surfing the Globe. The New Yorker (2012)

Moln, L.D., Takahashi, N., Peterson, G.: Risk and Trust in Social Exchange. The American Journal of Sociology 105(5), 1396–1427 (2000)

Molz, J.G., Gibson, S.: Mobilizing Hospitality. Ashgate Publishing Co., Burlington (2007)

Randall, C.: Theoretical Sociology. Harcourt, Brace, and Janovitch, San Francisco (1988)

Salganik, M.J., Dodds, P.S., Watts, D.J.: Experimental Study of Inequality and Unpredictability in an Artificial Cultural Market. Science 311, 854–856 (2006)

Simmel, G.: The Sociology of Georg Simmel. The Free Press, Glencoe (1950) edited by K.H. Wolff

Yamagishi, T., Cook, K.S., Watabe, M.: Uncertainty, Trust and Commitment Formation in the United States and Japan. American Journal of Sociology (104), 165–194 (1998)

Yamagishi, T., Yamagishi, M.: Trust and Commitment in the United States and Japan. Motivation and Emotion 18(2), 129–164 (1994)

Zigos, J.: CouchSurfing's Sex Secret, Business Insider (2013), `http://www.businessinsider.com/couchsurfing-the-best-hook-up-app-2013-12`

Appendix A: Proportional Odds Model Description

The standard ordered logistic model assumes a latent variable, Y*, that is connected to the observed dependent variable with three categories (Y) thorough a series of cut points (α):

$$Y_i^* = \alpha + x' \, \beta + \epsilon_i$$

$$\begin{aligned}
\text{and, } Y &= 1 \quad if \ Y^* < \alpha_1 \\
Y &= 2 \quad if \ \alpha_1 < Y^* \le \alpha_2 \\
Y &= 3 \quad if \ Y^* > \alpha_2
\end{aligned}$$

By assuming that the error terms are independently distributed and follow a logistic distribution, a proportional odds model can be defined as,

$$\Pr\left(Y \le y_j \big| x\right) = \frac{e^{(\alpha_j - x'\beta)}}{1 + e^{(\alpha_j - x'\beta)}} \tag{1}$$

and estimated using standard log-odds ratio and the logit link function. Proportional odds imply that the coefficients that describe the relationship between Pr (Y=1) and Pr (Y=2) are the same ones that describe Pr (Y=2) and Pr (Y=3). A strong benefit of such a model is that it produces a simple set of coefficients that can be easily interpreted. A major drawback is that the proportional odds assumption is a strong assumption that is seldom respected. When the proportional odds assumption does not hold, the model in [1] produces biased results. A compromise is to estimate a model where the proportional odds assumption is relaxed for the subset of coefficients that do not maintain it (t):

$$\Pr\left(Y \le y_j \big| x\right) = \frac{e^{(\alpha_j - x'\beta - t'\gamma)}}{1 + e^{(\alpha_j - x'\beta - t'\gamma)}} \tag{2}$$

In our analysis we start with a proportional odds model and then visually inspect the results by plotting the expected probabilities (analysis not shown). If the proportional odds assumption holds, the distance between the categories of the given covariate ought to remain the same across the levels of the dependent variable as specified in [1]. We performed the test by first normalizing all the coefficients to the lowest category of the dependent variable and use it as the reference category for the estimates across the other levels. The distance between the coefficients for the three covariates specified above in the text differs significantly so that, for example, the estimated probability of *Best Friend* differs for the two levels of *Exchange tie* (yes and no). As a result, we opted to selective relax the proportionality assumption as in [2].

Appendix B: Univariate Statistics

		Mean or proportion	std. dev.	N
Tie Strength:				
	Acquaintances	.72		1,564,853
	Friends	.249		541,106
	Best Friends	.03		66,007
Gender:				
	Female to female tie	.144		313,326
	Female to male tie	.503		1,091,738
	Male to male tie	.24		521,151
Mean age		24.44 years	6.29 years	
Age difference		6.01 years	6.25 years	
Difference join CS		13.88 months	12.58 months	
Report tie:				
	After a month	.392		850,355
	Within the same month	.608		1,321,611
Information:				
	No information	.063		136,851
	... about one partner	.063		137,594
	... about both partners	.874		1,897,521
Hospitality Exchange Tie:				
	No	.698		1,516,475
	Yes	.302		655,491
Month tie creation (since 1/2003)		81.43 month	16 month	
Friend in common		.136	.64	

Look into My Eyes & See, What You Mean to Me. Social Presence as Source for Social Capital

Katja Neureiter, Christiane Moser, and Manfred Tscheligi

ICT&S Center, University of Salzburg,
Sigmund-Haffner-Gasse 18, 5020 Salzburg, Austria
{katja.neureiter,christiane.moser,manfred.tscheligi}@sbg.ac.at
http://icts.sbg.ac.at

Abstract. Eye contact is presumed to be one of the most important non-verbal cues in human communication. It supports mutual understanding and builds the foundation for social interaction. In recent years, a variety of systems that support eye contact have been developed. However, research hardly focuses on investigating the impact of eye contact on social presence. In a study with 32 participants, we investigated the role of eye contact and gaze behavior with respect to social presence. Our results indicate that not only a system's capability to enable eye contact but also a user's consciously perceived eye contact are important to experience that the communication partner is 'there', i.e., social presence. Considering social presence as a source for social capital, i.e., valuable relationships that are characterized by trust and reciprocity, we discuss in what way social presence can serve as a contributing factor in video-mediated communication.

Keywords: Social Presence, Social Capital, Video-Mediated Communication.

1 Introduction

The variety of available communication systems (CS) have made it easier than ever to connect with people at almost any time and place. The potential benefits are numerous, ranging from easily exchanging information and sharing experiences to supporting a feeling of connectedness. Video-mediated communication (VMC) systems are presumed to be a 'rich' form of communication [5], allowing enhanced sensory stimulation. Head nods, smiles, or eye contact, for example, *'[...] give speakers and listeners information they can use to regulate, modify, and control exchanges'* [21, p.1125]. By providing a variety of non-verbal cues, they support a user's experience of social presence [2,32,33], the *'sense of being with another in a mediated environment'* [4, p.10]. VMC systems that allow for social presence can serve as a source for social capital [28], and therefore encourage a user to invest in social relationships.

One of the most important non-verbal cues in human communication is eye contact [1], as it is considered to build the foundation for social interaction

L.M. Aiello and D. McFarland (Eds.): SocInfo 2014, LNCS 8851, pp. 183–198, 2014.

[22]. Eye contact regulates, for example, information flow and provides insights into the relationship between the communication partners [20]. If eye contact is missing, people often do not experience full involvement in a conversation [1]. Thus, eye contact can play an important role in mediated communication, influencing the perception of a dialogue partner. Within the last years, gaze and eye contact have gained particular interest in research (e.g., [2,12,27]), but there are only a handful of studies that focus on the perception of eye contact in VMC (e.g., [7,18,35]).

The aim of this paper is to contribute to a better understanding in what way social presence can serve as a source for the development of *valuable* relationships, i.e., encourage a user to invest in social relationships (social capital). We carried out a user study in the lab where we investigated the role of non-verbal cues (eye contact, gestures) in VMC to better understand how these factors contribute to the experience of social presence. We applied an experimental study design with two conditions. The experimental condition (EC) allowed eye contact, the control condition (CC) did not. Participants (N=32) communicated twice via the given system, one time having eye contact and one time not. Our results provide insights on the role of eye contact with respect to social presence, which serve as a basis to discuss in what way the system holds potential for the development of valuable relationships.

2   Related Work

In order to gain a better understanding of the concept of social presence and it's contributing factors, the following section provides a brief overview on the concept and will point out studies that consider eye contact in VMC with respect to social presence. We will also give an overview on the theory of social capital, as it is relevant for the discussion of our findings and will describe in what way the concepts are interrelated with each other.

2.1   The Theory of Social Presence

In the late 1980s and early 1990s, researchers started studying the effects of computer mediated communication and came up with the concept of social presence to describe in what way VMC fosters interpersonal relationships. The concept raised substantial attention in the context of learning environments, focusing on how to support collaboration and interaction best (e.g., [34,31,16]).

First approaches for a definition of social presence originate from Short et al. [33], defining it as *'the degree of salience of the other person in a mediated communication and the consequent salience of their interpersonal interactions'* (p.65). Starting from the system's perspective, they conceptualized social presence as a medium's quality (e.g., being warm or cold, personal or impersonal).

Newer approaches focus on the individual perspective. Biocca and Harms [4], for example, define social presence as a *'sense of being with another in a mediated environment ... the moment-to-moment awareness of co-presence of a mediated body and the sense of accessibility of the other being's psychological, emotional, and intentional states'* (p.10). The authors describe three different levels of presence from a user perspective: the perceptual level, which is primarily the detection and awareness of the other's mediated body (i.e., co-presence), the subjective level, entailing the sense of the communication partner's emotional state or behavioral interaction, and the intersubjective level, which addresses reciprocal dynamics.

So far, a variety of contributing factors to social presence have been identified, ranging from a technology's ability to convey a variety of information (e.g., gaze, eye contact, or behavioral cues [3,26]) to user characteristics (e.g., user's perceptual or cognitive abilities [17]). Ijsselsteijn and colleagues [17] considered that content factors need to be taken into account when encompassing, for example, objects, actors, or even the context in which an activity takes place.

2.2   Eye Contact in VMC

The importance of non-verbal cues and, in particular, of eye contact in mediated communication has increasingly raised researchers' attention. Grayson and Monk [15], for example, investigated image size and camera position as influencing factors to discriminate mutual gaze. Chen [7] explored a user's sensitivity to eye contact, stressing that people are less sensitive towards eye contact when the user looks below the communication partner's eyes than when looking to the left or right side or above the eyes. Consequential, they suggest parameters for the design of video conferencing systems (e.g., maximum viewing distance). Eye contact has also been investigated in the context of video conferencing and has been identified to support collaboration between remote groups of people (e.g., [24,29,35]).

Unfortunately, there is no single definition of *'eye contact'* and the term is often used synonymously with mutual gaze, eye gaze or gaze awareness. Questions arising are whether eye contact is actually looking into someones eyes or simply gazing at someone's face, or whether it is a measurable variable or a subjective experience. Mc Nelley [24] reports, for example, that some people experience eye contact even when someone is looking somewhere in their face (e.g., chin or nose), preferring a so called more 'generalized eye contact', which might depend on ones personality or even cultural aspects.

Gale and Monk [11] note that gaze awareness is depending on the knowledge of what object in the environment someone is looking at. The authors consider the knowledge that someone is looking at you as mutual gaze or eye contact, *'...because it is only possible to know that someone is looking at your face if you are looking at theirs'* (p.585) and not somewhere in the environment or at another object. This definition refers to mutual awareness of communication partners during a conversation.

In the present study we use the term 'perceived eye contact' to describe our participants' experience of eye contact, even if the system did not support any eye contact (due to the vertical displacement of the cameras) and 'mutual gaze' to describe our participants' experience of eye contact when the system actually enabled eye contact.

2.3 Eye Contact and Social Presence in VMC

There is an increased interest in the relevance of eye contact in VMC and a trend of developing systems that support eye contact even in the context of communication with humanoid avatars (e.g., [2,12]). Nevertheless, only a handful studies consider investigating the role of eye contact with respect to social presence.

Bente et al. [2] investigated the effects of gaze on social presence in the context of an avatar-mediated communication. They found out that when gender-homogeneous female dyads communicate with each other, longer gaze duration positively correlates with higher levels of co-presence. However, this effect could not be reproduced with mixed-sex dyads.

Mukawa et al. [27] explored the impact of eye contact on users' behavior when making first contact via a VMC system. Their results indicate that participants, who communicated via the system supporting eye contact immediately had the awareness of visual connection. In contrast, participants, who used a system not allowing eye contact, needed confirmation through additional non-verbal cues (e.g., gestures like waving their hands) to make sure that their communication partner was aware of them. The authors did not explicitly asses social presence, but the awareness of the communication partner can be interpreted as indicator for a user's experience of (co-)presence [4].

2.4 Social Capital

Social presence has been identified as potential source for social capital, allowing the development of valuable relationships [28]. Social capital theory relates to resources that are inherent in the structure of social relationships [8]. It is *the aggregate of the actual or potential resources which are linked to the possession of a durable network of more or less institutionalized relationships of mutual acquaintance and recognition* [6, p.243]. According to Putnam [30], such relationships are characterized by norms of trustworthiness and reciprocity that arise from connections among individuals or social networks. Trustworthiness is the willingness to rely on a communication partner's actions. It is developing over time and is an important contributor for building up personal relationships [30]. Reciprocity characterizes the social interaction of giving and receiving [23], for example, as a response to a friendly action a person responds with a favor.

In the context of mediated communication, Garrison [13] points out that conditions such as trust or closeness need to be met in order to recognize collaboration (e.g., in a learning context) as valuable experience. The sense of closeness and trust can be achieved by establishing social presence, the experience that the communication partner is 'there'. It allows a user to immediately respond to a communication partner's action and can reduce perceived distance [36]. Moreover, enhanced sensory stimulation, i.e., non-verbal behavior (e.g., facing each other [25]) contributes to the awareness of the communication partner [27] and can positively influence the development of valuable relationships.

3   The Study

The study aims at investigating the role of eye contact and gaze behavior with respect to social presence in a mediated communication to better understand the role of eye contact in VMC and in what way such a system holds potential for the development of valuable relationships, i.e., is a source for social capital. Two central research questions were defined: 1) To what extent does participants' social presence differ when the system allows (no) eye contact? 2) How does participants' gaze behavior differ when they have (no) eye contact?

The system we used was developed within a research project that aims at facilitating enhanced communication and interaction among older adults. It consists of two screens and two cameras (see Figure 1a). The lower camera tracks facial expressions, the upper camera tracks gestures and postures. This arrangement allows an almost life-sized illustration of the user, which in turn enables enhanced sensory stimulation.

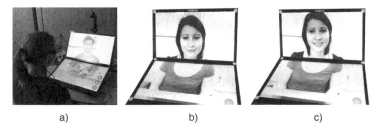

a)   b)   c)

Fig. 1. VMC System

However, the setup does not allow eye contact, due to the vertical displacement of the camera, mounted on the top of the screen, evoking the experience that the communication partner is looking at one's chest (see Figure 1b). In this setup eye contact can only be established for the communication partner when a user is directly looking into the lower camera, entailing a limited view on the communication partner.

In order to achieve eye contact in the experimental condition we added a third camera, installed at the confederate's device (see Figure 2a). It was positioned in the middle of the upper screen. Camera 2a and 2b could be activated individually and allowed either for eye contact (EC) (see Figure 1c and 2c) or not (CC) (see Figure 1b and 2b).

3.1   Study Design

A within design was chosen, to reduce error variance, associated with individual differences of the users. Participants were randomly assigned to the starting condition (CC, EC). Overall, 32 participants, aged between 22 and 66 years

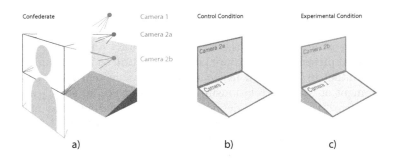

Fig. 2. Camera Positioning in the Experimental Setup

($M = 40.84, SD = 12.99$), took part in our study, 50% were male and 50% female. All had at least basic computer skills and were generally interested in new communication technologies. 81.3% regularly used VMC systems in daily life (e.g., Skype) and only a small amount of participants (18.8%) did not use any VMC systems. The recruitment of participants happened via email and telephone.

Methodological Approach. Social presence was assessed by means of a self-reporting questionnaire, based on the IPO-SPQ [9]. The original instrument consists of two parts: a semantic differential questionnaire and ten statements about system qualities, asking users to indicate to what extent they agree to the given statements (7 = totally agree, 6 = agree, 5 = rather agree, 4 = neither/nor, 3= rather disagree, 2 = disagree, 1= totally disagree). For the purpose of the study we only used the subjective attitude statements, allowing to gain more detailed information about participants' experience of social presence (e.g., 'It provides a great sense of realism.', 'It was just as though we were all in the same room.') (a full list of all statements on the system qualities are available in [9]).

To explore the determinants of gaze behavior, participants' eye movements were recorded using SMI eye tracking glasses[1]. For the analysis, four areas of interest (AOIs) were defined: 1) the head 2) the eyes 3) the lower screen, displaying gestures, and 4) the lower camera. Three eye tracking metrics were defined as dependent variables to quantify the attention on an AOI: dwell time (%) (summary of time spent within an AOI), fixation count (total number of fixations), and the fixation time (ms) (sum of fixation duration), indicators of visual attention [14]. Additionally, to the objective measures of social presence and gaze behavior we carried out an interview at the end of the study to gather detailed information about participants' experiences, e.g., with respect to social presence. Moreover, participants were asked to indicate if they were aware of any differences between the two conditions, and if yes, what the difference was.

---

[1] http://www.eyetracking-glasses.com

Procedure. When participants arrived, they were introduced to the confederate (to avoid that they felt uncomfortable when they make first contact via the device) and the test leader gave background information about the general procedure of the study and the tasks, they were asked to carry out, i.e., playing two rounds of a simple quiz.

Participants took place in front of the VMC system and the eye tracking glasses were calibrated. Before starting with the first round of the quiz a test-call was carried out, which was also intended as 'breaking the ice', to provide the participants first insights how the quiz will be played, and to avoid biases on participants' gaze behavior (as they did not know how the confederate will appear to them on the screen). This test-call also provided proof to the study leader that the eye tracking system was working properly.

Afterwards, participants started playing the quiz (round 1), either in the experimental condition or in the control condition. The confederate explained terms (e.g., job), to the participant, who was asked to guess. The terms were prepared beforehand by the study leader and the confederate always explained the same terms in the same order. If a participant guessed right, the confederate explained the next term. After 4 minutes, the quiz was terminated by the confederate and participants were asked to complete the questionnaire (IPO-SPQ), indicating to what extent they agreed to the given statements on social presence. Moreover, they were asked about their experiences when playing the quiz, i.e., how they 'perceived' their communication partner. After the second round they were additionally asked if they had experienced any difference in comparison to the first round.

At the end of the study, a brief interview was carried out, asking questions about participants' first impression when communicating via the device, their experience of social presence, if eye contact was important to them, and if they had experienced eye contact during both rounds playing the quiz.

To ensure that the confederate always behaved the same way (independently from the condition) he was trained how to explain the terms, was instructed to use similar gestures when explaining the terms and to look into the camera, which was placed at the middle of the upper screen (see Figure 2a, Camera 2b).

3.2   Data Analysis

The quantitative data (social presence questionnaires, eye tracking data) was analyzed, using SPSS Version 21. Both descriptive and interference statistical analyses (t-tests) were carried out. To test the reliability of the test scores (IPO-SPQ), Cronbach's Alpha was calculated to estimate the internal consistency, revealing a value of .90, indicating a good internal consistency. For the analysis of the eye tracking data it was required to map fixations of the individual video recordings of the participants on a defined reference image that represented the visual target area. Moreover, the defined AOIs were marked on the reference image. The 'mapped data' was exported and calculated using SPSS. Descriptive analyses were carried out on the dependent variables dwell time, fixation count, and fixation time.

4   Results

Data analysis revealed several interesting findings regarding the role of eye contact and gaze behavior with respect to social presence.

4.1   Participants' Experience of Social Presence

The analyses of the IPO-SPQ show high ratings on social presence in both conditions. In the experimental condition social presence was rated slightly higher ($M = 5.72, SD = 0.99$) than in the control condition ($M = 5.47, SD = 1.02$) (see Figure 3) indicating that participants in the experimental condition said that they *agreed* to the given statements, whereas participants in the control condition *rather agreed*. We could neither identify a significant correlation between age and social presence in the control condition ($r = 0.06, p \geq .05$) nor in the experimental condition ($r = 0.27, p \geq .05$). The same applies for gender and social presence (CC: $r = .01, p \geq .05$; EC: $r = .16, p \geq .05$). Age and gender were not associated with social presence.

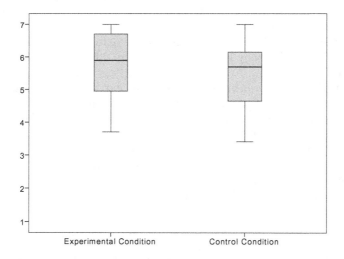

Fig. 3. Social Presence (N=32)

To investigate differences in social presence with respect to the two conditions we carried out a t-test, which revealed a small difference between the means, but no significance ($t = 1.84, p = .076$). However, with a relatively low probability of 7.6%, the mean difference occurred by chance (see Table 1). This finding encouraged us to further investigate differences in social presence with respect to the two conditions and in dependence to an other interesting finding we made during our pre-tests: even if the system did not allow eye contact (CC), some of our participants indicated that they perceived eye contact and that they were not aware of any difference between the two conditions when communicating via the given system. So we had a deeper look at our data, investigating to what extent social presence differed among those participants, who were aware that there was a difference between the two conditions and experienced mutual gaze within the experimental condition and lacked eye contact in the control condition.

Table 1. Paired Samples Test: Social Presence

Pair	Mean Difference	SD	t	df	Sig. (2-tailed)
SP EC/CC	-.25	.78	1.84	31	.076*

4.2 (Consciously Perceived) Difference, Eye Contact, and Social Presence

Slightly more than half of our participants (56.3%) consciously perceived a difference between the two conditions, indicating that they experienced mutual gaze in the experimental condition and lacked eye contact in the control condition. Almost half of the participants (43.8%) indicated that they did not recognize any difference. Consciously perceived difference cannot be equated with perceived eye contact, as the majority of participants who did not recognize any difference between the two conditions (85.7%), indicated that they perceived eye contact in both conditions. A possible explanation for this finding is that some people prefer a 'generalized eye contact', simply gazing towards someone's face but not explicitly into someone's eyes [24].

The group of participants, who recognized a difference between the two conditions encompassed 56.3% ($N = 18$). The age span ranged from 22 and 61 years ($M = 39.72, SD = 12.27$), 50% male and 50% female. The descriptive analysis of the IPO-SPQ shows that social presence differed between the two conditions. Participants in the experimental condition indicated more agreement to the given statements than participants in the control condition (see Figure 4). No effects were identified with respect to age (CC: $r = -.18, p \geq .05$; EC: $r = .09, p \geq .05$) and gender (CC: $r = .02, p \geq .05$; EC: $r = .21, p \geq .05$). T-tests reveal that the mean difference for social presence is 0.58, meaning that participants in the experimental condition rated on average social presence at 0.58 points higher than participants in the control condition. This difference is significant ($t = 3.02, p = .008$) and did not occur by chance (see Table 2). Participants of our study, who consciously perceived a difference in terms of eye contact between the two conditions, indicated to experience more social presence in the experimental condition than in the control condition where they lacked eye contact.

Almost half of the participants 43.8% ($N = 14$) were not aware of a difference between the two conditions. The age span ranged from 23 to 66 years ($M = 42.29, SD = 14.20$). Again no effects were identified with respect to age (CC: $r = -.13 p \geq .05$; EC: $r = .23, p \geq .05$) and gender (CC: $r = -.03, p \geq .05$;

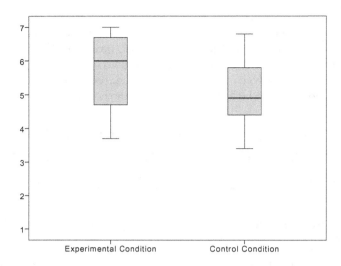

Fig. 4. Social Presence when consciously perceiving a difference (N=18)

Table 2. Paired Samples Test: Social presence when consciously perceiving a difference

Pair	Mean Difference	SD	t	df	Sig. (2-tailed)
SP EC/CC	-.58	.82	3.02	17	.008*

EC: $r = .29, p \geq .05$). The difference between the subjective experience of social presence in the control and experimental condition was small, negative (i.e., participants in the control condition experienced more social presence) and not significant ($t = -1.34, p = .204$). Thus, we assume that the difference occurred by chance and that participants' social presence does not differ with respect to the two conditions (see Table 3). However, it has to be considered, that most of the participants, who did not recognized any difference between the two groups indicated that they perceived eye contact in both conditions (85.7%). This indicates that with respect to social presence, not the system's capabilities to convey eye contact might have been important but participants' perceived eye contact (even if it was not possible to actually establish eye contact).

Table 3. Paired Samples Test: Social presence when not consciously perceiving a difference (N=14)

Pair	Mean Difference	SD	t	df	Sig. (2-tailed)
SP EC/CC	-.17	.48	-1.34	13	.204

4.3   Gaze Behavior and Social Presence

With respect to our second research question, to what extent participants' gaze behavior differs when they have (no) eye contact, we analyzed their gaze behavior, considering four different AOIs: 1) head, 2) eyes, 3) gestures, and 4) the camera. Three eye-tracking metrics were defined as dependent variables: dwell time (%), fixation count and the fixation time (ms). Unfortunately, eye tracking did not work out for all participants, which can be explained by various factors. Jacob et al. [19] report about a considerable minority of participants, who cannot be tracked reliably (about 10-20%). In our study, we could gather eye tracking data from 22 out of 32 participants (68.75%). Participants of this group were aged between 22 and 66 years ($M = 42.68, SD = 11.97$), 54.5% were male and 45.5% female.

Table 4. Descriptives for eye tracking metrics

AOI	Measures	Experimental Condition Mean (SD)	Control Condition Mean (SD)
Head	Dwell Time	78.27 (22.33)	79.14 (17.26)
	Fixation Time	124.27 (401.92)	127.17 (255.06)
	Fixation Count	463.00 (153.97)	466.05 (125.29)
Eyes	Dwell Time	17.69 (13.80)	17.39 (16.55)
	Fixation Time	33.67 (25.34)	32.73 (29.28)
	Fixation Count	140.27 (102.01)	132.45 (120.54)
Gestures	Dwell Time	1.88 (2.45)	2.91 (3.66)
	Fixation Time	3.50 (4.48)	5.10 (6.50)
	Fixation Count	20.96 (24.13)	30.91 (38.26)
Camera	Dwell Time	.04 (.11)	.04 (.07)
	Fixation Time	.11 (2.76)	.09 (.14)
	Fixation Count	.73 (1.78)	.68 (1.09)

Descriptive data show that participants attention with respect to the defined AOIs head, eye, gestures, and camera did not differ between the two conditions (see Table 4). Moreover, fixation count and time, indicators for user's interest in a certain AOI were similar in both conditions. With respect to the AOI gestures we could identify differences, indicating that participants tended to pay more attention on gestures when they had no eye contact. It has to be considered that we identified a high standard deviation, explaining a high variation of the data from the average. T-tests revealed a significant difference with respect to the fixation time, indicating that participants' attention on gestures differed. In the control condition, participants paid significantly more attention on gestures than in the experimental condition (see Table 5). This effect could also be identified among participants, who consciously perceived a difference between the two conditions (see Table 6). It indicates that participants' attention and interest on gestures differed with respect to the two conditions, meaning that participants paid significantly more attention on gestures when they lacked eye contact.

Table 5. Paired Samples Test: Gaze Behavior in the control and experimental condition

Pair	Mean Difference	SD	t	df	Sig. (2-tailed)
Fixation Count EC/CC	-3.05	151.19	-.09	21	.926
Dwell Time EC/CC	-.86	22.19	-.18	21	.857
Fixation Time EC/CC	127170.84	35507.49	16.80	21	.000*

Table 6. Paired Samples Test: Gaze Behavior in the control and experimental condition with respect to consciously perceived eye contact

Pair	Mean Difference	SD	t	df	Sig. (2-tailed)
Fixation Count EC/CC	44.46	42.12	3.81	12	.003*
Dwell Time EC/CC	3.64	7.44	1.76	12	.103
Fixation Time EC/CC	121851.59	39892.52	11.01	12	.000*

5   Discussion

In this paper, we present results of a study that aimed at investigating the role of eye contact and gaze behavior with respect to social presence. The study reveals three central findings we will now briefly discuss and reflect on with respect to social capital.

First, we identified a tendency that participants' social presence differed with respect to the two conditions. This would mean that the system's capabilities to convey eye contact had a positive effect on social presence, i.e., the experience that the communication partner is close. If the system allowed mutual gaze (EC), participants rather experienced social presence than when the system did not support mutual gaze. We would like to emphasize that this finding was *not* significant and has to be considered as tendency. Further investigations are required to verify this result.

Second, we found out that social presence significantly differed in the two conditions among participants, who consciously perceived a difference between the two conditions with respect to eye contact. This means, that participants, who were aware of a difference, experienced more social presence when they perceived mutual gaze than if they lacked eye contact in the control condition. We argue that the system's capability to convey eye contact was important with respect to social presence when participants actually were aware of the difference between the two conditions. For the group of participants, who did not recognize any difference between the two conditions we could not identify any differences with respect to social presence. Based on this finding we assume that not the system's capabilities to convey eye contact had an effect on their social presence during the mediated communication, but their *experience* that they had eye contact. As this group did not recognize any difference we assume that they preferred a 'generalized eye contact' and thus were not aware of the difference between the two conditions.

Third, participants tended to pay more attention on gestures of their communication partner in the control condition than in the experimental condition, indicating that they paid more attention on additional non-verbal cues (gestures) when the system did not allow for mutual gaze than if the system allowed mutual gaze.

We would like to point out that the results of this research need to be considered in light of some limitations. First, with respect to our eye tracking data, the quiz itself might have had an influence on participant's gaze behavior. Social attention, in our case, the attention on the confederate during the conversation, depends, for example, on whether a person is listening or talking to somebody else. Persons tend to look, for example, more often at their communication partner when being asked a questions than when answering a question [10]. Moreover, gaze is avoided when thinking in order to reduce the cognitive load. As we chose a quiz, this of course might have influenced participants' gaze behavior. The idea behind choosing the quiz was to trigger an informal atmosphere to avoid that participants feel uncomfortable when talking with a stranger. Second, our eye tracking data revealed a quite high standard deviation, which means that there were high variances among participants with respect to their gaze behavior. The identified effect needs to be verified. As a future work, we would like to further investigate participants' gaze behavior in VMC with respect to social presence.

5.1 Reflections on the Interrelation between Social Presence and Social Capital

The aim of this paper is to contribute to a better understanding in what way the present VMC system holds potential for social capital, the development of valuable relationships. By investigating the role of eye contact and gaze behavior with respect to social presence, we discuss in what way the system can support the development of relationships, that are characterized by norms of trustworthiness and reciprocity [30].

We consider two facets of social presence as relevant to better understand in what way systems that support social presence hold potential for social capital: co-presence and mutual understanding (see [28]).

Co-presence, the experience that the communication partner is 'there', allows a user to immediately respond to a communication partner's action, which is one aspect of reciprocity. In VMC, mutual understanding is mainly supplied by the variety of non-verbal cues a system provides [27]. For example, eye contact or gestures allow users to gather feedback on their communication partner's reactions, which is an important precondition for the development of trustworthy relationships.

Eye contact is considered as one of the most important cues in human communication, being proximate to face-to-face communication. With respect to social presence, our study revealed a quite diverse picture in terms of the role of eye contact respectively mutual gaze when communicating via the given system. Not only the system's capability to allow mutual gaze was important but also a user's perceived eye contact. Among those participants, who consciously perceived a difference between the two conditions, social presence was higher when they experienced mutual gaze than if they did not. Based on this, we assume that eye contact is an important cue with respect to social presence. Consequently, we can assume that mutual gaze facilitates social dynamics such as reciprocity and supports mutual understanding in mediated communication, important aspects of social capital.

Nevertheless, it needs to be considered that participants, who did not consciously recognize any difference did not indicate any differences with respect to their experience of social presence. The majority indicated that they perceived eye contact in both conditions and showed high ratings on social presence, meaning that not the systems's capabilities allowing for mutual gaze were important but participant's perceived eye contact. This in turn would mean that although eye contact, i.e., mutual gaze is presumed having an important role in mediated communication, influencing the perception of the dialogue partner, the present VMC system holds potential for the development of social capital, even if it does not support mutual gaze.

6   Conclusion

The goal of this paper is to better understand the role of eye contact with respect to social presence in VMC. We carried out a user study in the lab, applying an experimental study design with two conditions. The experimental condition allowed eye contact; the control condition did not. Our results indicate that a user's *perceived* eye contact was important to experience social presence and that even if mutual gaze was not supported by the system participants experienced eye contact and social presence during the mediated communication with the given system. Considering social presence, especially in terms of co-presence and mutual understanding as source for social capital we assume that the system supports the development of relationships that are characterized by reciprocity and trustworthiness and holds potential for social capital.

Acknowledgements. This research was enabled by the ConnectedVitality project and was supported by PresenceDisplays who provided the system. The financial support by the AAL JP and the support of all involved parties in the project is gratefully acknowledged.

References

1. Argyle, M., Dean, J.: Eye-contact, distance and affiliation. Sociometry, pp. 289–304 (1965)
2. Bente, G., Eschenburg, F., Aelker, L.: Effects of simulated gaze on social presence, person perception and personality attribution in avatar-mediated communication. In: Proc. PRESENCE 2007, pp. 207–14 (2007)
3. Biocca, F., Harms, C., Gregg, J.: The networked minds measure of social presence: Pilot test of the factor structure and concurrent validity. In: International Workshop on Presence (2001)

4. Biocca, F., Harms, C.: Defining and measuring social presence: Contribution to the networked minds theory and measure. In: Proc. PRESENCE 2002, pp. 1–36 (2002)

5. Bohannon, L.S., Herbert, A.M., Pelz, J.B., Rantanen, E.M.: Eye Contact and Video-Mediated Communication: A Review. Displays (2012)

6. Bourdieu, P.: The forms of capital. Handbook of theory and research for the sociology and education. Greenwood, New York (1986) Richardson, J.G

7. Chen, M.: Leveraging the asymmetric sensitivity of eye contact for videoconference. In: Proc. CHI 2002, pp. 49–56 (2002)

8. Coleman, J.S.: Social capital in the creation of human capital. American Journal of Sociology, S95–S120 (1988)

9. De Greef, P., Ijsselsteijn, W.A.: Social presence in a home tele-application. Cyber Psychology & Behavior 4(2), 307–315 (2001)

10. Freeth, M., Foulsham, T., Kingstone, A.: What affects social attention? Social presence, eye contact and autistic traits. PloS One 8(1) (2013)

11. Gale, C., Monk, A.F.: Where am i looking? the accuracy of video-mediated gaze awareness. Perception & Psychophysics 62(3), 586–595 (2000)

12. Garau, M., Slater, M., Bee, S., Sasse, M.A.: The impact of eye gaze on communication using humanoid avatars. In: Proc. CHI 2001, pp. 309–316 (2001), http://doi.acm.org/10.1145/365024.365121

13. Garrison, D.R.: Online collaboration principles. Journal of Asynchronous Learning Networks 10(1), 25–34 (2006)

14. Goldberg, J.H., Stimson, M.J., Lewenstein, M., Scott, N., Wichansky, A.M.: Eye tracking in web search tasks: design implications. In: Proc. ETRA 2002, pp. 51–58 (2002)

15. Grayson, D.M., Monk, A.F.: Are you looking at me? eye contact and desktop video conferencing. Trans. Comput.-Hum. Interact. 10(3), 221–243 (2003), http://doi.acm.org/10.1145/937549.937552

16. Gunawardena, C.N.: Social presence theory and implications for interaction and collaborative learning in computer conferences. International Journal of Educational Telecommunications 1(2), 147–166 (1995)

17. IJsselsteijn, W.A., de Ridder, H., Freeman, J., Avons, S.E.: Presence: concept, determinants, and measurement. In: Electronic Imaging, pp. 520–529. International Society for Optics and Photonics (2000)

18. Ishii, H., Kobayashi, M.: Clearboard: A seamless medium for shared drawing and conversation with eye contact. In: Proc. CHI 1992, pp. 525–532. ACM (1992)

19. Jacob, R.J., Karn, K.S.: Eye tracking in human-computer interaction and usability research: Ready to deliver the promises. Mind 2(3), 4 (2003)

20. Kendon, A.: Some functions of gaze-direction in social interaction. Acta Psychologica 26, 22–63 (1967)

21. Kiesler, S., Siegel, J., McGuire, T.W.: Social psychological aspects of computer-mediated communication. American Psychologist 39(10), 1123 (1984)

22. Kleinke, C.L.: Gaze and eye contact: a research review. Psychological Bulletin 100(1), 78 (1986)

23. Mauss, M.: The gift: forms and functions of exchange in archaic societies. Cohen & West, London (1966), trans. Cunnison

24. McNelley, S.: Immersive group telepresence and the perception of eye contact (2005)

25. Mehbabian, A.: Orientation behaviors and nonverbal attitude communication1. Journal of Communication 17(4), 324–332 (1967)

26. Mukai, S., Murayama, D., Kimura, K., Hosaka, T., Hamamoto, T., Shibuhisa, N., Tanaka, S., Sato, S., Saito, S.: Arbitrary view generation for eye-contact communication using projective transformations. In: Proc. VRCAI 2009, pp. 305–306 (2009)
27. Mukawa, N., Oka, T., Arai, K., Yuasa, M.: What is connected by mutual gaze?: user's behavior in video-mediated communication. In: CHI EA 2005, pp. 1677–1680 (2005)
28. Neureiter, K., Moser, C., Tscheligi, M.: Social Presence as influencing Factor for Social Capital. In: Challenging Presence Proceedings of the International Society for Presence Research, pp. 57–64 (2014)
29. Nguyen, D., Canny, J.: Multiview: Spatially faithful group video conferencing. In: Proc. CHI 2005, pp. 799–808 (2005), http://doi.acm.org/10.1145/1054972.1055084
30. Putnam, R.D.: Bowling alone: The collapse and revival of American democracy. Simon and Schuster Nova York (2000)
31. Richardson, J.C., Swan, K.: Examing social presence in online courses in relation to students' perceived learning and satisfaction. JALN 7(1), 68–88 (2003)
32. Rüggenberg, S., Bente, G., Krämer, N.: Virtual encounters. creating social presence in net-based collaborations. In: Proc. Int. Workshop on Presence (2005)
33. Short, J.W., Christie, B.: The social psychology of telecommunications. John Wiley and Sons (1976)
34. Tu, C.-H., McIsaac, M.: The relationship of social presence and interaction in online classes. The American Journal of Distance Education 16(3), 131–150 (2002)
35. Vertegaal, R., Weevers, I., Sohn, C., Cheung, C.: Gaze-2: conveying eye contact in group video conferencing using eye-controlled camera direction. In: Proc. CHI 2003, pp. 521–528 (2003)
36. Woods Jr, R.H., Baker, J.D.: Interaction and immediacy in online learning. International Review of Research in Open & Distance Learning 5(2) (2004)

From "I Love You Babe" to "Leave Me Alone" - Romantic Relationship Breakups on Twitter

Venkata Rama Kiran Garimella[1], Ingmar Weber[2], and Sonya Dal Cin[3]

[1] Aalto University
kiran.garimella@aalto.fi
[2] Qatar Computing Research Institute
iweber@qf.org.qa
[3] University of Michigan
sdalcin@umich.edu

Abstract. We use public data from Twitter to study the breakups of the romantic relationships of 661 couples. Couples are identified through profile references such as @user1 writing "@user2 is the best boyfriend ever!!". Using this data set we find evidence for a number of existing hypotheses describing psychological processes including (i) pre-relationship closeness being indicative of post-relationship closeness, (ii) "stonewalling", i.e., ignoring messages by a partner, being indicative of a pending breakup, and (iii) post-breakup depression. We also observe a previously undocumented phenomenon of "batch un-friending and being un-friended" where users who break up experience sudden drops of 15-20 followers and friends.

Our work shows that public Twitter data can be used to gain new insights into psychological processes surrounding relationship dissolutions, something that most people go through at least once in their lifetime.

Keywords: relationships, breakups, Twitter.

1 Introduction

The breakup of a romantic relationship is one of the most distressing experiences one can go through in life. It is estimated that more than 85% of Americans [2] go through this process at least once in their life time. Correspondingly, lots of research in psychology and other fields has investigated relationship breakups, looking at dimensions such as the impact of breakups on mental health [33], post-breakup personal growth [39], or the increased use of technology for the actual act of breaking up [42].

Through the advent of social media, it is possible to publicly declare one's relationship either using a dedicated functionality as provided by Facebook's "relationship status" or, as in the case of Twitter, stating a relation in one's public profile. For example, @user1 on Twitter might write "@user2 is the best boyfriend ever!!". In fact, updating one's social network information to mention a new partner has become almost synonymous with the beginning of a committed relationship, leading to the expression "Facebook official" [29].

L.M. Aiello and D. McFarland (Eds.): SocInfo 2014, LNCS 8851, pp. 199–215, 2014.
© Springer International Publishing Switzerland 2014

Given the scale and richness of data available on these social networks, they have proven a treasure trove for studying relationships and relationship breakups. Most of the existing work here that has not relied on small-scale survey data has used proprietary data such as coming from Facebook [1] or online dating sites [9,43]. In this work, we show that it is possible to study relationship breakups using public data from Twitter. Concretely, we analyze data for couples where at least one partner in their profile mentions the other one at the beginning of our study period (Nov. 4, 2013). We then periodically look for removals of this profile mention before Apr. 27, 2014 and take this as indication of a breakup, which we validate using CrowdFlower.

We use this data to address research questions related to (i) finding indicators of an imminent breakup in the form of changes in communication patterns, (ii) the connection between pre- and post-breakup closeness, (iii) evidence for post-breakup depression and its dependence on being either the rejector/rejectee, and (iv) the connection between "stonewalling" and relationship breakups.

Using Twitter data or other public social network data to address such questions comes with a number of advantages, including (i) ease of data collection, (ii) size of data, (iii) less self-reporting bias, (iv) timely collection around the moment of break-up, and (v) having social context in the form of network information. However, using this type of data also comes with a number of drawbacks including (i) noise of data, (ii) lack of well-defined variables, (iii) difficulties in observing psychological variables, (iv) limited power to determine *causal* links, and (v) privacy concerns. We discuss more of the limitations and challenges of our study in Section 5.

Our findings include:

- Using crowdsourcing we validate that it is possible to identify a large set of relationship breakups on Twitter.
- We observe changes in communication patterns as the breakup approaches, such as a decrease in the fraction of messages to the partner, and an increase in the fraction of messages to other users.
- We observe batch un-friending and being un-friended as indicated by the sudden loss of both 15-20 Twitter friends[1] and followers.
- We confirm that couples who breakup tend to be "fresher" when compared to couples that do not breakup.
- We observe an increased usage of "depressed" terms after the breakup compared to couples that do not breakup.
- We find a higher level of depressed term usage for likely "rejectees" compared to "rejectors", both before and after the breakup.
- Communication asymmetries, related to one-sided "stonewalling", are more likely for couples who will breakup.
- There are higher levels of post-breakup communication for couples who had higher pre-breakup levels of interaction.

---

[1] We use the term "friend" as Twitter terminology referring to another Twitter user that a user follows.

2   Data Collection

Twitter is an online social networking and micro blogging platform. It is one of the biggest social networks with around 270 million active users.[2]. Each Twitter user has a profile, also called bio, where they can describe themself. The content of this free text field is referred to as *profile description* in this study.

Terminology – Messages vs. Mentions: We define that a Twitter user @user1 has sent a *message* to @user2 if a tweet by @user1 starts with '@user2'. An example message from @user1 to @user2 could be: "@user2 can I come over to your place?". These *public* messages are not to be confused with direct messages, which can only be sent to followers, are always private and can not be accessed via the Twitter API. A user @user1 is said to *mention* @user2 in a tweet if @user2 occurs anywhere in the tweet. An example could be "I love @user2 soo much!".[3] Note that all messages as well as all retweets are special kinds of mentions.

Our data collection starts with a 28 hour snapshot of Twitter containing about 80% of all public tweets in late July 2013 (provided by GNIP[4] as part of a free trial). Each tweet in this data set comes with meta data that includes the user's profile description at the time the tweet was created. We searched this meta data for profiles of users containing mentions of other users and along with terms such as "boyfriend", "girlfriend", "love", "bf", "gf", or "taken". For example, the user profile of @user1 containing "I am taken by @user2, the love of my life" would be considered because it mentions another user "@user2" and contains the word "love" (as well as "taken"). We removed profiles mentioning other accounts of the same person such as on Instagram, Facebook, Vine, etc. by looking at simple word matches like 'ig', 'instagram', 'vine', 'fb', etc, or if the user being mentioned is the same as the actual user. e.g. Profiles like "I love football. Follow me on instagram: @user2" would be removed. We also had a few thousands of profiles mentioning popular celebrities, especially @justinbieber and @katyperry. Many of these seemed to indicate one-sided, para-social relationships where people claimed @justinbieber as their "boyfriend" or their "love".

In the end, we had 78,846 users (39,423 pairs) with at least one of the users in the pair mentioning the other in their profile, tentatively indicating a romantic relationship. We tracked these ~80k users starting from Nov. 4, 2013, till the end of Apr. 2014 (24 weeks). We obtained weekly snapshots of the tweets, user profiles (containing the profile description, their self-declared location, time zone, name, the number of followers/friends/tweets, etc.) and their mutual friendship relations (Does @user1 follow @user2? Does @user2 follow @user1?). Note that even though we started with a set of ~80k users, some of them deleted their accounts over the course of the study, some of them are private or made them private during the ~6 months of data collection. So by the end of the data collection, we were left with 73,868 users.

---

[2] https://about.twitter.com/company
[3] https://support.twitter.com/articles/14023-what-are-replies-and-mentions
[4] http://gnip.com/

For the current study, we limited ourselves to English-speaking countries to avoid cultural differences and difficulties in analyzing different languages, e.g., with respect to sentiment. Hence, we only kept users who had their profile location set to US, Canada or UK, identified using the profile timezone and their profile language set to English. This left us with 6,737 couples (13,474 users).

As our simplistic approach of identifying tentative romantic relationships gave some false positives, such as "Host for @SacrificeSLife All things video games. I love comic books...", we used Crowdflower[5] (an online crowd sourcing platform) to clean our data. Concretely, we asked three human judges to manually label if two users were involved in a romantic relationship in Nov. 2013, and again in Apr. 2014 by looking at the pair of Twitter profile descriptions at the relevant time. So each user couple was labeled for two snapshots in time.

The human judges had to decide on a simple "Yes/No" answer, indicating a romantic relationship or not for that snapshot. Since this is a potentially subjective task, the judges were asked to answer "No" unless it is very clear that the pair are in a romantic relationship. To ensure additional quality, we only used results where all the three human judges agreed on a label. All three judges agreed on the same label in 66% of the cases.

From this labeling, we can infer if a couple who were in a romantic relationship at the start of the study (Nov. 04, 2013), were still in a relationship at the end of the study (27 Apr. 2014). If they were in a relationship in Nov. and not in Apr., we assume that the couple broke up sometime during this period.

We also used Crowdflower to manually label the gender of the users given the name, profile description and profile picture, again using three human judges for each task. The judges had to pick one of "Male/Female/Cannot say" about the gender of the Twitter profile. Similar to the above task, we made sure that the labels were of good quality and picked only those users for whom all the judges agreed on a gender (80% inter-judge agreement). We also ignored the users who were labeled "Cannot say".

In the end, we obtained 1,250 pairs of users highly likely to have undergone a relationship breakup, as well as 2,301 pairs of users who were in a genuine romantic relationship but did not breakup. Para-social relationships with celebrities, mentioned above, were filtered out during this step as the *pair* was not labeled to be in a romantic relationship to begin with.

We decided to remove couples likely to be married, using a simple keyword search for "married", "wife", "husband", etc., as these groups have been observed to follow different relationship dynamics compared to casual dating relationships [6]. There were also a small fraction of same-sex couples which were removed as, again, they are likely to follow other dynamics [13]. This left us with a set of 661 pairs (1,322 users) which we refer to as BR. As a reference set, we also randomly sampled a set of 661 pairs of users who we knew were in a romantic relationship, but did not breakup over the course of our study. We refer to this set as NBR.

For the BR user pairs, we looked at their weekly profile description snapshots and identified the week when at least one user removed the mention of another

---

[5] http://www.crowdflower.com

user in their profile. We define this to be the week the two users broke up. Fig. 1a shows the distribution of breakups in our data over time. Though there is some temporal variation we did not break down the data further into, say, pre- and post-Christmas breakups. Still, to avoid temporal-specific peculiarities we also paired the 661 BR pairs with the 661 NBR pairs concerning the week of the breakup. This way we when we refer to "one week before the breakup" for a particular couple in our analysis, we use the very *same* week for the randomly paired NBR pair.

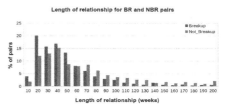

(a) Number of breakups in our data set over time

(b) Length of relationships (in weeks) for BR and NBR pairs

Fig. 1.

3 Results

3.1 Length of Relationship

It is known from literature that with an increase in relationship duration the breakup probability decreases [23]. This is consistent with observations from our data. Concretely, to estimate the length of the relationship between @user1 and @user2, we looked at the oldest tweet in our data set where one mentioned the other. We could do this as, even though we only started our study in Nov. 2013, we collected (up to) 3,200 historic tweets for each user in our study at that point. For the vast majority (80%) this covered all their tweets.

Using this occurrence of the first mention in a tweet as proxy for relationship duration, we find that the average relationship length for BR pairs in Nov. 2013 was 35 weeks whereas it was 60 weeks for NBR pairs. Figure 1b shows a histogram of the estimated relationship duration at the beginning of our study period.

3.2 Post-breakup Changes in Profile Description

The removal by one user of the mention in their profile description of the other user is, as described before, our definition of a breakup. However, we were interested in which *other* changes of the profile description would coincide with a breakup. To study this, we looked at the profile descriptions of BR users (a) at the start of the data collection - 04 Nov 2013, and (b) the week after their breakup. We then generated word clouds for these two sets of profiles. Figures 2a, 2b show the profiles before/after. (We removed the very frequent words "love" and "follow" from both before and after as they were at least 75% more frequent than the next most frequent word before breakup and hence distorting the distribution.)

There are several clearly visible differences and these reconfirm that our data set really does contain actual relationship breakups. For example, the terms "taken" or "baby" both lose in relative importance compared to the other terms. Note, however, that terms such as "taken" do not disappear completely which relates to a temporal difference in when the two partners update their profiles. See Section 3.8 for details.

To rule out the influence of background temporal changes due to, say, Christmas or Valentine's Day we also generated similar word clouds using users from NBR. For this set, we could not observe any differences over time and the figures, very similar to the "before" cloud, are omitted here.

Though the relative loss of "taken" is expected, we were also interested in which terms would *gain* in relative importance and, in a sense, come to replace the former reference to the partner. To quantify the change in relative importance, we ranked the words before/after by frequency and looked at those words which increased in terms of rank the most. Concretely, we weighted terms by the formula $(before\_rank - after\_rank)/after\_rank$, which gives more weight to terms moving to the top, rather than moving up from, say, rank 100 to rank 80. The top gainers are, in descending order, "im", "god", "dont", "live", "single", "dreams", "blessed", "fuck". One story that potentially emerges from this is that people (i) become more self-centered, (ii) find stability in religion and spirituality, but also (iii) curse life for what has happened. A positive impact of spirituality on post-breakup coping has also been observed before [15].

(a)   (b)

Fig. 2. (a) Word cloud of the profile descriptions before breakup, at the beginning of our study. (b) Word cloud of the profile descriptions one week after the breakup.

3.3   Changes in Communication Styles with (Ex-)Partner

As Twitter is used for many purposes, including sharing factual information, we were interested to see if there would be a noticeable change in tone when one partner would message the other, either before or after the breakup. As simple analysis tools, we generated word clouds of 4-grams of words from conversations (messages) between pairs of users breaking up. Figures 3a and 3b show the shifts in personal communication patterns.

The change is roughly from "I love you so ..." to "I hate when you ...", indicating a (to us) surprising amount of *public* fighting and insulting happening after the breakup. For future analysis it might be interesting to quantify which relationships "turn sour", e.g., as a function of pre-breakup closeness.

(a)   (b)

Fig. 3. (a) Word cloud of the 4-grams from messages exchanged between BR users before breakup. (b) Word cloud of the 4-grams from messages exchanged between BR users after breakup.

3.4   Changes in Communication Patterns around Breakups

Apart from looking for expected before/after changes, we were interested to see if there were any gradual changes in communication patterns as people gradually edged towards a breakup. For this, we considered only those users who had at least four weeks of data before and two weeks after the breakup (1,070 users). We then looked at changes in (i) the fraction of tweets that contain a message to the partner, (ii) the fraction of tweets that are messages to non-partner users, and (iii) the fraction of tweets that are "original", i.e., non-retweet tweets. In all cases, these were then macro-averaged such that each couple, independent of their number of tweets, contributed equally to the average.

The trends are noticeable and consistent: as the breakup approaches – and beyond – (i) the number of messages to the partner decreases, (ii) the number of messages to other users increases, and (iii) the fraction of original tweets goes down. Though not the goal of this study, these observations could potentially lead to "early breakup warning" systems.

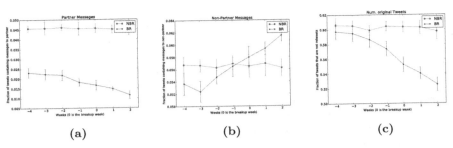

Fig. 4. Comparison of various features using data from four weeks before, during and two weeks after the breakup. (a) Fraction of the total tweets containing mentions of the partner. (b) Fraction of the total tweets containing direct messages to someone other than the partner. (c) Fraction of the total tweets that are not retweets. Error bars indicate standard errors.

3.5 Breakup-Induced Batch Un-friending and Being Un-friended

After the breakup, we were expecting partners to potentially unfollow each other but, apart from that, we were expecting "business as usual" as far as the social network was concerned. However, when we tried to quantify our hypothesis that there should not be ripple effects affecting other connections, we found evidence for the opposite.

Concretely, we monitored the number of friends and followers of each of our users over time. To be able to quantify the temporal changes, we only considered users who had at least two weeks of data before and two weeks after the breakup (1,156 users). As can be seen in Table 1, for the BR pairs there is a sudden drop of about 20 followers/friends on average and 16 in the median. Unfortunately, we do not have data on *who* is being unfollowed or stops following and we can only speculate that these are former mutual friends (We had to remove an outlier user @tatteddarkskin (249k followers, 270k friends), who changed this Twitter id to @iammald, the week he broke up).

Table 1. Average number of friends/followers for BR and NBR for two weeks before and two weeks after the breakup. The numbers in parentheses are for the median.

	T-2	T-1	friends T0	T+1	T+2
BR	579 (294)	582 (295)	562 (280)	564 (281)	577 (285)
NBR	588 (273)	591 (273)	596 (275)	598 (275)	601 (276)
			followers		
BR	683 (328)	689 (329)	669 (313)	672 (314)	675 (316)
NBR	778 (285)	780 (288)	785 (290)	787 (291)	788 (292)

3.6   Making Profiles Private

Given that relationship breakups can be traumatic experiences and that going through this in public can potentially be perceived as embarrassing, we wanted to see if breakups have effects on users' privacy settings.

On Twitter, by default all information is public and anyone can read your tweets and see your network information. However, Twitter users have the option to make their account private, which restricts access to their tweets to their followers where each follower now requires approval by the user. However, even for private profiles, the profile description and meta information such as the number of tweets, friends or followers remains accessible through the Twitter API. But the tweets' content or the identities of users in the user's social network are then hidden.

Out of 1,250 users that ever broke up, (excluding users who broke up in the very first and last week), 98 users eventually made their account private. Of these, 74 users made their account private within +/- one week of the breakup with 22 users already making this change *before* the week of the breakup. On the other hand, only 23 of the NBR users made their account private.

Put differently, BR users had a 7% probability of making their profiles private whereas this was 2% for NBR users (excluding first and last week). Interestingly, though there is work on privacy issues on Twitter [18,26], we are not aware of any study that looks at when and why users change their Twitter privacy settings. For Facebook on the other hand, a connection between relationship breakups and changes in privacy settings has been observed [24].

3.7   Evidence for Post-breakup Depression

Relationship breakups are known to be linked to depression [35]. As far as Twitter is concerned it has also been shown that tweets can give indications of depression [7,41]. Following features used as part of depression classifiers, we decided to use certain categories of the Linguistic Inquire and Word Count (LIWC) dictionary[6] [30]. Concretely, we combined terms from the "sad", "negemo" (= negative emotions), "anger" and "anxious" categories into a single category we call "depressed". The choice of these categories to be merged is inspired by [40] who find that "much of the research on uncertainty reduction theory (URT) has documented that high levels of uncertainty between romantic partners are correlated with greater feelings of anger, sadness, and fear, and that reduced uncertainty is accompanied by a decrease in the experience of negative emotion."

For the "depressed" category we then looked at the fraction of tweets during a particular week that contained at least one term from this category. These fractions were then first averaged for each partner of a couple and then averaged across couples. The resulting fractions over time are shown in Table 2.

For each week we find a statistically significant difference between BR and NBR couples ($p < .01$ using non-parametric bootstrap resampling) where BR

---

[6] http://www.liwc.net/

Table 2. Fraction of tweets containing depressing words, 2 weeks before and after the breakup

	T-2	T-1	T0	T+1	T+2
BR	0.124	0.129	0.132	0.143	0.149
NBR	0.105	0.106	0.107	0.104	0.107

pairs consistently have higher levels of these words. Moreover, their usage of these terms increases over time and the difference between T-2 and T+2 are significant ($p < .01$ using non-parametric bootstrap resampling).

3.8 Being Dumped Hurts More Than Dumping

When it comes to coping with relationship breakups previous work has found differences depending on whether a person is the "rejector", i.e., the initiator of the break-up or the "rejectee" [31,36]. To identify potential breakup initiators, we looked at BR pairs that initially had a mutual profile reference, but where one removed the mention of the other earlier, i.e., not in the same week. We hypothesize that the initiators are first to remove the reference of the former partner and label them "Rejectors" and the others were "Rejectees". There were 164 pairs where we observed such a behavior. Out of these, in 67% of the cases, women were the rejectors.

As far as word usage of the "depressed" terms is concerned, we found that rejectors feel less depressed compared to the rejectees (observed previously in [31] and [36]) as shown in Table 3. Again, the differences between pairs in the same week and the weeks T-2 and T+2 were tested for significance using bootstrap sampling and found to be significant at $p < .01$.

Table 3. Rejector's vs. rejectee's depression levels before and after breakup

	T-2	T-1	T0	T+1	T+2
Rejector	0.116	0.124	0.125	0.129	0.131
Rejectee	0.138	0.128	0.145	0.154	0.163

3.9 Pre-breakup Communication Asymmetries

Stonewalling is one of the "four horsemen of the apocalypse" defined by Gottman [12]. Stonewalling refers to ignoring the other partner and we quantify it by looking for communication asymmetries, where if only one side is "doing all the talking" there is evidence of stonewalling. Concretely, in each of the four weeks before the breakup, we looked at the number of messages exchanged between users. Here we looked if there were at least five times as many messages in one direction as the other direction and an absolute difference of at least five messages (to avoid cases where the difference was a mere one or two messages vs. zero messages). For BR couples, we found this kind of stonewalling in 224 out of the 585 couples (38%). For NBR, we only found it in 59 out of the 585 couples (10%).

3.10 Post-breakup Closeness

Existing work has looked at predicting post-breakup closeness using pre-breakup closeness [38] and found a positive connection between the two. We show that our data set can also be used to study this phenomenon by operationalizing these concepts as follows. We mark a pair as being "close" after a breakup if they both mention each other at least in two distinct weeks after they breakup (requiring a total of at least four tweets). 97 BR couples (16% of all BR couples with at least two post-breakup weeks) satisfy this condition for maintained, bi-directional communication and we call them PBC (for post-breakup closeness).

To quantify pre-breakup closeness, we looked at a user's fraction of all pre-breakup tweets that were messages to the partner. We did the same for their partner and then averaged this value for this couple, and then across all couples. This we did separately for the PBC set and the remaining BR pairs called NPBC. The same procedure of averaging pre-breakup tweets ratios was repeated for (i) the fraction of mentions to the partner and (ii) the fraction of retweets of the partner. We also obtained (up to) 3,200 of a user's favorites at the end of the study period and looked at the fraction of those that were for tweets by the partner. This value was again averaged across partners and then across couples. The results are presented in Table 4. For all four measures of "closeness" there is a significant difference between the PBC and the NPBC groups with higher levels of pre-breakup communication and interaction for couples who stay in touch after the breakup, confirming results in [38]. We also found the same trend when looking at the pre-study relationship duration (see Section 3.1) and the average relationship duration was 47 weeks for PBC, but only 34 for NPBC.

Table 4. Difference between various interaction related features for PBC and NPBC. All PBC and NPBC values are statistically significantly different ($p < .01$ using bootstrap resampling).

	Messages	Mentions	Retweets	Favorites
PBC	0.0559	0.0842	0.011	0.0897
NPBC	0.0296	0.0551	0.006	0.0505

4 Related Work

The only work we know of on studying romantic relationships on Twitter is Clayton et al. [3]. Using answers to specific questions (from surveys) from a few hundred users, they look at how Twitter mediates conflict between couples. They find evidence that "active Twitter use leads to greater amounts of

Twitter-related conflict among romantic partners, which in turn leads to infidelity, breakup, and divorce".

Currently, we are using Twitter merely as a data source to study relationship breakups per se. However, one could also study the more intricate relationship between technology use and personal relationships. Weisskirch, et al. [42] look at the attachment styles of couples involved in a relationship breakup online. It is the only work that we are aware of that looks at the act of breaking up through technology. Manual inspection of tweets around breakup revealed a few instances of actual breakups through public (!) tweets in our dataset too.

Apart from facilitating breakups, increased importance of technology in romantic relationships [29] potentially has other negative impact on romantic relationships such as jealousy, or surveillance [40,4,8]. On the positive side, researchers have looked at if technologies such as video chat can positively affect long-distance relationships by making it easier to feel connected [14,28].

Hogerbrugge et al. [16] study the importance of social networks in the dissolution of a romantic relationship. They define certain factors such as the overlap of networks of partners or social capital and study how these factors affect breakup. Though we did not collect data for the Twitter social *network*, or its changes over time, it would be possible to validate their findings on Twitter using our approach of identifying breakups.

Backstrom et al. [1] used the network structure of an individual's ego network to identify their romantic partner. Note that a social tie on Facebook is not the same as one on Twitter, mainly because, (i) Twitter network is directed, (ii) the use of Facebook and Twitter may be different. Still, the notion of 'dispersion' defined in their paper might be related to the loss of friends/followers in our study (see Section 3.5). Lefebvre et al. [24] study relationship dissolution on Facebook, mainly focusing on the phases and behavior of users who go through breakups on Facebook. There is evidence of limiting profile access in order to manage the breakup which is similar to our findings in Section 3.6.

Researchers conducting retrospective [10] and diary [35] studies of emotional adjustment following a breakup have found evidence of negative emotional responses including sadness and anger. In contrast to the current findings, Sbarra et al. found no difference between rejectors and rejectees in the extent of negative emotion following a breakup, and suggested that this might reflect difficulties in accurately identifying who initiated the breakup. Though imperfect, the current approach of identifying the first person to remove a profile mention as the "initiator" or "rejector" may provide a good proxy for being the person who is more ready to terminate the relationship or who feels more control over the breakup; this latter feature of controllability has been found to predict better adjustment post breakup [10,36]. It may also be that the larger sample size in the current study provides more statistical power to detect these effects than has been available in smaller survey studies.

Researchers studying close relationships have identified a number of factors that predict longevity and dissolution of non-marital relationships, including duration of the relationship, commitment, closeness, conflict, inclusion of other in the self, and the availability of alternatives [23]. Our analyses were informed by these extant findings, and we attempted to identify proxies for several of these important predictors, e,g., that with bi-directional profile mentions might be a sign of greater commitment than unidirectional mentions, and our findings tend to support those of the meta-analysis. However, we note that some relationship features may be more easily extracted from Twitter data than others. Factors like duration of the relationship and conflict might emerge clearly in the Twitter data (e.g., Figures 1b, 3b). Others, such as commitment or inclusion of other in the self (IOS), are typically assessed using multi-item self-report question-naires, and are not directly observable in tweets (at least not with any degree of frequency). Therefore, computational social scientists should pay particular attention to the need for studies that demonstrate the relations between the pat-terns they observe in data from online social networks and validated measures of relationship factors.

5   Discussion

Though our point of departure was a privileged data set, derived from a trial period for data access by GNIP, other ways to gather data are possible. For example, one could use services such as Followerwonk[7] to obtain a list of Twitter users with "boyfriend" in their profile description. For historic studies, one could use from the 1% "Spritzer" sample of public tweets on the Internet Archive[8] to find a sample of such users.

As with most similar, observational studies reasoning about causal links is tricky. For example, the increased usage of depressed terms (see Section 3.7) after the breakup could be a consequence of the breakup itself, or it could indicate that relationships are more likely to end when someone is about to undergo increased levels of depression.

Despite being limited when it comes to detecting causal links, observational studies such as ours are useful to validate existing models and theories as well as to provider pointers as to where a more in-depth study could be promising. For example, the observation that there is a sudden loss both in the number of friends and followers (see Section 3.5) is worth following up on. Were those to-be-removed friends only added due to social pressure in the first place? Or were they actual "friends" but maintaining communication with them would have been too emotionally taxing?

Most existing work on post relationship breakup behavior is based on surveys conducted long after the breakup, where people typically recall the experiences they have been through. This method has serious flaws as pointed out in [34]. Fortunately, we can collect data right around the time of the breakup.

Undoubtedly, couples in our data set are *not* representative of all heterosexual dating relationships in the United States, UK and Canada. Manually inspecting the data indicates an over-representation of teenagers. However, even the set of

---

[7] http://followerwonk.com/bio/
[8] https://archive.org/details/twitterstream

teenage dating relationships make up a significant part of relationships and are worth studying, especially as they seem to follow established patterns when it comes to the effect of the duration of relationship on the breakup probability (c.f. Section 3.1 and [23]), the occurrence of post-breakup depression (c.f. Section 3.7 and [35]), or communication asymmetries in the form of "stonewalling" (c.f. Section 3.9 and [12]).

Even though in this study we ignored para-social relationships with teen celebrities like @justinbieber and @katyperry, we could have looked at how these relationships evolve over time [5]. Maybe a "breakup" with Justin Bieber exerts just as much emotional stress as a breakup with a real boyfriend.

For the current study, we only looked at one-time relationship dissolutions. We did not try to identify cases where a couple got together again or cases where a partner "moved on" and entered a new relationship (even though the latter is easy to identify from our data set). Having a larger and periodically updated set of couples to monitor could allow studying such phenomena as well.

Arguably, Twitter could be used more as a type of "information network" than a "social network" [21], questioning its use as a data source for interpersonal relations. However, Myers et al. found that "from an individual user's perspective, Twitter starts off more like an information network, but evolves to behave more like a social network". Also, in our study we only consider people who at least partly use it as a social network for personal relations in the first place.

For this study we built a data set with a "high precision" approach, at the potential expense of recall. To be considered a "couple in a relationship", each pair of users underwent a sequence of filtering steps, including crowd labeling. The scale of our study could be improved by turning to machine-learned classifiers to detect romantic relationships even when partners are not mentioning each other in their profile descriptions. This is similar to work that looks into "when a friend in Twitter is a friend in life" [44] and work that classifies pairs of communicating users on Twitter into friends or foes [25].

So far we have only looked at basic measures of communication styles, such as the fraction of tweets mentioning a partner. However, there has been a body of work on inferring *personality traits* from Twitter usage [32,17,37,11]. This work could potentially be applied to our data set to look more into which types of personalities undergo what types of relationship breakups.

Not focusing on romantic relationships, there is research looking at unfollowing on Twitter [27,19,20,22]. Though unfollowing could be seen as a "mini-breakup" we observed that, maybe surprisingly, 44% of couples still follow each other two weeks after the week of the breakup and in another 32% of cases one of the partners still follows the other one at this point. For comparison, initially 96% of couples mutually follow each other.

6   Conclusions

We used public Twitter data to analyze the dissolution of 661 romantic relationships on Twitter during the period of Nov. 2013 to Apr. 2014. We compared the behavior of the users involved with those of 661 couples to those that did not breakup during the same period. Our analysis confirmed a number of existing hypotheses such as: (i) the breakup probability decreases with length of the relationship, (ii) post-breakup usage of "depressed" terms increases, (iii) rejectees have higher levels of usage of depressed terms compared to rejectors, (iv) communication asymmetries and one-sided stonewalling is indicative of breakups, and (v) higher levels of post-breakup closeness for couples who also have a higher pre-breakup closeness.

We also found evidence of the, to our knowledge, undocumented phenomenon of "batch un-friending and being un-friended" at the end of a relationship. Concretely, we observed sudden drops of size 15-20 for both the number of friends and followers a user has around the time of the breakup.

Though our data set is undoubtedly not representative of all relationship breakups we believe our study still shows the huge potential that public social media offers with respect to studying sociological and psychological processes in a scalable and non-obtrusive manner.

References

1. Backstrom, L., Kleinberg, J.: Romantic partnerships and the dispersion of social ties: A network analysis of relationship status on facebook. In: Proceedings of the 17th ACM Conference on Computer Supported Cooperative Work & Social Computing, pp. 831–841. ACM (2014)
2. Battaglia, D.M., Richard, F.D., Datteri, D.L., Lord, C.G.: Breaking up is (relatively) easy to do: A script for the dissolution of close relationships. Journal of Social and Personal Relationships 15(6), 829–845 (1998)
3. Clayton, R.B.: The third wheel: The impact of twitter use on relationship infidelity and divorce. Cyberpsychology, Behavior, and Social Networking (2014)
4. Clayton, R.B., Nagurney, A., Smith, J.R.: Cheating, breakup, and divorce: Is facebook use to blame? Cyberpsy., Behavior, and Soc. Networking 16(10), 717–720 (2013)
5. Cohen, J.: Parasocial break-up from favorite television characters: The role of attachment styles and relationship intensity. Journal of Social and Personal Relationships 21(2), 187–202 (2004)
6. Cupach, W.R., Metts, S.: Accounts of relational dissolution: A comparison of marital and non-marital relationships. Communications Monographs 53(4), 311–334 (1986)
7. De Choudhury, M., Gamon, M., Counts, S., Horvitz, E.: Predicting depression via social media. In: ICWSM 2013 (2013)
8. Drouina, M., Millera, D.A., Dibbleb, J.L.: Ignore your partners current facebook friends; beware the ones they add! Computers in Human Behavior 35, 483–488 (2014)
9. Fiore, A.T., Donath, J.S.: Homophily in online dating: When do you like someone like yourself? In: CHI, pp. 1371–1374 (2005)
10. Frazier, P.A., Cook, S.W.: Correlates of distress following heterosexual relationship dissolution. Journal of Social and Personal Relationships 10(1), 55–67 (1993)
11. Golbeck, J., Robles, C., Edmondson, M., Turner, K.: Predicting personality from twitter. In: SocialCom/PASSAT, pp. 149–156 (2011)

12. Gottman, J.: Why marriages succeed or fail: And how you can make yours last. Simon and Schuster (1995)
13. Gottman, J.M., Levenson, R.W., Gross, J., Frederickson, B.L., McCoy, K., Rosenthal, L., Ruef, A., Yoshimoto, D.: Correlates of gay and lesbian couples' relationship satisfaction and relationship dissolution. Journal of Homosexuality 45(1), 23–43 (2003)
14. Hassenzahl, M., Heidecker, S., Eckoldt, K., Diefenbach, S., Hillmann, U.: All you need is love: Current strategies of mediating intimate relationships through technology. ACM Trans. Comput.-Hum. Interact. 19(4), 30:1–30:19 (2012)
15. Hawley, A.R.: The roles of spirituality and sexuality in response to romantic breakup. Ph.D. thesis, Bowling Green State University (2012)
16. Hogerbrugge, M.J., Komter, A.E., Scheepers, P.: Dissolving long-term romantic relationships: Assessing the role of the social context. Journal of Social and Personal Relationships 30(3), 320–342 (2013)
17. Hughes, D.J., Rowe, M., Batey, M., Lee, A.: A tale of two sites: Twitter vs. facebook and the personality predictors of social media usage. Computers in Human Behavior 28(2), 561–569 (2012)
18. Jin, S.A.A.: Peeling back the multiple layers of twitter's private disclosure onion: The roles of virtual identity discrepancy and personality traits in communication privacy management on twitter. New Media & Society 15(6) (2013)
19. Kivran-Swaine, F., Govindan, P., Naaman, M.: The impact of network structure on breaking ties in online social networks: unfollowing on twitter. In: CHI, pp. 1101–1104 (2011)
20. Kwak, H., Chun, H., Moon, S.B.: Fragile online relationship: a first look at unfollow dynamics in twitter. In: CHI, pp. 1091–1100 (2011)
21. Kwak, H., Lee, C., Park, H., Moon, S.B.: What is twitter, a social network or a news media? In: WWW, pp. 591–600 (2010)
22. Kwak, H., Moon, S.B., Lee, W.: More of a receiver than a giver: Why do people unfollow in twitter? In: ICWSM (2012)
23. Le, B., Dove, N.L., Agnew, C.R., Korn, M.S., Mutso, A.A.: Predicting nonmarital romantic relationship dissolution: A meta-analytic synthesis. Personal Relationships 17(3), 377–390 (2010)
24. LeFebvre, L., Blackburn, K., Brody, N.: Navigating romantic relationships on facebook: Extending the relationship dissolution model to social networking environments. Journal of Social and Personal Relationships p. 0265407514524848 (2014)
25. Liu, Z., Weber, I.: Predicting ideological friends and foes in twitter conflicts. In: WWW, pp. 575–576 (2014)
26. Mao, H., Shuai, X., Kapadia, A.: Loose tweets: An analysis of privacy leaks on twitter. In: WPES, pp. 1–12 (2011)
27. Moon, S.: Analysis of twitter unfollow: How often do people unfollow in twitter and why? In: SocInfo, p. 7 (2011)
28. Neustaedter, C., Greenberg, S.: Intimacy in long-distance relationships over video chat. In: CHI, pp. 753–762 (2012)
29. Papp, L.M., Danielewicz, J., Cayemberg, C.: "Are we facebook official?" implications of dating partners' facebook use and profiles for intimate relationship satisfaction. Cyberpsychology, Behavior, and Social Networking 15, 85–90 (2012)
30. Pennebaker, J.W., Chung, C.K., Ireland, M., Gonzales, A., Booth, R.J.: The development and psychometric properties of liwc 2007. LIWC. Net, Austin (2007)
31. Perilloux, C., Buss, D.M.: Breaking up romantic relationships: Costs experienced and coping strategies deployed. Evolutionary Psychology 6(1), 164–181 (2008)

32. Quercia, D., Kosinski, M., Stillwell, D., Crowcroft, J.: Our twitter profiles, our selves: Predicting personality with twitter. In: SocialCom/PASSAT, pp. 180–185 (2011)
33. Rhoades, G.K., Kamp Dush, C.M., Atkins, D.C., Stanley, S.M., Markman, H.J.: Breaking up is hard to do: The impact of unmarried relationship dissolution on mental health and life satisfaction. Journal of Family Psychology 25(3), 366 (2011)
34. Sakaluk, J.K.: Problems with recall-based attachment style priming paradigms: Exclusion criteria, sample bias, and reduced power. Journal of Social and Personal Relationships p. 0265407513508728 (2013)
35. Sbarra, D.A., Emery, R.E.: The emotional sequelae of nonmarital relationship dissolution: Analysis of change and intraindividual variability over time. Personal Relationships 12(2), 213–232 (2005)
36. Sprecher, S., Felmlee, D., Metts, S., Fehr, B., Vanni, D.: Factors associated with distress following the breakup of a close relationship. Journal of Social and Personal Relationships 15(6), 791–809 (1998)
37. Sumner, C., Byers, A., Boochever, R., Park, G.J.: Predicting dark triad personality traits from twitter usage and a linguistic analysis of tweets. In: ICMLA (2), pp. 386–393 (2012)
38. Tan, K., Agnew, C.R., VanderDrift, L.E., Harvey, S.M.: Committed to us: Predicting relationship closeness following nonmarital romantic relationship breakup. Journal of Social and Personal Relationships, 0265407514536293 (2014)
39. Tashiro, T., Frazier, P.: I'll never be in a relationship like that again: Personal growth following romantic relationship breakups. Personal Relationships 10(1), 113–128 (2003)
40. Tong, S.T.: Facebook use during relationship termination: Uncertainty reduction and surveillance. Cyberpsychology, Behavior, and Social Networking 16(11), 788–793 (2013)
41. Tsugawa, S., Mogi, Y., Kikuchi, Y., Kishino, F., Fujita, K., Itoh, Y., Ohsaki, H.: On estimating depressive tendencies of twitter users utilizing their tweet data. In: VR, pp. 1–4 (2013)
42. Weisskirch, R.S., Delevi, R.: Its ovr b/n u n me: Technology use, attachment styles, and gender roles in relationship dissolution. Cyberpsychology, Behavior, and Social Networking 15(9), 486–490 (2012)
43. Xia, P., Jiang, H., Wang, X., Chen, C.X., Liu, B.: Predicting user replying behavior on a large online dating site. In: ICWSM (2014)
44. Xie, W., Li, C., Zhu, F., Lim, E.P., Gong, X.: When a friend in twitter is a friend in life. In: WebSci, pp. 344–347 (2012)

The Social Name-Letter Effect
on Online Social Networks

Farshad Kooti[1,*], Gabriel Magno[2,*], and Ingmar Weber[3]

[1] USC Information Sciences Institute, Marina del Rey, USA
[2] Universidade Federal de Minas Gerais, Belo Horizonte, Brazil
[3] Qatar Computing Research Institute, Doha, Qatar

Abstract. The Name-Letter Effect states that people have a preference for brands, places, and even jobs that start with the same letter as their own first name. So Sam might like Snickers and live in Seattle. We use social network data from Twitter and Google+ to replicate this effect in a new environment. We find limited to no support for the Name-Letter Effect on social networks. We do, however, find a very robust Same-Name Effect where, say, Michaels would be more likely to link to other Michaels than Johns. This effect persists when accounting for gender, nationality, race, and age. The fundamentals behind these effects have implications beyond psychology as understanding how a positive self-image is transferred to other entities is important in domains ranging from studying homophily to personalized advertising and to link formation in social networks.

1 Introduction

According to the *Name-Letter Effect* (NLE), people have a preference for partners, brands, places, and even jobs that share the first letter with their own name. Correspondingly, a Sarah would be more likely to marry a Sam, go to Starbucks, move to San Francisco, and work in sales. This phenomena has been replicated in numerous settings [18,19,9,8,2,20,1] and is part of text books in psychology [13]. Some researchers have, however, questioned the validity or at least the generality of such studies [22,23,14,17,5]. By its supporters, the NLE is usually attributed to "implicit egotism" [20] with people preferring situations that reflect themselves.

We turn to data from online social networks, Twitter and Google+, to see if the NLE can be replicated in a large online setting. Concretely, we seek evidence for or against the NLE in choosing social connections (Sarah following Sam) and in expressing brand interest (Peter following Pepsi). Our findings here are mixed and, depending on the exact setting, we find statistically significant evidence both for and against the NLE.

Extending the NLE and the idea of implicit egotism, we look for a *Same-Name Effect* (SNE) where a Sarah is more likely to follow another Sarah and Tom Cruise is in particular popular among Toms. Here, we observe the presence of the SNE in different settings. We show that the SNE exists for both genders and in different countries. We also show that the SNE affects linking both to celebrities and to normal users and affects

---

* This work was done while the first two authors were interns at Qatar Computing Research Institute.

L.M. Aiello and D. McFarland (Eds.): SocInfo 2014, LNCS 8851, pp. 216–227, 2014.
© Springer International Publishing Switzerland 2014

both strong and weak ties. Finally, we show that there is an anti-correlation between the number of friends and the extent of link preference bias caused by the SNE.

To the best of our knowledge, this is the first time that the Name-Letter Effect and the Same-Name Effect have been studied in an online setting. It is also the largest study of its kind with more than a million connections analyzed. Our analysis quantifies a factor that affects link formation in online social networks. Understanding the processes governing which links are established is crucial for areas such as information diffusion or link prediction. Moreover, the strength of the NLE or the SNE for an individual could be an estimate of the person's positive self-image. Understanding this could help in understanding homophily, and it could also be used in personalized advertising.

2   Related Work

The NLE was first observed by Nuttin in 1985 [18]. The effect was studied by asking volunteers to pick their favorite letter from pairs or triads of letters where only one of them belonged to the participant's first or last name. Nuttin showed that independent of visual, acoustical, semantic, and frequency characteristics, letters belonging to own first and last name are preferred over other letters. The most popular explanation for the NLE is "implicit egotism" [21]. People have positive feelings about themselves and these feelings are associated implicitly to places, events, and objects related to the self [20].

Later, the presence of the effect was tested in different languages and cultures. It has been shown that the NLE exists in twelve European languages [19]. Also, Hoorens et al., showed that the NLE exists across languages, i.e., participants picked the letters in another language that were either visually or acoustically similar to the letters in their names in their own language [9].

After the discovery of the NLE, many studies verified the existence of the effect in a wide range of decision making situations: People are disproportionately more likely to live in cities and take jobs that are similar to their name [21]. Also, brands that have the same initial as a person's name are preferred by that person [2] and there is a higher chance of donation when the name of the solicitor is similar to the name of the contacted person [1]. Studies have even found that NLE affects marriage; people are more likely to marry a person with a similar name [11,23]. On the other hand, the NLE was not observed in choosing favorite foods and animals [8].

Besides many studies providing evidence for an NLE, there have also been papers questioning the presence of the effect in different areas or the reason for the effect. E.g., in [17], the authors show that a wrong statistical test was conducted in an earlier work on verifying the existence of the NLE in the initial of a baseball player and number of strike outs by him. Also, other works had shown different biases that might create the same results as a NLE [22,23,14]. For example, in the study that showed people are more likely to live in cities with the same initial as theirs, one explanation might be that people in those cities named their babies with such names. Although there are some papers challenging the existence of NLE, the critics are usually concerned about the way a particular study was done, and the main effect is still generally accepted.

3   The Name-Letter Effect on OSNs

In this section, we first test the generality of the NLE on Twitter and Google+ in different domains, such as preferred brands, celebrities, and news media. Then, we investigate the NLE in the social context. Concretely, do users follow other users with the same initial disproportionately more than users with a different initial? Here we use the term "follow" to refer both to Twitter following and to Google+ "has added to a circle". In both case, the acting user expresses an interest in the updates of the user acted upon.

3.1   Data Description

Twitter: Most of the analyses in this work is done on a large Twitter social network gathered in [3]. The network contains all the 52 million users who joined Twitter by September 2009 and all the 1.9 billion links among them. We also used users' location information from [12], which uses both location and time zone fields for inferring a user's country.

   Google+: The Google+ dataset was created by collecting public information available in user profiles in the network. The data collection ran from March 23rd of 2012 until June 1st of 2012. In total we were able to retrieve information from 160 million profiles . With the social links of the users, we have constructed a directed graph that has 61 million nodes and 1 billion edges. Details of the Google+ platform and a data characterization of an early version of the dataset is discussed in a previous work [15].

3.2   NLE and Brand Preference

For testing the NLE on Twitter and Google+, we considered a variety of domains and we picked a pair of popular Twitter and Google+ accounts from each domain. Then, we gathered all the followers of each account as of May 2013 (or a large 1 million uniform, random sample of them) in Twitter, and all the followers of each account in Google+ as in the time of the data collection (2012).

   We examine the brand NLE by performing the Pearson's chi-squared test of independence. We do this by counting the followers of each account who have the same initial to see if there are disproportionately many followers for the brands and users with the same initial. For each pair of brands, we create a 2×2 table showing the number of followers for each account whose initial is the same as initial of either of brands. Since both the popularity of the brands and the frequency of name initials are not necessarily the same across the world, in all the analyses in this section we only consider followers in the US. To filter the users in Twitter we used the location field from the users' accounts and only picked users who had one of the top 20 most populated US cities, *United States*, or *USA* in their location field. The location filter in Google+ was done by extracting the geographic coordinates of the last location present on the *Places lived* field, picking only the users from USA.

   Table 1 shows an illustrating example of the 2×2 tables. A represents the number of users who follow Brand 1 and have the same initial as Brand 1. Similarly, D is the number of followers of Brand 2 who have the same initial as Brand 2. For testing the

NLE, first, we calculate the expected values for the cells that the initial of the followers matches the brand's initial (here A and D). The expected value, is the value that the fields would have if, given the total values, the followers were split uniformly and without any preference. Here the expected value of A would be $\frac{(A+C)*(A+B)}{A+B+C+D}$ and the expected value of D would be $\frac{(B+D)*(C+D)}{A+B+C+D}$. Then, expected values smaller than the observed values for A and D indicate the existence of the NLE.

Table 2 shows all the considered Twitter and Google+ accounts and whether a significant NLE exists or not. We picked these pairs mainly because these accounts have high number of followers. Moreover, the pairs presented here and in the rest of the paper are all the pairs that we did the analysis on, and we are not "cherry-picking" the results. Out of the eight considered domains in Twitter, shown in Table 2, only three of them show a statistically significant NLE, three cases imply NLE but the results are not statistically significant, and the remaining two pairs exhibit a negative NLE. The results suggest that the NLE exists only in some special cases and it is not a generalizable concept for following brands on Twitter. This analysis was done by considering the first name of the user. We repeated the analysis using the Twitter handle (i.e., screen name) of the users. For 61% of the users the initial of the actual name matches the initial of their Twitter handle. Due to this high overlap, testing the NLE by using the handles yields very similar results to using their declared names: in only two cases the results are statistically significant, for the game consoles and the actors, and in these cases the effect is much smaller than the NLE with actual names (3.5% and 1% respectively). In Google+, none of the three pairs of brands/celebrities had statistically significant results, with two of the pairs exhibiting low positive and one negative NLE.

3.3 NLE and Social Link Preference

In this section, we test the NLE in the context of friend link preference. This means that we check if users prefer to establish links to other users with the same initial. To have two sets of users with the same initials for testing the NLE on link preference, we first picked the four most popular names on Twitter that have pairs of same initials: "Michael", "Matthew", "Jason", and "James". Since these names are used in many countries, considering all users might falsely show the NLE: say "Michael" and "Matthew" are popular in a particular country, but not in others, in this case there will be lots of links from "Matthew" to "Michael", but not to "Jason". This could create an

Table 1. Illustration of testing the NLE. If link formation is independent of the initials of the brands, the observed value would be close to the expected value for A, namely, $\frac{(A+C)*(A+B)}{A+B+C+D}$. Larger than expected observed values for A and D indicate the existence of the NLE.

	Brand 1	Brand 2	Total
Brand 1 initial	A	B	$A + B$
Brand 2 initial	C	D	$C + D$
Total	$A + C$	$B + D$	

Table 2. The Twitter and Google+ accounts considered for the brand NLE and the average percentage of preference for the brands with the same initial. There is no significant NLE for most of the brands.

Twitter			
Account 1	Account 2	NLE	p-value
Sega	Nintendo	9%	< 0.001
Jim Carrey	Tom Cruise	4%	< 0.001
Firefox	Internet Explorer	5%	< 0.1
Canon	Nikon	5%	—
Puma	Adidas	0.9%	—
CNN	New York Times	0.4%	—
Nokia	Samsung	-1.3%	—
Pepsi	Coca-Cola	-1.7%	—

Google+			
Account 1	Account 2	NLE	p-value
Sergey Brin	Larry Page	1%	—
Nokia	Samsung	-16%	—
Pepsi	Coca-Cola	1%	—

Table 3. Results of the NLE on link preference. Effect sizes are shown in the parentheses. In Twitter, users with same initials have negative effect size, contradicting the NLE ($p - values <$ 0.001). Google+ results were not statistically significant.

Twitter			
	Michael	Jason	Total
Matthew	**6,455** (-2%)	4,285 (+4%)	10,740
James	12,016 (+1%)	**7,236** (-2%)	19,252
Total	18,471	11,521	

Google+			
	Mark	James	Total
Michael	**3,605** (0%)	1,829	5,434
John	3,213	**1,598** (-1%)	4,811
Total	6,818	3,427	

apparent NLE in the results, that might not actually exist, or at least not due to implicit egotism. To overcome this issue, we limited ourselves to users in the US.

Table 3 shows the results of the number of times "Matthews" and "Jameses" follow "Michaels" and "Jasons" for Twitter. Surprisingly, the results show a slight, statistically significant[1] *negative* NLE ($\chi^2(1) = 15.58$). This analyses was repeated with a pair of female names ("Melissa" and "Jennifer") following a pair of male names ("Michael" and "Jason") and vice versa. Again in both cases a negative NLE existed, but this time not statistically significant. The results clearly show that the NLE does not exist for general social link preference. The same analyses were done for the Google+ dataset, using the two most popular pairs of same initials: "Michael", "John", "Marks" and "James". Again, there was a negative NLE, but not a statistically significant one.

3.4   NLE and Location, Job, and Hobbies

Earlier studies have shown people prefer to live in the cities with the same initials and also choose occupations that have the same initial as their name [21]. We tried to replicate these findings using our data. For Twitter we gathered the profile information of more than 4 million random users and used their location field to see the effect of NLE in the city that people choose to live. For Google+ we retrieved the city from the

---

[1] In this work, we consider $p - value < 0.001$ as statistically significant, unless explicitly specified.

"Places lived" field. We tested the NLE for the top ten largest city in the US[2]. The ten largest cities in the US have seven unique initial letters, which leads to 21 (seven choose two), pairs of letters for checking NLE. In Twitter, out of the 21 pairs, 8 pairs show statistically significant results, with 6 of them showing positive NLE. In Google+, 7 pairs were statistically significant, with 6 of them showing positive NLE.

Similarly for the occupations, we consider the following jobs: engineer, cashier, waiter(ess), teacher, and nurse. In Twitter we search the users' bios for the corresponding strings. The "bio" is the field in the profile that users introduce themselves in and they often include their occupation. In Google+, we examine the "Occupation" field, and looked for the same set of strings. Both in Twitter and Google+ we find only one statistically significant result out of the ten (five choose two) possible pairs of letters, and this single statistically significant pair has negative NLE.

We also test the NLE for the hobbies of the users. More specifically, we look for popular sports in the bio of the users in Twitter and in the "Introduction" field of Google+. We consider football, basketball, baseball, lacrosse, soccer, volleyball, tennis, and hockey. We test the NLE again for the all 21 possible pairs of initials of the sports and the names. Only four of the pairs show a statistically significant result, with only one positive NLE in Twitter and two positive NLE in Google+.

Overall, our findings therefore question the existence or at least the general scope and strength of the NLE as we failed to replicate earlier claims in this new setting.

4   The Same-Name Effect on OSNs

In this section, we test another effect in link creation preference in a more restricted case where both users have the exact same first name, rather than just the same initial. Since all letters of the users' names are involved, this effect should be stronger than the NLE. We call this effect *same-name effect* (SNE). In other words, are Michaels disproportionately more likely to follow other Michaels compared to other users? A similar idea was tested in an earlier study, where it was shown that people are more likely to marry others with the same *last* name [11]. Here, we analyze linking between users with the same name and show that there is a strong SNE that is robust to many variations.

First, we test the SNE by considering the gender of users as the first name typically identifies the gender. Since men (women) might be more likely to follow other men (women) [16], considering both groups together might cause a false indication of a SNE. So, we perform the SNE test within each gender. Also, as mentioned earlier, having users from different countries might introduce a bias in the results, so again we are considering only users in the US.

We pick the five most popular male names on Twitter among users from the US: "Michael", "John", "David", "Chris", and "Brian". Then, we count the number of times each of the users with these names follows other users with these names. Table 4 (appendix) shows the resulting 5 × 5 table and the effect sizes of 4-13% on Twitter. We calculate the effect size of each name as the average of pairwise preference of that

---

[2] http://en.wikipedia.org/wiki/List_of_United_States_cities_by_population

name over other names in the table. This same analysis is repeated in Google+, and the results are the same: male users significantly preferred to follow other users with the same name 7-30% more than expected. We also tested the SNE with the five most popular female names in the US on Twitter and Google+: "Jennifer", "Jessica", "Ashley", "Sarah", and "Amanda". The results were similar to the previous case and even stronger: female users significantly preferred to follow others with the same name 30-45% more than expected in Twitter, and 10-29% in Google+.

An alternative explanation for the observed preference could be the fact that different first names are popular in different ethnicities and races. To address this concern, we repeated the analysis for all male first names in the US with more than 10,000 users (56 names in Twitter, 58 names in Google+). We tested the SNE pairwise for these names and the SNE existed for all 1,540 pairs of names with an average effect size of 19% in Twitter, and for all 1,653 pairs of names with an average effect size of 28% in Google+. The fact that the SNE exists for all of the pairs suggests that the preference is not just because of homophily because for at least some cases the names would be associated with the same particular race or ethnicity.

Moreover, we used last names as a proxy of the ethnicity. We used 1990 census data to gather last names that are prominent for only one race in the US.[3] We gathered the top 1,000 last names in each of the five races of white, African-American, Asian, Hispanic, and native American natives.[4] For each race we considered only the last names that are in the top 500 of a particular race and do not occur among the top 1,000 names for any of the other race's lists. Then we tested the SNE within each race for the pairwise combination of the top 50 popular first names, 1,225 pairs, though not all of these 50 first names were found for all of the five races. Table 5 shows that for all five races a strong and consistent SNE exists in Twitter. In Google+, most of the results were not statistically significant, although implying positive SNE.

To account for age, we use offline data from social security statistics[5]. We focus on the common ages of 20-30 years old on Twitter at the time the data was collected (2009), which corresponds to users born between 1979-1989. We use the records of social security to gather the most popular boy baby names during the mentioned years. Then, we pick all the names that were in the top five at least once: "Michael", "Jason", "Christopher", "Matthew", "David", "James", and "Daniel". We conduct a similar analysis to the previous section on these names. A statistically significant SNE again existed with 12-17% preference in Twitter, and 5-23% statistically significant preference in Google+. We also try the same experiment with the most popular girl baby names during 1979-1989. Again, a significant SNE is observable with a 16-24% preference in Twitter and 10-106% preference in Google+.

Finally, to see if the SNE exists in different languages and cultures, we picked three countries with different languages: Brazil, Germany, and Egypt. Then, we picked the most popular names in each of those countries and tested the SNE. We found that a statistically significant SNE exists in all three countries, both for Twitter and Google+.

---

[3] http://names.mongabay.com/

[4] Note that in later census the race/ethnicity has been treated differently and that "Hispanic" can now be of any race according to the census terminology.

[5] http://www.ssa.gov/oact/babynames/top5names.html

The effect sizes for Brazilian users range from 13-22% in Twitter and from 16-22% in Google+. Similarly, in Germany and Egypt users significantly preferred to follow other users with the same name (6 - 101%).

5 Discussion

We have focused on testing and observing the NLE and the SNE rather than on explaining them. When using implicit egotism as an explanation the crucial assumption is that users are free to choose the brands they like or the members of their social network. This basic assumption is arguably flawed as people can only connect to people (or brands) they know. But as the distribution of names is not homogeneous across all parts of society this creates implicit selection biases. For example the name Emma was very popular for girls born during 2002-2012[6] but less popular earlier which, in turn, means that an Emma would be more likely to go to school with another Emma and hence have a chance to connect. Similarly, the name DeShawn is popular among African Americans [6] which means a DeShawn growing up in a predominantly black neighborhood would again have a higher than expected chance of connecting to another DeShawn. In fact, previous research has shown that mere familiarity with a name correlates with likeability [4,7].

We tried to avoid obvious pitfalls, such as selecting names associated with a particular demographic groups, and we looked at names that were popular during a certain period. Additionally, the fact that for testing the NLE and the SNE on link preference we only used the network of early adopters of Twitter (up to September 2009) and Google+ (less than a year after the launch) helps to further homogenize the user set across age and income. Also, we have used users' last names to test the SNE within one race. Still, naming conventions within a family, where family members are given the same first name, could explain part of the observed the SNE.

It is also not clear what fraction of users use their real name in online social networks. We believe this is the case for the majority of the users, especially for Google+, since Google explicitly asked users to use their real name and banned the accounts of users with fake names[7]. There might be much less use of real names on Twitter, but the fact that our findings for Twitter and Google+ are very consistent suggests that there is no dramatic difference between Twitter and Google+ in the way people chose their name. And even if the majority of the names are not real, we still found the SNE, which might have a different explanation than the implicit egotism. Also, note that for testing SNE, we tested the effect on common English names, so we are not analyzing users completely fake names like "cowboy".

6 Conclusions

The Name-Letter Effect (NLE) states that people prefer the letters in their own names over other letters. We investigated the existence of the NLE in the context of Twitter

---

[6] http://www.ssa.gov/oact/babynames/top5names.html
[7] http://gawker.com/5824622/names-banned-by-google-plus

and Google+. Our findings question at least the generality of the NLE. Going beyond the NLE, we analyzed users' linking behavior for a same-name effect (SNE), where instead of comparing the initials we compared the whole name. In this stronger version, we observe a robust effect, even when accounting for gender, age, race, and location.

Besides the psychological aspects of NLE, there are some real-world implications. E.g., one study has showed that using NLE can increase the chance of donation made by people [1]. In recent years, the Coca-Cola *share a coke*[8] campaign has proven to be very successful by increasing sales[9].

References

1. Bekkers, R.: George gives to geology jane: The name letter effect and incidental similarity cues in fundraising. International Journal of Nonprofit and Voluntary Sector Marketing 15(2), 172–180 (2010)
2. Brendl, C.M., Chattopadhyay, A., Pelham, B.W., Carvallo, M.: Name letter branding: Valence transfers when product specific needs are active. Journal of Consumer Research 32(3), 405–415 (2005)
3. Cha, M., Haddadi, H., Benevenuto, F., Gummadi, K.: Measuring User Influence in Twitter: The Million Follower Fallacy. In: ICWSM (2010)
4. Colman, A.M., Hargreaves, D.J., Sluckin, W.: Psychological factors affecting preferences for first names. Names: A Journal of Onomastics 28(2), 113–129 (1980)
5. Dyjas, O., Grasman, R.P., Wetzels, R., van der Maas, H.L., Wagenmakers, E.J.: What's in a name: a bayesian hierarchical analysis of the name-letter effect. Frontiers in Psychology 3 (2012)
6. Fryer, R.G., Levitt, S.D.: The causes and consequences of distinctively black names. The Quarterly Journal of Economics 119(3) (2004)
7. Hargreaves, D.J., Colman, A.M., Sluckin, W.: The attractiveness of names. Human Relations 36(4) (1983)
8. Hodson, G., Olson, J.M.: Testing the generality of the name letter effect: Name initials and everyday attitudes. Personality and Social Psychology Bulletin 31(8), 1099–1111 (2005)
9. Hoorens, V., Nuttin, J.M., Herman, I.E., Pavakanun, U.: Mastery pleasure versus mere ownership: A quasi-experimental cross-cultural and cross-alphabetical test of the name letter effect. European Journal of Social Psychology 20(3) (1990)
10. Johnson, B., Eagly, A.: Effects of involvement on persuasion: A meta-analysis. Psychological Bulletin 106, 290–314 (1989)
11. Jones, J.T., Pelham, B.W., Carvallo, M., Mirenberg, M.C., et al.: How do i love thee? let me count the js: Implicit egotism and interpersonal attraction. Journal of Personality and Social Psychology 87, 665–683 (2004)
12. Kulshrestha, J., Kooti, F., Nikravesh, A., Gummadi, P.K.: Geographic dissection of the twitter network. In: ICWSM (2012)
13. Leary, M.R., Tangney, J.P. (eds.): Handbook of Self and Identity, 2nd edn. The Guilford Press (2011)
14. Lebel, E.P., Paunonen, S.V.: Sexy but often unreliable: The impact of unreliability on the replicability of experimental findings with implicit measures. Personality and Social Psychology Bulletin 37(4), 570–583 (2011)

---

[8] http://www.coca-cola.co.uk/faq/products/share-a-coke.html
[9] 3MM P4W Consumption Oct-Dec 2011 B3 Survey Australia

15. Magno, G., Comarela, G., Saez-Trumper, D., Cha, M., Almeida, V.: New kid on the block: exploring the google+ social graph. In: Proceedings of the 2012 ACM Conference on Internet Measurement Conference, pp. 159–170. ACM, New York (2012)
16. Magno, G., Weber, I.: International gender differences and gaps in online social networks. In: SocInfo (2014)
17. McCullough, B., McWilliams, T.P.: Baseball players with the initial "k" do not strike out more often. Journal of Applied Statistics 37(6), 881–891 (2010)
18. Nuttin, J.M.: Narcissism beyond gestalt and awareness: The name letter effect. European Journal of Social Psychology 15(3), 353–361 (1985)
19. Nuttin, J.M.: Affective consequences of mere ownership: The name letter effect in twelve european languages. European Journal of Social Psychology 17(4), 381–402 (1987)
20. Pelham, B.W., Carvallo, M., Jones, J.T.: Implicit egotism. Current Directions in Psychological Science 14(2), 106–110 (2005)
21. Pelham, B.W., Mirenberg, M.C., Jones, J.T.: Why susie sells seashells by the seashore: implicit egotism and major life decisions. J. Pers. Soc. Psychol. 82(4), 469–487 (2002), http://www.biomedsearch.com/nih/Why-Susie-sells-seashells-by/11999918.html
22. Simonsohn, U.: In defense of diligence: a rejoinder to pelham and carvallo. Journal of Personality and Social Psychology 101(1), 31–33 (2011)
23. Simonsohn, U.: Spurious? name similarity effects (implicit egotism) in marriage, job, and moving decisions. Journal of Personality and Social Psychology 101(1), 1–24 (2011)

A Appendix

Table 4. Results for the SNE on popular names in the US on Twitter and Google+. The effect sizes are positive for all five names showing the SNE ($p - values < 0.001$).

Twitter	Michael	John	David	Chris	Brian	Total
Michael	28,587 (+5%)	36,590	29,051	25,928	15,093	135,249
John	28,393	42,417 (+4%)	31,540	27,823	16,906	147,079
David	24,303	33,713	29,388 (+5%)	24,441	14,255	126,100
Chris	22,632	31,383	25,107	25,999 (+6%)	14,089	119,210
Brian	15,394	20,974	16,676	15,636	11,383 (+13%)	80,063
Total	119,309	165,077	131,762	119,827	71,726	

Google+	Michael	David	John	Chris	James	Total
Michael	5,949 (+7%)	4,492	4,659	5,349	1,829	22,278
David	4,739	5,375 (+10%)	4,526	4,858	1,657	21,155
John	4,590	3,979	5,154 (+9%)	4,687	1,598	20,008
Chris	3,971	3,431	3,659	4,791 (+9%)	1,410	17,262
James	2,349	1,980	2,329	2,556	1,287 (+30%)	10,501
Total	21,598	19,257	20,327	22,241	7,781	

Table 5. The SNE test for users with a race-specific last name. The "# of last names" indicates the number of race-specific last names found. "Could be tested" is the number of first name pairs where each first name had a non-zero count for the race-specific last name. There is a large number of statistically significant positive effects, and only a single first name pair with a significant *negative* effect.

Race	# of last names	Could be tested	Median SNE	Sig. positive	Sig. negative	Non-sig. positive	Non-sig. negative
			Twitter				
White	35	1225	45%	986	0	237	2
Asian	394	1193	59%	433	1	645	114
Hispanic	341	1073	86%	350	0	541	146
Native American	72	345	100%	78	0	221	46
African-American	64	263	100%	56	0	241	66

Race	# of last names	Could be tested	Median SNE	Sig. positive	Sig. negative	Non-sig. positive	Non-sig. negative
			Google+				
White	35	80	43%	0	0	79	1
Asian	394	608	47%	61	0	494	53
Hispanic	341	95	60%	0	0	95	0
Native American	72	1	-17%	0	0	0	1
African-American	64	16	18%	0	0	16	0

A.1 SNE and Social Tie Strength

We also investigated the correlation of the SNE and the strength of the tie between users. Concretely, are users' strong ties more affected by the SNE than their weak ties? Again, we limited this analysis to users from the US and the mentioned popular names of American users on Twitter.

We eliminated all the super-users to better capture the strong and weak ties among normal users. For all normal users, for the link from user A to B, we looked at the Jaccard similarity of the friends of A and B as a measure of the strength of the tie. Then, we considered half of the links with the lower strength as weak links and the other half as strong ties ($threshold = 0.008$). First, we tested the SNE by only considering weak ties, and then by only considering strong ties. In both cases, the SNE was statistically significant. For weak links the preference ranged from 13% to 17% and for strong ties from 10% to 13%, and for all the five names the SNE was slightly stronger for the weak ties. Our results suggest that people are more affected by SNE when they are establishing a weak link. This is in contrast with an earlier study that has found the NLE only affects people's important decisions, such as choosing a job or place to live, and not the more trivial decisions like favorite animals or foods [8]. This observation was explained by an earlier finding that the NLE is a type of implicit egotism and implicit egotism is boosted under stress [10]. However, we do not find evidence to support this finding in Twitter. Though, the results on Google+ are not consistent with these findings and the preference for weak ties ranged from 7-33% and for strong ties

12-45%. Further investigation of differences between Twitter and Google+ is needed to figure out the root of the mentioned inconsistency.

A.2   SNE and Number of Friends

Finally, we examined the correlation between the SNE and the number of friends (followees) of users. The aim is to see if the SNE differs for users with more compared to users with less friends. Similar to before, we considered only users from the US and the mentioned popular names on Twitter. Then, we grouped users based on their number of friends logarithmically, up to 64 friends and a group for users with more than 64 friends. The resulting groups are fairly balanced, with the smallest group (one friend) containing 8% of considered users and the largest group (between 16 and 32 friends) 20% of them. We also use the same group sizes for Google+.

We tested the SNE in each of the groups by only considering the links going out from users of that group and then taking the average of the SNE for the five considered names. Figure 1 shows that there is a noticeable reverse correlation between the number of friends and the SNE. Users with fewer friends are more likely to follow other users with the same name compared to the users with a higher number of friends.

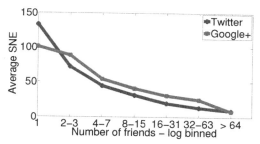

Fig. 1. The average SNE of users grouped by the number of friends

TweetCred: Real-Time Credibility Assessment of Content on Twitter

Aditi Gupta[1], Ponnurangam Kumaraguru[1], Carlos Castillo[2], and Patrick Meier[2]

[1] Indraprastha Institute of Information Technology, Delhi, India
{aditig,pk}@iiitd.ac.in
[2] Qatar Computing Research Institute, Doha, Qatar
chato@acm.org, pmeier@qf.org.qa

Abstract. During sudden onset crisis events, the presence of spam, rumors and fake content on Twitter reduces the value of information contained on its messages (or "tweets"). A possible solution to this problem is to use machine learning to automatically evaluate the credibility of a tweet, i.e. whether a person would deem the tweet believable or trustworthy. This has been often framed and studied as a supervised classification problem in an off-line (post-hoc) setting.

In this paper, we present a semi-supervised ranking model for scoring tweets according to their credibility. This model is used in *TweetCred*, a real-time system that assigns a credibility score to tweets in a user's timeline. *TweetCred*, available as a browser plug-in, was installed and used by 1,127 Twitter users within a span of three months. During this period, the credibility score for about 5.4 million tweets was computed, allowing us to evaluate *TweetCred* in terms of response time, effectiveness and usability. To the best of our knowledge, this is the first research work to develop a real-time system for credibility on Twitter, and to evaluate it on a user base of this size.

1 Introduction

Twitter is a micro-blogging web service with over 600 million users all across the globe. Twitter has gained reputation over the years as a prominent news source, often disseminating information faster than traditional news media. Researchers have shown how Twitter plays a role during crises, providing valuable information to emergency responders and the public, helping reaching out to people in need, and assisting in the coordination of relief efforts (e.g. [9, 12, 18]).

On the other hand, Twitter's role in spreading rumors and fake news has been a major source of concern. Misinformation and disinformation in social media, and particularly in Twitter, has been observed during major events that include the 2010 earthquake in Chile [12], the Hurricane Sandy in 2012 [10] and the Boston Marathon blasts in 2013 [9]. Fake news or rumors spread quickly on Twitter and this can adversely affect thousands of people [16]. Detecting credible or trustworthy information on Twitter is often a necessity, especially

L.M. Aiello and D. McFarland (Eds.): SocInfo 2014, LNCS 8851, pp. 228–243, 2014.
© Springer International Publishing Switzerland 2014

Fig. 1. Screenshot of timeline of a Twitter user when *TweetCred* browser extension is installed

during crisis events. However, deciding whether a tweet is credible or not can be difficult, particularly during a rapidly evolving situation.

Both the academic literature, which we survey on Section 2, and the popular press,[1] have suggested that a possible solution is to automatically assign a score or rating to tweets, to indicate its trustworthiness. In this paper, we introduce *TweetCred* (available at `http://twitdigest.iiitd.edu.in/TweetCred/`), a novel, practical solution based on ranking techniques to assess *credibility* of content posted on Twitter in real-time. We understand credibility as "the quality of being trusted and believed in," following the definition in the Oxford English Dictionary. A tweet is said to be credible, if a user would trust or believe that the information contained on it is true.

In contrast with previous work based on off-line classification of content in a post-hoc setting (e.g. [8, 12] and many others), *TweetCred* uses only the data available on each message, without assuming extensive historical or complete data for a user or an event. Also in contrast with previous work, we evaluate *TweetCred* with more than a thousand users who downloaded a browser extension that enhanced their Twitter timeline, as shown in Figure 1.

The main contributions of this work are:

– We present a semi-supervised ranking model using SVM-rank for assessing credibility based, on training data obtained from 6 high impact crisis events of 2013. An extensive set of 45 features is used to determine the credibility score for each of the tweets.
– We develop and deploy a real time system, *TweetCred*, in the form of a browser extension, web application, and REST API. The *TweetCred* extension

---

[1] `http://www.huffingtonpost.com/dean-jayson/twitter-breaking-news_b_2592078.html`

was installed and used by 1,127 Twitter users within a span of three months, computing the credibility score for about 5.4 million tweets.

– We evaluate the performance of *TweetCred* in terms of response time, effectiveness and usability. We observe that 80% of the credibility scores are computed and displayed within 6 seconds, and that 63% of users either agreed with our automatically-generated scores or disagreed by 1 or 2 points (on a scale from 1 to 7).

This paper is organized as follows: Section 2 briefly reviews work done around this domain. Section 3 describes how we collect labeled data to train our system, and Section 4 how we apply a learning-to-rank framework to learn to automatically rank tweets by credibility. Section 5 presents the implementation details and a performance evaluation, and Section 6 the evaluation from users and their feedback. Finally, in the last section we discuss the results and future work.

2   Survey

In this section, we briefly outline some of the research work done to assess, characterize, analyze, and compute trust and credibility of content in online social media.

Credibility Assessment. Castillo et al. [4] showed that automated classification techniques can be used to detect news topics from conversational topics and assess their credibility based on various Twitter features. They achieved a precision and recall of 70-80% using a decision-tree based algorithm. Gupta and Kumaraguru [7] in their work on analyzing tweets posted during the terrorist bomb blasts in Mumbai (India, 2011), showed that the majority of sources of information are unknown and have low Twitter reputation (small number of followers). The authors in a follow up study applied machine learning algorithms (SVM-rank) and information retrieval techniques (relevance feedback) to assess credibility of content on Twitter [8], finding that only 17% of the total tweets posted about the event contained situational awareness information that was credible. Another, similar work was done by Xia et al. [19] on tweets generated during the England riots of 2011. They used a supervised method based on a Bayesian Network to predict the credibility of tweets in emergency situations. O'Donovan et al. [15] focused their work on finding indicators of credibility during different situations (8 separate event tweets were considered). Their results showed that the best indicators of credibility were URLs, mentions, retweets and tweet length.

Credibility Perceptions. Morris et al. [14] conducted a survey to understand users' perceptions regarding credibility of content on Twitter. They found that the prominent features based on which users judge credibility are features visible at a glance, for example, the username and picture of a user. Yang et al. [21] analyzed credibility perceptions of users on two micro-blogging websites: Twitter in the USA and Weibo in China. They found that location and network overlap features had the most influence in determining the credibility perceptions of users.

Credibility of Users. Canini et al. [3] analyzed the usage of automated ranking strategies to measure credibility of sources of information on Twitter for any given topic. The authors define a credible information source as one which has trust and domain expertise associated with it. Ghosh et al. [6] identified topic-based experts on Twitter using features obtained from user-created list, relying on the wisdom of Twitter's crowds.

System. Ratkiewicz et al. [17] introduced Truthy,[2] a system to study information diffusion on Twitter and compute a trustworthiness score for a public stream of micro-blogging updates related to an event. Their focus is to detect political smears, astroturfing, and other forms of politically-motivated disinformation campaigns.

To the best our knowledge, the work presented in this paper is the first research work that describes the creation and deployment of a practical system for credibility on Twitter, including the evaluation of such system with real users.

3   Training Data Collection

TweetCred is based on semi-supervised learning. As such, it requires as input a *training set* of tweets for which a credibility label is known.

To create this training set, we collect data from Twitter using Twitter's streaming API,[3] filtering it using keywords representing six prominent events in 2013: (i) the Boston Marathon blasts in the US, (ii) Typhoon Haiyan/Yolanda in the Philippines, (iii) Cyclone Phailin in India, (iv) the shootings in the Washington Navy Yard in the US, (v) a polar vortex cold wave in North America, and (vi) the tornado season in Oklahoma, US. These events affected a large population and generated a high volume of content in Twitter. Table 1 describes the characteristics of the data collected around the events we used to build a training set.

In order to create ground truth for building our model for credibility assessment, we obtained labels for around 500 tweets selected uniformly at random from each event. The annotations were obtained through crowdsourcing provider CrowdFlower.[4] We selected only annotators living in the United States. For each tweet, we collected labels from three different annotators, keeping the majority among the options chosen by them.

The annotation proceeded in two steps. In the first step, we asked users if the tweet contained information about the event to which it corresponded, with the following options:

—R1. The tweet contains information about the event.

—R2. The tweet is related to the event, but contains no information.

—R3. The tweet is not related to the event.

—R4. None of the above (skip tweet).

---

[2] http://truthy.indiana.edu/

[3] https://dev.twitter.com/docs/api/streaming

[4] http://www.crowdflower.com/

Table 1. Number of tweets and distinct Twitter users from which data was collected for the purposes of creating a training set. From each event, 500 tweets were labeled.

Event	Tweets	Users
Boston Marathon Blasts	7,888,374	3,677,531
Typhoon Haiyan / Yolanda	671,918	368,269
Cyclone Phailin	76,136	34,776
Washington Navy yard shootings	484,609	257,682
Polar vortex cold wave	143,959	116,141
Oklahoma Tornadoes	809,154	542,049
Total	10,074,150	4,996,448

Along with the tweets for each event, we provided a brief description of the event and links from where users could read more about it. In this first step, 45% of the tweets were considered informative (class R1), while 40% were found to be related to the event for which they were extracted, but not informative (class R2), and 15% were considered as unrelated to it (class R3).

In the second step, we selected the 45% of tweets that were marked as informative, and annotated them with respect to the credibility of the information conveyed by it. We provided a definition of credibility ("the quality of being trusted and believed in"), and example tweets for each option in the annotation. We asked workers to score each tweet according to its credibility with the following options:

—C1. Definitely credible.
—C2. Seems credible.
—C3. Definitely incredible.
—C4. None of the above (skip tweet).

Among the informative tweets, 52% of tweets were labeled as *definitively credible*, 35% as *seems credible*, and 13% as *definitively incredible*.

4   Credibility Modeling

Our aim is to develop a model for ranking tweets by credibility. We adopt a semi-supervised learning-to-rank approach. First, we perform feature extraction from the tweets. Second, we compare the speed and accuracy of different machine learning schemes, using the training labels obtained in the previous section.

4.1   Feature Extraction

Generating feature vectors from the tweets is a key step that impacts the accuracy of any statistical model built from this data. We use a collection of features from previous work [1, 4, 8, 22], restricting ourselves to those that can be derived from single tweets in real-time.

Table 2. Features used by the credibility model

Feature set	Features
Tweet meta-data	Number of seconds since the tweet; Source of tweet (mobile / web/ etc); Tweet contains geo-coordinates
Tweet content (simple)	Number of characters; Number of words; Number of URLs; Number of hashtags; Number of unique characters; Presence of stock symbol; Presence of happy smiley; Presence of sad smiley; Tweet contains 'via'; Presence of colon symbol
Tweet content (linguistic)	Presence of swear words; Presence of negative emotion words; Presence of positive emotion words; Presence of pronouns; Mention of self words in tweet (I; my; mine)
Tweet author	Number of followers; friends; time since the user if on Twitter; etc.
Tweet network	Number of retweets; Number of mentions; Tweet is a reply; Tweet is a retweet
Tweet links	WOT score for the URL; Ratio of likes / dislikes for a YouTube video

A tweet as downloaded from Twitter's API contains a series of fields in addition to the text of the message.[5] For instance, it includes meta-data such as posting date, and information about its author at the time of posting (e.g. his/her number of followers). For tweets containing URLs, we enriched this data with information from the Web of Trust (WOT) reputation score.[6] The features we used can be divided into several groups, as shown in Table 2. In total, we used 45 features.

4.2 Learning Scheme

We tested and evaluated multiple learning-to-rank algorithms to rank tweets by credibility. We experimented with various methods that are typically used for information retrieval tasks: Coordinate Ascent [13], AdaRank [20], RankBoost [5] and SVM-rank [11]. We used two popular toolkits for ranking, RankLib[7] and SVM-rank.[8]

Coordinate Ascent is a standard technique for multi-variate optimization, which considers one dimension at a time. SVM-rank is a pair-wise ranking technique that uses SVM (Support Vector Machines). It changes the input data, provided as a ranked list, into a set of ordered pairs, the (binary) class label for every pair is the order in which the elements of the pair should be ranked. AdaRank trains the model by minimizing a loss function directly defined on the performance measures. It applies a boosting technique in ranking methods. RankBoost is a boosting algorithm based on the AdaRank algorithm; it also runs for many iterations or rounds and uses boosting techniques to combine weak rankings.

Evaluation metrics. The two most important factors for a real-time system are correctness and response time, hence, we compared the methods based on

---

[5] https://dev.twitter.com/docs/api/1.1/get/search/tweets
[6] The WOT reputation system computes website reputations using ratings received from users and information from third-party sources. https://www.mywot.com/
[7] http://sourceforge.net/p/lemur/wiki/RankLib/
[8] http://www.cs.cornell.edu/people/tj/svm_light/svm_rank.html

two evaluation metrics, NDCG (Normalized Discounted Cumulative Gain) and running time. NDCG is useful to evaluate data having multiple grades, as is the case in our setting. Given a query q and its rank-ordered vector V of results $\langle v_1, \ldots, v_m \rangle$, let label($v_i$) be the judgment of v_i. The discounted cumulative gain of V at document cut-off value n is:

$$DCG@n = \Sigma_{i=1}^{n} \frac{1}{log_2(1+i)}(2^{label(v_i)} - 1) .$$

The normalized DCG of V is the DCG of V divided by the DCG of the "ideal" (DCG-maximizing) permutation of V (or 1 if the ideal DCG is 0). The NDCG of the test set is the mean of the NDCGs of the queries in the test set.

To map the training labels from Section 3 to numeric values, we used the following transformation: 5=Informative and definitively credible (class R1.C1), 4=Informative and seems credible (R1.C2), 3=Informative and definitively incredible (R1.C3), 2=Not informative (R2), 1=Not related (R3). From the perspective of quality of content in a tweet, a tweet that is not credible, but has some information about the event, is considered better than a non-informative tweet.

Evaluation. We evaluated the different ranking schemes using 4-fold cross validation on the training data. Table 3 shows the results. We observe that AdaRank and Coordinate Ascent perform best in terms of $NDCG@n$ among all the algorithms; SVM-rank is a close second. The gap is less as we go deeper into the result list, which is relevant given that Twitter's user interface allow users to do "infinite scrolling" on their timeline, looking at potentially hundreds of tweets.

The table also presents the learning (training) and ranking (testing) times for each of the methods. The ranking time of all methods was less than one second, but the learning time for SVM-rank was, as expected, much shorter than for any of the other methods. Given that in future versions of *TweetCred* we intend to re-train the system using feedback from users, and hence need short training times, we implemented our system using SVM-rank.

Table 3. Evaluating ranking algorithms in terms of Normalized Discounted Cumulative Gain (NDCG) and execution times. Boldface values in each row indicate best results.

	AdaRank	Coord. Ascent	RankBoost	SVM-rank
NDCG@25	**0.6773**	0.5358	0.6736	0.3951
NDCG@50	**0.6861**	0.5194	0.6825	0.4919
NDCG@75	0.6949	**0.7521**	0.6890	0.6188
NDCG@100	0.6669	**0.7607**	0.6826	0.7219
Time (training)	35-40 secs	1 min	35-40 secs	9-10 secs
Time (testing)	<1 sec	<1 sec	<1 sec	<1 sec

The top 10 features for the model of credibility ranking built using SVM-Rank are: (1) tweet contains *via*, (2) number of characters, (3) number of unique characters, (4) number of words, (5) user has location in profile, (6) number of retweets, (7) age of tweet, (8) tweet contains a URL, (9) ratio number of statuses/followers of the author, and (10) ratio friends/followers of the author. We observe that majority of the top features for assessing credibility of content were tweet based features rather user attributes.

5   Implementation and Performance Evaluation

In order to encourage many users to interact with *TweetCred*, we provided it in a way that was easy to use, as a browser extension. We also provided access to *TweetCred* as a web-based application and as an API, but the browser extension was much more commonly used.

5.1   Implementation

The implementation includes a back-end and a front-end which interact over RESTful HTTP APIs.

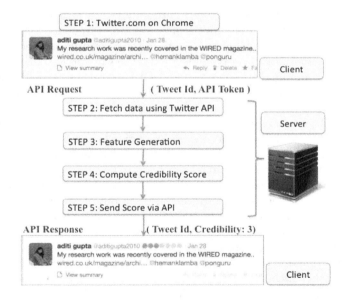

Fig. 2. Data flow steps of the *TweetCred* extension and API

Back-end. Figure 2 shows the basic architecture of the system.

The flow of information in *TweetCred* is as follows: A user logs on to his/her Twitter account on http://twitter.com/, once the tweets starts loading on the webpage, the browser extension passes the IDs of tweets displayed on the page to our server on which the credibility score computation module is done. We do not scrape the tweet or user information from the raw HTML of web page and merely pass the tweet IDs to web server. The reason is that what the server needs to compute credibility is more than what is shown through Twitter's interface.

From the server a request is made to Twitter's API to fetch the data about an individual tweet. Once the complete data for the tweet is obtained, the feature vectors are generated for the tweet, and then the credibility score is computed using the prediction model of SVM-rank. This score is re-scaled to a value in the range from 1 to 7 using the distribution of values in our training data. Next, this score is sent back to the user's browser. Credibility scores are cached for 15 minutes, meaning that if a user requests the score of a tweet whose score was requested less than 15 minutes ago, the previously-computed score is re-used. After this period of time, cached credibility scores are discarded and computed again if needed, to account for changes in tweet or user features such as the number of followers, retweets, favorites and replies.

All feature extraction and credibility computation scripts were written in *Python* with *MySQL* as a database back-end. The RESTful APIs were implemented using *PHP*. The hardware for the backend was a mid-range server (Intel Xeon E5-2640 2.50GHz, 8GB RDIMM).

Front-end. The Chrome browser currently enjoys the largest user base by far among various web browsers,[9] and hence was our target for the first version of the browser extension. In order to minimize computation load on the web browser, heavy computations were offloaded to the web server, hence the browser extension had a minimalistic memory and CPU footprint. This design ensures that the system would not result in any performance bottleneck on client's web browser.

In an initial pilot study conducted for *TweetCred* with 10 computer science students that are avid Twitter users, we used the *Likert Scale* of score 1–5 for showing credibility for a tweet.[10] We collected their feedback on the credibility score displayed to them via personal interviews. The users found it difficult to differentiate between a high credibility score of 4 and a low credibility score of 2, as the difference in values seemed too small. Eight out of the ten participants felt that the scale of rankings should be slightly larger. They were more comfortable with a scale of 1–7 ranking, which we adopted.

TweetCred displays this score next to a tweet in a user's timeline, as shown in Figure 1. Additionally, the user interface includes a feedback mechanism. When end users are shown the credibility score for a tweet, they are given the option to provide feedback to the system, indicating if they agree or disagree with the credibility score for each tweet. Figures 3(a) shows the two options given to the user upon hovering over the displayed credibility score. In case the user disagrees

---

[9] As of August 2014, Chrome has 59% of market share, more than doubling the 25% of the second place, Firefox
http://www.w3schools.com/browsers/browsers_stats.asp

[10] http://www.clemson.edu/centers-institutes/tourism/documents/sample-scales.pdf

(a)   (b)

Fig. 3. Users can provide feedback to the system. Figure (a) shows how users can push the agree ("thumbs up") button to agree with a rating, the case for the disagree ("thumbs down") button is analogous. Figure (b) shows how users can provide their own credibility rating for a tweet.

with the credibility rating, s/he is asked to provide what s/he considers should be the credibility rating, as shown in Figure 3(b). The feedback provided by the user is sent over a separate REST API endpoint and recorded in our database.

5.2   Response Time

We analyzed the response time of the browser extension, measured as the elapsed time from the moment in which a request is sent to our system to the moment in which the resulting credibility score is returned by the server to the extension. Figure 4 shows the CDF of response times for 5.4 million API requests received. From the figure we can observe that for 82% of the users the response time was less than 6 seconds, while for 99% of the users the response time was under 10 seconds. The response time is dominated by the requests done to Twitter's API to obtain the details for a tweet.

6   User Testing

We uploaded *TweetCred* to the Chrome Web Store,[11] and advertised its presence via social media and blogs. We analyzed the deployment and usage activity of *TweetCred* on the three-months period from April 27th, 2014 to July 31st, 2014. A total of 1,127 unique Twitter users used *TweetCred*. They constitute a diverse sample of Twitter users, from users having very few followers to one user having 1.4 million followers. Their usage of *TweetCred* was also diverse, with two users computing the credibility scores of more than 50,000 tweets in his/her timeline, while the majority of users computed credibility scores for less than 1,000 tweets.

Table 4 presents a summary of usage statistics for *TweetCred*. In total 5,451,961 API requests for the credibility score of a tweet were made.

---

[11] http://bit.ly/tweetcredchrome

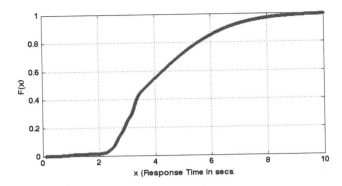

Fig. 4. CDF of response time of *TweetCred*. For 82% of the users, response time was less than 6 seconds and for 99% of the users, the response time was under 10 seconds.

Table 4. Summary statistics for the usage of *TweetCred*

Date of launch of *TweetCred*	27 Apr, 2014
Credibility score seen by users (total)	5,438,115
Credibility score seen by users (unique)	4,540,618
Credibility score requests for tweets (Chrome extension)	5,429,257
Credibility score requests for tweets (Browser version)	8,858
Unique Twitter users	1,127
Feedback was given for tweets	1,273
Unique users who gave feedback	263
Unique tweets which received feedback	1,263

We received feedback from users of our system in two ways. First, the users could give their feedback on each tweet for which a credibility score was computed. Secondly, we asked users to fill a usability survey on our website.

6.1   User Feedback

Out of the 5.4 million credibility score requests served by *TweetCred*, we received feedback for 1,273 of them. When providing feedback, users had the option of either agreeing or disagreeing with our score. In case they disagreed, they were asked to mark the correct score according to them. Table 5 shows the breakdown of the received feedback. We observed that for 40% of tweets for which user's provided feedback agreed with the credibility score given by *TweetCred*, while 60% disagreed—this can be partially explained by self-selection bias due to cognitive dissonance: users are moved to react when they see something that does not match their expectations.

Credibility Rating Bias. For the approximately 60% tweets for which users disagreed with our score, for 49% of the tweets the users felt that credibility score

Table 5. Feedback given by users of *TweetCred* on specific tweets ($n = 1,273$)

	Observed	95% Conf. interval
Agreed with score	40.14	(36.73, 43.77)
Disagreed with score	59.85	(55.68, 64.26)
Disagreed: score should be higher	48.62	(44.86, 52.61)
Disagreed: score should be lower	11.23	(9.82, 13.65)
Disagreed by 1 point	8.71	(7.17, 10.50)
Disagreed by 2 points	14.29	(12.29, 16.53)
Disagreed by 3 points	12.80	(10.91, 14.92)
Disagreed by 4 points	10.91	(9.17, 12.89)
Disagreed by 5 points	6.52	(5.19, 8.08)
Disagreed by 6 points	6.59	(5.26, 8.16)

should have been higher than the one given by *TweetCred*, while for approximately 11% thought it should have been lower. This means *TweetCred* tends to produce credibility scores that are lower than what users expect. This may be in part due to the mapping from training data labels to numeric values, in which tweets that were labeled as "not relevant" or "not related" to a crisis situation were assigned lower scores. To test this hypothesis, we use keyword matches to sub-sample, from the tweets for which a credibility score was requested by users, three datasets corresponding to crisis events that occurred during the deployment of TweetCred: the crisis in Ukraine ($3,637$ tweets), the Oklahoma/Arkansas tornadoes ($1,362$ tweets), and an earthquake in Mexico ($1,476$ tweets).

Figure 5 compares the distribution of scores computed in real-time by Tweet-Cred for the tweets on these three crisis events against a random sample of all tweets for which credibility scores were computed during the same time period. We observe that in all crisis events the credibility scores are higher than in the background distribution. This confirms the hypothesis that *TweetCred* gives higher credibility scores to tweets that are related to a crisis over general tweets.

6.2 Usability Survey

To assess the overall utility and usability of the *TweetCred* browser extension, we conducted an online survey among its users. An unobtrusive link to the survey appeared on the right corner of Chrome's address bar when users visited Twitter.[12] The survey link was accessible only to those users who had installed the extension, this was done to ensure that only actual users of the system gave their feedback. A total of 67 users participated. The survey contained the standard 10 questions of the *System Usability Scale* (SUS) [2]. In addition to SUS questions, we also added questions about users' demographics such as gender, age, etc. We obtained an overall SUS score of 70 for *TweetCred*, which is considered above

---

[12] http://twitdigest.iiitd.edu.in/TweetCred/feedback.html

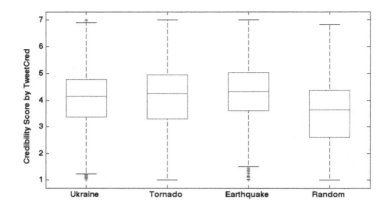

Fig. 5. Distribution of credibility scores. We observe that during crisis events larger percentage of tweets have higher credibility than during non-crisis.

average from a system's usability perspective.[13] In the survey, 74% of the users found *TweetCred* easy to use (agree/strongly agree); 23% of the users thought there were inconsistencies in the system (agree/strongly agree); and 81% said that they may like to use *TweetCred* in their daily life.

User Comments. *TweetCred* system was appreciated by majority of users for its novelty and ease of use. Users also expressed their desire to know more about the system and its backend functionality. One recurring concern of users was related to the negative bias of the credibility scores. Users expressed that the credibility score given by *TweetCred* were low, even for tweets from close contacts in which they fully trust. For instance, one of the user of *TweetCred* said: *"People who I follow, who I know are credible, get a low rating on their tweets"*. Such local friendships and trust relationships are not captured by a generalized model built on the entire Twitter space. Other comments we received about *TweetCred* in the survey and from tweets about *TweetCred* were:

– "I plan on using this to monitor public safety situations on behalf of the City of [withheld]'s Office of Emergency Management."

– "Very clever idea but Twitter's strength is simplicity - I found this a distraction for daily use."

– "It's been good using #TweetCred & will stick around with it, thanks!"

– "It's unclear what the 3, 4 or 5 point rating mean on opinions / jokes, versus factual statements."

7   Conclusions and Future Work

We have described *TweetCred*, a real-time web-based system to automatically evaluate the credibility of content on Twitter. The system provides a credibility

---

[13] http://www.measuringusability.com/sus.php

rating from 1 (low credibility) to 7 (high credibility) for each tweet on a user's Twitter timeline. The score is computed using a semi-supervised automated ranking algorithm, trained on human labels obtained using crowdsourcing, that determines credibility of a tweet based on more than 45 features. All features can be computed for a single tweet, and they include the tweets content, characteristics of its author, and information about external URLs.

Future Work. Our evaluation shows that both in terms of performance, accuracy, and usability, it is possible to bring automatic credibility ratings to users on a large scale. At the same time, we can see that there are many challenges around issues including personalization and context. With respect to personalization, users would like to incorporate into the credibility ratings the fact that their trust some of their contacts more than others. Regarding context, it is clear from the user feedback and our own observations, that there are many cases in which it may not be valid to issue a credibility rating, such as tweets that do not try to convey factual information. In future, we would also like to study the intersection between the psychology literature about information credibility and the credibility of content in Twitter.

TweetCred's deployment stirred a wide debate on Twitter regarding the problem and solutions for the credibility assessment problem on Twitter. The browser extension featured in many news websites and blogs including the Washington Post,[14] the New Yorker[15] and the Daily Dot[16] among others, generating debates in these platforms. We can say that social media users expect technologies that help them evaluate the credibility of the content they read. *TweetCred* is a first step towards fulfiling this expectation.

Acknowledgments. We would like to thank Nilaksh Das and Mayank Gupta in helping us with the web development of *TweetCred* system. We would like to express our sincerest thanks to all members of Precog, Cybersecurity Education and Research Centre at Indraprastha Institute of Information Technology, Delhi, and Qatar Computing Research Institute for their continued feedback and support.[17] We would like to thank the Government of India for funding this project.

References

[1] Aggarwal, A., Rajadesingan, A., Kumaraguru, P.: Phishari: Automatic realtime phishing detection on Twitter. In: 7th IEEE APWG eCrime Researchers Summit, eCRS (2012)

---

[14] http://wapo.st/1pWEOWd
[15] http://newyorker.com/online/blogs/elements/2014/05/can-tweetcred-solve-twitters-credibility-problem.html
[16] http://www.dailydot.com/technology/tweetcred-chrome-extension-addon-plugin/
[17] http://precog.iiitd.edu.in/
http://cerc.iiitd.ac.in/
http://www.qcri.org.qa/

[2] Brooke, J.: SUS: A quick and dirty usability scale. In: Jordan, P.W., Weerdmeester, B., Thomas, A., Mclelland, I.L. (eds.) Usability evaluation in industry. Taylor and Francis, London (1996)

[3] Canini, K.R., Suh, B., Pirolli, P.L.: Finding credible information sources in social networks based on content and social structure. In: SocialCom (2011)

[4] Castillo, C., Mendoza, M., Poblete, B.: Information credibility on Twitter. In: Proc. WWW, pp. 675–684. ACM (2011),
http://doi.acm.org/10.1145/1963405.1963500

[5] Freund, Y., Iyer, R., Schapire, R.E., Singer, Y.: An efficient boosting algorithm for combining preferences. J. Mach. Learn. Res. 4, 933–969 (2003)

[6] Ghosh, S., Sharma, N., Benevenuto, F., Ganguly, N., Gummadi, K.: Cognos: crowdsourcing search for topic experts in microblogs. In: Proc. SIGIR (2012)

[7] Gupta, A., Kumaraguru, P.: Twitter explodes with activity in Mumbai blasts! a lifeline or an unmonitored daemon in the lurking? Tech. Rep. IIITD-TR-2011-005, IIIT, Delhi (2011)

[8] Gupta, A., Kumaraguru, P.: Credibility ranking of tweets during high impact events. In: Proc. 1st Workshop on Privacy and Security in Online Social Media, PSOSM 2012, pp. 2:2–2:8. ACM (2012),
http://doi.acm.org/10.1145/2185354.2185356

[9] Gupta, A., Lamba, H., Kumaraguru, P.: $1.00 per rt #bostonmarathon #prayforboston: Analyzing fake content on Twitter. In: Proc. Eighth IEEE APWG eCrime Research Summit (eCRS), p. 12. IEEE (2013a)

[10] Gupta, A., Lamba, H., Kumaraguru, P., Joshi, A.: Faking sandy: characterizing and identifying fake images on Twitter during hurricane sandy. In: Proc. WWW Companion. International World Wide Web Conferences Steering Committee, pp. 729–736 (2013b)

[11] Joachims, T.: Optimizing search engines using clickthrough data. In: Proc. KDD, pp. 133–142. ACM (2002), http://doi.acm.org/10.1145/775047.775067

[12] Mendoza, M., Poblete, B., Castillo, C.: Twitter under crisis: can we trust what we rt? In: Proc. First Workshop on Social Media Analytics, SOMA 2010, pp. 71–79. ACM (2010), http://doi.acm.org/10.1145/1964858.1964869

[13] Metzler, D., Croft, W.B.: Linear feature-based models for information retrieval. Information Retrieval 10(3), 257–274 (2007)

[14] Morris, M.R., Counts, S., Roseway, A., Hoff, A., Schwarz, J.: Tweeting is believing?: Understanding microblog credibility perceptions. In: Proc. CSCW. ACM (2012), http://doi.acm.org/10.1145/2145204.2145274

[15] O'Donovan, J., Kang, B., Meyer, G., Hšllerer, T., Adali, S.: Credibility in context: An analysis of feature distributions in Twitter. ASE/IEEE International Conference on Social Computing, SocialCom (2012)

[16] Oh, O., Agrawal, M., Rao, H.R.: Information control and terrorism: Tracking the mumbai terrorist attack through Twitter. Information Systems Frontiers (March 2011), http://dx.doi.org/10.1007/s10796-010-9275-8

[17] Ratkiewicz, J., Conover, M., Meiss, M., Gonçalves, B., Patil, S., Flammini, A., Menczer, F.: Truthy: mapping the spread of astroturf in microblog streams. In: Proc. WWW 2011 (2011), http://dx.doi.org/10.1145/1963192.1963301

[18] Vieweg, S., Hughes, A.L., Starbird, K., Palen, L.: Microblogging during two natural hazards events: what Twitter may contribute to situational awareness. In: Proc. SIGCHI, CHI 2010, pp. 1079–1088. ACM (2010),
http://doi.acm.org/10.1145/1753326.1753486

[19] Xia, X., Yang, X., Wu, C., Li, S., Bao, L.: Information credibility on Twitter in emergency situation. In: Proc. Pacific Asia Conference on Intelligence and Security Informatics, PAISI 2012 (2012)

[20] Xu, J., Li, H.: Adarank: A boosting algorithm for information retrieval. In: Proc. SIGIR, pp. 391–398. ACM, New York (2007), http://doi.acm.org/10.1145/1277741.1277809

[21] Yang, J., Counts, S., Morris, M.R., Hoff, A.: Microblog credibility perceptions: Comparing the usa and china. In: Proc. CSCW, pp. 575–586 (2013), http://doi.acm.org/10.1145/2441776.2441841

[22] Yardi, S., Romero, D., Schoenebeck, G., Boyd, D.: Detecting spam in a Twitter network. First Monday 15(1) (January 2010), http://firstmonday.org/htbin/cgiwrap/bin/ojs/index.php/fm/article/view/2793

Can Diversity Improve
Credibility of User Review Data?

Yoshiyuki Shoji, Makoto P. Kato, and Katsumi Tanaka

Department of Social Informatics
Graduate School of Informatics, Kyoto University
Kyoto, Japan
{shoji,kato,tanaka}@dl.kuis.kyoto-u.ac.jp

Abstract. In this paper, we propose methods to estimate the credibility of reviewers as an individual and as a group, where the credibility is defined as the ability of precisely estimating the quality of items. Our proposed methods are built on two simple assumptions: 1) a reviewer who has reviewed many and diverse items has high credibility, and 2) a group of reviewers is credible if the group consists of many and diverse reviewers. To verify the two assumptions, we conducted experiments with a movie review dataset. The experimental results showed that the diversity of reviewed items and reviewers was effective to estimate the credibility of reviewers and reviewer groups, respectively. Therefore, yes, the diversity does improve the credibility of user review data.

1   Introduction

The rapid growth of the World Wide Web and Internet shopping services has enabled users to select from a huge number of commercial products on the Internet. Thus, the importance of user review data has increased, as it provides opinions and impressions that help users choose a quality item. There are many reviews for a variety of items on the Web, some of which are authored by professionals and others that are authored by non-professionals. Since professional reviews are available only for a limited number of items, even non-professional reviews are also useful for users to help making a decision.

However, there is a problem of *credibility* in utilizing reviews of general users. Since user reviews can be posted by any kinds of users including experts, novices, and even spammers, each review and aggregation of reviews can be biased and different from what the general public feels. Even users familiar with a particular domain cannot always produce a widely acceptable review, as they can be highly accustomed and accordingly biased to the domain. For example, users who have watched many Science Fiction(SF) movies might be likely to give a lower score to a SF movie than ordinary users, since they know more high-quality SF movies and use them as the basis for evaluating the other SF movies.

In this paper, we focus particularly on the credibility of reviewers, where the credibility of reviewers is defined as the ability of precisely estimating the item quality. This ability is defined for a single reviewer, as well as a group of reviewers

L.M. Aiello and D. McFarland (Eds.): SocInfo 2014, LNCS 8851, pp. 244–258, 2014.
© Springer International Publishing Switzerland 2014

where the quality of items is estimated by aggregated reviews (e.g. the mean of their review scores). Thus, two problems regarding credibility are addressed in this paper: 1) estimating the credibility of a single reviewer, and 2) estimating the credibility of a group of reviewer.

We tackle the first problem to discover *experts* based on their review experience approximated by the number of reviews, as well as diversity of reviewed items. Although the credibility of a reviewer possibly correlates to the number of reviews that he has posted, many reviews do not always guarantee high credibility of a reviewer. As we discussed earlier, users who have reviewed only a specific category of items might post highly biased reviews. Therefore, we also consider the diversity of reviewed items to accurately estimate the reviewer credibility, assuming that a reviewer who has reviewed in diverse categories has higher credibility. For example, we expect that users who reviewed a wide variety of movies have a higher ability to evaluate the quality of movies than those who reviewed only SF movies.

We tackle the second problem to precisely estimate the quality of items by aggregating reviews of a reviewer group. Even if the credibility of individuals is low, it is possible to achieve high credibility when their reviews are aggregated. This phenomenon is known as *the wisdom of crowds* [12], in which one of the key criteria to obtain quality results is diversity of opinions. Thus, our proposed method to estimate the credibility of a reviewer group stands on diversity of reviewers, with an assumption that a group of more diverse reviewers has higher credibility.

To verify the two assumptions mentioned above, we conducted experiments with a movie review dataset. The credibility of reviewers was measured by the similarity between their review score and a *true* score, which was approximated by the score given by a well-known professional reviewer. Our experimental results showed that the diversity of reviewed items and reviewers in a group was effective to estimate the credibility of a reviewer and a group of reviewers, respectively. Therefore, yes, the diversity does improve the credibility of user review data. The rest of this paper is organized as follows. Section 2 describes the related work. In Section 3, we introduce methods of estimating the credibility of reviewers based on diversity. Section 4 describes our experiments, and Section 5 evaluates our method in light of the experimental results. We conclude this paper in Section 6.

2   Related Work

This section introduces research on finding experts and its application to recommendation in Section 2.1, and research on diversity in Section 2.2.

2.1   Expert Detection and Its Application to Recommendation

Finding experts has a long history and has been recently conducted in consumer generated media(CGM) sites. One of the representative examples is expert finding in community-based question and answering (CQA) sites. Liu and Koll [5] proposed a method to find experts from CQA sites by focusing on the past answers given by users. In this work, experts are defined as users who can answer a certain kind of questions. The basic assumption used in their method is that users are able to answer a question if they have answered similar questions in the past.

There is some previous work on discovering experts to improve the accuracy of recommendations. One of the assumptions in this line of work is that an item evaluated as high-quality by experts is likely to be high-quality for many other users. Amatriain et al. [2] proposed a recommendation method that utilizes only the *nearest experts*, which are defined as users who posted a sufficient number of reviews, and are the most similar to a user who receives a recommendation. The performance of the proposed method was comparable to traditional collaborative filtering algorithms, even when a small expert set was used. Their expert detection method was based solely on the number of reviews, and the method did not take into account reviewed items. In our work, we utilize the diversity of reviewed items to find experts, and propose a method to aggregate reviews to precisely estimate the quality of items.

Sha et al. [9] proposed a method of seeking two different kinds of experts from an online photo sharing community: *trend makers* and *trend spotters*, and recommending trends in the community esitimated by these experts.

McAuley and Leskovec proposed [7] a method to find domain experts by using their review experience. Users are expected to become more professional in a domain if they work on the domain for a longer time. This work pointed out two important perspectives of expertise: 1) a user becomes an expert if s/he has been engaged in a domain for a long time, and 2) the evaluations done by novices tends to be diverse, while those by experts tends to be focused.

2.2   Measuring Diversity

Our proposed method incorporates a diversity-based measure to find experts and evaluate the credibility of a group of reviewers. There have been various studies on diversity specialized for different problems.

Collective intelligence has been actively discussed, as the collaboration on Web sites became a popular activity. Surowiecki [12] presented in his book some conditions of data under which the wisdom of crowds work correctly: *diversity of opinion*, *independence*, and *decentralization*. Once the three requirements are satisfied, useful knowledge can be built from the data by means of *aggregation*.

Diversity has been extensively used in the field of information retrieval. One of the most active research topics is diversification of Web search results [1,13,3]. For example, maximal marginal relevance [4] was used to diversify search results by decreasing the score of the pages similar to ones ranked in higher positions.

The research areas that focus on diversity are not limited to computer sciences, but include sociology, ecology, life science, economics, etc. Many diversity measures have been proposed especially in the biology area. Stirling [11] summarized three key factors regarding categorical diversity: *variety*, *balance*, and *disparity*. Biodiversity has recently received attention, and is measured by

Shannon-Wiener index [8], which was developed based on Shannon entropy. The index highly correlates to the number of breeds and balance across different breeds. Another diversity index, Simpson's diversity index [6], is defined as the probability of breed coincidence of two randomly-selected individuals. An alternative to these diversity measures was proposed in our previous work [10].

Since the diversity is a multi-faceted concept as can be seen in the earlier discussion, the optimal design of a diversity measure highly depends on its application domain. In this paper, we use two different kinds of diversity measures for reviewer groups, namely, entropy-based and variance-based diversity measures. The former measures the variety and balance, while the latter measures the disparity of reviewers. These two measures were compared in our experiments.

3 Method

This section introduces methods to estimate the credibility of a reviewer and a reviewer group based on diversity measures. Our methods are designed to be applicable to a wide variety of user review data such as movies, hotels, books, restaurants, etc.

3.1 User Review Data

User review data can be modeled by a tripartite graph with a category hierarchy. The tripartite graph consists of reviewers, items, categories, as well as reviewer-item and item-category edges. The category hierarchy is a set of category-category edges. More specifically, user review data D is defined as follows:

$$D = (U, R, I, B, C, H), \tag{1}$$

where U is a set of reviewers, I is a set of items, C is a set of categories. A set of edges $R \subset U \times I$ represents reviews of reviewers for items, e.g. $(u, i) \in R$ indicates that reviewer u reviewed item i. A set of edges $B \subset I \times C$ represents categories of items, e.g. $(i, c) \in B$ indicates that item i belongs to category c. A set of edges $H \subset C \times C$ represents *is-a* relationships between pairs of categories, e.g. $(c_j, c_k) \in H$ indicates that category c_j is a sub-category of category c_k.

Category tree $T = (C, H)$ is a rooted tree whose root is $c_{\text{root}} \in C$. Children of c_{root}, i.e. $M = \{c \mid c \in C \wedge (c, c_{\text{root}}) \in H\}$, are called *main categories* and distinguished from the other categories.

Some variables used in our proposed methods are defined below. The number of reviews given by user u is defined as follows:

$$n_u = |\{i \mid i \in I \wedge (u, i) \in R\}|. \tag{2}$$

The number of items that belong to category c is defined as follows:

$$n_c = |\{i \mid i \in I \wedge (i, c) \in B\}|. \tag{3}$$

Finally, we define the number of items that belong to category c and have been reviewed by user u as follows:

$$n_{u,c} = |\{i \mid i \in I \wedge (u, i) \in R \wedge (i, c) \in B\}|. \tag{4}$$

3.2 Estimating the Credibility of a Reviewer

The first problem we tackle is to estimate the credibility of each reviewer. Recall that the credibility is the ability of precisely estimating the quality of items. Our method is based on the assumption that a reviewer who reviewed many and diverse items has high credibility. The reason why we came up with this assumption is explained as follows. Suppose that there are two reviewers: one reviewed 10 movies, while another reviewed 100 movies. According to the assumption, the latter reviewer is more credible, as his expertise is expected to be higher than the former reviewer. Then suppose that there are another pair of reviewers: one reviewed 100 SF movies, while another reviewed 100 a wide variety of movies. We assume that the latter is more credible since his review is expected to be unbiased compared to the former reviewer.

The following formula is derived if we follow the assumption on the credibility of individual reviewer:

$$\text{Credibility}(u) = \alpha n_u \text{Div}(u), \tag{5}$$

where α is a parameter, n_u is the number of items reviewed by user u, and $\text{Div}(u)$ is the diversity of items reviewed by user u. We then model the diversity of reviewed items based on the idea of Shannon-Wiener index [8], which measures the diversity by the entropy over species. Regarding main categories as species in our case, Shannon-Wiener index is defined as follows:

$$H(u) = -\sum_{c \in M} p_u(c) \log p_u(c), \tag{6}$$

where $p_u(c)$ is the probability that user u reviews an item that belongs to category c. This probability can be estimated by the number of items of category c reviewed by user u divided by the number of items reviewed by user u: $p_u(c) = n_{u,c}/n_u$.

One of the problems of Shannon-Wiener index is that it is agnostic about the prior category distribution. Suppose that there are 10 horror and 100 SF movies. Although the maximum entropy is achieved by reviewing 10 horror and 10 SF movies, this reviewer is considered as biased to horror movies, as he reviewed all the horror movies despite the small number of horror ones. Therefore, we slightly modify Shannon-Wiener index by taking into account the prior category distribution. More specifically, we measure the diversity by the difference of the category distribution of a reviewer from the prior category distribution, i.e. Kullback-Leibler divergence of the two distributions. Letting $p(c)$ be the prior category probability, Kullback-Leibler divergence is defined as follows:

$$\text{KL}(u) = -\sum_{c \in M} p_u(c) \log \frac{p_u(c)}{p(c)}, \tag{7}$$

where $p(c)$ is the number of items of category c divided by the number of items: $p(c) = n_c/|I|$.

Finally, we define the diversity of a reviewer as follows:

$$\mathrm{Div}(u) = \exp(-\mathrm{KL}(u)). \tag{8}$$

Note that the exponential function is not essential, but is applied to make the diversity function $\mathrm{Div}(u)$ positively correlate to the diversity. This diversity function becomes larger when the category distribution of a reviewer and prior category distribution are closer. Thus, a reviewer who has evenly reviewed items is considered as credible, as he is considered as unbiased to any category.

3.3   Estimating the Credibility of a Group of Reviewers

The second problem we address is to estimate the credibility of a group of reviewers. Even if the credibility of individual reviewers is not so high, the credibility of a group of reviewers can be high when their reviews are aggregated. For example, the average review score of a group can be close to true quality of items, even if no individual reviewer can precisely estimate the quality.

According to the previous studies on collective intelligence [12], the diversity of members in a group is an important factor to obtain a high-quality result from the group by means of aggregation. For example, there are two groups: one includes ten SF maniacs, while another includes five SF and five horror maniacs. Given an item to each group, the average review score given by the former group might be more biased than the latter, as the aggregated score may reflect only a specific preference of the homogeneous group.

Therefore, we propose methods to estimate the credibility of a reviewer group based on the diversity of the reviewers. Our assumption for this problem is that a group of many and diverse reviewers has high credibility. As the diversity can be measured by three types of aspects, namely, balance, variety, and disparity [11], we propose two diversity measures that take into account different aspects, i.e. entropy-based and variance-based diversity measures.

The entropy-based diversity measure is similar to the one we used in estimating the credibility of individual reviewers, and takes into account the balance and variety of reviewers[1]. A high entropy-based diversity measure indicates that there are more types of reviewers in a group and the distribution of reviewers is balanced across the types. On the other hand, the variance-based diversity measure reflects the disparity aspect of diversity, and becomes high if reviewers in a group are dissimilar each other.

To compute the two diversity measures briefly explained above, it is necessary to model the similarity between reviewers in some way. To this end, we opted to characterize reviewers by using their expertise estimated by their reviews, with an assumption that a reviewer who has reviewed diverse items in a category has high expertise in the category. For instance, a reviewer who have watched and reviewed all of space opera, cyberpunk, and science fantasy movies is expected

---

[1] Balance and variety are simultaneously measured since they are not divisible in many cases.

to have more knowledge in the SF category than one who have reviewed only space opera movies.

In a similar way to the diversity computation for a single reviewer, the expertise of user u in main category c is measured by Kullback-Leibler divergence of the sub-category distribution of a reviewer and prior sub-category distribution:

$$\mathrm{KL_{sub}}(u, c) = - \sum_{s \in \mathrm{Sub}(c)} p_u(s|c) \log \frac{p_u(s|c)}{p(s|c)}, \tag{9}$$

where $\mathrm{Sub}(c)$ is a set of sub-categories of main category c (i.e. $\mathrm{Sub}(c) = \{s \mid s \in C \wedge (s, c) \in H\}$), $p_u(s|c)$ is the probability that user u reviews an item of category s conditioned by category c ($p_u(s|c) = p_u(s)/p_u(c)$), and $p(s|c)$ is the prior probability of category c conditioned by category c ($p(s|c) = p(s)/p(c)$).

As the Kullback-Leibler divergence negatively correlates to the expertise in a main category, we apply an exponential function in the same way as the diversity computation for a single reviewer, and define the expertise of user u in main category c as follows:

$$e_{u,c} = \exp(-\mathrm{KL_{sub}}(u, c)). \tag{10}$$

Below, we explain the two diversity measures in the details.

Entropy-Based Diversity Measure

The entropy-based diversity measure is the entropy of the expertise distribution of a group as a whole with consideration of the prior expertise distribution. We first model the expertise of group $G \subset U$ in category c by aggregating the expertise of reviewers in the group:

$$e_{G,c} = \frac{1}{|G|} \sum_{u \in G} e_{u,c}. \tag{11}$$

We then model the *prior* expertise in category c:

$$e_c = \frac{1}{|U|} \sum_{u \in U} e_{u,c}. \tag{12}$$

The prior expertise can be interpreted as the average expertise in all the reviewers. Although these expertise scores do not represent a probability, we could normalize the expertise scores to treat them as probabilities:

$$p_G^e(c) = \frac{1}{|G|} \sum_{u \in G} e_{u,c}, \tag{13}$$

$$p_c^e = \frac{1}{|U|} \sum_{u \in U} e_{u,c}. \tag{14}$$

Kullback-Leibler divergence of the expertise distribution of a reviewer group and the prior expertise distribution is defined as follows:

$$KL^e(G) = -\sum_{c \in M} p_G^e(c) \log \frac{p_G^e(c)}{p^e(c)}. \tag{15}$$

This divergence represents the closeness between the expertise of a group and prior expertise, and becomes smaller if the group expertise is more evenly distributed against the prior expertise.

Entropy-based diversity measure EDiv is then defined as follows:

$$EDiv(G) = \exp(-KL^e(G)). \tag{16}$$

Note that the exponential function is not essential again.

The entropy-based diversity measure increases as the expertise of a group as a whole is evenly distributed in each category. Note that this measure does not take into account the diversity of each reviewer in a group, and becomes high in both of the following cases: 1) all the reviewers in the group have balanced expertise in each category, and 2) the expertise distribution of the group is close to the prior expertise distribution, even though the expertise distribution of each reviewer is far from the prior expertise distribution.

Variance-Based Diversity Measure

As computing the variance-based diversity measure requires the dissimilarity between reviewers, we first map reviewers on a $|M|$-dimensional space, where each dimension represent the expertise in a main category. A vector of reviewer u is denoted by $\mathbf{v}_u$ and defined as follows:

$$\mathbf{v}_u = (e_{u,c_1}, e_{u,c_2}, \ldots, e_{u,c_{|C|}}), \tag{17}$$

where $e_{u,c}$ is the expertise of reviewer u in category c.

Variance-based diversity measure VDiv, which is the average dissimilarity between individual reviewers and the mean of the reviewers in the group, is defined as follows:

$$VDiv(G) = \frac{1}{|G|} \sum_{u \in G} \|\mathbf{v}_u - \bar{\mathbf{v}}_G\|, \tag{18}$$

where $\bar{\mathbf{v}}_G$ is the mean of reviewer vectors of group G, i.e. $\bar{\mathbf{v}}_G = \frac{1}{|G|} \sum_{u \in G} \mathbf{v}_u$.

In summary, we proposed diversity measures to estimate the credibility of reviewers as an individual and as a group. A variant of Shannon-Wiener index was proposed to measure the diversity for both of the cases, and a variance-based diversity measure was used only for a reviewer group. Note that the entropy-based and variance-based diversity measures correlate to some extent, but behave differently in some cases. For example, the entropy-based diversity measure becomes high if reviewers in a group have similar expertise in a wide variety of categories, whereas the variance-based diversity measure does not. In the next section, we demonstrate the correlation between the credibility and diversity measured by the proposed methods.

4   Experiment

To clarify the effectiveness of our diversity measures for estimating the credibility of reviewers, we conducted experiments by using movie review data taken from Yahoo! Movies. Through the experiments, we tested the validity of the two assumptions: 1) a reviewer who has reviewed many and diverse items has high credibility, and 2) a group of reviewers is credible if the group consists of many and diverse reviewers.

4.1   Dataset

The movie review data was taken from Yahoo! Movies[2], which is one of the biggest movie communities in Japan. We collected 27,516 movies and 158,385 reviewers. There are 1,124,555 reviews and 38 categories in this dataset.

Since some real review data including ours do not contain explicit hierarchy information in categories, we applied a heuristic method to construct a hierarchy. Our method first extracted existing categories as main categories (e.g. 38 categories in our data), and then generated sub-categories by combining any pair of co-occurring main categories. More precisely, letting M be a set of main categories, we define a set of sub-categories as $S = \{c_j \oplus c_k \mid i \in I \wedge (i, c_j) \in B \wedge (i, c_k) \in B\}$, where $\oplus$ is an operator to concatenate two category names. We let the resultant set of sub-categories belong to main categories from which the sub-categories were generated, e.g. edges (c, c_j) and (c, c_k) were added to H for $c = c_j \oplus c_k$. For example, "Star Wars" belongs to two main categories SF and $adventure$. We created a sub-category SF - $adventure$ and let it belong to SF and $adventure$. Finally, a set of categories is defined as $C = M \cup S$.

Note that we created a special sub-category indicating that a movie belongs to only a main category and does not belong to any sub-category. Given a movie belonging only to main category c, we added subcategory $c' = c \oplus c$ to the entire category set, and edge (c', c) to H. This special type of sub-categories was added because movies without any sub-category are not taken into account in the expertise estimation. For instance, the movie "Blade Runner" belongs only to SF category. This movie was assigned to a SF - SF sub-category.

Tables 1 and 2 show the detailed statistics of reviewers and movies in our dataset, from which we can find many reviewers who posted a review only once, and movies with a few reviews.

4.2   Evaluating the Credibility of a Reviewer

The first assumption is that a reviewer who has reviewed many and diverse items has high credibility. To test this assumption, we compared the correlation

---

[2] http://movies.yahoo.co.jp/

Table 1. Statistics of reviewers

# of reviewers	
Reviewed only 1 movie	140,180
Reviewed less than 10 movies	204,178
Reviewed 1,000+ movies	39
Reviewed 2,000+ movies	7
Total	158,385

# of reviews per reviewer	
Arithmetic mean	6.35
Mode	1
Median	1
Max	5,301

Table 2. Statistics of movies

# of movies	
Reviewed by only 1 Reviewer	6,326
Reviewed by 10+ Reviewers	8,877
Reviewed by 1000+ Reviewers	158
Reviewed by 2000+ Reviewers	29
Total	27,514

# of reviewers per movie	
Arithmetic mean	40.82
Mode	1
Median	4
Max	6,304

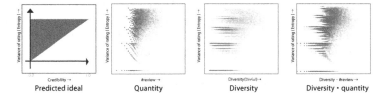

Fig. 1. Quantity, diversity, and their combination vs. review score entropy

between the credibility and following measures: quantity (n_u in Equation 2), diversity (Div(u) in Equation 10), and both diversity and quantity (Credibility(u) in Equation 5 ($\alpha = 1$)).

Before testing the first assumption, we start with illustrating the characteristics of these three measures. Figure 1 shows how well the three measure distinguish expert reviewers from the others, where the horizontal axis represents the value of each measure, and the vertical axis represents the entropy of review scores. Each point in the figures represents the value of a measure and review score entropy of a reviewer. According to McAuley and Leskovec's work, experienced reviewers have a higher review score entropy, while novice reviewers cannot take full advantage of the range of scores, and are likely to evaluate items in a narrow and biased manner. For example, novice reviewers may use only three or four even if they are asked to evaluate movies at a five-point scale. Thus, the review score entropy can be a good indicator of experts.

In the ideal case, points in the figures should converge towards the upper right corner: some novice reviewers gave a wide or a narrow range of scores, while the most expert reviewers gave a wide range of scores. It can be seen from Figure 1 that both of the quantity and diversity can distinguish experts (reviewers with high review entropy) from the others. The diversity measure shows a slightly better discriminative power as reviewers broadly spread along the horizontal axis.

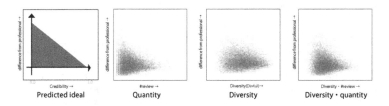

| Predicted ideal | Quantity | Diversity | Diversity · quantity |

Fig. 2. Quantity, diversity, and their combination vs. RSS to professional scores

To test the first assumption, it is necessary to obtain *true* quality of each item. Since it is hard to get exact true quality score, we approximated it by the score given by a well-known professional reviewer. We extensively compared reviewers who rated many and diverse movies, and carefully selected one who gives a widely acceptable score. Finally, we decided to use reviews authored by Yuichi Maeda, and manually collected his reviews from his Web site[3]. He is a Japanese professional critic and movie journalist who has written 1,832 reviews since 2003 to 2014 on his site. We found 1,689 movies included in both of his and our review data. As the range of his review scores was different from ours, we converted them to a five-point scale and used the scores as true quality of items.

The credibility of a reviewer was estimated by the residual sum of squares (RSS) between his score and a score of reviewer u:

$$\mathrm{RSS}(u) = \frac{1}{|I_u \cap P|} \sum_{i \in I_u \cap P} (\mathrm{score}(u, i) - \mathrm{score}_{\mathrm{pro}}(i))^2, \qquad (19)$$

where P is a set of movies reviewed by the professional, I_u is a set of movies reviewed by user u ($I_u = \{i \mid i \in I \wedge (u, i) \in R\}$), $\mathrm{score}(u, i)$ is a review score of u for movie i, and $\mathrm{score}_{\mathrm{pro}}(i)$ is a review score of the professional for movie i.

Figure 2 demonstrates that a reviewer becomes more similar in rating to the professional reviewer if the reviewer has reviewed more and more diverse movies.

4.3   Evaluating the Credibility of a Group of Reviewers

To test the second assumption regarding the credibility of a group of reviewers, we compared following measures: quantity ($|\{u \mid u \in U \wedge (u, i) \in R\}|$ for item i), entropy-based diversity measure ($\mathrm{EDiv}(G)$ in Equation 16), and variance-based diversity measure ($\mathrm{VDiv}(G)$ in Equation 18). The absolute error between the score of the professional and the average score of group G for item i is defined as follows:

$$\mathrm{AE}(G, i) = \left| \frac{1}{|G|} \sum_{u \in G} \mathrm{score}(u, i) - \mathrm{score}_{\mathrm{pro}}(i) \right|. \qquad (20)$$

If our second assumption is probable, the absolute error from large and diverse groups is smaller than that of smaller and/or more homogeneous groups.

---

[3] http://movie.maeda-y.com/

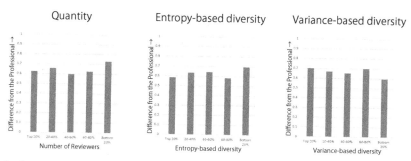

Fig. 3. Quantity, entropy-based, and variance-based diversity measure vs. RSS to professional scores (for all the groups)

Figure 3 shows the average absolute error of groups in each *bin*. We sorted all the groups based on one of the three measures, and categorized them into five bins based on the order of groups. For example, the leftmost bin of each figure includes groups ranked within top 20% when they are sorted in descending order of each measure. Thus, the left bins of each graph contain reviewer groups that are estimated as more credible, whereas the right bins contain reviewer groups that are estimated as less credible.

In the ideal case, the bars would slant upward to the right: the absolute error to the professional should become bigger for smaller or more homogeneous groups, while the error should be smaller for bigger or more diverse groups. of a group whose members are many or diverse is close to it. The bars of the quantity and entropy-based diversity measure show slightly similar trends to the ideal case, though they are not conclusive. The graph of the variance-based diversity measure does not show a trend similar to the ideal case. When we compare the leftmost bins, which contains the most diverse groups (top 20%), it can be seen that the entropy-based diversity measure outperforms the quantity-based measure in finding the most credible reviewers.

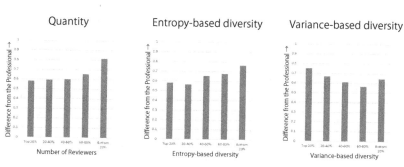

Fig. 4. Quantity, entropy-based, and variance-based diversity measure vs. RSS to professional scores (for groups with less than 100 reviewers)

As we have observed from Figure 3, there is much absolute error difference between groups with different diversity. We hypothesized that the absolute error to the professional can be small enough if plenty of reviews are available for each movie, and investigated a case where a limited number of reviews are available. Figure 4 shows the average absolute error of groups with less than 100 reviews. In this case, the entropy-based diversity measure and number of reviewers can more accurately estimate the credibility of reviewer groups.

5   Discussion

Our first experiment was successful in evaluating the credibility of a reviewer, supporting our hypothesis that reviewers who see diverse movies and reviewers who see many movies are reliable. We learned that these reviewers are characterized by a more even spread among their review scores and a amaller difference in rating with professional reviewers.

The reason why the difference of opinion of amatures and of professional does not converge to 0 is the difference of average; the professional's average rating is 3.3, and amature's is 3.6. Professionals are sometimes forced to see and to rate unfavorite movies at the job. Amateur can choose their favorite movies to review.

From the second experiment, we established that the entropy-based diversity and reviewer group size are good barometers to measure the credibility of a group. In contrast, Variance-based diversity does not work well.

The entropy-based diversity can measure the credibility of a group especially in case the number of reviewer is lower than 100. It's interesting to note that, when the number of members is small, the diversity of members is important, but when it is large, this is not the case. Generally, when the size of the a group is large enough, the most group is reliable when it is likely that the credibility of the group is saturated, we don't need to consider the size and diversity of the group. Figure 5 shows the relationship between the effect of the entropy-based diversity and the size of a group. The horizontal axis lists the groups binned by size. Each bin contains same number of groups. Groups were classified into high-diversity groups and low-diversity groups by their median entropy-based diversity. The vertical axis shows the average distance between the rating of the professional and that of the group. When the number of reviewers is less or equal to 440, a high diversity of reviewers minimized the score difference with the professional review. This fact supports our proposition. Contraly, in cases where the number of reviewers exceeds 440, the diversity of reviewers did not affect the score difference. Naturally, a larger group will be more credible because of the law of large numbers. The accuracy of the average score, however, trended down for the cluster of movies that assumed 119 to 182 reviews. This can be attributed to two possible causes. The movie reviewed by many reviewers is a popular movie, who tend to attract an audience of persons unfamiliar with movies. Their opinions are not very credible as evidenced by professional reviewers often shooting down popular movies. It refrects a characteristic of the review dataset; online user review are not implicit data, but intentional data.

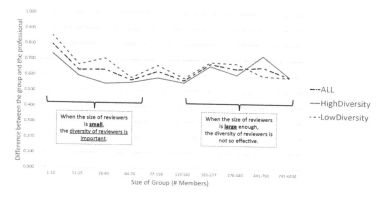

Fig. 5. Effectiveness of the Entropy-based Diversity

The variance-based diversity does not work well, regardless of the group size. One reason could be a biased group (i.e. a community of specialists) providing a correct opinion. Another cause could be generalists. They are simillar to each other. A group that consists of non-diverse generalists can rate movies accurately.

6   Conclusion

In this paper, we proposed a method to estimate credibility of individuals and reviewer groups. We proposed two simple assumptions: a reviewer who has reviewed many and diverse items has a high credibility, and a group of reviewers is credible if the group consists of many and diverse reviewers. We modeled a general user review structure with a category tree and proposed diversity-based measurement calculations. Through experiments using a real dataset of movie reviews, the effectiveness of the assumption 1 was confirmed; a reviewer, who reviews many and diverse movies has a high credibility. The effectiveness assumption 2 was partially confirmed; when the number of members is small, the entropy-based diversity is a good indicator to measure the credibility of a group.

Acknowledgments. This work was supported in part by the following projects: Grants-in-Aid for Scientific Research (Nos. 24240013) from MEXT of Japan.

References

1. Agrawal, R., Gollapudi, S., Halverson, A., Ieong, S.: Diversifying search results. In: Proceedings of the Second ACM International Conference on Web Search and Data Mining, WSDM 2009, pp. 5–14. ACM, New York (2009)

2. Amatriain, X., Lathia, N., Pujol, J.M., Kwak, H., Oliver, N.: The wisdom of the few: A collaborative filtering approach based on expert opinions from the web. In: Proceedings of the 32nd International ACM SIGIR Conference on Research and Development in Information Retrieval, SIGIR 2009, pp. 532–539. ACM, New York (2009)

3. Capannini, G., Nardini, F.M., Perego, R., Silvestri, F.: Efficient diversification of web search results. Proc. VLDB Endow. 4(7), 451–459 (2011)

4. Carbonell, J., Goldstein, J.: The use of mmr, diversity-based reranking for reordering documents and producing summaries. In: Proceedings of the 21st Annual International ACM SIGIR Conference on Research and Development in Information Retrieval, SIGIR 1998, pp. 335–336. ACM, New York (1998)

5. Liu, X., Croft, W.B., Koll, M.: Finding experts in community-based question-answering services. In: Proceedings of the 14th ACM International Conference on Information and Knowledge Management, CIKM 2005, pp. 315–316 (2005)

6. Magurran, A.E.: Measuring biological diversity. African Journal of Aquatic Science 29(2), 285–286 (2004)

7. McAuley, J.J., Leskovec, J.: From amateurs to connoisseurs: Modeling the evolution of user expertise through online reviews. In: Proceedings of the 22nd International Conference on World Wide Web, WWW 2013, pp. 897–908. International World Wide Web Conferences Steering Committee, Republic and Canton of Geneva (2013)

8. Nolan, K.A., Callahan, J.E.: Beachcomber biology: The shannon-weiner species diversity index. In: Proc. Workshop ABLE, vol. 27, pp. 334–338 (2006)

9. Sha, X., Quercia, D., Michiardi, P., Dell'Amico, M.: Spotting trends: The wisdom of the few. In: Proceedings of the Sixth ACM Conference on Recommender Systems, RecSys 2012, pp. 51–58. ACM, New York (2012)

10. Shoji, Y., Tanaka, K.: Diversity-based HITS: Web page ranking by referrer and referral diversity. In: Jatowt, A., Lim, E.-P., Ding, Y., Miura, A., Tezuka, T., Dias, G., Tanaka, K., Flanagin, A., Dai, B.T. (eds.) SocInfo 2013. LNCS, vol. 8238, pp. 377–390. Springer, Heidelberg (2013)

11. Stirling, A.: A general framework for analysing diversity in science, technology and society. Journal of the Royal Society Interface 4(15), 707–719 (2007)

12. Surowiecki, J.: The wisdom of crowds. Anchor (2005)

13. Wang, J., Zhu, J.: Portfolio theory of information retrieval. In: Proceedings of the 32nd international ACM SIGIR Conference on Research and Development in Information Retrieval, SIGIR 2009, pp. 115–122. ACM, New York (2009)

Social Determinants of Content Selection in the Age of (Mis)Information

Alessandro Bessi[1,2], Guido Caldarelli[2,3], Michela Del Vicario[2], Antonio Scala[3], and Walter Quattrociocchi[2,4]

[1] IUSS Institute for Advanced Study, Piazza della Vittoria 5, 27100 Pavia, Italy
[2] IMT Alti Studi Lucca, Piazza S. Ponziano 6, 55100 Lucca, Italy
[3] ISC-CNR Uos "Sapienza", 00185 Roma, Italy
[4] Laboratory for the Modeling of Biological and Socio-technical Systems, Northeastern University, Boston, MA 02115 USA

Abstract. Despite the enthusiastic rhetoric about the so called *collective intelligence*, conspiracy theories – e.g. global warming induced by chemtrails or the link between vaccines and autism – find on the Web a natural medium for their dissemination. Users preferentially consume information according to their system of beliefs and the strife within users of opposite worldviews (e.g., scientific and conspiracist) may result in heated debates. In this work we provide a genuine example of information consumption on a set of 1.2 million of Facebook Italian users. We show by means of a thorough quantitative analysis that information supporting different worldviews – i.e. scientific and conspiracist news – are consumed in a comparable way. Moreover, we measure the effect of 4709 evidently false information (satirical version of conspiracist stories) and 4502 debunking memes (information aiming at contrasting unsubstantiated rumors) on polarized users of conspiracy claims.

Keywords: misinformation, collective narratives, crowd dynamics, information spreading.

1 Introduction

The large availability of data from online social networks (OSN) allows for the study of mass social dynamics at an unprecedented level of resolution. Along this path, recent studies have pointed out several important results in the emerging field of computational social science [1, 2] ranging from the influence-based contagion, up to the emotional contagion, passing through the virality of false claims [3–5]. In particular in [5, 6] it has been shown that massive digital misinformation permeates online social dynamics creating viral phenomena even on intentional parodistic false information. Social interaction, healthcare activity, political engagement and economic decision-making are influenced by digital hyper-connectivity – i.e. the increasing and exponential rate at which people, processes and data are connected and interdependent [7–16]. Everyone can produce and access a variety of information actively participating in the diffusion and reinforcement of worldviews and narratives. Such a process has been dubbed

L.M. Aiello and D. McFarland (Eds.): SocInfo 2014, LNCS 8851, pp. 259–268, 2014.

as *collective intelligence* [17, 18]. However, despite the enthusiastic rhetoric about the ways in which digital technologies have burst the interest in debating political or social relevant issues, their role in enforcing informed debates and shaping the public opinion still remain unclear. A large body of literature from political science focused on the socio-cognitive aspects of citizens participating in the political discussion. As pointed out by [19] individuals can be uninformed or misinformed. The role of corrections in the diffusion and formation of biased beliefs have been addressed in [20]. In this work we address such a challenge by accounting for the consumption of information belonging to different worldviews on online social media. The World Economic Forum listed *massive digital misinformation* as one of the main risks for the modern society [21]. Conspiracy theories as alternative explanations to complex phenomena (e.g., globalization or climate change) find on the Web a natural medium for their dissemination and, not rarely, they are used as argumentation for policy making and foment collective debates [22]. Conspiracy theses tend to reduce the complexity of reality by explaining significant social or political aspects as plots conceived by powerful individuals or organizations. Since these kinds of arguments can sometimes involve the rejection of science, alternative explanations are invoked to replace the scientific evidence. For instance, people who reject the link between HIV and AIDS generally believe that AIDS was created by the U.S. Government to control the African American population [23]. The spread of misinformation in such trusted networks can be particularly difficult to detect and correct because of the social reinforcement – i.e. people are more likely to trust an information originating from within their network or someway consistent with their system of beliefs [24–34, 15, 35]. Since unsubstantiated claims are proliferating over the Internet, what would happen if they were used as the basis for policy making? Such a scenario makes crucial the quantitative understanding of the social determinants related to content selection, information consumption, and beliefs formation and revision. Misinformation is pervasive and as a first reaction we noticed the emergence of blogs and pages devoted to debunk false claims, namely *debunkers*. Meanwhile, the strong polarization of users with respect to one or another worldview (fomented by the possibility to ban and to write negative comments) triggered the proliferation of satirical pages producing demential imitation of conspiracy theses (e.g., chemtrails containing sildenafil citratum – i.e. the active ingredient of Viagra^{TM}– or the political action committee to abolish the thermodynamic laws), namely *trolls*. In this work we provide a genuine example of robust generative patterns about information consumption on the Italian Facebook on a sample of 1.2 million of individuals. In particular, we show, through a thorough quantitative analysis, similar consumption patterns of information supporting different (and opposite) worldviews. Then, we measure the social response of polarized users of alternative news to 4709 satirical version of conspiracy theses and to 4502 debunking memes (information aiming at correcting the diffusion of unsubstantiated claims) for increasing level of user engagement on the preferred category of information (scientific news and conspiracy news). We find that polarized users of conspiracy-like claims interacting with debunking or parody of conspiracy claims are more likely to interact again with conspiracy rumors.

2   Data Collection

In order to define the space of our investigation, we were helped by Facebook groups very active in the debunking of conspiracy theses (see acknowledgments section). The resulting dataset is composed of 73 public Facebook pages divided in scientific news and conspiracist news for which we downloaded all the posts (and their respective users interactions) in a timespan of 4 years (2010 to 2014). In addition, we consider 6 pages very active in debunking conspiracy information, namely hoax-busters, and 2 pages satirizing conspiracy theories by diffusing intentional false information as a satirical imitation of conspiracy theses. These latter have produced information that went viral despite their evident satirical taste. Among these, the OGM yellow tomatoes and the violet carrots created by industries to satisfy aesthetic needs (notice that the first tomatoes arrived in Europe were yellow) or the wonderful anti-hypnotic effects of lemon (such a post received more than 45.000 shares). The entire data collection process is performed exclusively with the Facebook Graph API [36], which is publicly available and which can be used through one's personal Facebook user account. The exact breakdown of the data is presented in Table 1. The first category includes all pages diffusing conspiracist information – pages which disseminate controversial information, most often lacking supporting evidence and sometimes contradictory of the official news. The second category is that of scientific dissemination including scientific institutions and scientific press having the main mission to diffuse scientific knowledge. We focus our analysis on the interaction of users with the public posts – i.e. likes, shares, and comments. Each of these actions has a particular meaning. A *like* stands for a positive feedback to the post; a *share* expresses the will to increase the visibility of a given information; and *comment* is the way in which online collective debates take form. Comments may contain negative or positive feedbacks with respect to the post.

Table 1. Breakdown of Facebook dataset. The number of pages, posts, comments and likes for all category of pages.

	Total	Science	Conspiracy	Hoaxbusters	Troll
Pages	81	34	39	6	2
Posts	271, 296	62, 705	208, 591	4, 502	4, 709
Likes	9, 164, 781	2, 505, 399	6, 659, 382	67, 324	40, 341
Comments	1, 017, 509	180, 918	836, 591	17, 883	58, 686
Unique Comments	279, 972	53, 438	226, 534	5, 115	42, 910
Unique Likes	1, 196, 404	332, 357	864, 047	12, 427	16, 833

3   Results and Discussion

3.1   Information Consumption

We start our analysis by characterizing information consumption patterns by focusing on the behavior of usual consumers of conspiracy and scientific news. Through a thresholding strategy we select the most active users in a specific category according to their liking activity on the posts of the two categories. As we assume *likes* to be positive feedbacks with respect to the information reported on the post [37], a user is labeled as polarized in one category if the 95% of his likes is given on posts published on pages of such a category. We are label $255,225$ users polarized in science and $790,899$ users polarized in conspiracy. In Figure 1(a) we show the empirical complementary cumulative distribution function (CCDF) of the users' persistence rate, namely r, intended as the mean time interval (in hours) between likes of a user on posts of their preferred category of information (scientific or conspiracy news). Usual consumers of conspiracy and scientific news present a very similar information consumption patterns.

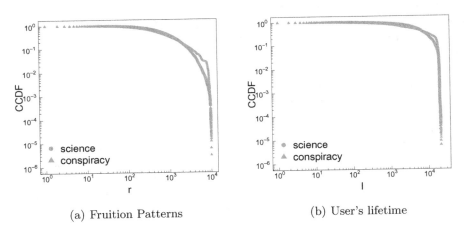

(a) Fruition Patterns   (b) User's lifetime

Fig. 1. Panel a) **Fruition Patterns**: Empirical CCDF of the mean time interval (in hours) between likes for each user. The two distributions are indicating a similar behavior in the rate of persistence of the users. Panel b) **User's lifetime**. Empirical CCDF of user's lifetime (in hours) – i.e. the time interval between the first and the last like of each polarized user in the category which he belongs to.

In Figure 1(b) it is shown the CCDF of users' lifetime, namely l – i.e., the time interval (in hours) between the first and the last like of the users on posts of the category which they are assigned to. These results show that information belonging to different (and opposite) worldviews are consumed in a similar way by their respective users.

3.2   Engagement and Interaction with External Information

We continue our analysis by addressing the relationship between the exposition to external information and the level of engagement of a user preferred kind of content. We use information from a) *hoaxbusters* pages aiming at debunking and correcting the diffusion of false claims (mainly conspiracy theses) such as the link between vaccines and autism or the astonishing medical powers of soursop and b) *troll* pages intentionally posting satirical and imitations of conspiracy theses. In particular, we analyze how the activity (comments and likes) of polarized users on troll and debunking posts changes as a function of θ – i.e. the engagement degree, intended as the number of likes of a polarized user in the category which he/she belongs to. In Figure 2 we show the number of polarized users as function of the threshold θ.

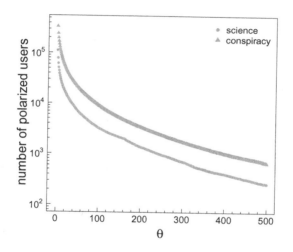

Fig. 2. Users Engagement. Number of polarized users as a function of the engagement degree θ.

In Figure 3 we show the activity (number of likes and comments) of polarized users of scientific and conspiracy news on respectively, 4,502 debunking (panel a) and 4709 troll (panel b) information as a function of θ. On the one hand, consumers of scientific news are more active in liking and commenting debunking posts. On the other hand, consumers of conspiracist posts are more prone to like (and not to comment) satirical imitation of the story they are usually exposed to. Such a trend of polarized users increases with their level of commitment and engagement.

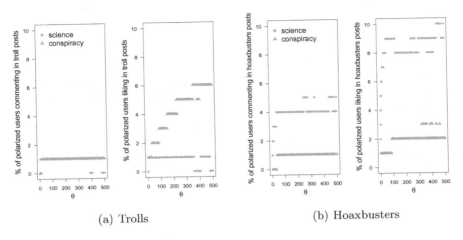

(a) Trolls   (b) Hoaxbusters

Fig. 3. Users on external contents. Users activity (likes and comments) as a function of the engagement degree θ of conspiracy and scientific news on troll (panel a) and hoaxbusters (panel b) posts.

The results of Figure 3 suggest that conspiracists are interested in diffusing their stories; their tendency to avoid scrutiny [38–40] allows for the mixing of conspiracy news and their satirical imitation. On the other hand, also polarized users of scientific news tend to like and comment information that are consistent with their worldview (debunking of unsubstantiated claims). Such results are a warning on the effectiveness of online debunking activities since they are mainly fruited by users of scientific pages and are not considered by consumers of conspiracy information. Coherently with [5], high levels of commitment in conspiracy theses decrease the level of interest in official and main stream information and increases the possibility to interact with unsubstantiated rumors even if these are statirical.

3.3   Conspiracy News within Online Debunking and Trolls

We want to understand if debunking posts are effective in changing the tendency of engaged conspiracy users to interact with unsubstantiated claims. Hence, we measure the survival probability of conspiracy users who commented (active interaction) either posts from debunking pages or false information as a function of the level of user engagement θ. More precisely, we compute the probability that a user's lifetime – i.e. the temporal distance between the first and the last like of the user in the category which he belongs to – is greater than some specified temporal distance t. Let define the random variable T with cumulative distribution function $F(t)$ on the interval $[0, \infty)$. Then the probability that a user's lifetime is not greater than a specific t is given by the cumulative probability distribution $F(t) = Pr(T \leq t)$. Hence, the survival function is the probability that a user will continue to like posts supporting the narrative in which he is polarized on beyond a given time t given by $S(t) = Pr(T > t) = 1 - F(t)$. To compute such a measure we use the *Kaplan Meier estimate* [41]. Let n_t denote the number of users that are still liking posts supporting the narratives in which they are polarized on, just before time t; and let d_t denote the number of users that stop liking at time t. Then the estimated survival probability after time t is $(n_t - d_t)/n_t$. Assuming that the times t are independent, the Kaplan Meier estimate of the survival function at time t is defined by $\hat{S}(t) = \prod^t \left(\frac{n_t - d_t}{n_t} \right)$. Figure 4 shows in panel (a) the quantile discretization, for different levels of engagement θ, of the survival probability distribution of usual consumers of conspiracy news which interacted with troll posts; and in panel (b), as a control, the quantile discretization, for different levels of engagement θ, of the survival probability distribution of polarized users not exposed to intentional false claims.

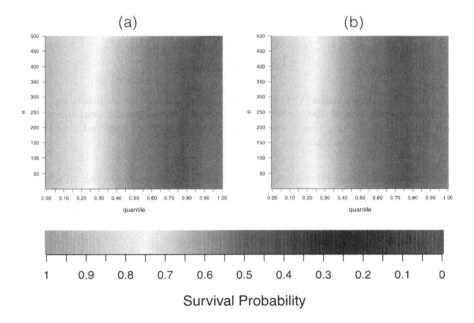

Fig. 4. Survival probability of conspiracists exposed to troll posts. Heatmap of the quantile discretization of the survival probability distribution of conspiracy users against their level of engagement θ exposed (panel a) and not exposed (panel b) to satirical and demential imitation of the story they are usually exposed to.

Similar results hold for the reaction to information having the goal to persuade users of the unsubstantiated nature of conspiracy theses. Figure 5 shows the quantile discretization of the survival probability distribution for increasing level of users engagement θ of usual consumers of conspiracy news exposed (panel a) and not exposed (panel b) to debunking memes.

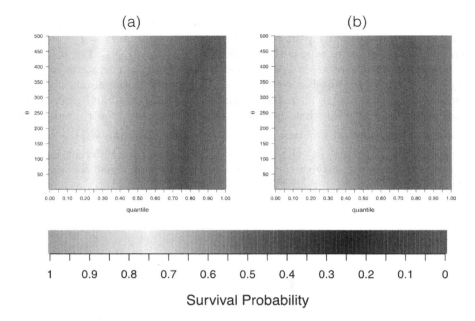

Fig. 5. Survival probability of conspiracists exposed to debunking posts. Quantile discretization of the survival probability distribution of conspiracy users against their level of engagement θ exposed (panel a) and not exposed (panel b) to posts debunking conspiracy theses.

These results suggest that with the increasing of the engagement of a user in conspiracy stories, the more the exposure to external information reinforce the user's consumption pattern. Despite the size effect (this is not a controlled experiment), users exposed and user not exposed with the same level of engagement present a different behavior.

Acknowledgments. Funding for this work was provided by EU FET project MULTIPLEX nr. 317532 and SIMPOL nr. 610704. The funders had no role in study design, data collection and analysis, decision to publish, or preparation of the manuscript. We want to thank "Protesi di Protesi di Complotto", "Scientificast", "Simply Humans" and "Semplicemente me" page staff. as well as Titta, Sandro Forgione, Salvatore Previti, Fabio Petroni, and Elio Gabalo for precious suggestions and discussions.

References

1. Lazer, D., Pentland, A., Adamic, L., Aral, S., Barabasi, A.-L.L., Brewer, D., Christakis, N., Contractor, N., Fowler, J., Gutmann, M., Jebara, T., King, G., Macy, M., Roy, D., Van Alstyne, M.: Social science. computational social science. Science 323(5915), 721–723 (2009)

2. Conte, R., Gilbert, N., Bonelli, G., Cioffi-Revilla, C., Deffuant, G., Kertesz, J., Loreto, V., Moat, S., Nadal, J.-P., Sanchez, A., Nowak, A., Flache, A., Miguel, M.S., Helbing, D.: Manifesto of computational social science. European Physical Journal Special Topics EPJST 214 (2012)

3. Kramer, A.D.I., Guillory, J.E., Hancock, J.T.: Experimental evidence of massive-scale emotional contagion through social networks. Proceedings of the National Academy of Sciences 111(24), 8788–8790 (2014)

4. Aral, S., Muchnik, L., Sundararajan, A.: Distinguishing influence-based contagion from homophily-driven diffusion in dynamic networks. Proceedings of the National Academy of Sciences 106(51), 21544–21549 (2009)

5. Mocanu, D., Rossi, L., Zhang, Q., Karsai, M., Quattrociocchi, W.: Collective attention in the age of (mis)information. CoRR, abs/1403.3344 (2014)

6. Bessi, A., Coletto, M., Davidescu, G.A., Scala, A., Caldarelli, G., Quattrociocchi, W.: Science vs conspiracy: collective narratives in the age of (mis)information. Technical report, IMT Lucca (2014)

7. Lotan, G., Graeff, E., Ananny, M., Gaffney, D., Pearce, I., Boyd, D.: The revolutions were tweeted: Information flows during the 2011 tunisian and egyptian revolutions. International Journal of Communications 5, 1375–1405 (2011)

8. Lewis, K., Gonzalez, M., Kaufman, J.: Social selection and peer influence in an online social network. Proceedings of the National Academy of Sciences 109(1), 68–72 (2012)

9. Leskovec, J., Huttenlocher, D., Kleinberg, J.: Signed networks in social media. In: Proceedings of the SIGCHI Conference on Human Factors in Computing Systems, CHI 2010, pp. 1361–1370. ACM, New York (2010)

10. Kleinberg, J.: Analysis of large-scale social and information networks. Philosophical Transactions of the Royal Society A: Mathematical, Physical and Engineering Sciences 371 (2013)

11. Kahn, R., Kellner, D.: New media and internet activism: from the 'battle of seattle' to blogging. New Media and Society 6(1), 87–95 (2004)

12. Howard, P.N.: The arab spring's cascading effects. Miller-McCune (2011)

13. Gonzalez-Bailon, S., Borge-Holthoefer, J., Rivero, A., Moreno, Y.: The dynamics of protest recruitment through an online network. Scientific Report (2011)

14. Bond, R.M., Fariss, C.J., Jones, J.J., Kramer, A.D.I., Marlow, C., Settle, J.E., Fowler, J.H.: A 61-million-person experiment in social influence and political mobilization. Nature 489(7415), 295–298 (2012)

15. Quattrociocchi, W., Caldarelli, G., Scala, A.: Opinion dynamics on interacting networks: media competition and social influence. Scientific Reports (May 4, 2014)

16. Aral, S., Walker, D.: Identifying influential and susceptible members of social networks. Science 337(6092), 337–341 (2012)

17. Levy, P.: Collective Intelligence: Mankind's Emerging World in Cyberspace (2000)

18. Malone, T.W., Klein, M.: Harnessing collective intelligence to address global climate change. Innovations: Technology, Governance, Globalization 2(3), 15–26 (2007)

19. Kuklinski, J.H., Quirk, P.J., Jerit, J., Schwieder, D., Rich, R.F.: Misinformation and the currency of democratic citizenship. Journal of Politics 62(3), 790–816 (2000)

20. Nyhan, B., Reifler, J.: When corrections fail: The persistence of political misperceptions. Political Behavior 32(2), 303–330 (2010)

21. Howell, L.: Digital wildfires in a hyperconnected world. In: Report 2013. World Economic Forum (2013)

22. Bessi, A., Coletto, M., Davidescu, G.A., Scala, A., Quattrociocchi, W.: Misinformation in the loop: the emergence of narratives in osn. Technical report, IMT Lucca (2014)
23. Sunstein, C.R., Vermeule, A.: Conspiracy theories: Causes and cures*. Journal of Political Philosophy 17(2), 202–227 (2009)
24. McKelvey, K., Menczer, F.: Truthy: Enabling the study of online social networks. In: Proc. CSCW 2013 (2013)
25. Meade, M.L., Roediger, H.L.: Explorations in the social contagion of memory. Memory & Cognition 30(7), 995–1009 (2002)
26. Mann, C., Stewart, F.: Internet Communication and Qualitative Research: A Handbook for Researching Online (New Technologies for Social Research series). Sage Publications Ltd. (September 2000)
27. Garrett, R.K., Weeks, B.E.: The promise and peril of real-time corrections to political misperceptions. In: Proceedings of the 2013 Conference on Computer Supported Cooperative Work, CSCW 2013, pp. 1047–1058. ACM, New York (2013)
28. Buckingham Shum, S., Aberer, K., Schmidt, A., Bishop, S., Lukowicz, P., Anderson, S., Charalabidis, Y., Domingue, J., Freitas, S., Dunwell, I., Edmonds, B., Grey, F., Haklay, M., Jelasity, M., Karpištšenko, A., Kohlhammer, J., Lewis, J., Pitt, J., Sumner, R., Helbing, D.: Towards a global participatory platform. The European Physical Journal Special Topics 214(1), 109–152 (2012)
29. Carletti, T., Fanelli, D., Grolli, S., Guarino, A.: How to make an efficient propaganda. Europhysics Letters 2(74), 222–228 (2006)
30. Centola, D.: The spread of behavior in an online social network experiment. Science 329(5996), 1194–1197 (2010)
31. Paolucci, M., Eymann, T., Jager, W., Sabater-Mir, J., Conte, R., Marmo, S., Picascia, S., Quattrociocchi, W., Balke, T., Koenig, S., Broekhuizen, T., Trampe, D., Tuk, M., Brito, I., Pinyol, I., Villatoro, D.: Social Knowledge for e-Governance: Theory and Technology of Reputation. Roma: ISTC-CNR (2009)
32. Quattrociocchi, W., Conte, R., Lodi, E.: Opinions manipulation: Media, power and gossip. Advances in Complex Systems 14(4), 567–586 (2011)
33. Quattrociocchi, W., Paolucci, M., Conte, R.: On the effects of informational cheating on social evaluations: image and reputation through gossip. IJKL 5(5/6), 457–471 (2009)
34. Bekkers, V., Beunders, H., Edwards, A., Moody, R.: New media, micromobilization, and political agenda setting: Crossover effects in political mobilization and media usage. The Information Society 27(4), 209–219 (2011)
35. Quattrociocchi, W., Amblard, F., Galeota, E.: Selection in scientific networks. Social Network Analysis and Mining 2(3), 229–237 (2012)
36. Facebook. Using the graph api. Website, 8 (2013) (January 1, 2014)
37. Viswanath, B., Mislove, A., Cha, M., Gummadi, K.P.: On the evolution of user interaction in facebook. In: Proceedings of the 2nd ACM Workshop on Online Social Networks, WOSN 2009, pp. 37–42. ACM, New York (2009)
38. Hogg, M.A., Blaylock, D.L.: Extremism and the Psychology of Uncertainty. Blackwell/Claremont Applied Social Psychology Series. Wiley (2011)
39. Fine, G.A., Campion-Vincent, V., Heath, C.: Rumor Mills: The Social Impact of Rumor and Legend. Social problems and social issues (Transaction publishers)
40. Bauer, M.: Resistance to New Technology: Nuclear Power, Information Technology and Biotechnology. Cambridge University Press (1997)
41. Kaplan, E.L., Meier, P.: Nonparametric estimation from incomplete observations. Journal of the American Statistical Association 53(282), 457–481 (1958)

How Hidden Aspects Can Improve Recommendation?

Youssef Meguebli[1], Mouna Kacimi[2], Bich-liên Doan[1], and Fabrice Popineau[1]

[1] SUPELEC Systems Sciences (E3S), Gif sur Yvette, France
{youssef.meguebli,bich-lien.doan,fabrice.popineau}@supelec.fr
[2] Free University of Bozen-Bolzano, Italy
mouna.kacimi@unibz.it

Abstract. Nowadays, more and more people are using online news platforms as their main source of information about daily life events. Users of such platforms discuss around topics providing new insights and sometimes revealing hidden aspects about topics. The valuable information provided by users needs to be exploited to improve the accuracy of news recommendation and thus keep users always motivated to provide comments. However, exploiting user generated content is very challenging due its noisy nature. In this paper, we address this problem by proposing a novel news recommendation system that (1) enrich the profile of news article with user generated content, (2) deal with noisy contents by proposing a ranking model for users' comments, and (3) propose a diversification model for comments to remove redundancies and provide a wide coverage of topic aspects. The results show that our approach outperforms baseline approaches achieving high accuracy.

Keywords: News recommendation, Opinion mining, Diversification.

1 Introduction

News Media platforms play a crucial role in covering daily life topics ranging from social to political issues. Such platforms often allow users to publish their reactions to the published information and freely express their opinions. The editorial content is generated using a top down approach where the provided information follows the publisher plan and target specific aspects that are made explicit in the editorial content. By contrast, user generated content follows a bottom up approach where users start discussing some specific issues forming debates around a given topic. Consequently, they reveal hidden topic aspects which are not confined to any predefined plan and thus extend information by continuously bringing new insights. This calls for an effective strategy for news recommendation that would provide users news articles that match with their interests and on which they are willing to comment. The willingness to comment on a news article is driven by the kind of aspects discussed by users around the topic. For this reason, it is important to capture that information when recommending an article to a user. A straightforward way to achieve this goal is to enrich the content of news articles with user comments for a more effective recommendation. User generated content is a free source of information which can be subject to a lot of noise. Thus, it is important to select only prominent comments using ranking strategy. Moreover, these comments have to be representative which require the application of diversification techniques to

L.M. Aiello and D. McFarland (Eds.): SocInfo 2014, LNCS 8851, pp. 269–278, 2014.
© Springer International Publishing Switzerland 2014

capture a wide set of aspects. Our proposed approach goes beyond existing techniques [1,20,26,4,26] that employ user generated content for search and recommendation in several ways. First, Ganesan et al., [4] use product reviews and assume that comments belong to an already known set of aspects. In our work, we are interested in aspects about daily life topics reported by news articles. These aspects are not classified but we extract them automatically using an unsupervised approach. Second, comments on news sites usually contain a lot of noise, thus unlike the approach by Yee et al. [26], we do not use all comments to enrich the content of news article but we select only the topk comments. Additionally, we perform diversification on those comments to have a large coverage of new aspects. Our work aims at providing an effective news recommendation to facilitate the access of users to published news stories and more importantly, to motivate readers to comment on the news articles of interests and get involved in discussions with other users. We first propose an unsupervised technique for aspect extraction from user generated content and editorial content. Second, we propose a novel recommendation approach that (1) enriches the content of news articles with user generated content to improve the effectiveness of recommendation, (2) ranks user comments to select only prominent content and filter noise, and (3) proposes a comment diversification model based on authorities, semantic and sentiment diversification. Third, we test our approach on four datasets.

2 Related Work

The emergence of Web 2.0 has led to a rapid growth of user generated content (UGC), such as product, movie, and hotel reviews, and comments on news stories. Due to its richness and insightfulness, user generated content was exploited by several studies [9,19,11,22,22,14,16,23,15,17] for different purposes including blog summarization [9], community detection for predicting the popularity of online content [19], spam detection [11], comments volume prediction [22], comments rating prediction [14], comments ranking [23,15], and identification of political orientation of users [17]. A key point for exploiting user generated content is to extract interesting and useful knowledge from it. Hence, some approaches [25,27] have focused on aspect extraction from annotated data. For instance, Wang et al., [25] identifies the main aspects of reviews by starting from few seed keywords which are fed into a bootstrapping-based algorithm. Most of these approaches are domain-specific, or usually highly dependent on the training data. In this paper, we employ an unsupervised approach to extract hidden aspects of news articles from their related users' comments. Another key point when exploiting user generated content is how to find the most useful or helpful information. To address this issue, several approaches have focused on ranking user reviews [8,12,24,3]. Danescu et al., [3] show, through extensive experiments, that exploiting relationships between reviews can significantly improve ranking quality. Litva et al., [15], propose to use PageRank to rank comments in news sites. In our work, we use this last technique to rank user comments due to its simplicity, domain-independence, and effectiveness. Directly related approaches to our work employ user generated content for search and recommendation [1,20,26,4,26,4]. Shmueli et al., [20] analyze the co-commenting patterns of users for recommending news articles to users who will likely comment them.

The closest works to ours are by Yee et al., [26,4] and Ganesan et al., [26,4] which exploit users' comments to enrich the content of documents. Yee et al., [26,4] prove that the potential of Youtube users' comments in the search index yields up to a 15% improvement in search accuracy compared to user-supplied tags or video titles. Similarly, Ganesan et al., [4] use the content of customer reviews to represent entities (hotels and cars) in the context of entity ranking. They measure the score of entities based on how their reviews match with users' keyword preferences. Two main points make the difference between our work and these approaches. First, Ganesan et al., [4] use product reviews which belong to an already known set of aspects. In our work, we are interested in aspects about daily life topics reported by news articles. These aspects are not classified but we extract them automatically using an unsupervised approach. Second, comments on news sites usually contain a lot of noise, thus unlike the approach by Yee et al., [26] we do not use all comments to enrich the content of news articles but we select only the topk comments. Additionally, we perform diversification on those comments to have a larger coverage of new aspects.

3 Aspects Extraction

We describe here how aspects are extracted from user comments and news article content. Note that the same extraction method is used for both types of content, with the sole difference that the computation of aspects scores depends either on the corpus of comments or on one of the articles.

3.1 Generation of Candidate Aspects

To extract aspects from the comments of user u_i, we first identify the sentences[1] expressed in all his comments. Then, we rank their contained terms using $tf * idf$ scoring function. In our work, tf represents the term frequency in the set of sentences of user u_i, and idf represents the inverted document frequency in the set of sentences of all users in the platform. The idea is to select highly scored unigrams as a base for generating candidate aspects. Similarly, for a given article a_j, we use the same unigram extraction from its content however this time tf represents the term frequency in the set of sentences of article a_j and idf represents the inverted document frequency in the set of sentences of all news articles in the platform. From the selected unigrams, we generate bi-grams, then we take the bi-grams as input and we build a set of n-grams by concatenating bi-grams that share an overlapping word. At each step we take the topk n-grams based on the score of their composed unigrams[2]. We check the redundancy of the generated candidates using Jaccard similarity [18]. If two n-grams have a similarity higher than a defined threshold, we would discard one of them. In our work, we have set the maximum length of the n-grams to 3 since there were no meaningful n-grams of a higher length.

---

[1] Using OpenNLP http://opennlp.sourceforge.net/
[2] In this work we have set k=500.

3.2 Selection of Promising Aspects

Generating n-grams that have high *tf* ∗ *idf* scores is not enough to identify the aspects discussed in users' comments and articles content. It is important for the words in the generated n-grams to be strongly associated within a sentence in the original text to avoid covering incorrect information. To capture this association, we use *pointwise mutual information* [21] (PMI) of words in n-grams based on its alignment to the narrow comments of each user (or article content). Formally, suppose $m_i = w_1...w_n$ is a generated n-grams. We define the $Score_n$ as follows:

$$S_{PMI}(w_1...w_n) = \frac{1}{n} \sum_{i=1}^{n} pmi_{local}(w_i) \tag{1}$$

where $pmi_{local}(w_i)$ is a local pointwise mutual information function defined as:

$$pmi_{local}(w_i) = \frac{1}{2C} \sum_{j=i-C}^{i+C} pmi'(w_i, w_j), i \neq j \tag{2}$$

where C is a contextual window size. The $pmi_{local}(w_i)$ measures the average strength of association of a word w_i with all its C neighboring words (on the left and on the right). When this is done for each $w_i \in m$, this would give a good estimate of how strongly associated the words are in m. We used a modified PMI scoring [4] referred to as pmi' and is defined as:

$$pmi'(w_i, w_j) = \log_2 \frac{p(w_i, w_j) \cdot c(w_i, w_j)}{p(w_i) \cdot p(w_j)} \tag{3}$$

where $c(w_i, w_j)$ is the frequency of two words co-occurring in a sentence from the original text within the context window of C and $p(w_i, w_j)$ is the corresponding joint probability. The co-occurrence frequency, $c(w_i, w_j)$ is integrated into our PMI scoring to reward frequently occurring words from the original text. By adding $c(w_i, w_j)$ into the PMI scoring, we ensure that low frequency words do not dominate and that moderately associated words with high co-occurrences have relatively high scores.

4   Comments Ranking

We adopt the opinion ranking approach proposed by Litva et. al., [15] because of its simplicity, domain-independence, and effectiveness. For each article a_j, we take all its related comments and build a graph where each node is a comment. An edge is created between two comments if their cosine similarity exceeds a given threshold[3]. Once we have the comments graph, we apply the **PageRank** algorithm to compute a score for each comment. The topk comments are then used to enrich the content of the news article a_j. We recall that the **PageRank** algorithm models use behavior in a hyperlink graph, where a random surfer visits a web page with a certain probability based on the page's PageRank. The probability that the random surfer clicks on one link is solely

---

[3] In our implementation we set the threshold to 0.5.

given by the number of links on that page. So, the probability for the random surfer reaching one page is the sum of probabilities for the random surfer following links to this page. It is assumed that the surfer does not click on an infinite number of links, but gets bored sometimes and jumps to another page at random. Besides its interpretation, the random jump is used to avoid dead-ends and spider traps in the graph. Formally, the **PageRank** algorithm is given by:

$$PR(A) = (1 - d) + d \left(\frac{PR(T_1)}{C(T_1)} + ... + \frac{PR(T_n)}{C(T_n)} \right)$$

where $PR(A)$ is the PageRank of page A, $PR(T_i)$ is the **PageRank** of pages T_i which links to page A, $C(T_i)$ is the number of outgoing links of page T_i, and d is a damping factor which can be set between 0 and 1. By replacing pages by comments, and hyperlinks by similarity edges, we can directly apply **PageRank** to our comment graph.

5 Comments Diversification

In this section, we introduce the technique used to diversify comments on news sites which was inspired by the work in [10]. By diversifying comments, we aim to remove redundancies and thus to provide a wide coverage of topic aspects. We are given a set of comments $C = \{c_1, c_2, ..., c_n\}$ where $n \geq 2$. Our goal is to select a subset $L_k \subseteq C$ of comments that is diverse. We assume three main components that define the diversity of a set of comments : *authority*, *semantic diversity*, and *sentiment diversity*. Naturally, before discussing whether a set is diverse or not, it should first contains comments with high authority scores. Note that the *authority* of each comment is given by the **PageRank** score described in the previous section. To diversify a set of comments, we need to give more preference to dissimilar comments. We assume that two comments are dissimilar if (1) they discuss different aspects, and/or (2) they exhibit different sentiments about the news article topic, including positive, negative, and neutral sentiments. To satisfy these two requirements, we define two distance functions. The first one is a *semantic distance* function $d : C \times C \rightarrow R^+$ between comments, where smaller the distance, the more similar the two comments are. The second one is a *sentiment distance* function $s : C \times C \rightarrow R^+$ between comments, where the smaller the distance, the closer in sentiments the two comments are. We formalize a set selection function $f : 2^C \times h \times d \times o \rightarrow R^+$, where we assign scores to all possible subsets of C, given an authority function $h(.)$, a semantic distance function $d(.,.)$, a sentiment distance function $s(.,.)$, and a given integer $k \in Z^+ (k \geq 2)$. The goal is to select a set $L_k \subseteq D$ of comments such as the value of f is maximized. In other words, the goal is to find:

$$L_k^* = \text{Max}_{L_k \subseteq D, |L_k| = k} f(L_k, h(.), d(.,.), s(.,.))$$

where all arguments other than L_k are fixed inputs to the function. The goal of this model is to maximize the sum of the authority, the semantic dissimilarity, and the sentiment dissimilarity of the selected set. The function we aim at maximizing can be formalized as follows:

$$f(L) = \alpha(k-1) \sum_{a \in L} h(a) + 2\beta \sum_{a,b \in L} d(a,b) + 2\gamma \sum_{a,b \in L} s(a,b)$$

where $|L| = k$, and $\alpha, \beta, \gamma > 0$ are parameters specifying the trade-off between relevance, semantic diversity, and sentiment diversity[4]. The model allows to put more emphasis on relevance, on semantic diversity, on sentiment diversity, or on any mixture of these measures. Note that we need to scale up the three terms of the function. The authority scores are computed based on **PageRank** and the semantic distance is computed based on Jaccard similarity function. As for sentiment distance $s(a, b)$, it equals to 0 when a and b have the same sentiment, 1 otherwise. The sentiment orientation includes *positive*, *negative*, and *neutral* sentiments. The problem of diversifying search results is NP-hard [5,2]. However, there exist a well-known approximation algorithm to solve it [6], which works well in practice [10]. Gollapaudi et al. [6] show that their Max-sum diversification objective can be approached to a facility dispersion problem, known as the MaxSumDispersion problem [7,13]. In our work, we follow the same principle and model our diversification problem as a MaxSumDispersion problem having the following objective function: $f'(L) = \sum_{a,b \in L} d'(a, b)$ where d'(.,.) is a distance metric. We show in the following that f' is equivalent to our f function. Thus, we define the distance function $d'(a, b)$ as follows:

$$d'(a, b) = \begin{cases} 0, & \text{if a=b;} \\ \alpha(r(a) + r(b)) + 2\beta d(a, b) + 2\gamma s(a, b) & \text{otherwise.} \end{cases}$$

Considering the binary sentiment function, we claim that if d(.,.) is a metric then d'(.,.) is also a metric (proof skipped). We replace $d'(.,.)$ by its definition in $f'(L)$, disregarding pairwise distances between identical pairs, thus we obtain:

$$f'(L) = \alpha(k-1) \sum_{a \in L} r(a) + 2\beta \sum_{a,b \in L} d(a, b) + 2\gamma \sum_{a,b \in L} s(a, b)$$

we can easily see that each $r(a)$ is counted exactly $(k-1)$ times. Hence, the function f' is equivalent to our function f. Given this mapping, we can use a 2-approximation algorithm as proposed in [7,13].

6   Experiments

We have crawled four real datasets based on the activities of 645 users on four news sites, namely **CNN, The Telegraph, The Independent** and **Al-Jazeera**.[5] The choice of these users was based on two key-properties: the number of users' comments and whether they follow the four news sites or not. More precisely, we start, by selecting the most active users on each news site based on the number of comments posted and then we choose users that have posted comments on the four news sites. This process results in the selection of four datasets, the first one contains the activities of 150 users which are a subset of the most active users on **CNN**, the second dataset contains the activities of 180 users which are a subset of the most active users on **The Telegraph**,

[4] In our implementation we have set $\alpha = \beta = \gamma = 1$.
[5] http://www.cnn.com/, http://www.telegraph.co.uk/, http://www.independent.co.uk/ and http://www.aljazeera.com/

the third dataset contains the activities of 164 users which are a subset of the most active users on The Independent and the last dataset contains the activities of 151 users which are a subset of the most active users on Al-Jazeera. For each of those users, we have collected the details of his comments in the four news sites mentioned earlier (content, published time, etc.). Additionally, we have collected the details of all the commented news articles (e.g., news title, content, opinions, published time, etc.) from May 2010 to December 2013. Statistics about the number of commented articles and the number of comments for each dataset are shown in Table 1. To evaluate our approach, we have randomly selected 233 users among the most active users in the four news platforms described above. For each user we performed recommendation at different time points $t_1, t_2, ..t_n$. The reason behind time dependent evaluation is twofold: (1) to take into account profile updates since users continuously post comments bringing new information about their interests, and (2) to use data before time point t_i for recommendation and data starting from time point t_i for assessment, as described later. The time points $t_1, t_2, ..t_n$ are chosen in such a way that between t_{i-1} and t_i, there is at least m news articles commented by the user. For each user u_i, we have chosen $m = \frac{N_i}{10}$ where N_i is the total number of commented news articles by the user u_i. This setting resulted in 2330 rounds of recommendation.

Table 1. Dataset statistics

	Dataset1 (CNN Seed)		Dataset2 (Telegraph Seed)	
	#articles	#comments	#articles	#comments
CNN	41, 245	12, 056, 789	665	874, 879
Telegraph	1, 908	1, 257, 645	56, 527	10, 704, 741
Independent	1, 412	987, 437	7, 999	1, 608, 665
Al-Jazeera	801	102, 254	451	62, 835
	Dataset3 (Independent Seed)		Dataset4 (Al-Jazeera Seed)	
CNN	528	421, 542	2, 233	1, 652, 875
Telegraph	23, 272	6, 710, 580	1, 126	894, 710
Independent	27, 012	2, 985, 412	394	54, 760
Al-Jazeera	303	48, 058	9, 313	531, 452

To assess the effectiveness of our approach we have used an automatic evaluation to avoid the subjectivity of manual assessments. We have considered the action of commenting on an article to be an indicator that the article fits the interests of the user. Based on this assumption, we check the list of recommended articles. The one that user has commented on are considered relevant. Note that it is probable that we have missing information. A person might well be interested in an article even though he does not comment on it. So, the actual results are most probably higher than our findings. We have used two baseline approaches and tested several variations of our proposed technique. We have used the following strategies:(1) **NoEnrich** is the first baseline and its a simple content filtering approach based solely on the content of news articles. (2) **Yee** is the second baseline and its the closest works to ours which exploit all the set of user comments to enrich the content of documents (news articles in our case). (3)

Authority_k where we use our approach to enrich news articles with the topk authoritative comments related to it, selected as described in section 5. In our experiments we have used $k = 5$, $k = 10$, and $k = 20$. (4) **Diversity_k** where we use our approach to enrich news articles with the most diverse topk comments related to it, as described in section 6. In our experiments we have used $k = 5$, and $k = 10$. To compare the results of the different methods, we use Precision at k ($P@k$). The $P@k$ is the fraction of recommended articles that interest the user in question considering only the top-k results. The results of our experiments are shown in table 2. We can clearly see that our approach outperforms the baseline approaches by a significant margin. The improvement goes up to 17% in precision@5 compared to NoEnrich and 21% compared to Yee which is substantial. Having a closer look at the results, we can see that relying only on the content of news articles does not provide good performance. Even worse, when trying to enrich the content by all user comments, the precision decreases. By applying ranking, the precision improves but the gains are small ranging from 1% to 4%. However, when we apply diversification to the top 100 comments, the top5 and top10 diversified comments give the best results. These results meet our expectations since they perfectly reflect the role and the nature of comments in news platforms. Relying only on the content of articles does not perform well because user profiles built from comments focus on some aspects that might be different from the ones provided by the news article. Taking all comments into account is not a good idea either since comments are subject to noise and some of them might even deviate from the topic of interest, and thus this approach had the worst performance. Selecting the topk comments to be included in the article content is a good idea but due to redundancies this method loses its effect especially when k increases, which is the case of *Authority_20*. Finally, diversifying comments before enriching the content of articles provides a high gain in precision. This is because of the wider coverage of aspects. If the aspects discussed in the comments are explicit in the news article, then their weight is increased, otherwise they are added which increase the chance of more users getting interested in the article. For example, the aspects extracted from the CNN news article *British couple to be deported from Australia for living in wrong suburb* are too generic with NoEnrich and Yee strategies. They are mainly about *Australian Live*. By contrast, the aspects become more focused with comment ranking and talk for example about *Australian Visa* and *Deportation*. Then, we see that diversification extracts more aspects such as *Australia tax* and *people contracts*.

Table 2. Overall performance of our approach

	P@1	P@3	P@5	P@10	P@20
NoEnrich	0.424	0.494	0.481	0.513	0.540
Yee [26]	0.393	0.474	0.445	0.453	0.503
Authority_5	0.439	0.510	0.509	0.534	0.558
Authority_10	0.454	0.535	0.530	0.550	0.565
Authority_20	0.439	0.530	0.521	0.553	0.559
Diversity_5	0.484	0.575	0.587	0.595	0.586
Diversity_10	**0.575**	**0.646**	**0.654**	**0.640**	**0.607**

7   Conclusions

In this paper, we addressed the problem of recommendation in the context of news sites. In particular, we employed different ways to leverage user generated content on articles for refining the list of recommended news stories. Two approaches were proposed: (i) employing only relevant comments using comments ranking strategy, and (ii) using diverse comments. Our study on an extensive set of experiments showed that diverse comments achieve the best results compared to baseline approaches. As future work, we aim at exploring the impact of co-comments patterns. To this end, we plan to extend our model to a hybrid recommender model in which we employ collaborative filtering recommendation techniques.

References

1. Abbar, S., Amer-Yahia, S., Indyk, P., Mahabadi, S.: Real-time recommendation of diverse related articles. In: Proceedings of the 22nd International Conference on World Wide Web, WWW 2013, Republic and Canton of Geneva, Switzerland, pp. 1–12. International World Wide Web Conferences Steering Committee (2013)
2. Agrawal, R., Gollapudi, S., Halverson, A., Ieong, S.: Diversifying search results. In: WSDM, pp. 5–14 (2009)
3. Danescu-Niculescu-Mizil, C., Kossinets, G., Kleinberg, J., Lee, L.: How opinions are received by online communities: a case study on amazon.com helpfulness votes. In: Proceedings of the 18th international conference on World Wide Web, WWW 2009, pp. 141–150. ACM, New York (2009)
4. Ganesan, K., Zhai, C.: Opinion-based entity ranking. Inf. Retr. 15(2), 116–150 (2012)
5. Gollapudi, S., Sharma, A.: An axiomatic approach for result diversification. In: WWW, pp. 381–390 (2009)
6. Gollapudi, S., Sharma, A.: An axiomatic approach for result diversification. In: Proceedings of the 18th International Conference on World Wide Web, WWW 2009, pp. 381–390. ACM, New York (2009)
7. Hassin, R., Rubinstein, S., Tamir, A.: Approximation algorithms for maximum dispersion. Operations Research Letters 21, 133–137 (1997)
8. Hong, Y., Lu, J., Yao, J., Zhu, Q., Zhou, G.: What reviews are satisfactory: novel features for automatic helpfulness voting. In: Proceedings of the 35th International ACM SIGIR Conference on Research and Development in Information Retrieval, pp. 495–504. ACM, New York (2012)
9. Hu, M., Sun, A., Lim, E.-P.: Comments-oriented document summarization: Understanding documents with readers' feedback. In: Proceedings of the 31st Annual International ACM SIGIR Conference on Research and Development in Information Retrieval, pp. 291–298. ACM, New York (2008)
10. Kacimi, M., Gamper, J.: Diversifying search results of controversial queries. In: Proceedings of the 20th ACM International Conference on Information and Knowledge Management, CIKM 2011, pp. 93–98. ACM, New York (2011)
11. Kant, R., Sengamedu, S.H., Kumar, K.S.: Comment spam detection by sequence mining. In: Proceedings of the Fifth ACM International Conference on Web Search and Data Mining, WSDM 2012, pp. 183–192. ACM, New York (2012)
12. Kim, S., Pantel, P., Chklovski, T., Pennacchiotti, M.: Automatically assessing review helpfulness. In: Proceedings of the 2006 Conference on Empirical Methods in Natural Language Processing, pp. 423–430 (2006)

13. Korte, B., Hausmann, D.: An analysis of the greedy heuristic for independence systems. Annals of Discrete Mathematics 2, 65–74 (1978)
14. Lin, C., He, Y.: Joint sentiment/topic model for sentiment analysis. In: Proceedings of the 18th ACM Conference on Information and Knowledge Management, CIKM 2009, pp. 375–384. ACM, New York (2009)
15. Litvak, M., Matz, L.: Smartnews: Bringing order into comments chaos. In: Proceedings of the International Conference on Knowledge Discovery and Information Retrieval, KDIR, vol. 13 (2013)
16. Meguebli, Y., Kacimi, M., Doan, B.-L., Popineau, F.: Building rich user profiles for personalized news recommendation. In: Proceedings of 2nd International Workshop on News Recommendation and Analytics (2014)
17. Meguebli, Y., Kacimi, M., Doan, B.-L., Popineau, F.: Unsupervised approach for identifying users' political orientations. In: de Rijke, M., Kenter, T., de Vries, A.P., Zhai, C., de Jong, F., Radinsky, K., Hofmann, K. (eds.) ECIR 2014. LNCS, vol. 8416, pp. 507–512. Springer, Heidelberg (2014)
18. Real, R., Vargas, J.M.: The probabilistic basis of jaccard's index of similarity. Systematic Biology 45(3), 380–385 (1996)
19. Rendle, S., Freudenthaler, C., Gantner, Z., Schmidt-Thieme, L.: Bpr: Bayesian personalized ranking from implicit feedback. In: Proceedings of the Twenty-Fifth Conference on Uncertainty in Artificial Intelligence, UAI 2009, Arlington, Virginia, United States, pp. 452–461. AUAI Press (2009)
20. Shmueli, E., Kagian, A., Koren, Y., Lempel, R.: Care to comment?: Recommendations for commenting on news stories. In: Proceedings of the 21st International Conference on World Wide Web, WWW 2012, pp. 429–438. ACM, New York (2012)
21. Terra, E., Clarke, C.L.A.: Frequency estimates for statistical word similarity measures. In: Proceedings of the 2003 Conference of the North American Chapter of the Association for Computational Linguistics on Human Language Technology, NAACL 2003, pp. 165–172. Association for Computational Linguistics, Stroudsburg (2003)
22. Tsagkias, M., Weerkamp, W., de Rijke, M.: Predicting the volume of comments on online news stories. In: Proceedings of the 18th ACM Conference on Information and Knowledge Management, CIKM 2009, pp. 1765–1768. ACM, New York (2009)
23. Tsagkias, M., Weerkamp, W., de Rijke, M.: News comments:Exploring, modeling, and online prediction. In: Gurrin, C., He, Y., Kazai, G., Kruschwitz, U., Little, S., Roelleke, T., Rüger, S., van Rijsbergen, K. (eds.) ECIR 2010. LNCS, vol. 5993, pp. 191–203. Springer, Heidelberg (2010)
24. Tsur, O., Rappoport, A.: Revrank: A fully unsupervised algorithm for selecting the most helpful book reviews. In: International AAAI Conference on Weblogs and Social Media (2009)
25. Wang, H., Lu, Y., Zhai, C.: Latent aspect rating analysis on review text data: A rating regression approach. In: Proceedings of the 16th ACM SIGKDD International Conference on Knowledge Discovery and Data Mining, KDD 2010, pp. 783–792. ACM, New York (2010)
26. Yee, W.G., Yates, A., Liu, S., Frieder, O.: Are web user comments useful for search. In: Proc. LSDS-IR, pp. 63–70 (2009)
27. Zhuang, L., Jing, F., Zhu, X.-Y.: Movie review mining and summarization. In: Proceedings of the 15th ACM International Conference on Information and Knowledge Management, CIKM 2006, pp. 43–50. ACM, New York (2006)

The Geography of Online News Engagement

Martin Saveski[1], Daniele Quercia[2], and Amin Mantrach[2]

[1] MIT Media Laboratory, Cambridge, MA, USA
msaveski@mit.edu
[2] Yahoo Labs, Barcelona, Spain
dquercia@acm.org, amantrac@yahoo-inc.com

Abstract. Geographical processes might well impact online engagement in big countries like the USA. Upon a random sample of 200K news articles and corresponding 41M comments posted on the Yahoo! News in that country, we show that nearby individuals tend to comment and engage with similar news articles more than distant individuals do. Interestingly, at state level, topics one reads about are associated with specific socio-economic conditions and personality traits.

1 Introduction

Online actions whose geographic processes have been well-studied include not only posting status updates on Twitter [34,12,14], but also uploading pictures on Flickr [7,27], and visiting Foursquare venues [26,24].

Despite their importance, the geographic processes of online engagement on news platforms have not been widely studied. To partly fix that, we consider a dataset containing articles and user comments posted on the Yahoo! News site for more than two years, and we make two main contributions:

- We find that users engage with each other (i.e., they comment on the same articles) depending on where they live (Sections 4 and 5).
- Since one's interests have been linked to one's socio-economic conditions and personality traits, we test whether this is also the case at geographic level, and we do so by combing our online data with census data (Section 6). We find that those in states with high levels of education and well-being comment articles about research&technology but not those about politics, gossips, or sport. Instead, those in states with high levels of crime and unemployment comment on articles about sports, but not on those about economy or research&technology. Also, as for personality traits, users from states that tend to have residents low in Neuroticism (emotionally stable) comment on articles about music, those in Open and Extravert states on articles about sports, and Conscientious states on articles about economics.

2 Related Work

The main goal of this work is to study the influence of geographic processes on user engagement with online news. Next, we review work related to this topic.

L.M. Aiello and D. McFarland (Eds.): SocInfo 2014, LNCS 8851, pp. 279–289, 2014.

Influence of Time on Our Actions Online. Golder and Macy [11] examined how the use of emotion words by Twitter users changed over the course of one day, and they found that it was regularly shifted along time zones. That is similar to what Mislove *et al.* [25] independently reported when contrasting the usage of Twitter in the west coast with that in the east coast.

News in Tweets and Geographic Spread on Twitter. Kwak *et al.* [23] found that reciprocal relations on Twitter (75% of them) tend to be between users who live no more than three time zones away, hinting that the geographical distance may be related to the interest similarity. Recent studies have also examined the geographic spread of topics on Twitter by investigating the adoption of hashtags across locations around the world [20]. They found that physical distance between locations constrained the spreading of hashtags: the adoption of the same hashtag by two locations was inversely proportional to their geographical distance.

User Engagement in Online News Platforms. Jones and Altadonna [16] examined the introduction of badges (i.e., awards for users with frequent posting) to encourage user engagement on the Huffington Post website. They found that longer threads do not come from badges, but from the desirability of news articles. Diakopoulos and Naaman [8] studied the relationships between news comment topicality, temporality, sentiment, and quality. They found that some topics aroused more deleted comments (by the moderators), and correlation between the negative sentiment and the fraction of deleted comments. They also found that the frequency at which users comment is correlated with the negativity of the comments.

From this brief literature review, one concludes that we hitherto lack a detailed understanding of how geography impacts the engagement on news platforms. We thus set out to partly fix that by studying how geographical processes impact user engagement on news articles (Sect. 4).

3 Initial Analysis

3.1 Data Description

Our dataset consists of a random sample of 200K news articles and corresponding 41M comments, published from August 2010 to February 2013. Yahoo! News features articles from a variety of news publishers including: Reuters, ABC News, Associated Press, The Atlantic Wire and other. For each article, we know its publication time and comments. Each comment comes with a timestamp, the commenter's anonymous user identifier and *IP* address (which we translate into the corresponding city name using the Yahoo! Places Web service).

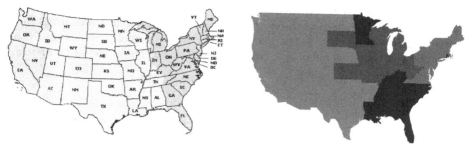

Fig. 1. (Left) USA Map of time zones. (Right) US Map of like-minded states that engage on the same articles (four different clusters of like-minded states emerge).

3.2   State Commenting Graph

To understand whether any geographical process shapes user engagement, we build a graph whose nodes are US states and whose links are weighted with the number of times two users in states i and j comment on the same article. To see the extent to which different states show similar commenting patterns (whether they are like-minded, in that, they tend to engage with the same articles), we apply a community detection algorithm on the graph. We use the Louvain community detection algorithm [5], whose main advantages are both the automatic detection of the optimal number of communities (no need to set that number a priori) and its high clustering accuracy [9]. After running the algorithm, four main clusters of like-minded states are detected and mapped in Figure 1 (right). Interestingly, we see that the four detected groups are geographically clustered (i.e., cover contiguous regions). Furthermore, one readily sees a similarity between this map and the USA Map of time zones (left panel of Figure 1).

4   The Time Zone Effect

To quantify whether time zone affects engagement, we test the hypothesis:

[H1] *Users in the same time zone preferentially engage with the same articles, while users in different time zones engage with different articles.*

To this end, we perform an experiment in three steps (which we shall detail): (1) We measure the observed engagement among users in the same time-zone, 1-time, ..., k time zones apart; (2) By keeping all factors constant except the time zone that are randomly permuted, we measure again the user engagement due to chance; and (3) we compare both engagement measures to assess if the time zone affects engagement.

(1) Engagement in k-Time Zone Apart. To measure engagement, we associate users with their time zones[1] and count the number of times users from k-time zone apart engage in the same articles. More formally, we measure the probability p_k that two users in k time zones apart engage in the same article:

$$p_k = \frac{\sum_{i \in S} \sum_{j \in S} I_k(i, j) \cdot interaction_{ij}}{n},$$

where S is the set of all states; I_k is an indicator function that equals to 1 if states i and j are k time zones apart, or 0 otherwise; $interaction_{ij}$ is the number of times users from states i and j have engaged in the same article; and n normalizes the numerator for the total number of interactions across all time zones.

(2) Engagement due to Chance. To test whether what we observe is not due to chance, we resort to a null (random) model [30]. We reshuffle the assignment of time zones by associating each user to a random zone, and repeat this procedure 2000 times to obtain accurate estimates. The random model removes the time zone effect and keeps all other factors constant. Thus, the difference between the engagement values that are observed and those in the random model depend only on effects strictly related to time zones. If there is no difference, then what we observe does not depend on time zone. As one may expect, if the time zones associated with each user are shuffled, the probability of engagement between two users is approximately the same (i.e., ≈ 0.27) regardless of the time difference.

(3) Compare the Two Engagements. By comparing the observed engagement with the engagement under the random model (Figure 2), we find that users in the same time zone and (to a lesser extent) those one time zone away engage with the same articles (first two dark bars) more than expected by chance (light bars). By contrast, those in three and four-time zone away engage less than chance. We perform a t-test to verify whether the differences between observed values and those in the random model are statistically significant. We find that all differences are significant at p-value less than 0.001.

5   The Geography of News Engagement

We have just ascertained that users who live in the same time zone interact with each other more than what people in different time zones do. Since our null model is oversimplified, we now adopt a geographic notion that is finer grained than that of time zones. We do so by resorting to a widely-used spatial interaction model called "the gravity model" [35]. In analogy to the gravitational interaction between planetary bodies, the model posits that the interaction between two

---

[1] States that belong to more than one time zone are assigned to the time zone in which the majority of the territory belongs to. We considered only the continental states, Alaska and Hawaii have been excluded from the analysis.

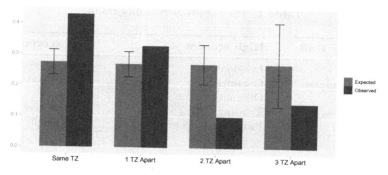

Fig. 2. The probability that two users who are k time zone (TZ) apart engage on the same article. Light bars show the expected engagement in a random model (suppressing the time zone effect), and the dark bars show the observed levels of engagement.

places (e.g., two states) is proportional to their mass (e.g., their population) and inversely proportional to their distance. Despite some criticisms [31], the model has been successfully used to describe 'macro scale' interactions (e.g., between cities, and across states), using both road and airline networks [4,18] and its use has extended to other domains, such as the spreading of infectious diseases [3,33], cargo ship movements [19], and to model intercity phone calls [22].

Here we posit that a gravity model can be used to estimate user engagement on the same articles at the *inter*-state level. The model takes the form:

$$F_{i,j}^{est} = g \frac{m_i m_j}{d_{i,j}^2} \qquad (1)$$

where $F_{i,j}^{est}$ is the estimated engagement, or number of comments users living in states i and j make on the same articles, g is a scaling constant fitted to the data, and $d_{i,j}$ is the distance between the two states, for which we use the Euclidean distance between the two centroids of i and j. Engagement between areas with large number of users and at short distances are predicted to be large, whereas engagement at longer distances or between areas with low mass are predicted to be small. Overall, the correlation between the observed number of comments and gravity model estimates, measured with the Pearson Correlation Coefficient, is as high as .70, which suggests that overall the gravity model provides a good description of user engagement between states, but also that there is still a significant amount of variation not accounted for by the model. We posit that this unexplained portion is due to prevailing socioeconomic factors.

6 The Socio-economic Factors of Engagement

To begin with, we assign topics to both articles and comments. Since we need explicit topic labels (previously we just needed to compute similarity measures),

Table 1. The big five personality traits

Personality trait	High scorers	Low scorers
Openness	Imaginative	Conventional
Conscientiousness	Organized	Spontaneous
Extraversion	Outgoing	Solitary
Agreeableness	Trusting	Competitive
Neuroticism	Prone to stress and worry	Emotionally stable

we cannot use unsupervised techniques (e.g., topic modeling). Instead, we opt for studying a subset (13.8%) of the articles that have been editorially labeled with topical categories from the IPTC news subject taxonomy[2]. The taxonomy consists of 1400 topics and is organized into three levels, according to the specificity of the topics. To have the finest-grained topical view, we use the lowest level of the taxonomy. The number of labels associated with each article ranges from 1 to 25, where the average number of labels per article is 5. We aggregate these topics at state level by considering the number of times users from a given state commented on articles with a certain tag, and the number of times the tag appears in the data set (to avoid the bias of dominant topics).

The Big Five Personality Traits. The five-factor model of personality, or the big five, is the most comprehensive, reliable and useful set of personality concepts [6,10]. An individual is associated with five scores that correspond to the five main personality traits and that form the acronym of *OCEAN* (Table 1 collates a brief explanation). Imaginative, spontaneous, and adventurous individuals are high in **O**peness. Ambitious, resourceful and persistent individuals are high in **C**onscientiousness. Individuals who are sociable and tend to seek excitement are high in **E**xtraversion [2,32]. Those high in **A**greeableness are trusting, altruistic, tender-minded, and are motivated to maintain positive relationships with others [15]. Finally, emotionally liable and impulsive individuals are high in **N**euroticism [17,21].

These big five traits have been studied not only at individual level but also at geographic level [28]. Rentfrow *et al.* [29] have examined the personality scores of half a million US residents and found clear patterns of regional variation across the country, and they have also strong relationships between state-level personality and socioeconomic indicators.

We now correlate state-level personality scores with engagement with articles about specific topics (Figure 3, right). Economy is popular in states with conscientious residents ($r = 0.42$), and unpopular in states with residents who tend to be agreeable ($r = -0.61$) and open ($r = -0.42$). Sport is popular in states whose residents tend to be both extroverts ($r = 0.49$) and open to new experiences ($r = 0.50$). As one might expect, agreeable states avoid articles about religion ($r = -0.53$) and war&unrest ($r = -0.63$). The latter category is also avoided

---

[2] http://www.iptc.org/site/NewsCodes/View_NewsCodes/

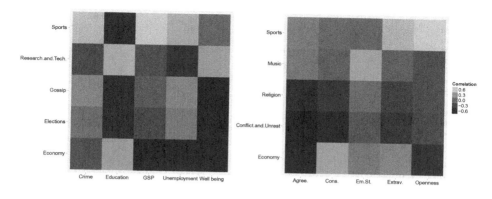

Fig. 3. Correlation between state's topics of interest and: socioeconomic indicators (left panel) and personality traits (right panel)

by conscientious states ($r = -0.49$). States with prevalence of neuroticism (emotional instability) tend to avoid article about music&theater ($r = 0.44$). Finally, states with low levels of neuroticism (i.e., emotional stability) show interest in diverse topics ($r = -0.44$).

Socioeconomic Indicators. We analyze the correlations between a state's assigned topics and the five most studied socioeconomic indicators: well-being index[3], crime level[4], rate of unemployment[5], Gross State Product[6], and education level[7] (number of people with higher education).

As illustrated in Fig. 3 (left), states with high levels of well-being (satisfaction with life) do not engage with articles about economy&business&finance ($r = -0.50$), about elections ($r = -0.53$), or about gossip&celebrities ($r = -0.53$). Economy is also not popular in states with unemployment ($r = -0.46$). Sport, instead, is popular in states with high levels of crime ($r = 0.48$), unemployment ($r = 0.39$), and low gross state product ($r = 0.52$); it is, instead, not very popular in states with high levels of education ($r = -0.43$) whose residents prefer to engage with articles about research&technology ($r = 0.43$) and avoid those on celebrities ($r = -0.40$). States with high levels of education also tend to be interested in diverse topics (i.e., those states have topical vectors with high Shannon diversity, which are correlated with education with an $r = 0.44$).

Putting All Together. In the previous section, we have found that the gravity model explains 70% of the variability of user engagement. We have now shown that socio-economic variables matter and, as a result, they might well explain

---

[3] http://www.thewellbeingindex.com
[4] http://www.ucrdatatool.gov
[5] http://www.bls.gov/web/laus/lauhsthl.htm
[6] http://www.usgovernmentspending.com
[7] http://www.census.gov

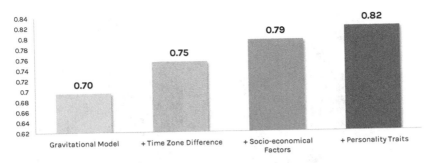

Fig. 4. Adjusted R^2 as predictors are incrementally added to the linear model

Table 2. Linear regression of comments on the same articles from different States. Significance: *** $p < 0.0001$, ** $p < 0.001$, * $p < 0.01$.

Variable	β	t-value	p-value	Variable	β	t-value	p-value
Gravitational Model	0.694	43.947	***	Bachelor	0.057	5.440	***
Time zone difference	0.855	11.893	***	SAT Scores	0.029	4.574	***
Well-being	1.181	9.220	***	Extraversion	0.002	7.383	***
Crime	-0.031	-1.365		Agreeableness	0.987	0.994	
Unemployment	0.000	0.045		Conscientiousness	-7.299	-6.038	***
GSP	-0.071	-3.734	***	Neuroticism	8.247	7.936	***
High Education	0.000	0.749		Openness	2.226	2.987	**

part of the remaining variability. To test the extent to which that is true, we build a linear regression predicting the number of user comments on the same articles from different states. By having not only the gravity model but also the socio-economic variables as predictors, the percentage of variability explained goes indeed up to 82% (Figure 4), which suggests that the linear model effectively predicts the observed user engagement (Figure 5). Table 2 reports the beta coefficients of the individual predictors in detail.

7   Discussion

Our study suffers from two main limitations. First, we have used the users' IP addresses to localize them. So users on the move might be associated with different IP addresses and consequently with different locations. While it might happen to associate the same user to different cities, we found that it had been extremely rare to associate them to different states. Second, our study does not establish any casual relationship. To that end, one would need to apply our methodology to different snapshots over a long period of time.

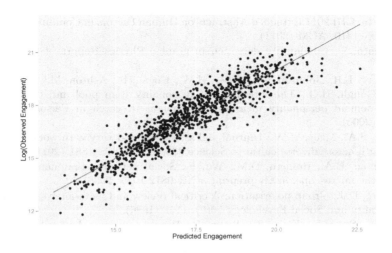

Fig. 5. Observed engagement versus the linear model's predictions

Based on our results, one might well wonder whether like-minded users comment on the same articles, creating fertile ground for group polarization [13]: as a by-product of commenting together (i.e., of engaging with each other), those like-minded users, the theory goes, might develop views that are more extreme than their initial inclinations [1]. For the future, it might be beneficial to explore how geo-temporal patterns of news engagement impact a country's opinion formation.

References

1. An, J., Quercia, D., Crowcroft, J.: Partisan Sharing: Facebook Evidence and Societal Consequences. In: ACM Conference on Online Social Networks (COSN) (2014)
2. Anderson, C., John, O.P., Keltner, D., Kring, A.M.: Who attains social status? effects of personality and physical attractiveness in social groups. Journal of personality and social psychology 81(1), 116 (2001)
3. Balcan, D., Colizza, V., Gonçalves, B., Hu, H., Ramasco, J.J., Vespignani, A.: Multiscale mobility networks and the spatial spreading of infectious diseases. Proceedings of the National Academy of Sciences 106(51), 21484–21489 (2009)
4. Barrat, A., Barthélemy, M., Pastor-Satorras, R., Vespignani, A.: The architecture of complex weighted networks.. Proceedings of the National Academy of Sciences of the United States of America 101(11), 3747–3752 (2004)
5. Blondel, V.D., Guillaume, J.L., Lambiotte, R., Lefebvre, E.: Fast unfolding of communities in large networks. Journal of Statistical Mechanics: Theory and Experiment 2008(10), 10008 (2008)
6. Costa, P.T., McCrae, R.R.: The revised neo personality inventory (neo-pi-r). The SAGE handbook of personality theory and assessment 2, 179–198 (2008)
7. Cox, A.M., Clough, P.D., Marlow, J.: Flickr: a first look at user behaviour in the context of photography as serious leisure. Information Research 13(1), 5 (2008)

8. Diakopoulos, N., Naaman, M.: Topicality, time, and sentiment in online news comments. In: CHI 2011 Extended Abstracts on Human Factors in Computing Systems, pp. 1405–1410. ACM (2011)

9. Fortunato, S.: Community detection in graphs. Physics Reports 486(3), 75–174 (2010)

10. Goldberg, L.R., Johnson, J.A., Eber, H.W., Hogan, R., Ashton, M.C., Cloninger, C.R., Gough, H.G.: The international personality item pool and the future of public-domain personality measures. Journal of Research in Personality 40(1), 84–96 (2006)

11. Golder, S.A., Macy, M.W.: Diurnal and seasonal mood vary with work, sleep, and daylength across diverse cultures. Science 333(6051), 1878–1881 (2011)

12. Huberman, B.A., Romero, D.M., Wu, F.: Social networks that matter: Twitter under the microscope. arXiv preprint arXiv:0812.1045 (2008)

13. Isenberg, D.J.: Group polarization: A critical review and meta-analysis. Journal of Personality and Social Psychology 50(6), 1141 (1986)

14. Java, A., Song, X., Finin, T., Tseng, B.: Why we twitter: An analysis of a microblogging community. In: Zhang, H., Spiliopoulou, M., Mobasher, B., Giles, C.L., McCallum, A., Nasraoui, O., Srivastava, J., Yen, J. (eds.) WebKDD 2007. LNCS, vol. 5439, pp. 118–138. Springer, Heidelberg (2009)

15. Jensen-Campbell, L.A., Graziano, W.G.: Agreeableness as a moderator of interpersonal conflict. Journal of personality 69(2), 323–362 (2001)

16. Jones, J., Altadonna, N.: We don't need no stinkin'badges: examining the social role of badges in the huffington post. In: Conference on Computer Supported Cooperative Work, pp. 249–252 (2012)

17. Jong, G.D., Sonderen, E.V., Emmelkamp, P.: A comprehensive model of stress: the roles of experience stress and Neuroticism in explaining the stress- distress relationship. Psychotherapy and Psychosomatics 68 (1999)

18. Jung, W., Wang, F.: Gravity model in the Korean highway. EPL (Europhysics Letters) 81 (2008)

19. Kaluza, P., Kölzsch, A., Gastner, M.T., Blasius, B.: The complex network of global cargo ship movements.. Journal of the Royal Society, Interface the Royal Society 7(48), 1093–103 (2010)

20. Kamath, K.Y., Caverlee, J., Lee, K., Cheng, Z.: Spatio-temporal dynamics of online memes: a study of geo-tagged tweets. In: Proceedings of the 22nd International Conference on World Wide Web, pp. 667–678. International World Wide Web Conferences Steering Committee (2013)

21. Karney, B.R., Bradbury, T.N.: The longitudinal course of marital quality and stability: A review of theory, methods, and research. Psychological bulletin 118(1), 3 (1995)

22. Krings, G., Calabrese, F., Ratti, C., Blondel, V.D.: Urban gravity: a model for inter-city telecommunication flows. Journal of Statistical Mechanics: Theory and Experiment 2009(07), L07003 (2009)

23. Kwak, H., Lee, C., Park, H., Moon, S.: What is twitter, a social network or a news media? In: Proceedings of the 19th international conference on World wide web, pp. 591–600. ACM (2010)

24. Lindqvist, J., Cranshaw, J., Wiese, J., Hong, J., Zimmerman, J.: I'm the mayor of my house: examining why people use foursquare-a social-driven location sharing application. In: Proceedings of the SIGCHI Conference on Human Factors in Computing Systems, pp. 2409–2418. ACM (2011)

25. Mislove, A., Lehmann, S., Ahn, Y.Y., Onnela, J.P., Rosenquist, N.: Pulse of the Nation: U.S. Mood Throughout the Day inferred from Twitter (2010), `www.ccs.neu.edu/home/amislove/twittermood/`

26. Noulas, A., Scellato, S., Mascolo, C., Pontil, M.: An empirical study of geographic user activity patterns in foursquare. ICWSM 11, 70–573 (2011)

27. Nov, O., Naaman, M., Ye, C.: Analysis of participation in an online photo-sharing community: A multidimensional perspective. Journal of the American Society for Information Science and Technology 61(3), 555–566 (2010)

28. Quercia, D.: Don't worry, be happy: The geography of happiness on facebook. In: Proceedings of the 5th Annual ACM Web Science Conference, pp. 316–325. ACM (2013)

29. Rentfrow, P.J., Gosling, S.D., Potter, J.: A theory of the emergence, persistence, and expression of geographic variation in psychological characteristics. Perspectives on Psychological Science 3(5), 339–369 (2008)

30. Sheskin, D.J.: Handbook of Parametric and Nonparametric Statistical Procedures, 4th edn. Chapman and Hall (2007)

31. Simini, F., González, M.C., Maritan, A., Barabási, A.L.: A universal model for mobility and migration patterns. Nature, 8–12 (2012)

32. Swickert, R.J., Rosentreter, C.J., Hittner, J.B., Mushrush, J.E.: Extraversion, social support processes, and stress. Personality and Individual Differences 32(5), 877–891 (2002)

33. Viboud, C., Bjornstad, O.N., Smith, D.L., Simonsen, L., Miller, M.A., Grenfell, B.T.: Synchrony, waves, and spatial hierarchies in the spread of influenza. Science 312(5772), 447–451 (2006)

34. Zhao, D., Rosson, M.B.: How and why people twitter: the role that micro-blogging plays in informal communication at work. In: Proceedings of the ACM 2009 International Conference on Supporting Group Work, pp. 243–252 (2009)

35. Zipf, G.K.: The P 1 P 2/D hypothesis: On the intercity movement of persons. American sociological review 11(6), 677–686 (1946)

On the Feasibility of Predicting
News Popularity at Cold Start

Ioannis Arapakis, B. Barla Cambazoglu, and Mounia Lalmas

Yahoo Labs, Barcelona, Spain
{arapakis,barla,mounia}@yahoo-inc.com

Abstract. We perform a study on cold-start news popularity prediction
using a collection of 13,319 news articles obtained from Yahoo News. We
characterise the online popularity of news articles by two different metrics
and try to predict them using machine learning techniques. Contrary to
a prior work on the same topic, our findings indicate that predicting
the news popularity at cold start is a difficult task and the previously
published results may be superficial.

Keywords: News popularity prediction, cold-start prediction.

1 Introduction

So far, some research effort has been made to address the problem of news popularity prediction relying on early-stage measurements and user-generated content
associated with the articles. The cold-start prediction scenario has been investigated, for the most part, in the context of recommender systems. To our knowledge, the only exception is the recent, widely cited work of Bandari et al. [2],
who investigate the problem using exclusively content-based features available at
cold start. The performance results reported by the authors imply that cold-start
popularity prediction may be feasible.

Our work challenges the positive interpretation of high accuracy values reported in [2]. To this end, we first try to reproduce the performance results
reported in [2] by following their experimental setting and methodology. We
then improve their methodology and integrate the right performance metrics
in a step-by-step fashion. Our work introduces a large number of new features
(including those used in [2]) which may further help predicting future article
popularity. As the popularity metric, in addition to tweet counts (the metric
used in [2]), we also use the view counts of article pages.

Although we could mostly reproduce the findings of [2] and obtain similar
results, our final findings, which are obtained after a more rigorous evaluation
and interpretation, indicate that predicting the popularity of news articles at
cold start is not really a viable task with the existing techniques. We point at
the high skewness in the popularity distribution as the source of the problem (i.e.,
large number of unpopular articles and very few popular articles). We show that
the techniques are biased to predict the large class of unpopular articles more

L.M. Aiello and D. McFarland (Eds.): SocInfo 2014, LNCS 8851, pp. 290–299, 2014.

accurately than the small class of popular articles (a common phenomenon in machine learning). Therefore, popular articles, which are more important to detect early, cannot be predicted and surfaced to a large extent.

2   Related Work

Some research efforts have addressed the cold-start prediction problem in the context of recommender systems. In [6], the authors present an approach to identify representative users and items using representative-based matrix factorisation. The authors of [5] discuss a hotel recommender system that employs context-based features. The authors overcome the cold-start problem by mining contextual information and analysing it for common traits per context group. In [3], the authors demonstrate how cold-start book recommendations based on social-tags can be combined with traditional collaborative filtering methods to improve performance. Finally, the work in [8] addresses the problem of cold-start social event recommendation, using the home location of the mobile phone users and the social events they have attended in the past.

To the best of our knowledge, the only work that has tackled the cold-start popularity prediction problem in the context of online news is the work of Bandari et al. [2]. In their work, the authors use a measure of popularity based on the number of times a news article is shared on Twitter. They devise a machine learning framework using some basic features including news source, genre, subjectivity of the language, and entities in the articles. The performance results reported by the authors suggest that popularity prediction is possible using only the limited information available before a news article is published. In this work, we reproduce the experimental results of [2] and demonstrate, using a more uniform dataset and a larger set of features, that predicting the news popularity at cold start is not a viable task with the existing techniques. Contrary to the findings of Bandari et al., we show that an article's popularity cannot be accurately estimated, solely on the basis of content features without incorporating any early-stage popularity information.

3   Data, Setup, and Characterisation of Metrics

Our analysis was conducted on a dataset consisting of $13,319$ news articles taken from Yahoo News. We opted for a single news portal to be able to extract features that are consistent across all articles. The dataset was constructed by crawling news articles over a period of two weeks. During the crawling period, we connected to the RSS feed API of the news portal every 15 minutes and fetched newly published articles. Each article was identified by its unique URI and stored in a database, along with meta-data like genre (e.g., politics, sports, crime), publication date, and article's HTML content at the time of publication.

To quantify the online popularity of news articles, we opted for two different metrics: the number of times an article was posted or shared on Twitter (`Tweets`) and the number of times an article page was viewed by the users (`Pageviews`).

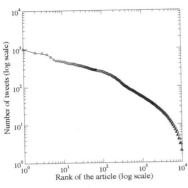

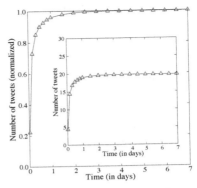

Fig. 1. Tweet counts of articles   **Fig. 2.** Tweet counts over time

The choice of `Tweets` was informed by the fact that, nowadays, an increasing number of users are interacting with social media applications and exchanging content. Online communities, such as Twitter, serve as conduits for information flow and can thereby help to assess the virality of online content. We also include `Pageviews` as a metric since it is commonly used as a proxy for website engagement and online content popularity.

In our setup, every request to the RSS feed API was followed by a request to a public Twitter API to collect sample values for the `Tweets` metric across time. For all articles stored in the database, the metric values were sampled every half an hour, over a period of one week after the articles' publication. This resulted in 337 observations per article. In addition, we sampled data about the page views, again every half an hour, from the access logs of the news site.

Fig. 1 shows the tweet counts of articles in decreasing order of counts. As expected, the distribution is heavily skewed, i.e., most articles are tweeted a few times while very few articles are tweeted many times. Consequently, the problem of identifying soon-to-be-popular news articles becomes a challenging task as we will see in Section 5. Fig. 2 shows the increase in `Tweets` over time (the values are averages over all articles). We report both the original values (inner plot) and normalized values (outer plot). According to the figure, the increase in the `Tweets` metric saturates after two days. About 90% of the tweets happen in the first 12 hours after an article is published.[1]

4   Features

We use a larger number of features that we extract from the content of the news articles as well as external sources. Our features are categorised under ten main headings, depending on how or where the feature is obtained from. In what follows, we explain each feature category separately.

---

[1] `Pageviews` were omitted in Figs. 1 and 2 due to the confidential nature of this metric.

Time. We use features related to the time of news publication. Our choice is motivated by [1,7], where the authors successfully employ date and time information as features for their prediction tasks.

News Source. Similar to [2], we use the news source as a feature. In our study, the articles are obtained from five news distributors. A large portion of the articles are delivered by two major distributors, Reuters and Associated Press, while the share of the remaining agencies in the news volume is much smaller.

Genre. In [2], Bandari et al. use meta-data about the article category (i.e., genre) as one of the features. The authors observe that news related to certain genres have a more prominent presence in their dataset and most likely in the social media as well. Based on their results, we use specific genres as features.

Length. Our length features include the number of characters, words, and sentences in the body of the articles. The first two types of features are also computed for the titles of the articles.

NLP. We also use linguistic features which may have an effect on the online popularity of news articles. Our approach involves computing the distribution of nouns, adverbs, and verbs in the title and body of news articles. Our motivation for applying text analysis, even at this basic level, is that linguistic features can provide insights into certain aspects of the textual meaning or the impact on the reading experience.

Sentiment Analysis. For sentiment analysis, we use SentiStrength, a lexicon-based sentiment analysis tool [9]. We compute a sentimentality score and a polarity score for an article by averaging the positive/negative sentiment scores returned by SentiStrength for the individual sentences in the article (as described in [4]). We compute the two scores also for the article's title, treating it as a single sentence.

Entity Extraction. Similar to [2], we use an in-house software to extract named entities from the news articles. Here, in particular, we were interested in observing if the number of named entities in a news article affects its popularity. In general, we observed that articles that mention at least one entity are more likely to become popular than articles that do not mention any entity.

Wikipedia. For each named entity extracted from the article, we retrieve the popularity of the entity in Wikipedia.[2] Title- and body-level popularity values are then computed by summing the popularity values of all entities extracted from the title and article body, respectively. Other aggregation techniques, like averaging, yield inferior performance.

---

[2] http://www.mediawiki.org/wiki/API:Main_page

Twitter. To determine the short-term popularity of articles, we compute the popularity of named entities in Twitter. For each entity, we track the volume of tweets referring to the entity starting one hour, one day, and one week before the article's publication date.[3]

Web Search. We repeat the same technique on a large sample of queries submitted to the front-end of a popular web search engine and compute the frequency of entities in the sample. Again, the popularity of an entity is computed at three different time intervals and the aggregate search popularity for an article is determined as before.

5 Experiments

We start our experiments by reproducing the classification results presented in [2] by Bandari et al. for `Tweets`. To this end, we split two weeks of articles (13,319 articles) into three classes based on their tweet counts: `A` (low popularity), `B` (medium popularity), and `C` (high popularity). Adopting the choice made in [2], the tweet count ranges are set to $[1, 20]$, $(20, 100]$, and $(100, \infty)$ for the `A`, `B`, and `C` classes, respectively. Articles that are not tweeted are removed from the data and not included in set `A`. We experiment with the same classifiers used in [2]: naive Bayesian (`NB`), bagging (`Bagging`), decision trees (`J48`), and support vector machines (`SVM`). Moreover, for comparison purposes, we include a baseline classifier `Baseline` that always predicts the majority class in the training data. We make predictions for one hour, one day, and one week after an article is published. We perform log transformation on features exhibiting a skewed distribution.

Despite our efforts to create a similar setup, there are two minor differences between our setup and the setup in [2]. First, the articles used in [2] (10,000 articles) are obtained from a large number of news sites while our collection is obtained from a single, relatively major news site. Second, in [2], the popularity of articles are assumed to saturate after four days. In our case, as the closest value, we can use the popularity values obtained after one week. Nevertheless, since the features used in our study form a more powerful superset of the features used in [2], we expect to attain better or at least similar classification performance.

In Table 1, we report the performance in terms of the accuracy metric, i.e., the fraction of test articles whose class is correctly predicted by the classifier.[4] The reported results are obtained by performing cross-validation with ten folds, again adopting the choice made in [2]. According to the table, for the (`Tweets`, `Week`) case, the best performing classifier (`SVM`) achieves an accuracy of 79.7%, which is a bit lower than the best accuracy value (83.96%) reported in [2] (achieved by `Bagging`). However, when we observe the relative improvement with respect to the baseline (79.7% − 70.3% = 9.4%), we find it to be slightly higher in our case. Although it is not directly reported in [2], the relative improvement in their case

---

[3] We use Topsy's Otter API, available at http://code.google.com/p/otterapi/

[4] We do not report the classification accuracies for the `Pageviews` metric as this may reveal confidential information about the distribution of page views.

Table 1. Accuracy (ten-fold cross validation, without zero-popularity articles)

Technique	Tweets		
	Hour	Day	Week
Baseline	0.840	0.710	0.703
NB	0.693	0.581	0.574
Bagging	0.858	0.749	0.741
J48	0.856	0.781	0.775
SVM	0.859	0.802	0.797

Table 2. Accuracy (training/test
split, without zero-popularity articles)

Technique	Tweets		
	Hour	Day	Week
Baseline	0.839	0.706	0.698
NB	0.735	0.589	0.584
Bagging	0.858	0.737	0.74
J48	0.852	0.779	0.774
SVM	0.861	0.803	0.798

can be estimated as $83.96\% - 76\% = 7.96\%$ using the data the authors provided in
Tables 5 and 6. In general, the reported results are comparable, and we believe
that we were able to reproduce the results reported in [2] to a certain degree.

Next, we repeat the same experiment using a training/test split in the time
dimension instead of cross-validation. This is because, in the latter approach,
the classifiers are allowed to use future information. In a real-life setting, this is
not meaningful since a model would be trained at a fixed point in time using
features extracted from previously seen articles and then it would be applied to
predict the popularity of new articles. Hence, we repeat the previous experiment
by splitting the data into training and test sets. The training set contains articles published in the first week and the test set contains articles published in the
following week. The two sets are roughly equal in size. According to Table 2, the
classification performance is somewhat similar to that in the previous experiment, i.e., there was no positive bias in results due to the use of cross-validation.
In the remaining experiments, we use the setup with a training/test split.

Another issue that we observe in the methodology followed in [2] is the artificial manipulation of the original news collection. Before conducting their experiments, the authors remove from the data every article that is not tweeted at all
after it was published. This manipulation may lead to unfair results because, in
a real-life setting, it is not possible to know whether an article will be tweeted
or not before it is published. Hence, herein, we repeat the previous experiment
after including zero-popularity articles in the A class. The results are reported
in Table 3. We observe that the classification problem is now easier than before
as the accuracy of the best performing classifier has increased in all scenarios.
In particular, the best accuracy increases from 79.8% to 82.5% in case of the
(Tweets, Week) scenario. On the other hand, the performance gap between the
best performing classifier and Baseline gets smaller. As an example, in case of
(Tweets, Week), the improvement drops from 10.0% to 8.5%.

All results reported so far indicate high classification accuracy. But, how
meaningful or useful are these results in practice? Can we really distinguish article popularity through classification? The answer lies in the surprisingly good
performance of the Baseline classifier, which always predicts the label of the majority class in the training data. This implies that high accuracy values could be
due to the highly skewed nature of the popularity distribution and the resulting
class imbalance. In such scenarios, the classifiers are biased to learn and predict
the majority class, leading to superficial accuracies.

Table 3. Accuracy (training/test split, with zero-popularity articles)

Technique	Tweets		
	Hour	Day	Week
Baseline	0.871	0.746	0.740
NB	0.772	0.642	0.633
Bagging	0.886	0.780	0.769
J48	0.883	0.805	0.804
SVM	0.890	0.829	0.825

Table 4. Fraction of instances in each of the three popularity classes

Class	Tweets		
	Hour	Day	Week
A	0.871	0.746	0.740
B	0.125	0.227	0.231
C	0.004	0.027	0.029

Table 5. The confusion matrix for (Tweets, Week)

Actual	Predicted		
	A	B	C
A	4,698	247	0
B	728	812	0
C	98	96	0

Table 6. Root mean squared error (training/test split, with zero-popularity articles)

Technique	Tweets		
	Hour	Day	Week
BaselineR	1.701	1.931	1.950
LR	1.132	1.270	1.305
kNNR	1.537	1.720	1.753
SVM	1.135	1.278	1.315

But, how skewed is the class distribution at hand? In Table 4, we display the fraction of articles in the test set for each of the three classes (confirming Fig. 1). As we can see, the collection is dominated by the unpopular articles in class A. In all cases, class C (the class of most popular articles) constitutes less than 4% of the sample. In a real-life setting, it is much more important to distinguish the articles in class C from the rest. The question is then how good are we in predicting class C articles. To answer this question, one can look at the confusion matrices, showing the true and false positive rates per class. In Table 5, as a representative, we provide the confusion matrix for the (Tweet, Week) scenario (using the best performing classifier, SVM). According to the table, the classifier does quite well in correctly identifying class A articles. However, it fails to distinguish class C articles from class A and B articles. This result indicates that the accuracy numbers reported in [2] are very likely to be not useful either.

Given that classification does not yield meaningful performance, we turn our attention to regression and observe the performance in predicting the actual popularity values rather than the popularity class values. To this end, we evaluate three regression approaches: linear regression (LR), k-nearest neighbor regression (kNNR), and support vector machines (SVM). For comparison, we also use a simple baseline (BaselineR) that always predicts the mean value in the training data. We perform a logarithmic transformation on the target values before regression.

In Table 6, the regression performance is reported in terms of root mean squared error. According to the table, LR is the best performing technique. Overall, the calculated errors are low and also there is considerable improvement with respect to BaselineR. As we go from Hour to Week, the error increases due to the larger variation in popularity. However, the improvement relative to BaselineR also increases since predicting the late-stage popularity is easier.

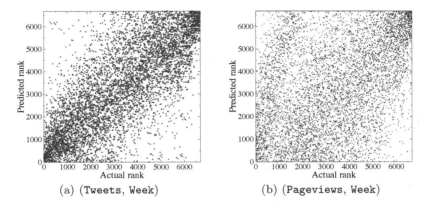

(a) (Tweets, Week)   (b) (Pageviews, Week)

Fig. 3. Actual versus predicted ranks

Although the regression results give an idea about the prediction quality, they still do not tell us whether the predictions are biased towards unpopular articles. Moreover, in practice, accurate ranking of articles (in decreasing order of popularity) is more important than accurate prediction of their popularity. That is, given a popular and an unpopular article, the difference between the predicted popularity values is not of high importance as long as we can correctly rank the popular article above the unpopular article.

To visualize the ranking performance, in Fig. 3, we display the actual versus predicted popularity ranks of the articles (e.g., the article with the highest popularity is ranked first). A stronger correlation is observed between the actual and predicted tweet counts. In case of Tweets, we observe a stronger correlation between the actual and predicted values compared to Pageviews. The Kendal Tau (KT) correlation values in Table 7 are consistent with the plots.

Finally, we evaluate the performance focusing on the top ranked articles. This is important because, as we mentioned before, only a small fraction of articles gain visibility due to the limited space in web pages and the limited attention span of users. Therefore, it is vital to get the popularity ranking right especially at the high ranks by correctly identifying the most popular articles. To evaluate the performance at top ranks, we define the recall@k (R@k) metric (this is different than the traditional recall metric in information retrieval). Our metric basically selects the articles that are placed in the top k ranks by the prediction algorithm and then computes what fraction of them also appear within the top k ranks in the actual popularity ranking. We report R@k values for $k \in \{10, 100, 1,000\}$ in Table 7. For $k = 1,000$, even with the best prediction scenario (i.e., Tweets), we observe that about 45% of the articles in the top 1,000 ranks are not ranked among the top 1,000 articles in the actual popularity ranking. The results are even worse as the k value decreases. These final results illustrate the real difficulty of the problem and indicate the superficial nature of the previous results obtained through classification and regression [2].

Table 7. Performance in terms of the Kendal Tau and recall@k metrics

Metric	Tweets			Pageviews		
	Hour	Day	Week	Hour	Day	Week
KT	0.551	0.569	0.561	0.078	0.286	0.287
R@10	0.000	0.000	0.000	0.000	0.000	0.000
R@100	0.240	0.110	0.090	0.010	0.020	0.060
R@1000	0.578	0.557	0.548	0.212	0.173	0.245

6   Conclusion

In this work, we investigated the cold-start news popularity prediction problem. We measured the popularity of articles in terms of their tweet counts as well as page views. Using the content of news articles and external sources, we engineered a large number of features that may indicate the future popularity of news articles. Our work revealed that predicting the news popularity at cold start is not a solved problem yet. We observed that classifiers were biased to learn unpopular articles due to the imbalanced class distribution. Hence, highly popular articles could not be accurately detected, rendering the predictions not useful in most practical scenarios. Our findings suggest that popularity is disconnected from the inherent structural characteristics of news content and cannot be easily modelled. News popularity may be more accurately predicted if early-stage popularity measurements are incorporated into the prediction models as features. In general, increasing the duration of such measurements will increase the accuracy of predictions but decrease their importance, leading to an interesting trade-off.

Acknowledgments. This work was supported by the MULTISENSOR project, partially funded by the European Commission, under the contract number FP7-610411.

References

1. Ahmed, M., Spagna, S., Huici, F., Niccolini, S.: A peek into the future: Predicting the evolution of popularity in user generated content. In: Proc. 6th ACM Int'l Conf. Web Search and Data Mining, pp. 607–616 (2013)
2. Bandari, R., Sitaram, A., Huberman, B.A.: The pulse of news in social media: Forecasting popularity. In: Proc. 6th Int'l Conf. Weblogs and Social Media (2012)
3. Givon, S., Lavrenko, V.: Predicting social-tags for cold start book recommendations. In: Proc. 3rd ACM Conf. Recommender Systems, pp. 333–336 (2009)
4. Kucuktunc, O., Cambazoglu, B.B., Weber, I., Ferhatosmanoglu, H.: A large-scale sentiment analysis for Yahoo! Answers. In: Proc. 5th ACM Int'l Conf. Web Search and Data Mining, pp. 633–642 (2012)
5. Levi, A., Mokryn, O., Diot, C., Taft, N.: Finding a needle in a haystack of reviews: cold start context-based hotel recommender system. In: Proc. 6th ACM Conf. Recommender Systems, pp. 115–122 (2012)
6. Liu, N.N., Meng, X., Liu, C., Yang, Q.: Wisdom of the better few: cold start recommendation via representative based rating elicitation. In: Proc. 5th ACM Conf. Recommender Systems, pp. 37–44 (2011)

7. Marujo, L., Bugalho, M., da Silva Neto, J.P., Gershman, A., Carbonell, J.: Hourly traffic prediction of news stories. In: 3rd Int'l Workshop on Context-Aware Recommender Systems (2011)
8. Quercia, D., Lathia, N., Calabrese, F., Lorenzo, G.D., Crowcroft, J.: Recommending social events from mobile phone location data. In: 2010 IEEE 10th Int'l Conf. Data Mining, pp. 971–976 (2010)
9. Thelwall, M., Buckley, K., Paltoglou, G.: Sentiment strength detection for the social Web. J. Am. Soc. Inf. Sci. Technol. 63(1), 163–173 (2012)

A First Look at Global News Coverage of Disasters by Using the GDELT Dataset

Haewoon Kwak* and Jisun An

Qatar Computing Research Institute, Doha, Qatar
{hkwak,jan}@qf.org.qa

Abstract. In this work, we reveal the structure of global news coverage of disasters and its determinants by using a large-scale news coverage dataset collected by the GDELT (Global Data on Events, Location, and Tone) project that monitors news media in over 100 languages from the whole world. Significant variables in our hierarchical (mixed-effect) regression model, such as population, political stability, damage, and more, are well aligned with a series of previous research. However, we find strong regionalism in news geography, highlighting the necessity of comprehensive datasets for the study of global news coverage.

Keywords: GDELT, global news coverage, news geography, regionalism, theory of newsworthiness, international news agency, foreign news, disaster.

1   Introduction

A wide news network woven by international news agencies helps to transform a remote disaster into an international crisis. Even though a majority of disasters remains unreported [9], the reported ones evoke compassion, and this potentially leads to various charitable acts, such as fund-raising to provide monetary support. These days it is not uncommon to expect help from the world when a tragedic disaster happens. In this sense, global news coverage of a disaster is a sufficient condition for worldwide public action. Then, a central question naturally arises: which disasters are covered and which are not? The systematic approach to address this question requires a comprehensive dataset of news media sites in different countries over long period, which remains to be unexplored in traditional media research. We revisit previous research based on a single country or a region [11,19] and examine whether it also holds globally.

In this work we use a large-scale news media coverage dataset collected by the GDELT (Global Data on Events, Location, and Tone) project. GDELT project monitors news media in over 100 languages from the whole world [14]. With large-scale data of 195 thousand disasters happening from April 2013 to July 2014 [1], we examine which disasters receive a great deal of attention from foreign news media.

---

* A long version of this work, "Understanding News Geography and Major Determinants of Global News Coverage of Disasters", will be available on arXiv.

L.M. Aiello and D. McFarland (Eds.): SocInfo 2014, LNCS 8851, pp. 300–308, 2014.

2   Theoretical Orientation

We clarify the theoretical orientation to address our research questions by reviewing previous literature. We begin with the theory of newsworthiness proposed in 1965 and proceed to following studies.

2.1   Theory of Newsworthiness

Foreign news coverage is the outcome of a news selection process [8]. Making a decision on which news items to report is essential because news is delivered through physically limited channels, such as pages in newspapers and minutes in TV news.

On one hand, one of the seminal studies that examines the factors of news selection was conducted by Galtung and Ruge [11]. They propose the theory of newsworthiness that is based on psychology of individual perception and explain which factors influence newsworthiness of an event. The suggested factors are frequency, intensity, unambiguity, meaningfulness, consonance, unexpectedness, continuity of an event, and some characteristics (e.g. identity) of an actor involved in the event. Some critics argue that applying their theory to the global news flow between nations is insufficient owing to the lack of systematic determinants based on the power structures of the world [21].Nevertheless, the theory of newsworthiness has provided a foundation of subsequent news flow studies with a few variations of some factors [12]. We consider several factors among their suggestions, such as unexpectedness and intensity of an event and the identity of an actor extracted from the GDELT dataset.

The significance of the number of people killed by a disaster in predicting its news coverage is still debatable. Gaddy and Tanjong reported its importance [10], while others found no significance [5]. Yet, the number of victims is a good proxy to reflect the intensity of a disaster, especially when the other measures for the damage of a disaster is unavailable. We thus include the number of victims of a disaster in our study and validate its importance.

2.2   Effect of National Attributes on Foreign News Coverage

While determinants of global news flow, in terms of the amount and its direction, have been repeatedly investigated based on a single or few countries for decades [16], results are not consistent between countries, mainly due to cultural difference [17].

On the other hand, one of the earliest studies reporting general patterns of global news flow was conducted with the 'Foreign News' dataset that contains news media of 46 countries for two weeks in 1995 [18]. The findings are relatively stable, and confirmed by subsequent studies. We note that these studies are usually conducted from the view of a guest and a host country relationship. The guest country is where the event happens, and the host country is where the news media exist. In this view, the problem of global news flow is transformed into the problem of dyadic relationship between countries, like whether an event in a certain guest country is covered in a particular host country. The general factors to affect the news coverage of a host country can be divided into two categories. One is the attributes of a guest country, and the other is proximity between two countries. We focus only on the former, the national attributes of the guest country, because our aim is to capture the global view and thus does not necessarily assume the dyadic relationship led by a specific host country.

A wide range of national attributes affecting news coverage is found across the studies [6,16], such as GNP per capita, territorial size, GDP, defense budget, population density, share in world trade, press freedom index, number of scientific publications, and Internet use. We consider all these variables. We also include some variables such as world giving index, to consider the humanitarian view of the disasters.

3   The GDELT Project

GDELT (Global Data on Events, Location, and Tone) is a recently developed event dataset containing more than 200 millions geolocated events with global coverage since 1979 [14]. GDELT began with monitoring a wide range of international news sources, including AfricaNews, Agence France Presse, Associated Press Online, Associated Press Worldstream, BBC Monitoring, Christian Science Monitor, Facts on File, Foreign Broadcast Information Service, United Press International, and the Washington Post. Now in cooperation with Google, it has expanded its sources to cover non-English news media. Today it tracks news media in over 100 languages from the whole world. After the first release of GDELT, several studies have confirmed that the GDELT dataset performs as well or better than the previously widely-used datasets, such as ICEWS (Integrated Conflict Early Warning System) due to its large coverage and the improved automatic coding system [7].

GDELT provides two types of datasets. One is called the Event Database, coded by CAMEO taxonomy since 1979, and the other is the Global Knowledge Graph (GKG), an expanded dataset about 'every person, organization, company, location, and over 230 themes and emotions from every news report' since 2013. We use the GKG dataset because it offers various fields to describe the characteristics of natural and man-made disasters, such as the type of a disaster, the number of news articles reporting the disaster, the number of victims, the location where the disaster occurred, and etc.

4   News Geography of Disasters

We first concentrate on news geography, the extent to which countries are represented in international disaster news. We show how different their news geographies are and then move on to examine the representativeness of each region for *global* attention. It directly relates with the external validity of previous studies about foreign news rooted on a single or several countries.

We divide the world into seven regions according to the World Bank: East Asia & Pacific, Europe & Central Asia, Latin America & Carribean, Middle East & North Africa, North America, South Asia, and Sub-Saharan Africa. The division mainly reflects geographical proximity. We map 10,009 news media into one of the seven regions according to the classification of Alexa, which is based on the nationality of website visitors. A list of news media falling in each region becomes a basis to construct news geography seen by each region.

We define the attention of a region, r_i, to a country, c_j, as the number of the disasters occurred in c_j covered by news media of r_i. We use the notation, $N_{r_i \Rightarrow c_j}$, for representing the attention of r_i to c_j. Then, we define news geography seen by r_i as $\mathbf{N}_{r_i} = \{N_{r_i \Rightarrow c_1}, N_{r_i \Rightarrow c_2}, N_{r_i \Rightarrow c_3}, ..., N_{r_i \Rightarrow c_K}\}$ where K is the number of the countries.

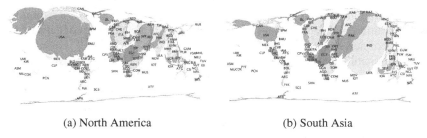

(a) North America   (b) South Asia

Fig. 1. News geography seen by each region

Figure 1 shows the news geography seen by North America and South Asia, respectively, through Cartogram, an intuitive visualization method of illustrating the territory of a country that is proportional to the assigned value. In the figure, the size of a territory is proportional to $N_{r_i \Rightarrow c_j}$ in the news geography seen by r_i.

By visual inspection, we observe clear differences of the news geography across the region. Although we omit other regions due to lack of space, as similar with two visualizations in Figure 1, every region is overrepresented in the news geography seen by the corresponding region. For example, disasters occurring in Latin America & Caribbean are not frequently reported in other regions. Similarly, Indonesia is well-recognized in $\mathbf{N}_{East Asia}$, Serbia in $\mathbf{N}_{Europe}$, Saudi Arabia in $\mathbf{N}_{Middle East}$, USA in $\mathbf{N}_{North America}$, India in $\mathbf{N}_{South Asia}$, and Kenya in $\mathbf{N}_{Africa}$. This strong regionalism raises concerns about the external validity of studies of foreign news coverage based on a single country or region.

At the same time, Figure 1 poses an interesting question about the over-representation of a certain country (e.g. Syria in news geography seen by North America) that cannot be explained by regionalism. This relatively regionalism-free country could be explained by the proximity in another layer, such as politics, economy or culture, instead of geographic proximity. Since the scope of this work is investigating global attention to disasters, rather than attention of a certain region or country, we do not study this further here.

5   Determinants of Global News Coverage of Disasters

5.1   Methods

We build a hierarchical (mixed-effect) multiple regression model to examine what affects global news coverage of disasters. We choose this model to control a random effect driven by variation rooted on country-level differences. Previous studies have shown that international news coverage varies significantly by country [20].

We define global news coverage of a disaster as the number of countries reporting the disaster, and use it as the dependent variable in our model. For 10,009 news media appearing in the GKG dataset, we extract the origin country of each news media from Alexa. In our data, the range of global news coverage lies between 1 and 34. The median is 1 (mean: 1.78), indicating that a large fraction of disaster news is consumed within a single country.

We consider 26 independent variables as candidates that might exert influence on the global news coverage of a disaster according to our theoretical orientations. The variables can be organized into three broad categories: (1) the attributes of a nation which measures political and economical status of a nation; (2) the attributes of a disaster; and (3) logistics of news gathering.

We obtain national attributes for all the countries listed in the GDELT dataset from various sources, including the World Bank Open Data, which provides a wide range of up-to-date measures of 254 countries. Fifteen national variables are driven from it: GDP (gross domestic product) per capita, GNI (gross national income) per capita, military expenditure, population, land size, population density, merchandise exports (US$), merchandise imports (US$), number of scientific journal publications, unemployment rate, foreign aid received (US$), Internet use (per 100 people), mobile cellular subscriptions (per 100 people), and homicide rate (per 100,000 people). In addition, we create a trade variable as the sum of the magnitude of exports and imports. While some information is not up-to-date, most of variables are reported for 2013. We additionally consider the index of press freedom [2], the world giving index [4], and the political stability index [3].

We obtain attributes of disasters directly from the GKG dataset. It marks a disaster as either man-made, natural, or both. Also, a fine-grained subtype of a disaster, such as 'Flooding' and 'Landslide', is available. We select one representative subtype for the disaster by considering the frequency of subtypes across all the disasters, while GDELT can assign multiple subtypes to one disaster. For the matter of categorical coding, we select top 30 subtypes out of 274 subtypes, which account for 31.5% of all disasters. All the other types are coded as 'other'. To measure unexpectedness of a disaster (denoted as UE_disaster), we use the inverse of the frequency of the same-subtype disaster occurring in the corresponding country. In other words, the more frequently a disaster occurs, the more expected the disaster is. We add variables about the impact of disasters: the number of people involved in the disaster, the type of people's involvement in the disaster (denoted as count type), and the type of people affected by the disaster (denoted as object type). We also consider the country where the disaster occurs.

We finally add one binary variable to show whether a disaster is reported by any of international news agencies, denoted as INAs covered. This reflects logistics of news gathering, determining the news flow by gate-keeping. We focus on three global news agencies, as many previous literature does: Agence France-Presse (AFP), Associated Press (AP) and Reuters.

After considering the above variables, elimination of multicollinearity is a crucial step because multicollinearity distorts estimated coefficients of variables. We take three steps of analysis to select relevant variables. First, we build a linear regression for each of the independent variables to see its predictive power for global news coverage of disasters. In this step, we discard the homicide rate variable as it shows low significance in predicting the global coverage. Then, we compute the (Pearson) correlation coefficient between each pair of variables. We find that a few national variables are correlated with each other (i.e., there are high positive correlations (where $r > 0.60$) among GDP, GNI,

GNI, Internet use, life expectancy, the number of scientific journal articles, political stability, and index of press freedom). We retain "political stability" as it is the most predictive factor among those eight variables. By a variance inflation factor (VIF) test, we additionally remove the trade variable. Lastly, through stepwise variable selection using AIC, we get the final regression model with 14 independent variables and an additional control variable (location). We confirm no collinearity by a VIF test; all the remaining variables have VIF below than 5.3.

For the analysis, we use the GKG dataset provided by the GDELT project that incorporates 3,574,627 events happening in 205 countries from April 2013 to July 2014. We extract 666,150 natural or man-made disasters and filter out disasters if any variable is missing. As a result, we have 195,513 disasters to build the model.

5.2 Results

We move on to examine the explanatory power of variables in determining global news coverage of a disaster with our hierarchical multiple regression model. Seven national variables are entered in the first model. Then, six disaster attributes are added to the second model for determining their unique contribution while controlling for the national characteristics. Finally, the full model is tested, including the simultaneous examination of all variables. We discuss only the contribution of individual variables in the full model due to lack of space. We use a 0.05 level of statistical significance to evaluate the results of the regression analysis.

Table 1 shows the regression result with estimated coefficient and its statistical significance. In the first model, three national variables, which are log(population), mobile subscription, political stability, have a significant effect on the dependent variable and explain 3.1% of its variance. The figures for the second model, in which disaster variables are added to the first model, indicate that the characteristics of a disaster themselves explain an additional 4.3% of the variance. Together, the national and disaster variables explain 7.9% of the variance in global news coverage.

In the final step, the newly added variable, INAs covered, explains an additional 18% of the variance, resulting in a total of 25.4% of the variance in global news coverage. Its gain is much more than the amount of variance explained by the national and disaster variables. The explanatory power of our model is comparable with previous studies in this research area; in the study of finding systemic determinants of news coverage of 38 countries [22], 20 out of 38 models have lower R^2 than ours.

National Attributes. We find that out of 7 variables considered, only two are statistically significant. Population has a positive coefficient, but its effect is marginal. High population of a country commonly leads to more emigrants to other countries. For example, China and India are major nationalities of US immigrants, in spite of their remoteness to US. As Lacy et al. point out, news coverage of newspapers is influenced by audience demand [13]. Higher number of immigrants possibly explain more news coverage for them. This relationship would be clarified when a guest, a host country framework for foreign news coverage is adopted, when the number of immigrants can be counted.

Table 1. Hierarchical multiple regression predicting global news coverage (N = 195,513)

	Model 1	Model 2	Full model
Intercept	$-0.87\ (0.63)$	$0.33\ (0.64)$	$0.19\ (0.54)$
National attributes			
Mobile subscriptions	$0.00\ (0.00)^{**}$	$0.00\ (0.00)^{**}$	$0.00\ (0.00)^{**}$
log(Population)	$0.08\ (0.03)^{*}$	$0.07\ (0.03)^{*}$	$0.06\ (0.03)^{*}$
Political stability	$-0.17\ (0.06)^{**}$	$-0.17\ (0.05)^{**}$	$-0.13\ (0.04)^{**}$
Disaster attributes			
Manmade disaster		$-0.95\ (0.02)^{***}$	$-0.76\ (0.02)^{***}$
Natural disaster		$-1.06\ (0.01)^{***}$	$-0.83\ (0.01)^{***}$
# of affected people		$0.00\ (0.00)^{***}$	$0.00\ (0.00)^{***}$
UE of disaster		$-0.28\ (0.07)^{***}$	$-0.17\ (0.07)^{**}$
Count type			
Kill		$-0.33\ (0.04)^{***}$	$-0.22\ (0.04)^{***}$
Other		$-0.73\ (0.04)^{***}$	$-0.52\ (0.04)^{***}$
Protest		$-0.56\ (0.04)^{***}$	$-0.45\ (0.04)^{***}$
Wound		$-0.40\ (0.04)^{***}$	$-0.26\ (0.04)^{***}$
Object type			
Victims		$-0.47\ (0.22)^{*}$	$-0.44\ (0.20)^{*}$
Disaster subtype			
Radiation leak		$0.58\ (0.37)$	$0.65\ (0.33)^{*}$
Toxic waste		$0.37\ (0.17)^{*}$	$0.06\ (0.15)$
Aftershocks		$1.37\ (0.37)^{***}$	$1.08\ (0.34)^{**}$
Flooding		$0.56\ (0.14)^{***}$	$0.33\ (0.13)^{**}$
Heat wave		$0.33\ (0.17)^{*}$	$0.18\ (0.15)$
Ice		$0.58\ (0.14)^{***}$	$0.31\ (0.13)^{*}$
Landslide		$0.67\ (0.14)^{***}$	$0.30\ (0.13)^{*}$
Monsoon		$0.29\ (0.15)^{*}$	$0.14\ (0.13)$
Severe weatehr		$1.10\ (0.17)^{***}$	$0.07\ (0.15)$
Tsunami		$0.33\ (0.15)^{*}$	$0.20\ (0.13)$
Other		$0.38\ (0.14)^{**}$	$0.20\ (0.13)$
Logistics of news gathering			
INAs covered			$3.74\ (0.02)^{***}$
R^2	0.03090539	0.07369887	0.2536145
AIC	802417.49	793681.70	750501.26
BIC	802519.33	794241.77	751071.51
Log Likelihood	-401198.75	-396785.85	-375194.63
Num. groups	74	74	74

$^{***}p < 0.001,\ ^{**}p < 0.01,\ ^{*}p < 0.05$

The political stability is negatively correlated with global news coverage. In other words, disasters happening in politically unstable countries receive more global attention. This finding is aligned with the study by Masmoudi that reveals the Western news media intentionally cover crisis or conflicts of unstable countries so that the stereotype of them can be reinforced [15]. Even though we find the possibility of such a tendency from the aggregated news media, we do not attempt to quantify the contribution of Western news media to this. In-depth analysis is required to assess the universality of reinforcing stereotype.

Disaster Attributes. The negative coefficients of man-made disaster and natural disaster mean that a disaster tagged as both man-made and natural disasters is more likely to get global attention than when only one theme is tagged. It implies that when a natural disaster happens, a complex situation where human factors are involved is likely to be covered by news media.

We find that the number of affected people is statistically significant with a marginal effect. This supports the previous finding that the number of killed people is an important factor for the natural disaster news coverage [10]. In our study, not only killed people but also affected people have been taken into account and we find that it has significant explanatory power.

The number of affected people greatly varies across disasters. For example, sometimes a few tens of thousands of people are evacuated when serious flooding occurs. Given that the significance of the number of killed people is still debatable [5], we believe that follow-up investigations in various settings are vital for assessing the significance of the affected people in global news coverage.

Unexpectedness of disaster subtype is negatively correlated with global news coverage. This finding counters what had been found in the theory of newsworthiness [11]. We explain a possible mechanism in the discussion section.

The count type is a categorical variable. The beta coefficients of category show the relative predictive power of each type compared to the base type, which is Kidnap in our model. Among the count types we consider, we find that the Kidnap type tends to get the most global attention compared to other type: Kill (-0.22), Wound (-0.26), Protest (-0.45), Other (-0.52). The negative coefficient of the Other type means that fewer countries report the Other type than they report the Kidnap type.

Lastly, some disaster types are more favorable globally than others. Although marginal, five disaster types are statistically significant and positively related to the number of countries covering the disaster: Aftershocks (1.08), Radiation leak (0.65), Flooding (0.33), Ice(0.31), and Landslide(0.30).

Logistic of News Gathering. INAs covered has the largest beta coefficients (3.74). Its positive sign indicates that a disaster covered by INA is more likely to get global attention. This is expected; however, the extent of the effect is not expected. INAs covered alone explains 18% of the variance in global news coverage. This result is in line with previous work reporting that the presence of INAs is a primary predictor of the amount of news coverage about the country [22]. We also agree with Wu's argument that the most news media sites are dependents of INAs because the cost of managing correspondents to investigate foreign issues is higher than using news copy of INAs. We discover that the INAs still play a prominent role in expanding news coverage of a foreign disaster.

6 Discussion

A lot of US media are tracked by GDELT, as the strongest player in the media industry. Thus, the number of news articles about a disaster is readily influenced by US news media. To neutralize this possible bias, we define global attention to a disaster as the number of countries covering the disaster. This equalizes the contribution of news media in each country to global attention. While this simplifies the international power relationship, it also captures well the news geography and news flow among countries, which is essential to our research question.

References

1. All GDELT Global Knowledge Graph (GKG) Files, `http://data.gdeltproject.org/gkg/index.html` (accessed August, 8 2014)
2. Index of press freedom, `http://rsf.org/index2014/en-index2014.php` (accessed August 8, 2014)
3. Political Stability and Absence of Violence, The Worldwide Governance Indicators, `http://www.govindicators.org` (accessed August 8, 2014)
4. World giving index, `http://www.theguardian.com/news/datablog/2010/sep/08/charitable-giving-country#data` (accessed august 8, 2014)
5. Adams, W.C.: Whose lives count? tv coverage of natural disasters. Journal of Communication 36(2), 113–122 (1986)
6. Ahern, T.J.: Determinants of foreign coverage in us newspapers. In: Foreign News and The New World Information Order, pp. 217–236 (1984)
7. Arva, B., Beieler, J., Fisher, B., Lara, G., Schrodt, P.A., Song, W., Sowell, M., Stehle, S.: Improving forecasts of international events of interest. In: The 3rd Annual Meeting of the European Political Science Association, Barcelona, Spain (June 2013)
8. Chang, T.-K., Lee, J.-W.: Factors affecting gatekeepers' selection of foreign news: A national survey of newspaper editors. Journalism & Mass Communication Quarterly 69(3), 554–561 (1992)
9. Franks, S.: The carma report: western media coverage of humanitarian disasters. The Political Quarterly 77(2), 281–284 (2006)
10. Gaddy, G.D., Tanjong, E.: Earthquake coverage by the western press. Journal of Communication 36(2), 105–112 (1986)
11. Galtung, J., Ruge, M.H.: The structure of foreign news the presentation of the congo, cuba and cyprus crises in four norwegian newspapers. Journal of Peace Research 2(1), 64–90 (1965)
12. Harcup, T., O'neill, D.: What is news? galtung and ruge revisited. Journalism Studies 2(2), 261–280 (2001)
13. Lacy, S., Chang, T.-K., Lau, T.-Y.: Impact of allocation decisions and market factors on foreign news coverage. Newspaper Research Journal 10(4) (1989)
14. Leetaru, K., Schrodt, P.A.: Gdelt: Global data on events, location, and tone, 1979–2012. Paper presented at the ISA Annual Convention, vol. 2, p. 4 (2013)
15. Masmoudi, M.: The new world information order. Journal of Communication 29(2), 172–179 (1979)
16. Östgaard, E.: Factors influencing the flow of news. Journal of Peace Research 2(1), 39–63 (1965)
17. Peterson, S.: Foreign news gatekeepers and criteria of newsworthiness. Journalism Quarterly 56(1), 116–125 (1979)
18. Sreberny, A., Stevenson, R.: Comparative analysis of international news flow: an example of global media monitoring. In: International Media Monitoring, pp. 55–72 (1999)
19. Westerståhl, J., Johansson, F.: Foreign news: News values and ideologies. European Journal of Communication 9(1), 71–89 (1994)
20. Wilke, J., Heimprecht, C., Cohen, A.: The geography of foreign news on television a comparative study of 17 countries. International Communication Gazette 74(4), 301–322 (2012)
21. Wu, H.D.: Investigating the determinants of international news flow a meta-analysis. International Communication Gazette 60(6), 493–512 (1998)
22. Wu, H.D.: Systemic determinants of international news coverage: A comparison of 38 countries. Journal of Communication 50(2), 110–130 (2000)

Probabilistic User-Level Opinion Detection on Online Social Networks

Kasturi Bhattacharjee and Linda Petzold

Department of Computer Science,
University of California, Santa Barbara
{kbhattacharjee,petzold}@cs.ucsb.edu

Abstract. The mass popularity of online social networks, such as Facebook and Twitter, makes them an interesting and important platform for exchange of ideas and opinions. Accurately capturing the opinions of users from their self-generated data is crucial for understanding these opinion flow processes. We propose a supervised model that uses a combination of hashtags and n-grams as features to identify the opinions of Twitter users on a topic, from their publicly available tweets. We use it to detect opinions on two current topics: U.S. Politics and Obamacare. Our approach requires no manual labeling of features, and is able to identify user opinion with a very high accuracy over a randomly chosen set of users tweeting on each topic.

1 Introduction

Social networks have emerged as one of the most powerful means of communication today. From beginning as a medium through which people remained connected to friends and family, they have emerged to become a facilitator of social causes and revolutions. Facebook and Twitter proved to be effective mediums of communication for protesters during the Arab Spring, enabling them to coordinate and conduct a revolution [24,21]. More recently, social media has been instrumental in facilitating the protests in Ukraine [27]. The massive popularity of social networks has led to their extensive use in political campaigns as well [34]. Social and political organizations such as *MoveOn.org* [15] and *Avaaz.org* [3] have emerged as platforms through which people start online petitions to increase public awareness on a myriad of important social and political issues.

Knowing the opinions of people is useful not only for predicting the outcome of socio-political events, but also for viral marketing, advertising and market prediction [20,7,2]. Since the volume of social network posts generated on a daily basis is enormous, it is important to be able to perform opinion detection in an automated fashion.

In this work we focus on the detection of opinions of Twitter users on a given topic by extracting informative features from their publicly available tweets, using a supervised learning approach. We chose two topics for which users tend to have strong opinions: U.S. Politics (during the 2012 Presidential Election) and Obamacare.

L.M. Aiello and D. McFarland (Eds.): SocInfo 2014, LNCS 8851, pp. 309–325, 2014.
© Springer International Publishing Switzerland 2014

Twitter has gained popularity among researchers due to its emergence as one of the most widely used social networks, and also because it allows for the crawling of some of its data. However, this data also brings along with it a host of challenges. The short length of a tweet, the abundance of grammatical errors, misspelt words, informal language and abbreviations make it difficult to extract the opinion expressed through a tweet accurately.

To overcome the above issues, we adopt the following strategies. We begin by preprocessing the data to reduce the amount of noise as described in detail in Section 3. This is a non-trivial step especially when dealing with Twitter data. Because the opinions detected on the basis of a single *tweet* are unreliable, we focus instead on assessing the opinion of a *user* by aggregating the information in all of their tweets relating to the topic of interest over a given time period. We use a probabilistic approach, regularized to avoid overfitting [26], to classify the user opinions as positive or negative on a given topic. The selection of features is critical to this task. We found that combining the use of hashtags and n-grams was highly informative in detecting user opinion. It is to be noted here that our method requires no prior manual selection or labeling of features. To test the robustness of our methodology, we implemented it for the detection of political opinions on the 2012 U.S. Presidential election, and on the topic of Obamacare, and obtained a high level of accuracy.

The remainder of this paper is organized as follows. In Section 2 we discuss related research. Section 3 describes the Twitter data that we crawled, and the labeling of users for the training and test sets. In Section 4, we present the model and the features that we used for opinion detection. Section 5 describes the experiments conducted on our test dataset, and the results obtained. Finally, in Section 6, we present conclusions on our work.

2   Related Work

Research involving sentiment analysis or opinion mining on social networks may be divided into two areas: techniques that are based on lexicons of words, and techniques that are based on machine learning. The lexicon-based methods work by using a predefined collection (lexicon) of words, where each word is annotated with a sentiment. Various publicly available lexicons are used for this purpose, each differing according to the context in which they were constructed. Examples include the Linguistic Inquiry and Word Count (LIWC) lexicon [32,31] and the Multiple Perspective Question Answering (MPQA) lexicon [25,40,41]. The LIWC lexicon contains words that have been assigned into categories and matches the input text with the words in each category [23]. The MPQA lexicon is a publicly-available corpus of news articles that have been manually annotated for opinions, emotions, etc. These lexicons have been widely used for sentiment analysis across various domains, not just specifically for social networks [1,17,8]. Other popular sentiment lexicons that have been designed for short texts are SentiStrength [38] and SentiWordNet [13,4]. These lexicons have been extensively used for sentiment analysis of social network data, online posts, movie reviews, etc. [16,22,37,12]. However, as we will see in Section 5.3, they do not perform very well when applied to our problem of assessing user opinion.

Machine learning techniques for sentiment analysis include classification techniques such as Maximum Entropy, Naive Bayes, SVM [18], k-NN based strategies [11], and label propagation [39]. These usually require labeling of data for training, which is accomplished either by manually labeling posts [39], or through the use of features specific to social networks such as emoticons and hashtags [18,11]. Some of the existing research combines lexicon-based methods and machine-learning methods [36]. These papers address a different (but related) problem than ours in that they perform tweet-level as opposed to user-level sentiment analysis. In Section 5.5, we will compare our method to user-level sentiment generated via tweet-level sentiment obtained by the methods of [36] and [18].

The methods in [35,30,9] perform user-level sentiment analysis. The method in [30] uses features derived from four different types of information of a social network user: user profile, tweeting behavior, linguistic content of the messages and the user network. Our method focuses on extracting informative features from only a user's tweets, and can achieve high accuracies with a smaller number of features and a simpler model. The methods in [9] determine the political alignment of Twitter users using their tweets, as well as their retweet networks. The dataset is selected by first creating a set of politically discriminative hashtags that co-occur with the hashtags #p2 ("Progressives on Twitter 2.0") and #tcot ("Top Conservatives on Twitter"). The tweets selected for the dataset carry at least one of the discriminative hashtags. In contrast, we select our dataset via identification of users who use the generic keywords in Table 1 at least once, which does not require the determination of discriminative words or hashtags. Moreover, [9] does not conduct any study on using combinations of hashtags and n-grams as features, which we have found to yield the best performance in opinion detection across two different topics (as described in Section 5.4 of this paper). Thus the results are not directly comparable. In addition, our method performs automatic feature selection, which [9] does not address. In [35], user-level sentiment analysis is performed using the users' following/mention network information. Since our dataset consists of randomly chosen users, we do not have the entire neighborhood of any user.

3   Datasets

Data Collection. We focused on two current topics for which people were more likely to voice their opinions on social media: U.S. Politics and Obamacare. For each of the topics of interest, we randomly selected users and collected their tweets over a period of time using the Twitter REST API. For U.S. Politics, our tweets were collected over the period of January 2012 to January 2013. The time period of the data collection coincided with the political campaigns leading up to the November 2012 U.S. Presidential election. For the dataset on Obamacare, we crawled tweets for 6 weeks over the months of June and July 2013.

To extract topical tweets, we filtered out tweets that contained words related to the topic of interest. For instance, for political tweets, we used words related to political figures, parties, causes or issues, or commentators whose bias is well-known. This approach is similar to that used in [33]. Table 1 shows the list of keywords used to obtain both the datasets and the categories that they belong to. The political dataset thus obtained was composed of 672,920 tweets from 552,524 users. The Obamacare dataset consisted of 187,141 tweets from 65,218 users. For the purposes of training and testing, we randomly picked users from each of the datasets, and then assigned a positive or negative opinion label (definition of these opinions are provided in Section 4.1) to them by manually reading *all* of their tweets. We labeled only those users whose opinion could be unambiguously determined from their tweets. We randomly chose 490 users (222 positive and 268 negative) for our labeled dataset on U.S. Politics, and 201 users (90 positive and 111 negative) for our labeled dataset on Obamacare.

Table 1. Keywords used to filter out topical tweets

Keyword	Keyword Type	Dataset
obama	Political figure	U.S. Politics
democrat	Political Party	U.S. Politics
p2	Political Party	U.S. Politics
romney	Political figure	U.S. Politics
gop	Political party	U.S. Politics
tcot	Political party	U.S. Politics
obamacare	Term for affordable health care	Obamacare
koch	Industrialists who are against Obamacare	Obamacare
affordable care	Term for affordable health care	Obamacare

Data Cleaning and Preprocessing. Twitter data is inherently noisy and filled with abbreviations and informal words. We performed the following cleaning and pre-processing on the dataset to enable better extraction of features from it.

1. URL removal: In our method, URLs would not contribute to the feature extraction and were therefore removed.
2. Stop word removal: Stop words such as "a", "the", "who", "that", "of", "has", etc. were removed from the tweets before extracting n-grams, which is a common practice.
3. Punctuation marks and special character removal: Punctuation marks such as ":", ";" etc. and special characters such as "[ ]", "{ }", "|", etc. were removed before extracting n-grams.
4. Additional whitespace removal: Multiple white spaces were replaced with a single whitespace.
5. Conversion to lowercase: Tweets are not generally case-sensitive owing to the informal language used. For instance, for our method, the word "Obama" should be considered the same as "obama" when parsing through a tweet. Hence we converted the tweets to lowercase to preserve uniformity in feature extraction.
6. Tokenization: The tweets were tokenized into words to extract n-grams from them. We use Python's Natural Language Toolkit 3.0 [6] for this purpose.

4   Methods

Given a user's tweets over time on a predetermined topic, our goal was to predict her opinion as accurately as possible. Thus we sought to learn a predictive model for user opinion from features derived from the tweets. In this section, we describe the problem definition, the model we used to solve the problem, and the features used for extracting user opinion. The results obtained are reported in Section 5.4.

4.1   Problem Definition

We adopted a probabilistic view for the user opinion in that we assumed it to be a distribution over *positive* and *negative* types. On the topic of US politics, we arbitrarily defined *positive* to mean that the user is pro-Obama or anti-Romney, and *negative* to mean that she is anti-Obama or pro-Romney. On the topic of Obamacare, *positive* was again arbitrarily defined to be a pro-Obamacare opinion, and *negative* was defined to be an anti-Obamacare opinion.

The main challenges involved were: (1) to determine appropriate features that carry information about the user's opinion (2) to learn a model that, with a sufficiently high accuracy, predicts the probabilistic user opinion from the features.

Thus, the problem definition may be summarized as follows: *Given a user's tweets over time on a topic, we seek the probabilities of her having a positive or a negative opinion.*

4.2   Model

We cast the problem at hand as a supervised binary classification problem in which the classifier outputs the probabilities of the opinions that a user can have. Logistic regression is a well-known and widely used probabilistic machine learning tool for classification. Given a binary output variable y and a set of features X, logistic regression estimates the conditional distribution $P(y = 1|X; \theta)$, where θ represents the parameters that determine the effect of the features on the output.

Logistic regression utilizes the following transfer function between X and y:

$$P(y = 1|X, \theta) = h_\theta(X) = \frac{1}{1 + \exp(-\theta^T X)}. \tag{1}$$

To estimate the parameter θ of the logistic model, we use Maximum Likelihood Estimation. Assuming that we have m i.i.d training samples $(y^i, X^i), i = 1, \ldots, m$, the log likelihood is given by

$$\log P(y|X, \theta) = \sum_{i=1}^{m} [y^i \log(h_\theta(X^i)) + (1 - y^i) \log(1 - h_\theta(X^i))]. \tag{2}$$

The loss function, which is the negative log-likelihood, being convex, we can minimize it to estimate the optimum θ, given by $\hat{\theta}$. We add a regularization to the loss function to avoid overfitting, as discussed in Sections 5.1 and 5.2.

Thus, given a set of features X and a set of known outputs y in the training data, the logistic regression model learns the parameter θ that determines the relationship between X and y. Once the model has been learned, it can then be used to predict the outcomes of the test data, given their features X.

4.3   Features for Classification

Deriving features from the tweets is a crucial step for successfully determining a user's opinion. The features must be such that they would reflect the opinion conveyed through the user's tweets, because if a human annotator were to determine the opinion of a user (which is the baseline we are comparing with), she would read the user's tweets to reach a conclusion.

Hashtags have become a very popular feature in Twitter and other social media sites. A hashtag is essentially a word that is prefixed with a # symbol that can be generated by a user and used in their tweets. *#followfriday*, *#mtvstars*, *#ipad*, #glee are examples of some popular hashtags on Twitter. The concept of hashtags was introduced in order to index tweets of a similar topic together, to make it easier for users to start a conversation with each other.

Apart from highlighting the topic of a tweet, hashtags have been found to carry some additional information regarding the bias of the tweet itself [11,39]. For example, hashtags such as *#ISupportStaceyDash*, *#iloveapple*, *#twilightsucks* all carry information about the topic of the tweet and also clearly exhibit the bias of the user. A manual inspection of our dataset suggested that hashtags might be used to provide information about the bias of the tweet. For example, hashtags such as *#romneyshambles*, *#gopfail*, *#defundobamacare* were more likely to occur in tweets in which the user portrays a negative opinion towards the topic. Similarly, hashtags such as *#iloveobama*, *#istandwithobama*, *#getcovered* occurred most often with tweets that carried a positive opinion towards the respective topic. For this reason, our first choice for features to use was hashtags.

Although hashtags are powerful carriers of sentiment information, sometimes they may not be sufficient to convey the bias hidden in the tweet. For instance, hashtags may just refer to a political party without seemingly carrying any bias, in which case the information we seek may be carried by the text of the tweet. Here is an example of such a tweet:

*"@MittRomney's refusal to release details of, well, anything, prove his cowardice & unfitness for the presidency. **#connecttheleft #gop**"*

In the above tweet, the hashtags used are *#gop* ("Grand Old Party") and *#connecttheleft* (a hashtag designed to connect the Democrats). Used together, these hashtags carry no information on the user's opinion. However, a human annotator would be able to identify the opinion by reading the entire text of the tweet. Hence, in order to augment the information obtained by using hashtags alone, we incorporated information from the tweet as well.

For this purpose, we use the n-gram model which is considered a powerful tool for sentiment extraction [5]. n-grams are essentially contiguous sequences of n words extracted from text. The n-gram model was developed as a probabilistic language model which predicts the occurrence of the next word in the sequence of words by modeling it as an $(n-1)$-order Markov process. In the domain of sentiment analysis, n-grams have been widely used since they help to capture phrases that carry sentiment expression [28,10].

We begin by using hashtags separately as features in the logistic regression model (as described further), and then use them in conjunction with n-grams to achieve better results.

Popular Hashtags. To eliminate the need for manual labeling of the hashtags, we extracted the most popular hashtags separately from each of the filtered datasets, by computing the total number of times each hashtag occurred in the respective dataset. For both the datasets, we used the 1000 most popularly used hashtags. We refer to these hashtags as *popular hashtags*. Not surprisingly, a manual inspection revealed that all of the popular political tags were related to politics either by representing names of the parties, their representatives, or political issues that gained importance during that time period. A similar pattern was observed for the popular Obamacare hashtags.

We then used the frequency of use of the popular hashtags as features in our model. Thus, in equation (3),

$$X_j^i = \text{ number of times popular hashtag j is used by user i.} \qquad (3)$$

Popular n-grams in Conjunction with Hashtags. As discussed previously, we used n-grams to augment the hashtag information. We used values of n = 1, 2 to extract out unigrams and bigrams from the tweets of each labeled user. Again, we picked the most popular n-grams from each dataset. For each dataset, we chose 2000 most popular unigrams and 2000 most popular bigrams. We combined the information we obtained from the hashtags with that obtained from the n-grams. This was done by performing logistic regression using multiple explanatory variables as follows

$$P(y=1|X,Z,\theta,\beta) = \frac{1}{1+\exp(-\theta^T X - \beta^T Z)}, \qquad (4)$$

where X and Z represent the hashtag-based features and the n-gram-based features respectively; θ and β represent the corresponding parameters. We tested each *type* of n-gram feature separately with the hashtags.

5   Experimental Results

In this section we outline in detail implementations of the proposed method with both l_1 and l_2 regularization, and the metrics we used to evaluate the results. Further, we describe the existing methods that we chose for comparison, and report the results obtained.

5.1   Logistic Regression with l_2 Regularization

Using both hashtags and n-grams yields a relatively large number of features (3000 for hashtags and bigrams). To avoid overfitting, we add a user-specified regularization term $\lambda\|X\|_2^2$ to our loss function, where $\lambda > 0$ is the regularization parameter [26]. The loss function thus becomes:

$$L(\theta) = -\log P(y|X, \theta) + \lambda\|\theta\|_2^2. \tag{5}$$

5.2   Logistic Regression with l_1 Regularization

We also explored the use of l_1-regularization [26]. This results in the loss function:

$$L(\theta) = -\log P(y|X, \theta) + \lambda\|\theta\|_1. \tag{6}$$

We used the open-source machine learning tool in Python, scikit-learn [29] to implement logistic regression with l_1 and l_2 regularizations. The selection of λ is discussed in Section 5.3.

Table 2. Metrics using l_2-regularization on U.S. Politics dataset

Feature type	Total Number of Features	Number of Selected Features	Mean Accuracy	Mean AUC	Mean F1-score	Mean Specificity
Popular hashtags	1000	288	86.32($\pm$0.043)	0.915	0.85	0.875
Popular hashtags, unigrams	3000	1488	86.12($\pm$0.031)	0.896	0.843	0.885
Popular hashtags, bigrams	3000	1398	**87.35($\pm$0.029)**	**0.909**	**0.858**	**0.895**
Popular hashtags, unigrams, bigrams	5000	2430	87.10	0.905	0.855	0.893

Table 3. Comparison of the proposed method with three state-of-the-art methods

Method	Accuracy(%)	Precision	Recall	Specificity
l_2 - regularized Logistic regression	**87.35**	**0.871**	**0.848**	**0.895**
SentiStrength	53.06	0.485	0.586	0.485
Maximum Entropy method	44.29	0.525	0.419	0.463
Combined method (SentiStrength and MaxEnt)	59.59	0.542	0.694	0.515

Table 4. Metrics using l_2-regularization on Obamacare dataset

Feature type	Total Number of Features	Number of Selected Features	Mean Accuracy	Mean AUC	Mean F1-score	Mean Specificity
Popular hashtags	1000	445	77.33($\pm$0.0466)	0.912	0.804	0.799
Popular hashtags, unigrams	3000	2295	87.30($\pm$0.022)	0.942	0.906	0.943
Popular hashtags, bigrams	3000	1506	87.54($\pm$0.025)	0.956	0.907	0.927
Popular hashtags, unigrams, bigrams	5000	3448	**90.8 ($\pm$0.033)**	**0.958**	**0.919**	**0.850**

5.3   Evaluation Metrics

To evaluate the performance of the model, we conducted hold-out cross validation by randomly splitting the data into 30% test set and 70% training set. On each run of the cross-validation, the best λ was learned from the validation error on the training set. The cross-validation was done 10 times, with the data being randomly shuffled each time. Our experiments showed that the best λ value did not vary much across the validation sets of the respective dataset. For the U.S. Politics dataset, we set $\lambda = 50.0$ for l_2-regularization, and for l_1-regularization, it was 0.01. For the Obamacare dataset, we set $\lambda = 25.0$ for the l_2-regularized model, and $\lambda = 0.0083$ for the l_1-regularized model. The average classifier metrics [14] such as ROC curves, AUC, accuracy, precision, recall, F1-score and specificity across the 10 sets is reported in Section 5.4. For the U.S. politics data, we tested on 147 users, and on 60 users for the Obamacare dataset. For each user, the class with the higher probability is assigned as the corresponding opinion label, with ties broken arbitrarily. There were no cases in either of the datasets in which ties were encountered.

5.4   Results

Table 2 and Figure 1(a) present the results obtained using logistic regression with l_2 regularization on U.S. Politics, and Table 4 and Figure 1(b) demonstrate the results on the Obamacare dataset. We ran this method using four combinations of features, as shown in the results. As can be observed, the values of each of the classifier metrics are excellent. The high values of precision and specificity indicate that the method could predict both positive and negative opinions accurately. The highest accuracy achieved by our classifier was **87.35%**

on U.S. Politics and **90.8%** on Obamacare. Figure 1(c) presents the ROC curves obtained using the l_2-regularized model on U.S. Politics and Obamacare.

Table 5 presents the results obtained with l_1-regularized logistic regression using the same kinds of features on the U.S. Politics dataset. It is to be noted that, using l_1 regularization, comparable accuracies were obtained with a much smaller number of features. For instance, using the combination of hashtags and bigrams, we were able to achieve a high accuracy of **86.10%** and an AUC of **0.916** from 32 features, as contrasted with using 1398 features and obtaining slightly higher accuracy of 87.35% and an AUC of 0.909 with l_2 regularization. Similarly, Table 6 shows the results using l_1-regularized logistic regression on the Obamacare dataset. A similar trend in results is observed in this case as well.

Selection of Informative Features. From Tables 5 and 6, we find that the l_1 regularizer yields excellent results with a small number of selected features. Table 7 shows a few of the features that the regularizer picked from either dataset as the most informative features. Thus the method results in automatic selection of the most useful features for opinion detection.

5.5   Comparison with Existing Methods

We compare our methods with three popularly used state-of-the-art methods that perform tweet-level sentiment analysis, and use their results to obtain opinion on a user level as described below. The following methods were tested on the U.S. Politics dataset.

SentiStrength. SentiStrength [38] is a lexicon-based method that was designed for use with short informal text including abbreviations and slang. It has been widely used by researchers for sentiment analysis of tweets, online posts, etc. (Section 2) . It uses a lexicon of positive and negative words which were initially annotated by hand, and later improved during a training phase. Given a sentence, the method assigns a sentiment score to every word in the sentence, and thereafter, the sentence is assigned the most positive score and the most negative score from among its words. According to [38], the algorithm was tested extensively for accuracy, and was found to outperform standard machine learning approaches. Hence we chose this as a baseline method to compare against.

Tweet-level Maximum Entropy classifier. The second method for comparison is a machine-learning method proposed in Section 3.3 of [18] which uses a Maximum Entropy based classifier trained on 1,600,000 tweets using emoticons as noisy labels. It uses the presence or absence of unigrams, bigrams and parts-of-speech tags as features for classification, and classifies a given tweet as positive or negative. The authors provide an online tool for this purpose [19], which we use for conducting our experiments. This method has also been widely used for sentiment analysis. It is to be noted that we used their pre-trained model that was trained on their annotated tweet set. We could not train the method on our labeled datasets because our datasets have labels on the user and not on the individual tweets, and it is non-trivial to transfer the user opinion to their tweets owing to the amount of noise per tweet. Moreover, we could not annotate

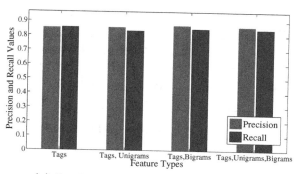

(a) Precision Recall values for U.S. Politics

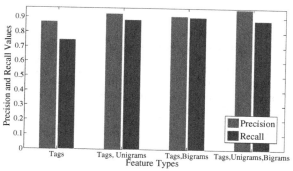

(b) Precision Recall values for Obamacare

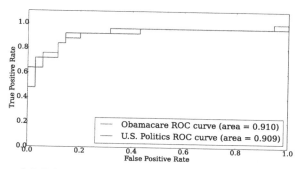

(c) ROC curves for U.S. Politics and Obamacare

Fig. 1. Classifier metrics with l_2 regularization

Table 5. Metrics using l_1 regularization on U.S. Politics dataset

Feature type	Total Number of Features	Number of Selected Features	Mean Accuracy	Mean AUC	Mean F1-score	Mean Specificity
Popular hashtags	1000	22	84.70($\pm$0.048)	0.896	0.823	0.82
Popular hashtags, unigrams	3000	34	85.67($\pm$0.025)	0.903	0.818	0.86
Popular hashtags, bigrams	3000	32	**86.10($\pm$0.030)**	**0.916**	**0.844**	**0.849**
Popular hashtags, unigrams, bigrams	5000	70	85.03	0.909	0.832	0.869

Table 6. Metrics using l_1 regularization on Obamacare dataset

Feature type	Total Number of Features	Number of Selected Features	Mean Accuracy	Mean AUC	Mean F1-score	Mean Specificity
Popular hashtags	1000	34	77.33($\pm$0.0466)	0.912	0.804	0.799
Popular hashtags and unigrams	3000	210	87.30($\pm$0.022)	0.942	0.906	0.943
Popular hashtags and bigrams	3000	132	87.54($\pm$0.025)	0.956	0.907	0.927
Popular hashtags, unigrams, bigrams	5000	372	**90.8 ($\pm$0.033)**	**0.958**	**0.919**	**0.850**

our datasets using emoticons because they are rarely used in our datasets (only 0.13% of the tweets used emoticons in the U.S. Politics dataset). Since they used emoticons to label the sentiment of a tweet and did not manually annotate them, theirs may be considered as a (partially) supervised method, as opposed to our fully supervised method.

Combined Method. The third method for comparison is a method described in Section 3.2 of [36] that combines the output of the lexicon-based method [38] and the tweet-level machine learning method [18]. The authors propose a way to combine the results of SentiStrength and the MaxEnt based method of [18] to perform a binary tweet-level sentiment classification with better accuracy than either of the individual methods.

Obtaining targeted user-level sentiment from tweet-level sentiment. We adopt the following strategies when comparing our method with the other three methods. First, to obtain a sentiment label for every tweet using *SentiStrength*, the most positive and most negative scores for every tweet were added up. If this sum was positive the tweet was labeled positive; if the sum was negative then it was labeled negative, and if the sum was zero the tweet was labeled neutral. This approach was proposed in Section 3.2 of [36].

Second, all of the methods described above determine whether a given tweet has an *overall positive or negative sentiment, irrespective of the target of the sentiment.* This varies from our definition of *positive* and *negative* as described in Section 4.1. Hence, to determine the sentiment of a tweet towards a target (Democrat or Republican), we selected a set of keywords that were associated with Democrats and another set for Republicans, with the objective of identifying targets for as many tweets as possible, and defined them as *positive targets* and *negative targets*, respectively. (The keywords used are given in Table 8). For **any** method that we compared with, given a tweet sentiment, we first computed a sum of the target words that the tweet contained, assigning +1 for a *positive target* and -1 for a *negative target*. If the sum was greater than 0 we assumed that the subject of the tweet was Democrats, in which case the sentiment remained unaltered. If the sum was less than 0 we assumed that the subject was Republicans. In this case, a positive sentiment towards Republicans would mean a *negative* sentiment according to our definition, and vice versa.

Table 7. Examples of features selected by l_1-regularization

Feature Type	Dataset	Sparse features
Hashtags	U.S. Politics	"tcot", "p2", "gop", "obama2012"
Bigrams	U.S. Politics	"tcot gop", "obama didnt", "mitt romney"
Hashtags	Obamacare	"obamacare", "tcot", "defundobamacare", "defund"
Bigrams	Obamacare	"defund obamacare", "shut down", "government over"

Third, to obtain *user-level* sentiment from the tweet-level sentiment output from **any** of the methods, we adopted the following strategy. For every user, we summed the (targeted) sentiments of all her tweets using +1 for *positive*, -1 for *negative* and 0 for neutral. The user output was considered positive if the sum was positive, negative if the sum was negative and was assigned randomly if the sum was zero. Table 3 represents the comparison of our method with the

Table 8. Keywords used to identify positive and negative targets in the U.S. Politics Dataset

Target Type	Keywords
Positive targets	"obama", "democrat", "p2", "barackobama", "barack", "democrats", "liberals", "obama2012", "dem", "p2b", "biden", "romneyshambles", "clinton", "releasethereturns", "forward2012", "obamabiden2012","connecttheleft", "ctl", "inobamasamerica", "obamawinning", "dnc", "dncin4words", "dnc2012", "150dollars", "repugnican"
Negative targets	"romney", "gop", "tcot", "mitt", "mittromney","republicans","teaparty", "imwithmitt", "mitt2012", "nobama2012", "romneyryan2012", "tlot", "webuiltit", "teaparty", "gop2012", "prolife", "romneyryan", "youdidntbuildthat", "obamaphone", "anndromney", "obamafail", "youjustpulledaromney", "nobama", "republican", "limbaugh", "paulryanvp"

existing methods. All of the classifier metrics clearly display that our method outperforms all the three methods.

6   Conclusion

In this paper we propose a method for detecting user-level opinion on a given topic from Twitter data. Our approach of performing user-level (as opposed to tweet-level) opinion detection using regularized logistic regression with hashtags and n-grams as features was found to produce excellent results. The l_2 and l_1 regularizations yielded comparable accuracy, however the l_1 regularization required far fewer features. Moreover, our method required no manual labeling of features. The method was applied to Twitter datasets on two different topics and yielded excellent results on both, which highlights its generalizability. The importance of informative features is evident in the results obtained; only a small percentage of the most informative features were required for accurate user opinion detection.

Acknowledgements. The research was supported by the Institute for Collaborative Biotechnologies through grant W911NF-09-0001 from the U.S. Army Research Office. The content of the information does not necessarily reflect the position or the policy of the Government, and no official endorsement should be inferred.

References

1. Akkaya, C., Wiebe, J., Mihalcea, R.: Subjectivity word sense disambiguation. In: Proceedings of the 2009 Conference on Empirical Methods in Natural Language Processing, vol. 1, pp. 190–199. Association for Computational Linguistics (2009)

2. Asur, S., Huberman, B.A.: Predicting the future with social media. In: 2010 IEEE/WIC/ACM International Conference on Web Intelligence and Intelligent Agent Technology (WI-IAT), vol. 1, pp. 492–499. IEEE (2010)
3. Avaaz (The World in Action) (2013), http://www.avaaz.org/en/highlights.php
4. Baccianella, S., Esuli, A., Sebastiani, F.: Sentiwordnet 3.0: An enhanced lexical resource for sentiment analysis and opinion mining. In: LREC. vol. 10, pp. 2200–2204 (2010)
5. Bespalov, D., Bai, B., Qi, Y.,, S.: Sentiment classification based on supervised latent n-gram analysis. In: Proceedings of the 20th ACM International Conference on Information and Knowledge Management, pp. 375–382. ACM (2011)
6. Bird, S., Klein, E., Loper, E.: Natural language processing with Python. O'Reilly Media, Inc. (2009)
7. Bollen, J., Mao, H., Zeng, X.: Twitter mood predicts the stock market. Journal of Computational Science 2(1), 1–8 (2011)
8. Bono, J.E., Ilies, R.: Charisma, positive emotions and mood contagion. The Leadership Quarterly 17(4), 317–334 (2006)
9. Conover, M.D., Gonçalves, B., Ratkiewicz, J., Flammini, A., Menczer, F.: Predicting the political alignment of twitter users. In: Privacy, security, risk and trust (passat), 2011 IEEE Third International Conference on Social Computing (SocialCom), pp. 192–199. IEEE Computer Society Press, Los Alamitos (2011)
10. Dave, K., Lawrence, S., Pennock, D.M.: Mining the peanut gallery: Opinion extraction and semantic classification of product reviews. In: Proceedings of the 12th International Conference on World Wide Web, pp. 519–528. ACM (2003)
11. Davidov, D., Tsur, O., Rappoport, A.: Enhanced sentiment learning using twitter hashtags and smileys. In: Proceedings of the 23rd International Conference on Computational Linguistics: Posters, pp. 241–249. Association for Computational Linguistics (2010)
12. Denecke, K.: Using sentiwordnet for multilingual sentiment analysis. In: IEEE 24th International Conference on Data Engineering Workshop, ICDEW 2008, pp. 507–512. IEEE (2008)
13. Esuli, A., Sebastiani, F.: Sentiwordnet: A publicly available lexical resource for opinion mining. Proceedings of LREC 6, 417–422 (2006)
14. Fawcett, T.: An introduction to ROC analysis. Pattern Recognition Letters 27(8), 861–874 (2006)
15. Galland,   A.:   Moveon.org   (2013),   http://front.moveon.org/thank-you-for-an-awesome-2013/#.Uty0RnmttFQ
16. Garas, A., Garcia, D., Skowron, M., Schweitzer, F.: Emotional persistence in online chatting communities. Scientific Reports 2 (2012)
17. Gilbert, E., Karahalios, K.: Predicting tie strength with social media. In: Proceedings of the SIGCHI Conference on Human Factors in Computing Systems, CHI 2009, pp. 211–220. ACM, New York (2009), http://doi.acm.org/10.1145/1518701.1518736
18. Go, A., Bhayani, R., Huang, L.: Twitter sentiment classification using distant supervision. CS224N Project Report, Stanford pp. 1–12 (2009(a))
19. Go, A., Huang, L., Bhayani, R.: Twittersentiment (2009b), http://www.sentiment140.com

20. Jansen, B.J., Zhang, M., Sobel, K., Chowdury, A.: Twitter power: Tweets as electronic word of mouth. Journal of the American Society for Information Science and Technology 60(11), 2169–2188 (2009)
21. Kassim, S.: How the arab spring was helped by social media (2012), http://www.policymic.com/articles/10642/twitter-revolution-how-the-arab-spring-was-helped-by-social-media
22. Kucuktunc, O., Cambazoglu, B.B., Weber, I., Ferhatosmanoglu, H.: A large-scale sentiment analysis for yahoo! answers. In: Proceedings of the Fifth ACM International Conference on Web Search and Data Mining, pp. 633–642. ACM (2012)
23. LIWC: LIWC software (2001), http://www.liwc.net/index.php
24. Marzouki, Y., Oullier, O.: Revolutionizing revolutions: Virtual collective consciousness and the arab spring (2012), http://www.huffingtonpost.com/yousri-marzouki/revolutionizing-revolutio_b_1679181.html
25. MPQA: MPQA (2005), http://mpqa.cs.pitt.edu/lexicons/
26. Ng, A.Y.: Feature election, l 1 vs. l 2 regularization, and rotational invariance. In: Proceedings of the Twenty-First International Conference on Machine Learning, p. 78. ACM (2004)
27. Onuch, O.: Social networks and social media in ukrainian "euromaidan" protests (2014), http://www.washingtonpost.com/blogs/monkey-cage/wp/2014/01/02/social-networks-and-social-media-in-ukrainian-euromaidan-protests-2/
28. Pak, A., Paroubek, P.: Twitter as a corpus for sentiment analysis and opinion mining. In: LREC (2010)
29. Pedregosa, F., Varoquaux, G., Gramfort, A., Michel, V., Thirion, B., Grisel, O., Blondel, M., Prettenhofer, P., Weiss, R., Dubourg, V., Vanderplas, J., Passos, A., Cournapeau, D., Brucher, M., Perrot, M., Duchesnay, E.: Scikit-learn: Machine learning in Python. Journal of Machine Learning Research 12, 2825–2830 (2011)
30. Pennacchiotti, M., Popescu, A.M.: A machine learning approach to twitter user classification. In: ICWSM (2011)
31. Pennebaker, J.W., Chung, C.K., Ireland, M., Gonzales, A., Booth, R.J.: The development and psychometric properties of liwc 2007, Austin, TX, LIWC. Net (2007)
32. Pennebaker, J.W., Francis, M.E., Booth, R.J.: Linguistic inquiry and word count: LIWC 2001 (2001)
33. Romero, D.M., Meeder, B., Kleinberg, J.: Differences in the mechanics of information diffusion across topics: idioms, political hashtags, and complex contagion on twitter. In: Proceedings of the 20th International Conference on World Wide Web, pp. 695–704. ACM (2011)
34. Rutledge, P.: How obama won the social media battle in the 2012 presidential campaign (2013), http://mprcenter.org/blog/2013/01/25/how-obama-won-the-social-media-battle-in-the-2012-presidential-campaign/
35. Tan, C., Lee, L., Tang, J., Jiang, L., Zhou, M., Li, P.: User-level sentiment analysis incorporating social networks. In: Proceedings of the 17th ACM SIGKDD International Conference on Knowledge Discovery and Data Mining, pp. 1397–1405. ACM (2011)
36. Tan, S., Li, Y., Sun, H., Guan, Z., Yan, X., Bu, J., Chen, C., He, X.: Interpreting the public sentiment variations on twitter. IEEE Transactions on Knowledge and Data Engineering (2012)
37. Thelwall, M., Buckley, K., Paltoglou, G.: Sentiment in twitter events. Journal of the American Society for Information Science and Technology 62(2), 406–418 (2011)

38. Thelwall, M., Buckley, K., Paltoglou, G., Cai, D., Kappas, A.: Sentiment strength detection in short informal text. Journal of the American Society for Information Science and Technology 61(12), 2544–2558 (2010)
39. Wang, X., Wei, F., Liu, X., Zhou, M., Zhang, M.: Topic sentiment analysis in twitter: a graph-based hashtag sentiment classification approach. In: Proceedings of the 20th ACM International Conference on Information and Knowledge Management, pp. 1031–1040. ACM (2011)
40. Wiebe, J., Wilson, T., Cardie, C.: Annotating expressions of opinions and emotions in language. Language Resources and Evaluation 39(2-3), 165–210 (2005)
41. Wilson, T.: Fine-grained subjectivity analysis. Ph.D. thesis, Doctoral Dissertation, University of Pittsburgh (2008)

Stemming the Flow of Information in a Social Network

Balaji Vasan Srinivasan[1], Akshay Kumar[2], Shubham Gupta[2], and Khushi Gupta[3]

[1] Adobe Research India Labs, Bangalore, India
[2] Adobe Systems India Pvt. Ltd, India
[3] Indian Institute of Technology, Kanpur, India
{balsrini,khgupta}@adobe.com, {kakshay,shubhamg}@iitk.ac.in

Abstract. Social media has changed the way people interact with each other and has contributed greatly towards bringing people together. It has become an ideal platform for people to share their opinions. However, due to the volatility of social networks, a negative campaign or a rumor can go viral resulting in severe impact to the community. In this paper, we aim to solve this problem of stemming the flow of a negative campaign in a network by observing only parts of the network. Given a negative campaign and information about the status of its spread through a few candidate nodes, our algorithm estimates the information flow in the network and based on this estimated flow, finds a set of nodes which would be instrumental in stemming the information flow. The proposed algorithm is tested on real-world networks and its effectiveness is compared against other existing works.

Keywords: social network, rumor source, rumor stemming, targeted influence.

1 Introduction

Social media is changing the way people communicate with each other and have become the preferred mediums of communication between friends and family. These sites serve as excellent platforms for sharing information and is being extensively used /leveraged for several business applications such as advertising, marketing, e-business and social campaigning. The structure of these networks is such that there is huge potential for a particular piece of information to go "viral". Marketers hope to exploit this and use it to their advantage when designing advertisements /campaigns. However, this also allows the possibility of a negative or inauthentic piece of information to become widely popular. Such pieces of incorrect or malicious information against individuals or brands by various agents, adversaries or competitors can have severe detrimental effects tarnishing hard built brand images.

There are several cases of documented social media disasters - both at the enterprise level as well at the political level. Rumors about the outbreak of swine flu resulted in a large online panic in 2009[4]. GreenPeace's extensive campaign against Nestle's misuse of the Indonesian rainforests lead to serious damage of the Swiss based corporation's brand image[17]. The ethical implications in these examples (attempts to handle inconvenient truths, etc.) are not addressed here and are beyond the scope of this paper. However, these examples also warrant for a method to limit the effect of an information deemed as negative, which will be the primary subject of our work.

L.M. Aiello and D. McFarland (Eds.): SocInfo 2014, LNCS 8851, pp. 326–335, 2014.
© Springer International Publishing Switzerland 2014

There are two stages to stem the spread of a misinformation. The spread needs to be estimated first, followed by approaches to limit it. The former problem has been studied extensively in the lights of disease spread [2,12,14]. The latter problem has been recently studied as the "Influence Limitation" problem[1,13]. In this paper, while we focus on the overall problem, our primary innovation is around the latter part of the problem.

Existing work in influence limitation removes the affected part of the network and maximizes the influence in the unaffected to limit the overall influence spread. However, real-world networks can seldom be differentiated into these 2 clear classes. To address this, we propose a methodology to score each users based on their "affectedness" to the negative campaign. We start with an estimate of the spread of negative information across the whole network and then determine the vulnerability of the nodes to the negative campaign to categorize them as *infected* (the negative information has already reached these nodes, the nodes are affected by the negative information and have started spreading them), *vulnerable* (the negative information may or may not have reached these nodes, but are most likely to move to the *infected* state in the near future) and *un-infected* (immune to the spread of the negative campaign) categories. Limiting the spread will then amount to targeting a "positive" campaign to the users in the vulnerable category (and may be the uninfected as well).

This paper is organized as follows. In Section 2, we discuss related prior work, highlighting their shortcomings and how we will overcome them. Section 3 provides an overview of our end-to-end approach for stemming the flow of negative information followed by the discussions of our algorithms to rank order the nodes based on their "vulnerability" in Section 4. We present the performances of our algorithm in Section 5 compared against existing work. Section 6 concludes the paper.

2   Related Work

The first stage of the problem is to estimate the spread of the information in a social network and could be cast as the "Rumor Source Identification" problem. Shah and Zaman [15] developed rumor-centrality to identify single-source in the epidemic. Prakash et al. [12] extended the source identification to address multiple sources via the use of Minimum Description Length and also developed mechanisms to identify the number of sources automatically. However, both these approaches require the complete snapshot of the spread in the network. To alleviate this issue, Seo et al [14] identify a few monitor nodes in the network and categorize them as positive and negative monitors on the basis of whether they have received the information or not. The rumor source identification is based on the intuition that the source should be close to positive monitors but far from the negative monitors. This approach is more practical because it requires the knowledge of the status of monitor nodes only unlike the entire network, although this may not be as accurate as the former approaches. Leskovic et al. [8] study this further by analyzing the placement of these monitor nodes to quickly detect rumor spreads.

The second stage of the problem requires identifying mechanisms to spread the positive word to the network. This is studied as the problem of Influence Limitation. This is a special case of the Influence Maximization problem, which was first studied by

Domingos and Richarson [5] and formalized by Kempe et al [7]. Influence maximization has since been extensively studied (e.g. [3,10]). Influence limitation in the context of social networks is comparatively lesser studied despite having a close resemblance to the disease spread models [2,12]. The problem of influence limitation in the presence of competing campaigns was first addressed by Budak et al [1]. Budak et al. show that this problem was NP-hard and in general does not follow the sub-modular property. They solved restricted versions of the problem that were proved to be submodular and provided approximation guarantees for greedy solutions. They compared the performance of their greedy algorithm against various heuristics like degree centrality. Premm et al [13] extended the work in [1] to solve the general influence limitation problem when the underlying functions are non submodular. They used the concept of Shapley Value from co-operative game theory in order to capture the marginal contribution of a node as the number of nodes it influences in the good campaign. They first calculate the effect of the negative campaign and remove the nodes affected after a delay r. The nodes with the highest Shapley Value among the remaining nodes are selected to seed the positive information. Tong et al. [18] take an alternative approach to influence limitation by effecting edge addition and deletion to limit the spread. This work is interesting, but will not work in an online social network where formation and deletion of edges are non-trivial. We take a similar but more feasible approach towards influence limitation. We first identify a set of nodes, who if inoculated, would ensure that the rest of the network does not hear about the information and target them with the positive information to limit the spread of the rumor.

3   Our Approach - Overview

The primary objective of our algorithm is to identify nodes vulnerable to a negative campaign that is already spreading in the network and target them with positive information to stem the flow/spread of the negative information.

To estimate the spread, we follow the approach in [14] and assume that the status of a few nodes in the network can be accurately known. These nodes known as the **monitors** could be a set of brand loyalists, evangelists or well-wishers in case of an enterprise who can accurately report about the negative campaign. Monitors who have received the information are termed as positive monitors and those who have not are negative monitors. We use the approach in [14] to estimate the source of rumor and use standard diffusion models to estimate the resulting spread [7].

Once the network infection status is approximately known, we score each node in the network in the order of their susceptibility to the negative information. We categorize the nodes into 3 types - *infected*, *vulnerable* and *uninfected* which are defined as below:

[INFECTED] Nodes that have seen the rumor and have started spreading it - these nodes are difficult to cure. We can neither influence them with our positive campaign nor use them for our positive campaign since they are sufficiently convinced about the negative campaign to spread them. It is difficult to cure these infected nodes with the positive information.

[VULNERABLE] Nodes that have either seen the rumor and are not spreading it or nodes that may see the rumor "soon" from one of the infected nodes. This set of nodes is critical to stemming the flow of rumor, because if not attended to, they may quickly get infected and start spreading the rumors. Therefore, these nodes must be reached out with the positive information.

[UNINFECTED] Nodes that are not infected and may not be infected soon. These are users not in the vicinity of the spread and can be ignored.

We identify the vulnerable nodes via the scores from two algorithms (described in Section 4 to compute the susceptibility score of the network. After identifying the "highly susceptible" targets, we identify influencers who can reach out to them efficiently. To effectively target the nodes we use the approach in [16] that extends the Maximum Influence Arborescence [3] to maximize influence on a specific set of targets. The complete flow of the proposed framework is shown in Figure 1.

Fig. 1. [color] Flow process of the algorithm

4   Vulnerability of Nodes

We categorize the nodes by calculating their **vulnerability** using 2 different algorithms: Algorithm to measure reachability of node from the estimated sources; and Algorithm to measure susceptibility of nodes who have not seen the negative campaign

4.1   Reachability of Nodes

This algorithm is based on the premise that the impact of a certain piece of information depends not only strength of the connection *via* which information is pervading but also on the relative prominence of nodes in a network. Our hypothesis here is that information is likely to propagate faster through an edge if it is from a prominent source than if it is from a less prominent source. We therefore consider two factors:

Information Diffusion: Probability of information diffusion from one node to another is generally captured as the weight of the edge between them. In our algorithm, we use the approach in [16] to determine these influence probabilities.

Importance: Relative importance of the users along the path of information transfer is measured using their page rank [11] in the entire graph. Page rank is a popular measure of centrality of the nodes in a network. A higher value of page rank indicates that the node is better connected and hence more prominent in the network.

Let Σ denote the set of paths between the rumor source R and a node n. Consider one such path $\sigma \in \Sigma$. Suppose the path is as follows: $R \to i_1 \to i_2 \to \cdots \to i_{n-1} \to i_n$. We define a reachability index RI to this path σ as the product of the diffusion probabilities in the entire path and the importance of the nodes along the path:

$$RI(\sigma) = w(R, i_1) \left(\prod_{j=1}^{k} pg(i_j)w(i_j, i_{j+1}) \right) \tag{1}$$

Here, $w(i, j)$ is the weight on the edge based on the diffusion probability between nodes $i\&j$ and $pg(i)$ is the page rank of the node i. The idea is to account for the path of information flow based not only on the diffusion probabilities but also the prominence of the nodes along the path. The reachability index associated with this path is thus the product of probabilities of influence along the edges and page ranks of the nodes encountered on this path. The reachability index associated with all the paths between a pair of nodes are computed and the highest reachability index is chosen.

However, calculating all the paths between two nodes in a graph is NP-hard. To scale this scoring, we utilize the following modification:

1. The weight on the edges w_{ij} are modified as the $- \log w_{ij}$.
2. Every node n of the original graph G is converted into a directed edge $n_1 \to n_2$ in a new graph G' with a weight as $- \log pg(n)$. All the incoming edges into node n would now go into n_1 whereas all the outgoing edges from n would now emanate from n_2. In effect, n_1 has only incoming edges barring the outgoing edge $n_1 \to n_2$ edge and n_2 has all outgoing edges except for the incoming edge $n_1 \to n_2$ edge.
3. The transformation of an edge $n \to m$ is shown in Fig. 2(a).

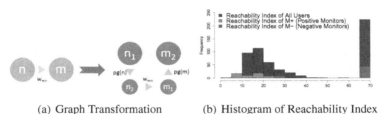

(a) Graph Transformation   (b) Histogram of Reachability Index

Fig. 2. Fig. 2(a): Transformation of the graph edges to account for page rank. ; **Fig. 2(b):** Histogram showing reachability index of nodes. The ones at the extreme end of the histogram are those nodes which are not connected to rumor source.

With these modifications, we can apply Dijkstra's Algorithm or any other shortest path finding algorithm to find the shortest path between the start node R and a node n to compute the susceptibility index of the node n. This algorithm is summarized in Algorithm 1. The $REACH(n)$ is defined as,

$$REACH(n) = \exp^{-RI(n)} \tag{2}$$

The exponential scaling is done to negate the logarithmic scaling in our modification for Djikstra's algorithm.

Figure 2(b) shows the histogram for $REACH(i)$ *i.e.* $e^{-RI(i)}$ where i is a node in the graph for a sample graph with a randomly chosen rumor source. All the positive monitors have a higher susceptibility than the negative monitors, $\min_{i \in M^+} REACH(i) > \max_{i \in M^-} REACH(i)$, where M^+ indicates the set of positive monitors and M^- indicates the set of negative monitors. This is used to define a threshold for node categorization into infected or uninfected. The threshold θ_1 is defined as the geometric mean between the maximum of positive monitors and minimum of negative monitors, given by $\theta_1 = \sqrt{\left(\min_{i \in M^+} REACH(i) \right) \left(\max_{i \in M^-} REACH(i) \right)}$. Nodes having

---

Algorithm 1. Algorithm to measure reachability of node from the estimated sources

---

Require: $G = (V, E, w), R$ where w is the probability of influence along the edges and R is the rumor source

1. $G' = (V', E', w')$: new directed weighted graph
2. **for** i in V **do**
3.   Add i_1 and i_2 to V'
4.   Add (i_1, i_2) to E'
5. $w'_{i_1, i_2} = -\log pg(i)$ where $pg(i)$ is the pagerank of node i in G
6. **end for**
7. **for** (i, j) in E **do**
8.   Add (i_2, j_1) to E'
9. $w'_{i_2, j_1} = -\log w_{i,j}$
10. **end for**
11. **for** i in V **do**
12.   **if** R and i_1 are connected in G' **then**
13. $RI(i) = shortest\_path\_length(G', R, i_1)$
14.   **else**
15. $RI(i) = \infty$
16.   **end if**
17. **end for**
18. **for** i in V **do**
19. $REACH(i) = e^{-RI(i)}$
20. **end for**
21. **return** REACH

REACH greater then θ_1 are termed as *uninfected* whereas the ones having reachability index less then θ_1 are further classified as *infected* and *vulnerable*. The second level categorization is obtained by defining a boundary θ_2 at $1 - \sigma$ (standard deviation) from the mean of the reachability index score of all nodes. Nodes with reachability index between θ_1 and θ_2 are categorized as *vulnerable* and rest as *infected*.

4.2 Susceptibility of the Node

The objective of this algorithm is to determine the susceptibility of a node based on its neighborhood. The hypothesis here is that vulnerable nodes have a heavily *affected* neighborhood. However, certain nodes that do not have a heavily affected neighborhood need to considered as well if they have the potential to affect a large part of unaffected nodes when and if they get affected. To encapsulate these conditions, we consider the following factors for the susceptibility score:

Susceptibility: We calculate the susceptibility using the predicted snapshot of the network at time r. For the nodes which will get infected at time $r + 1$, we look at all the incoming edges. The susceptibility is calculated as the sum of those edge weights which come from *affected* nodes, $sus(v)^r = \sum_{(u,v) \in E \,\wedge\, Affect(u)=1} w(e_{uv})$ We add a decay factor to the susceptibility to account for the recency in the infected neighborhood, $sus(v)^{r+k} = sus(v)^r e^{-\mu(k)}$ where μ is the decay factor.

Contagiousness: For each node that has not been affected at time r we calculate how contagiousness the node is by calculating the fraction of shortest paths from affected to

---

Algorithm 2. Vulnerability Score
___

Require: $G = (V, E, w)$

1. $G' = (V', E, w)$: Status of Graph G after delay r
2. **for** k in $1 \rightarrow \infty$ **do**
3. $G' = (V', E, w)$: Status of Graph G after time $r + k$
4.   **for** $i \in V$ and i is not $infected$ **do**
5. $sus(i) = \sum_{(u,v) \in E \,\wedge\, Affect(u)=1} w(e_{uv}) e^{-\mu(k)}$
6. $bc(i) = \sum_{\substack{\forall (s,t) \in E \\ s \in Affected \\ t \in Unaffected}} \frac{\sigma_{st}(v)}{\sigma_{st}}$
7.   **end for**
8. **end for**
9. **return** $sus + bc$
___

unaffected nodes that pass through the node. This is a modified version of the popular *between-ness centrality* $bc(v) = \sum_{\substack{\forall (s,t) \in E \\ s \in Affected \\ t \in Unaffected}} \frac{\sigma_{st}(v)}{\sigma_{st}}$ where $\sigma_{s,t}(v)$ is the number of shortest paths from node s to node t that pass through v and $\sigma_{s,t}$ is the total number of shortest paths from node s to node t. This measure is akin to percolation centrality introduced in [6]. This measures how critical is the current node to the percolation of information from the infected to the uninfected parts of the network. A node with high contagiousness though not immediately vulnerable must be protected in order to prevent serious information spread.

The final susceptibility score is the sum of the susceptibility at r and the contagiousness of a node. The algorithm is briefed in Algorithm 2. The nodes that are not connected to any infected nodes are not infected at all and belong to Category 3 of nodes. Among the rest, the categorization is done by choosing the top $50-$percentile as infected and the rest as vulnerable.

4.3 Reachability and Susceptibility

Reachability score considers the position of a node with respect to the affected parts and the path to the affected part of the network to determine its vulnerability. Susceptibility score on the other hand accounts for the node's presence in the paths from the affected parts of the network. Both the approaches account for the factor of delay after which the information has reached a particular node. In the reachability, this is implicit in the product of weights along the path whereas the susceptibility score has an explicit exponential time decay factor.

One difference between the two Algorithms (Algorithm 1 and 2) is that Algorithm 1 finds targets who may or may not have seen the rumor/negative campaign and are vulnerable, whereas Algorithm 2 finds the scores based on connectedness to already infected nodes. However, both the algorithms can be used to identify vulnerable nodes that have to be targeted with the positive campaign.

Reachability is loosely related to the popular Independent Cascade Model (ICM)[7] since it considers the independent paths between the nodes. Susceptibility on the other hand is loosely related to the Linear Threshold Model (LTM) since it evaluates the simultaneously affected neighborhood of a node.

We put nodes categorized as vulnerable by either of the algorithm into the *vulnerable* set. From the remaining nodes, those that are classified as infected by either of the approaches are put into *infected* set and the rest form the uninfected part.

5   Experiments

All our experiments were performed on a data set collected by querying for the followers of a specific twitter handle. For each follower, we obtain the information about their connections and tweets using standard Twitter APIs. The final graph is built from the connections between all the followers. An edge is defined between every node and its set of followers. The edges are weighted using the approach in [16]. The network had 571 nodes and 5747 edges. We use the rumor source identification in [14] with randomly chosen monitors in all our experiments. We used 25 (5% of the network size) monitors for our experiments.

Each of our experiment is repeated across 100 trials by randomly selecting a source and simulating the information flow based on the edge weights and considering this as our ground truth. Information flow is simulated using a Independent Cascade Model (ICM) [7]. The monitors were then selected uniformly at random for each trial. Since the ground truth is generated via simulation, we present our results with this dataset only, however, we have observed that the performance holds for other networks as well.

We determine the nodes using our approach in Section 4.3 and target them by identifying the influencers using the algorithm in [16]. We compare our performance with the approach in [13] and compare the spread of the positive information using the Multi-Campaign Independent Cascade Model (MCICM)[1].

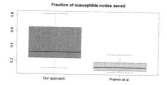 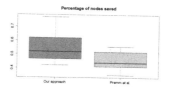

(a) Fraction of nodes saved   (b) Percentage of nodes saved

Fig. 3. [color] Performance comparison across configurations

In our first experiment, we calculate the fraction of network saved by our algorithm. We estimate the spread the rumor without any positive campaign in the network using the standard ICM and then compare it with the case had there been a positive campaign propagating using the MCICM model [1] and calculate the number of nodes saved from the infection (nodes that received the information in the former setup and did not receive the information in the latter). Figure 3(a) shows the fraction of susceptible nodes saved in each of the configuration. Our approach performs significantly better than the Shapely value based approach in [13] in best/worst/average cases as evidenced by the box plot. The Mann-Whitney statistic [9] for the results from the two approaches had a

p-value of 0.000367 (less than 0.05) indicating that they are statistically significant and the performance is not by chance.

In our next experiment, we find the percentage reduction in the infected nodes via the use of MCICM [1]. The resulting performance is shown in Fig 3(b), where again our approach performed better than the existing model from [13] in most cases. The Mann-Whitney statistic [9] for this experiment was observed to be 0.00235 again indicating statistical significance.

6   Conclusion

In this work, we have devised a mechanism for containing the spread of rumor in a network. The problem consists of two stages, identifying the spread of the rumor and then limiting it. We have proposed a comprehensive framework and also a new algorithm to score users vulnerable to the spread. We have compared our work with existing approach in [13], and the results are encouraging.

References

1. Budak, C., Agrawal, D., El Abbadi, A.: Limiting the spread of misinformation in social networks. In: Proceedings of the 20th International conference on World Wide Web, pp. 665–674. ACM (2011)
2. Capasso, V., Capasso, V.: Mathematical structures of epidemic systems, vol. 88. Springer (1993)
3. Chen, W., Wang, C., Wang, Y.: Scalable influence maximization for prevalent viral marketing in large-scale social networks. In: Proceedings of the 16th ACM SIGKDD International conference on Knowledge Discovery and Data Mining, pp. 1029–1038. ACM (2010)
4. Chew, C., Eysenbach, G.: Pandemics in the age of twitter: content analysis of tweets during the 2009 H1N1 outbreak. PloS one 5(11) (2010)
5. Domingos, P., Richardson, M.: Mining the network value of customers. In: Proceedings of the 7th ACM SIGKDD International Conference on Knowledge Discovery and Data Mining, pp. 57–66. ACM (2001)
6. Gönci, B., Németh, V., Balogh, E., Szabó, B., Dénes, Á., Környei, Z., Vicsek, T.: Viral epidemics in a cell culture: novel high resolution data and their interpretation by a percolation theory based model. PloS one 5(12) (2010)
7. Kempe, D., Kleinberg, J., Tardos, É.: Maximizing the spread of influence through a social network. In: Proceedings of the Ninth ACM SIGKDD International Conference on Knowledge Discovery and Data Mining, pp. 137–146. ACM (2003)
8. Leskovec, J., Krause, A., Guestrin, C., Faloutsos, C., Van Briesen, J., Glance, N.: Cost-effective outbreak detection in networks. In: Proceedings of the 13th ACM SIGKDD International Conference on Knowledge Discovery and Data Mining, pp. 420–429. ACM (2007)
9. Mann, H.B., Whitney, D.R.: On a test of whether one of two random variables is stochastically larger than the other. The Annals of Mathematical Statistics, 50–60 (1947)
10. Narayanam, R., Narahari, Y.: A shapley value-based approach to discover influential nodes in social networks. IEEE Transactions on Automation Science and Engineering (99), 1–18 (2010)
11. Page, L., Brin, S., Motwani, R., Winograd, T.: The pagerank citation ranking: bringing order to the web (1999)

12. Prakash, B.A., Vreeken, J., Faloutsos, C.: Spotting culprits in epidemics: How many and which ones? In: IEEE International Conference on Data Mining, vol. 12, pp. 11–20 (2012)
13. Premm Raj, H., Narahari, Y.: Influence limitation in multi-campaign social networks: A shapley value based approach (2012)
14. Seo, E., Mohapatra, P., Abdelzaher, T.: Identifying rumors and their sources in social networks. In: SPIE Defense, Security, and Sensing. pp. 83891I–83891I. International Society for Optics and Photonics (2012)
15. Shah, D., Zaman, T.: Rumors in a network: Who's the culprit? IEEE Transactions on Information Theory 57(8), 5163–5181 (2011)
16. Srinivasan, B., Anandhavelu, N., Dalal, A., Yenugula, M., Srikanthan, P., Layek, A.: Topic-based targeted influence maximization. In: Social Networking Workshop, 2014 Sixth International Conference on Communication Systems and Networks (COMSNETS), pp. 1–6 (January 2014)
17. Steel, E.: Nestlé takes a beating on social-media sites. The Wall Street Journal 29 (2010)
18. Tong, H., Prakash, B.A., Eliassi-Rad, T., Faloutsos, M., Faloutsos, C.: Gelling, and melting, large graphs by edge manipulation. In: Proceedings of the 21st ACM International Conference on Information and Knowledge Management, pp. 245–254. ACM (2012)

Is Twitter a Public Sphere for Online Conflicts?
A Cross-Ideological and Cross-Hierarchical Look

Zhe Liu[1,*] and Ingmar Weber[2]

[1] College of Information Sciences and Technology,
The Pennsylvania State University, University Park, Pennsylvania 16802
zul112@ist.psu.edu
[2] Qatar Computing Research Institute,
PO Box 5825, Doha, Qatar
iweber@qf.org.qa

Abstract. The rise in popularity of Twitter has led to a debate on its impact on public opinions. The optimists foresee an increase in online participation and democratization due to social media's personal and interactive nature. Cyber-pessimists, on the other hand, explain how social media can lead to selective exposure and can be used as a disguise for those in power to disseminate biased information. To investigate this debate empirically, we evaluate Twitter as a public sphere using four metrics: equality, diversity, reciprocity and quality. Using these measurements, we analyze the communication patterns between individuals of different hierarchical levels and ideologies. We do this within the context of three diverse conflicts: Israel-Palestine, US Democrats-Republicans, and FC Barcelona-Real Madrid. In all cases, we collect data around a central pair of Twitter accounts representing the two main parties. Our results show in a quantitative manner that Twitter is not an ideal public sphere for democratic conversations and that hierarchical effects are part of the reason why it is not.

Keywords: public sphere, social stratification, conflict, political communication, twitter.

1   Introduction

With the rapid growth of Twitter, it has become one of the most widely adopted platforms for online communication. Besides using it for relationship formation and maintenance, many people also regularly engage in discussions about controversial issues [1]. On one hand, this increasing adoption of Twitter for online deliberation inevitably creates a perfect environment for open and unrestricted conversations. On the other hand, individuals on Twitter tend to associate more with like-minded others and to receive information selectively. This leads the cyber-pessimist to emphasize the vital role of opinion leaders in shaping others' perceptions during a conflict and to foresee the online environment as a disguise for those in higher social hierarchy to disseminate information. In order to empirically understand whether Twitter creates a public sphere for democratic debates we ask questions like: How do people on

L.M. Aiello and D. McFarland (Eds.): SocInfo 2014, LNCS 8851, pp. 336–347, 2014.
© Springer International Publishing Switzerland 2014

different sides of ideological trenches engage with each other on Twitter? How much does social stratification matter in this process? And how universal are such patterns across different types of polarized conflicts?

For our study, we choose three conflicts of very different nature: the Palestine-Israel conflict, the Democrat-Republication political polarization, and the FC Barcelona-Real Madrid football rivalry. Our analysis is guided by four assessment metrics for the democratic public sphere introduced by [2], namely, (i) equality, (ii) diversity, (iii) reciprocity, and (iv) quality. We find that in general Twitter is not an idealized space for democratic, rational cross-ideological debate, as individuals from the bottom social hierarchy not only interact less but also provide lower quality comments in inter-ideological communication. We believe our results advance the understanding of opportunities and limitations provided by Twitter in online conflicts. It is also of relevance for the design and development of conflict intervention tools or procedures as we paint a detailed picture of cross-ideological communication.

2   Related Work

The notion of public sphere is defined by Habermas as democratic space for open and transparent communication among publics [3]. In his view, a public sphere was conceived as a space in which: first, communicators are supposed to disregard their social status, so that better argument could win out over social hierarchy. Second, debates should focus on issues of common concerns and should discursively formulate core values. Third, everyone should be able to access and take part in the public debates.

With the advent of the Internet, some optimistic researchers viewed it as a better public sphere than traditional media considering its high reach [4, 5], anonymity [6], diversity and interactivity [2]. In contrast, pessimistic scholars claimed that online discourse oftentimes ends in miscommunication and cannot directly enhance democracy [7]. Also, individuals within the same deliberating group online usually end up at a more extreme position in the same general direction [8, 9] due to selective exposure [10, 11]. In addition, [8] rejected the claim that social stratification is leveled out by the "blindness" of cyberspace, and argued that even in online environment social hierarchy hindered the democratic process of inter-personal communication.

In recent years, the center of the debate has been changed from "Internet as public sphere" to "SNS as public sphere". Optimists argued that the features and tools provided by SNS facilitate communication between individuals, and may be a better means of achieving a true public sphere than anything that has come before it [12, 13]. In contrast, [14, 15] claimed that certain Facebook designs make it a difficult platform for public discourse. In addition, [16-19] noticed that individuals on SNS formed dense clusters that were ideologically homogeneous, although [20] proposed a completely different view, stating that Twitter users tend to share news without bias.

To have a more comprehensive understanding of the afore-mentioned works, in Appendix Table A1 we performed a classification of the existing literatures according to the type of platform being studied, as well as Habermas's criteria of public sphere. We colored the literatures to indicate whether it is in support of or against a public sphere. From that table we saw that: first, most of the existing works mainly focused on online selective exposure, which is just a subcomponent of a healthy public sphere according to Habermas's conception. Second, although comprehensive assessments have been conducted on blogs and forums as public spheres [2, 8, 21], we argue that one cannot simply map these findings onto SNS, due to its very different network structures and communication features. Last but not least, social hierarchy, as a very important criterion in evaluating public sphere, has rarely been addressed in prior literatures. Thus, in this study we want to determine among others, if social hierarchy has an effect on individual's participation in democratic communication on Twitter.

3   Research Questions

In this work, we aim to assess if Twitter is a public sphere for democratic debates. We used Habermas's conception as the theoretical framework for our analysis and evaluate each of those dimensions with the assessment metrics: equality, diversity, reciprocity and quality, proposed by Schneider [2]. Appendix Figure A1 depicts the research framework of this study.

— *Equality.* A democratic state requires all individuals, regardless of their social status, to engage and contribute equally in communication [2]. We quantify a user's engagement by the total number of mentions they make to ideological-friends or foes.

— *Diversity.* A healthy public sphere requires a diverse communication network, which suggests the flexibility of an individual in adapting to varied opinions and views [30]. We adopt the measurement called external-internal (E-I) index to measure the diversity of one's communication. The E-I index is calculated as:

$$EI_i = \frac{E_i - I_i}{E_i + I_i}$$

where E_i is number of unique ideological-foes user i has interacted with, I_i is the number of unique ideological-friends user i has interacted with. The E-I index ranges from -1 to 1. The closer the E-I index is to -1, the more an individual tends to only talk to members of their own group, suggesting a high degree of insularity.

— *Reciprocity.* High reciprocal interactions promote the dyadic exchange of information and resources among individuals, and thus ensure a democratic communication environment. To evaluate the reciprocity levels across ideologies and hierarchies, we adopted the maximum length of inter-ideological conversations as the measurement. We chose the maximum over average in order to avoid the bias introduced due to Twitter API's restriction of getting more than 3,200 tweets per user.

— *Quality.* High-quality communication requires participants to be polite to each another, even during disagreements. Besides, it also encourages participants to make rational arguments supported by logical explanations. High-quality political discourse is important to building democratic consensus. Quality is measured using a crowd-sourcing method, which we will discuss in more details in later sections

4 Method

4.1 Data Collection and Labeling

To automatically detect users with similar or different ideologies, we started with three pairs of opposing seed users: *@AlqassamBrigade* and *@IDFSpokesperson*, *@TheDemocrats* and *@GOP*, and *@FCBarcelona_es* and *@realmadrid*. We intentionally chose these accounts as seed nodes due to their key roles in well-known real-life conflicts which are also reflected on Twitter. For each of the seed nodes, we obtained up to 3,200 of its latest tweets using the Twitter API. For each tweet, we identified up to 100 of its retweeters and labeled them as likely supporters. We use retweet as a signal for ideological categorization by following [22], as retweet usually represents one's endorsements and preferences [23]. We removed mediators and neutral intervenors, such as peace movement organizations and journalists, from our data sets based on their distinct retweeting patterns by following the method introduced in [24].

Classification results were validated via CrowdFlower [25] by assigning 100 random users in each ideology to the HIT workers. By comparing user's pre-assigned ideology to the majority-voted label obtained from CrowdFlower, we found that our classification method yielded on average an accuracy of 96.2%. With the classified users, we extracted all mentions between them as interactions between ideological-friends and foes. Table 1 lists the descriptions of our collected data sets. In total we collected 226,239 Twitter users involved in all three conflicts. Among over 400 million of their daily tweets, we extracted only tweets containing cross- or within-ideological interactions from 56,024 unique users. While comparing the inter- and intra-party tweets, we noticed that they are far less interactions between ideological-foes than friends.

Table 1. Data sets Statistics

Conflict	# Users	#Intra-Mentions	#Inter-Mentions	#Intra-Retweets	#Inter-Retweets
PA – IL	9,937	42,471	3,772	135,784	1,057
DEM – REP	17,869	105,557	16,927	471,291	3,330
DEM – REP	17,869	105,557	16,927	471,291	3,330
FCB – RMCF	28,218	47,924	7,996	104,875	13,093

In addition to dividing the collected users into two camps for each data set, we also split them into four social hierarchical groups according to their number of followers, including: the top 1%, the 1% - 10%, 10% - 70%, and 70% - 100% users. The division is arbitrary, but we think that the number of followers at least partly indicates a person's degree of influence on the social network [26], even though it may not fully represent the social status of an individual in real world.

4.2   Analysis of Inter and Intra-ideological Communication

To test our first three hypotheses, regarding the equality, diversity and reciprocity across hierarchical levels, one-way ANOVA tests were conducted, with significance level set at 0.05. Post-hoc analyses were also carried out with Tamhane's T2 test due to non-homogenous variances. Prior to analysis, all data were checked for normality and non-normal data was transformed using the Log(x+1) method.

For the hypothesis of communication quality, we again relied on CrowdFlower. We analyzed the quality of inter-ideological conversations from two perspectives, including the openness of the communicator's attitude, and the rationality of his/her argumentation. To be more specific, for each combination of the four hierarchical levels, we randomly sampled 50 user pairs with cross-ideological conversations. Next, for each of the user pairs, we extracted one of their complete conversations and highlighted one tweet in it at random. We displayed the selected conversation to the workers. From reading the highlighted tweet, we asked them to label the user's attitude and rationality according to our pre-defined coding schemes as shown in Appendix Table A2. To provide more contexts, the user's profiles as well as their automatically detected ideologies are also displayed in the HIT.

5   Results

5.1   Descriptive Results of Cross-Hierarchical Communication

To examine the social hierarchical effect on cross and within-ideological communication patterns, we calculated the conditional probability of a communicator interacting with another, given their social hierarchies. Note that due to the conditional probabilities, different activity levels of the different tiers do not affect our results. Here we used the PA–IL conflict for illustration purpose and only reported findings that can be generalized to all three data sets. As shown in Figure 1, the horizontal bars depicted the four social hierarchical levels of the conversation starter / receiver. The width of the bar denoted the number of interactions existed within that level.

From the width of the horizontal bars in Figure 1, we saw that except the bottommost level, users from the other three hierarchies have about the same probabilities of being mentioned by their ideological-friends. However, under an inter-ideological context, we noticed that users in the topmost hierarchical level have the highest chance of receiving a mention initiated by their foes, which is even higher than the sum of the probabilities derived from rest three levels. This indicated that people are more willing to attack or challenge "authorities" in online conflict. Besides, under both conditions, there is very little chance that the bottom users will be mentioned by either their friends or foes. In addition, from viewing the width of all ribbons, we found that users from the bottommost hierarchical level maintain the highest probability of initiating a mention of the top 1% of users.

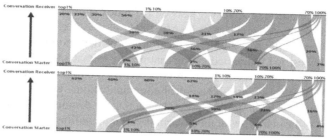

Fig. 1. Inter (below) and intra-ideological (above) communication across hierarchies for the PA-IL data set

5.2   Twitter as Public Sphere

This section presents the findings with regard to each of our proposed measurements. As similar patterns were observed for the two political data sets, only the analysis results from the PA – IL conflict would be shown below for illustration purpose.

Equality. For equality measurement, we categorized users into groups as introduced in [27] based on their number of inter- and intra-ideological mentions. We noticed from Table 1 that users in the upper hierarchical levels initiated more conversations with their ideological-friends than those from the lower levels. The ANOVA results further indicated that these differences were significant at the 5% level (PA-IL: F = 119.12, p = 0.00; DEM-REP: F = 530.34, p = 0.00). We assumed that this might be relevant to the political celebrities' intentions of maintaining their position and status, as well as to stay connected with their supporters, although this needs to be proved in future studies. FCB-RMCF data set revealed very different results, with only the bottom users initiated more conversations than users from the upper levels.

When analyzing the inter-ideological communications, we did not find such differences across social hierarchies within the PA-IL (F = 0.73, p = 0.41) and FCB-RMCF conflict (F = 0.59, p = 0.63). Although the ANOVA results on the DEM-REP data set was significant (F = 27.45, p = 0.00), from the results of the Tamhane's T2 test we further noticed that only users in the bottom group were involved in significantly less interactions with their ideological-foes. In that sense, we claim that at least from our experiments, Twitter allows individuals to disregard their social status in real world, and facilitates their equal participation in online political discourse.

Table 1. Equality of participation across social hierarchies (PA-IL)

Participation Type	#Users with Intra-ideological Mentions				#Users with Inter-ideological Mentions			
(# of mentions)	1%	1%-10%	10%-70%	70-100%	1%	1%-10%	10%-70%	70-100%
One time (1)	1 (1.1%)	35 (4.3%)	435 (9.5%)	359 (19.6%)	6 (19.4%)	32 (20.5%)	198 (20.9%)	110 (24.8%)
Light (2-5)	6 (6.8%)	95 (11.8%)	1002 (21.8%)	579 (31.7%)	8 (25.8%)	42 (26.9%)	292 (30.9%)	147 (33.2%)
Medium (6-20)	16 (18.2%)	220 (27.2%)	1307 (28.5%)	485 (26.5%)	8 (25.8%)	37 (23.7%)	239 (25.3%)	95 (21.4%)
Heavy (21-79)	31 (35.2%)	274 (33.9%)	1117 (24.4%)	307 (16.8%)	7 (22.6%)	30 (19.2%)	172 (18.2%)	66 (14.9%)
Very Heavy (80+)	34 (38.6%)	184 (22.8%)	726 (15.8%)	99 (5.4%)	2 (6.5%)	15 (9.6%)	45 (4.8%)	25 (5.6%)

Diversity. The one way ANOVA tests on E-I index showed significant differences for all three data sets (PA-IL: F = 25.29, p = 0.00; DEM-REP: F = 24.06, p = 0.00; and FCB-RMCF: F = 62.34, p = 0.00), with the second hierarchical group of both political data sets had the significantly lowest E-I index, indicating that people in that social hierarchy are more insular toward their ideological-foes. In contrast, the bottom hierarchy exhibited the highest tendency towards inter-ideological communications. Unlike the political data sets, our post-hoc analysis on the sports data set again demonstrated completely different patterns of insularity, with the bottom users more willing to interact within their own camps. Consistent with findings from prior studies [16-19], all E-I index were less than 0, indicating individual's preferences of talking to their ideological-friends.

Reciprocity. The ANOVA tests also indicated significant overall differences on the maximum length of intra-ideological conversations across hierarchies (PA-IL: F = 86.32, p = 0.00; DEM-REP: F = 807.15, p = 0.00; FCB-RMCF: F = 355.706, p = 0.00), with the maximum frequency of back-and-forth communications increased along with the level of the conversation starter's social hierarchy. In other words, when talking to friends with higher social status, people tended to show greater reciprocity.

However, when analyzing the reciprocity in cross-party debates, the ANOVA and post hoc tests showed no (DEM-REP: F = 27.56, p = 0.06) or almost no (PA-IL: F = 4.60, p = 0.00; FCB-RMCF: F = 10.24, p = 0.00) significant effect of social hierarchy on conversation reciprocity, with only conversation starters from the bottom hierarchy had significantly less back and forth exchanges in cross-ideological conversations.

Quality. Table 2 lists the annotation results on inter-ideological communications. The analysis results of the FCB-RMBC data set were not included here, as the majority of the inter-ideological conversations within that conflict are off-topic chit-chats. For the two political data sets, we found that "disagreement" tweets dominated all the inter-ideological discussions, accounting for more than 70% of all posts. "Insults or sarcasm" were the second most common communication type identified. About 8% of all arguments were personal attacks. 46.7% of all invective posts were from individuals in the bottom level. Inter-party agreements were fairly rare in our results.

Table 2. Statistics on inter-ideological communication types and rationality

Conflict	Agree	Insult	Neutral	Off-Topic	Unclear	Disagree		
						Highly-rational	Rational	Irrational
PA-IL	1.1%	7.2%	1.6%	1.1%	18.1%	4.1%	63.7%	3.0%
DEM-REP	3.4%	8.4%	1.7%	2.0%	9.2%	4.7%	67.6%	3.0%

Next, in our analysis of the argument rationality, we first noticed that the majority of people (89.9%) in inter-ideological discussions demonstrated at least some rational attempts to justify their viewpoints to opponents. Irrational arguments were detected in only 5.8% of all conversations. Highly rational statements were even rarer, accounting for only 4.3% of all annotated tweets. We noticed 31.6% of all statements with highly rational argument were from the top 1% of users.

Table 3. Type of rationality across social hierarchies

	1%	1-10%	10-70%	70-100%
Urls to Foes	19 (63.3%)	80 (54.1%)	430 (49.5%)	151 (22.2%)
Equal Urls	3 (10.0%)	27 (18.2%)	227 (26.2%)	158 (42.1%)
Urls to Friends	8 (26.7%)	41 (27.7%)	211 (24.3%)	66 (20.2%)

Assuming that rational individuals tend to rely on external resources to support their viewpoints, we also quantified users' rationality in this section by measuring the differences in the percentage of URL usages between inter- and intra-ideological mentions. Based on such percentage differences, we categorized all individuals into three groups: more URLs shared with ideological-friends, with ideological foes, and equal URLs shared with both ideological friends and foes. As shown in Table 3, we found that more than half of the individuals from the first two social hierarchical groups adopted more URLs when talking to ideological foes, whereas individuals in the bottom social hierarchy tended to be more rational to their ideological friends.

To further explore the differences in content between inter and intra-ideological conversations, we generated a word cloud in Figure 2 with the top 50 words with the largest relative differences in the usage probabilities. The font size in this word cloud correlates with the absolute difference of a word occurring with a higher probability in only one of the two classes. We colored words that appeared more in inter-ideological conversations blue and otherwise red. We found that, first, blue words are in general larger than the red ones, indicating that inter-ideological talks stick more to the controversial topics compared to the intra-ideological ones. Second, it is very clear that words adopted in inter-ideological conversations are more negative (e.g. "kill", "murder", "hate") in tone compared to words in intra-ideological talks (e.g., "thank", "great", "love").

Fig. 2. Tag cloud with relative differences in inter- vs. intra-ideological usage probabilities for the PA-IL data set

6   Discussion and Conclusion

Through our analyses on three data sets, we concluded that individuals demonstrated inconsistent communication behaviors in conflicts of different natures (political vs. sports).

Ideological and social status played important roles in shaping one's communication habits in political conflicts, and in the meanwhile posed a challenge for conducting democratic discussions on Twitter. First, our work also found selective exposure a problem in Twitter conflicts, as users are more willing to share and to communicate with their ideological-friends than foes. Second, we noticed that most of the cross-ideological mentions on Twitter were initiated toward political authorities in higher social hierarchies, whereas the general public in the bottom hierarchy were mostly ignored. Third, in general the duration of a within ideological conversation was longer than that of a cross-ideological one. Also, conversations initiated by the top users tended to last longer than those initiated by the bottom ones. Fourth, in our experiment more than 40% of cross-ideological tweets were disagreements. This leads us to think Twitter's failure in facilitating the establishment of cross-party agreements.

Although Twitter cannot be viewed as a public sphere for the above issues, we believe it still has a great potential in becoming a platform for resolving online conflicts. Through our analysis of equality we found that Twitter users disregard their social status, participated equally in cross-ideological communications. Additionally, to our surprise, there are very few insulting tweets labeled in our experiments. Most of the arguments on Twitter are claims based on rational viewpoint, though without referencing any external source, fact or data. We think this kind of logical argumentation can still help spread information or knowledge across-ideologies. As more and more communication happens online and publicly through social media, we deem such an analysis a valuable step towards understanding conflicts online. Understanding the online dynamics of such communication could, among other things, contribute to identifying appropriate mediators to resolve the conflict both online and offline.

References

1. Liu, Z., Weber, I.: Predicting ideological friends and foes in Twitter conflicts. In: WWW, Companion Volume (2014)
2. Schneider, S.M.: Expanding the Public Sphere through Computer-Mediated Communication: Political Discussion about Abortion in a Usenet Newsgroup, Massachusetts Institute of Technology (1997)
3. Habermas, J.: The structural transformation of the public sphere: An inquiry into a category of bourgeois society. MIT Press (1991)
4. Gerhards, J., Schäfer, M.S.: Is the internet a better public sphere? Comparing old and new media in the USA and Germany. New Media & Society 12(1), 143–160 (2010)
5. McKenna, K.Y., Bargh, J.A.: Plan 9 from cyberspace: The implications of the internet for personality and social psychology. Personality and Social Psychology Review 4(1), 57–75 (2000)
6. Blader, S.L., Tyler, T.R.: A four-component model of procedural justice: Defining the meaning of a "fair" process. Personality and Social Psychology Bulletin 29(6), 747–758 (2003)
7. Papacharissi, Z.: The virtual sphere The internet as a public sphere. New Media & Society 4(1), 9–27 (2002)
8. Dahlberg., L.: Computer-mediated communication and the public sphere: A critical analysis. Journal of Computer-Mediated Communication 7(1) (2001)

9. Sunstein, C.R.: The law of group polarization. Journal of Political Philosophy 10(2), 175–195 (2002)
10. Adamic, L.A., Glance, N.: The political blogosphere and the 2004 us election: divided they blog. In: 3rd International Workshop on Link Discovery, ACM (2005)
11. Munson, S.A., Resnick, P.: Presenting diverse political opinions: how and how much. In: SIGCHI Conference on Human Factors in Computing Systems, ACM (2010)
12. Semaan, B., et al.: Social media supporting political deliberation across multiple public spheres: Towards depolarization. In: CSCW. ACM (2014)
13. Westling, M.: Expanding the public sphere: The impact of Facebook on political communication (2007)
14. Robertson, S.P., Vatrapu, R.K., Medina, R.: Off the wall political discourse: Facebook use in the, US presidential election Information Polity 15(1), 11–31 (2008)
15. Robertson, S.P., Vatrapu, R.K., Medina, R.: The social life of social networks: Facebook linkage patterns in the 2008 us presidential election. In: Annual International Conference on Digital Government Research (2009)
16. Himelboim, I., McCreery, S., Smith, M.: Birds of a feather tweet together: Integrating network and content analyses to examine cross-ideology exposure on twitter. Journal of Computer-Mediated Communication (2013)
17. Conover, M.D., et al.: Political polarization on twitter. In: 5th Intl. Conference on Weblogs and Social Media (2011)
18. Conover, M.D., et al.: Predicting the political alignment of twitter users. In: IEEE 3rd International Conference on Social Computing (2011)
19. Paul, S.A., Hong, L., Chi, E.H.: Is Twitter a Good Place for Asking Questions? A Characterization Study. In: ICWSM (2011)
20. Morgan, J.S., Lampe, C., Shafiq, M.Z.: Is news sharing on twitter ideologically biased? In: CSCW, ACM (2013)
21. Papacharissi, Z.: The virtual sphere the internet as a public sphere. New Media & Society 4(1), 9–27 (2002)
22. Weber, I., V.R.K. Garimella, and A. Teka. Political Hashtag Trends. in ECIR. 2013.
23. Boyd, D., Golder, S., Lotan, G.: Tweet, tweet, retweet: Conversational aspects of retweeting on twitter. In: HICSS. IEEE (2010)
24. Liu, Z., Weber, I.: Cross-hierarchical communication in Twitter conflicts. In: Proceedings of the 25th ACM Conference on Hypertext and Social Media. ACM (2014)
25. http://www.crowdflower.com/
26. Kwak, H., et al.: What is twitter, a social network or a news media?. In: International Conference on World Wide Web (2010)
27. Jansen, H.J., Pundits, K.R.: Ideologues, and Ranters: The British Columbia Election Online. Canadian Journal of Communication, 2009 30(4), 613–632 (2009)
28. An, J., et al.: Media landscape in twitter: A world of new conventions and political diversity. In: ICWSM (2011)

Appendix

Table A1. Review of literatures on public sphere. Red indicates evidence in support of a public sphere, blue indicates evidence against.

Platform	Equality		Inclusiveness		Argument Rationality
	Disregard Hierarchies	**Equal Accessibility**	**Interaction Scale**	**Interaction Diversity**	
Website, Blog, Forum	[8]	[4] [6] [8] [21] [2]	[5] [21] [2]	[6] [2] [10] [11]	[21] [2]
SNS	[14]	[12] [13] [14]	[12] [13]	[20] [28] [18] [17] [16] [14] [15]	[15] [14]

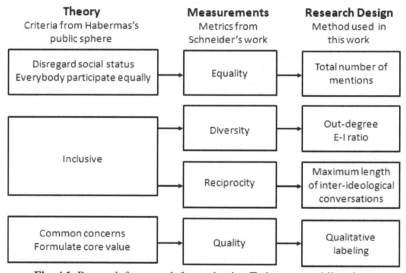

Fig. A1. Research framework for evaluating Twitter as a public sphere

Table A2. Coding scheme for communicator's attitude and rationality

Openness of Attitude	
Agree	A tweet that agrees with the other user or shows similar opinions on the covered material.
Neutral	A tweet that is neutral in nature, neither in obvious agreement or disagreement.
Disagree	A tweet that disagrees with (or critiques) the other user (or the party him/her supports) or shows different opinions on the covered material.
Insult or Sarcasm	A tweet that can be regarded as a derogatory message, such as curses, insults, personal abuse, sarcasm or words that indicated pejorative speak.
Off-Topic	A tweet that is totally unrelated to the conflict.
Unclear	A tweet that is does not fall into any of the above categories.
Rationality of Argument	
Highly rational	The user used information from external sources and with statements based on facts or data, etc.
Rational	The user claimed based on his/her viewpoint and with fair and logical argument to support the statement.
Irrational	The user claimed based on subjective arguments with-out any kind of validation or presentation of facts.

Distributions of Opinion and Extremist Radicalization: Insights from Agent-Based Modeling

Meysam Alizadeh[1] and Claudio Cioffi-Revilla[1, 2]

[1]Department of Computational Social Science and [2]Center for Social Complexity
Krasnow Institute for Advanced Study, George Mason University, Virginia 22030, USA

Abstract. We apply an agent-based opinion dynamics model to investigate the distribution of opinions and the size of opinion clusters. We use parameter sweeps to examine the sensitivity of opinion distributions and cluster sizes relative to changes in individuals' tolerance and uncertainty. Our results demonstrate that opinion distributions and cluster sizes are structurally unstable, not stationary, and have fat tails in most configurations of the model, rather than stable Gaussian distributions. Hence, extremist radical individuals occur far more frequently than "normally" expected. Opinion clusters, in addition to being fat-tailed, reveal a dynamic transition from lognormal to exponential distributions as parameters change.

1 Introduction

The availability of large data sets, such as election polls [1], has enabled researchers to empirically study the opinion of large groups of individuals. However, the accuracy of surveys decays over time, becoming useless in a short time-period. Events such as protests, debates, and personal experiences can be sources of opinion change, especially in the presence of intergroup conflict [2-4]. Moreover, policy makers often need to predict how distributional properties of opinions changes with influential events.

Social media data overcome some of the limitations of surveys. It can be used to make assessments of opinions in retrospect and real time [5], forecasting events [6], and assessing the impact of events on opinion [7]. Social media data, however, have their own shortcomings. A study by the Pew Research Center found that "reaction on Twitter to major political events often differs a great deal from public opinion measured by surveys" [8]. Other factors, such generational gaps [9] indicate that social media data are not always a reliable signal for analyzing public opinion distribution.

Computational models of opinion dynamics provide an alternative approach. One important contribution of this approach is the quantitative analysis of opinion. Issues such as the number of opinion clusters, emergence of extremists, and the size distribution of opinion clusters, are significant examples that are difficult to understand without the use of computational models. Here, we are interested in using an agent-based model to investigate distributional properties of opinion dynamics. We are especially interested in examining the emergence of "fat-" or "heavy-tailed" distributions: power laws, lognormal, and exponential distributions. We also examine how opinion distributions

L.M. Aiello and D. McFarland (Eds.): SocInfo 2014, LNCS 8851, pp. 348–358, 2014.

change with individuals' personality traits. Finally, we analyze the size distribution of clusters and its changes.

Identifying opinion distributions is important because knowing the specific distribution reveals how a population views a given issue and how individuals behave under different circumstances. Understanding opinion distributions is also important for research on extremism and radicalization, because distributions can be heavy-tailed. Similarly, the size distribution of opinion clusters is of interest, because it can identify the polarization of opinions, the number of isolated or unconnected individuals (e.g., so called "lone wolves"), and which opinions are discussed by individuals.

Another goal of this study is to better understand what to expect in terms of opinion distributions and clusters of opinion in a given population, based on specific social theories of human interactions and social influence—such as cognitive dissonance [10], social judgment [11], social identity [12], and sacred values [3], among others. Various computational opinion dynamics models have also been developed, based on these theories, using causal mechanisms such as social influence [13-14], homophily [15], differentiation [16-17], and striving for uniqueness [18]. Therefore, it is rewarding to explore distributional properties of opinions that emerge from these and similar models.

The rest of the paper is organized as follows. Section 2 reviews basic computational models of opinion dynamics. Section 3 presents the 2-dimensional "bounded confidence" opinion dynamics model [16] and our modification. Section 4 reports our simulation results and statistical analysis of opinion distributions and size distribution of opinion clusters. Section 5 provides a discussion of findings and main conclusions.

2 Computational Models of Opinion Dynamics

There are two main classes of opinion dynamics models. First are models from statistical physics, based on transition rates between different states of a system [19-22]. The second group consists of agent-based models, where emergent social behavior is studied through the interactions of autonomous actors with bounded rationality [23-24]. A prominent agent-based opinion dynamics model is the Bounded Confidence (BC) model developed independently by Deffuant, Weisbuch, and collaborators (DW) [15]; and Hegselmann and Krauze (HK) [25]. The two models are similar, differing mainly in their communication regime. While the DW model allows random pairwise encounters at each time step, the HK model allows agents to communicate with all other agents, adopting the average opinion of agents within their area of confidence.

For this study, we chose the model proposed by Huet et al [16] which is an extension of the DW model, for two reasons: (1) it captures two well-understood socio-psychological mechanisms, named homophily and differentiation; and (2) agents in the model have two opinions, which enables (but does not predetermine) the emergence of cognitive dissonance affecting agent decision-making. This model, and our modification as adapted for purposes of this study, is described in the next section.

3   Bounded Confidence with Rejection Model

Consider a set of N agents, each characterized by an opinion variables x_{1i}, $x_{2i} \in$ [-1, 1] and an opinion uncertainty variables $u_{1i}, u_{2i} \in [0, 1]$, where both variables are continuous and their interactions are pairwise. We assume that all agents have the same uncertainty U. Formally, if $|x_{1i}^t - x_{1j}^t| \leq U$ and $|x_{2i}^t - x_{2j}^t| \leq U$, then the opinion of the two agents falls within their bounded confidence interval. Thus, their opinions will converge, based on the following system of equations:

$$x_{1i}^{t+1} = x_{1i}^t + \mu \left(x_{1j}^t - x_{1i}^t\right) \tag{1}$$

$$x_{2i}^{t+1} = x_{2i}^t + \mu \left(x_{2j}^t - x_{2i}^t\right) \tag{2}$$

where μ is a constriction factor. Another possible state occurs when two agents are close in one opinion and far in another. Two cases arise in such a situation, depending on whether the difference is less than the "intolerance threshold" δ. (i) If $|x_{1i}^t - x_{1j}^t| \leq$ $(1 + \delta)U$, dissonance is insufficient to trigger rejection and the two agents ignore opinion 1 and converge on opinion 2 (equation 2). (ii) However, if $|x_{1i}^t - x_{1j}^t| > (1 + \delta)U$, then the conflict in opinions is sufficient to cause dissonance and agents will diverge in opinion 2 (equation 3) and ignore opinion 1.

$$x_{2i}^{t+1} = x_{2i}^t - \mu \, psign \left(x_{2j}^t - x_{2i}^t\right) \left(U - |x_{2i}^t - x_{2j}^t|\right) \tag{3}$$

Here *psign* is similar to *sign* function, except that it returns 1 if the argument is 0. The original model [16] placed upper and lower bounds on opinions, which creates two problems: it suppresses the emergent characteristic of the model and it pushes a significant portion of the population (approximately 25%) toward the boundaries of opinion (Fig. A.1). We fix these problems by removing bounds on opinions. The second row of Fig. A.1 displays results from a sample run by the original model, and a corresponding histogram from our modified model. Our modified model self-organizes opinion between values of approximately -1.5 and 1.5., which now become emergent upper and lower boundaries of opinion, respectively.

4   Simulation Results

4.1   Distribution of Opinions

In this study, we consider a population of 1,000 agents. As shown by the histogram in Fig. 1, a normal distribution seems appropriate. We fit the normal distribution on the results using maximum likelihood estimation (MLE). This shows that the normal distribution was not the best fit, although the histogram is bell-shape. Therefore, the hypothesis that opinions followed a normal distribution is rejected (Table A.1).

Next, lower and upper quantiles of the final opinion distribution were analyzed to assess whether it was heavy-tailed. The complementary cumulative density function (c.c.d.f.), defined as $Pr(X \geq x)$, was computed and each case was compared to the normal distribution (Figs. 1b and 1c). Results show that for $\delta = 1$ the upper tail of the

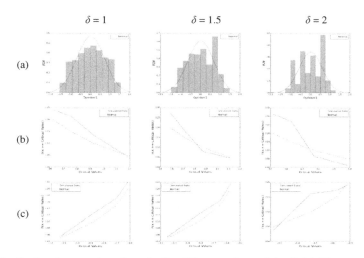

Fig. 1. Distributional properties of results for different values of the δ. (a) Histogram and fitted normal pdf. Rows (b) and (c): c.c.d.f.s for upper and lower tails, respectively.

data falls above the normal, except for the last critical value. Therefore, our simulated data are fat-tailed, although not long-tailed. For $\delta = 1.5$ and $\delta = 2$, the simulated data do not show fat-tail behavior. In the case of the lower tail, however, the simulated data are fat-tailed for almost all values.

The same procedure was followed for analyzing the distribution of opinions for different levels of U (Fig. 2). Results show that the range of emergent opinion increases (meaning an increase in opinion diversity) when increasing uncertainty U from 0.1 to 0.2. The histogram is irregular for $U = 0.3$. Chi-square values reject the normal distribution hypothesis for all levels of uncertainty (Table A.1). Results also show that the upper tail of opinion distribution is fat, but not as long as in the theoretical normal distribution, when $U = 0.1$ and $U = 0.2$ (Fig. 2b). Heavy-tailed behavior between $U = 0.1$ and $U = 0.2$ was also observed for the lower quantiles of the distribution.

4.2   Size Distribution of Opinion Clusters

We used Deffuant's algorithm [26] to identify opinion clusters, which defines a cluster in terms of minimal distance ε between agent opinions. In this study we used $\varepsilon = 0.05$. The c.c.d.f. were computed for three values of the δ, as before, and simulated distributions were compared with corresponding power law and lognormal distributions (Fig. 3). Results show that the upper tail of the simulated data is heavier than the power law for all three δ levels. For the lognormal, the upper tail of simulated data is heavier than the theoretical plot, except for the last critical value. The same results hold for different U, with one exception: when $U = 0.1$ the simulated c.c.d.f. has a thinner tail than the theoretical plot (Fig. 4). We conclude that the size distribution of opinion clusters is heavy-tailed and this property is independent of δ and U.

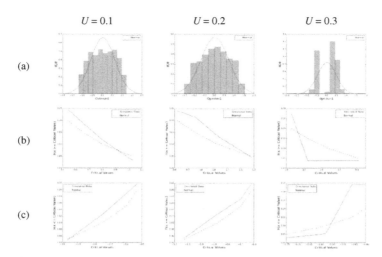

Fig. 2. Distributional properties of simulated data for different levels of U. (a) Histogram and fitted normal pdf. Plots (b) and (c): c.c.d.f. for upper and lower tails, respectively.

Given the finding of heavy tail distributions, the fit of power law, exponential, and lognormal distributions is examined next. In fitting a power law, the first step is to decide which portion of the data is to be fitted. Moreover, a minimum value must also be chosen, because a power law is undefined for $x = 0$, even when a data set is known or hypothesized to "scale" across its entire range [27]. Clauset et al's method [28] was used for finding the optimal value of x_{min}, by minimizing the Kolmogoroff-Smirnov statistic between data and theoretical power law values.

Fig. 5 shows the histogram of opinion cluster sizes, the fitted distributions, and log-log plots for the three δ values. After obtaining the final opinion values, a histogram of cluster sizes was obtained (Fig. 5a). The histogram was then log-log transformed to test whether data followed the log-linear function of a power law with scaling parameter α, given by the least-square regression estimate of the slope (Fig. 5b and c).

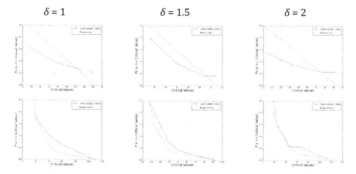

Fig. 3. Plots of c.c.d.f.s of the size distribution of opinion clusters for different thresholds of intolerance. *Top row*: Fitting a power law. *Bottom row*: Fitting a lognormal distribution.

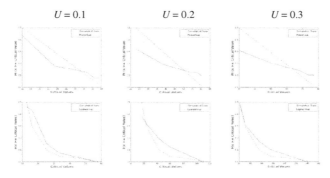

Fig. 4. Plots of c.c.d.f.s of opinion cluster size for different levels of opinion uncertainty. *Top row*: Fitting a power law. *Bottom row*: Fitting a lognormal distribution.

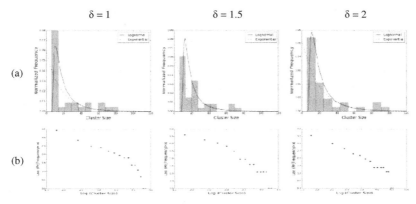

Fig. 5. Distributional properties of the simulated data of opinion cluster size for different values of intolerance threshold (a) Histogram and fitted distributions. (b) Log-log plot of the c.c.d.f.s

Fig. 5 shows the fitted plots, log-log plots, and c.c.d.f.s of cluster size for different δ values. Since the first bin of our data has high frequency and decays with increasing size, visually the data appear to fit the "many-some-rare" pattern typical of heavy-tailed distributions. Parameter estimates and chi-square values are reported in Table A.2, showing that a power law is not the best fit for any δ. The lognormal distribution is the best fit for $\delta = 1$. However, the exponential distribution is the best fit when $\delta = 1.5$. Finally, the lognormal is again the best fit when $\delta = 2$.

Fig. 6 shows the same results for three levels of U. We observe that both lognormal and exponential distributions appear plausible. Again, the power law is not the best fit across cases. The best fit is the lognormal distribution when $U = 0.1$ and $U = 0.2$, but the exponential distribution is better at $U = 0.3$. Hence, we conclude that the best-fitting heavy-tailed distribution depends on parameter combinations of δ and U.

Changes in estimated scale and shape parameter values of the best-fitting distributions are interesting. Those for power law and lognormal distributions vary with δ (Tables A.1). For the exponential distribution, the estimated location parameter is the

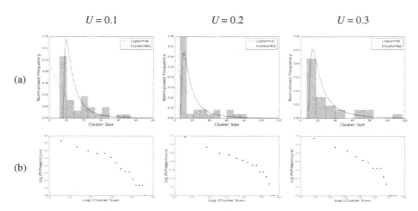

Fig. 6. Distributional properties of simulated data of opinion cluster size for different levels of U. (a) Histogram and fitted distributions. (b) Log-log plot of the c.c.d.f.s

same for all δ values, while the scale parameter varies. MLEs of location and scale parameters for power law and lognormal distributions vary by U (Table A.2). However, changes in scale parameters are more significant when U increases, compared to increasing δ. This finding suggests that size distribution parameters in opinion clusters are dynamically dependent on the stability of, or pattern of change in, individuals' personality traits.

5   Discussion and Conclusion

Simulation results from our Huet-modified agent-based opinion dynamics model showed that distributions of opinion are fat-tailed for different thresholds of intolerance and opinion uncertainty. *An immediate implication of this finding is that extremist opinions are not as rare as in bell-shaped distributions. That is, there will be more individuals with radical opinions, compared to when opinions are normally distributed, if individuals form their opinions based solely on homophily and differentiation.*

Analysis of the opinion clusters revealed that their size distribution is heavy-tailed for all values of δ and U. Our parameter estimation and goodness of fit tests demonstrate that changing the threshold for individuals' intolerance, or their levels of opinion uncertainty, leads to change in the distribution that fits best, fluctuating between lognormal and exponential. Such a system of opinion dynamics may be said to be structurally unstable, based on our distributional analysis.

Opinions have different underlying cognitive characteristics, thereby generating a spectrum of opinions. Some opinions are divisive, causing repulsive behaviors, whereas others require cooperation if individuals are to improve their best global outcome. These differences along with other factors such as occurrence of influential events and changes in personality traits produce different opinion distributions which in turn calls for different models of opinion dynamics (Table 1). Therefore, for example, the original Huet et al model seems inappropriate for modeling the opinion dynamics of brands, public events, popular subjects, hobbies, or professional topics, based on results from this study.

Heavy-tailed behavior is an indicator of critical phenomena in physical and social systems. Besides their intrinsic scientific interest, heavy-tailed properties can also be used for improving early warning, mitigation, and possibly prevention, such as in the case of emergent extremism. Our study offers a plausible framework for analyzing patterns across opinion distributions and associated social mechanisms.

Table 1. Implications of the Frequency Distribution of Opinion Cluster Size

Histogram	Fitted Dist.	Number of Opinion Clusters			Prob. of Radical Opinion	Isolated Individuals	Plausible Type of Opinion
		Small	Medium	Large			
	N.A.	None	None	2	High	Few	Any polarized opinion like political controversies [29] or sacred values [3]
	N.A.	Many	None	None	Rare	Many	Brands, public events, popular subjects [29]
	Normal	Rare	Some	Rare	Rare	Few	Hobbies and professional topics [29]
	Exp.	Many	Some	Few	≥ Normal	Many	Global media topics [29]
	Log normal	Many	Some	Few	≥ Exp.	Many	Global media topics [29]
	Power law	Many	Some	Few	≥ Log-norm	Many	Race Prejudice [30]

Acknowledgements. An earlier version of this paper was prepared by the first author (MA), who also wrote the original model code, for the Seminar on Complexity Theory in Computational Social Science, Department of Computational Social Science, George Mason University, Fall 2013, taught by the second author (CC-R). This version was extensively revised by both authors for presentation at the 6th International Conference on Social Informatics (SocInfo14), Barcelona, Spain, November 10–13, 2014. All errors are the responsibility of the authors, who declare no conflicts of interest. Funding for CC-R was provided grant no. Z878101 from the Office of Naval Research, and by the Center for Social Complexity of George Mason University. MA is funded by a Presidential Fellowship from the Provost's Office, GMU.

A   Appendix

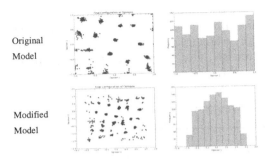

Fig. A.1 Final opinion configuration and distribution from the original and modified models

Table A.1. Estimated parameters of cluster size distributions and chi-square tests for different values of intolerance threshold δ

	$\delta = 1$				$\delta = 1.5$				$\delta = 2$			
	L	S	Chi2	p	L	S	Chi2	p	L	S	Chi2	p
Power law	-14	2.3	27.9	0.00	7.22	3.16	15.3	0.05	-13	2.12	26.3	0.00
Log Normal	6.1	10.1	10.9	0.2	5.16	9.2	6.6	0.58	5.29	10.3	2.36	0.97
Exponential	5.99	17.2	14.8	0.04	5.99	16.5	2.58	0.92	5.99	19.3	7.9	0.34

* Corresponding minimum values to fit power law: $x_{min1} = 6$, $x_{min2} = 21$, $x_{min3} = 9$

Table A.2. Estimated parameters of cluster size distribution and goodness of fit test for different levels of opinion uncertainty U

	$U = 0.1$				$U = 0.2$				$U = 0.3$			
	L	S	Chi2	p	L	S	Chi2	p	L	S	Chi2	p
Power law	3.48	3.56	28.4	0.00	-14	2.3	27.9	0.00	1.54	2.42	13.5	0.1
Log Normal	7.1	2.03	15.1	0.06	6.1	10.1	10.9	0.2	8.9	12.9	7.34	0.5
Exponential	5.99	4.12	13.8	0.05	5.99	17.2	14.8	0.04	6.99	23.1	6.18	0.52

* Corresponding minimum values to fit power law: $x_{min1} = 7$, $x_{min2} = 6$, $x_{min3} = 21$

References

1. Costa Filho, R.N., Almeida, M.P., Andrade, J.S., Moreira, J.E.: Scaling behavior in a proportional voting process. Physical Review E 60(1), 1067 (1999)
2. Brown, R., Condor, S., Mathews, A., Wade, G., Williams, J.: Explaining intergroup differentiation in an industrial organization. Journal of Occupational Psychology 59(4), 273–286 (1986)
3. Atran, S., Ginges, J.: Religious and Sacred Imperatives in Human Conflict. Science 336(6083), 855–857 (2012)
4. Alizadeh, M., Coman, A., Lewis, M., Cioffi-Revilla, C.: Intergroup Conflict Escalation Leads to more Extremism. Journal of Artificial Societies and Social Simulation 17(4), 4 (2014)

5. Zeitzoff, T.: Using Social Media to Measure Conflict Dynamics: An Application to the 2008–2009 Gaza Conflict. Journal of Conflict Resolution 6, 938–969 (2011)
6. Tufekci, Z., Wilson, C.: Social Media and the Decision to Participate in Political Protest: Observations From Tahrir Square. Journal of Communication 62(2), 363–379 (2012)
7. Elson, S.B., Yeung, D., Roshan, P., Bohandy, S.R., Nader, A.: Using Social Media to Gauge Iranian Public Opinion and Mood After the, Election. RAND Corporation Technical Report (2012)
8. Mitchell, A., Hitlin, P.: Twitter reaction to events often at odds with overall public opinion. Pew Research Center (2013)
9. Mitchell, A., Guskin, E.: Twitter News Consumers: Young, Mobile and Educated. Pew Research Center (2013)
10. Festinger, L.A.: Theory of Cognitive Dissonance. Stanford University Press, Stanford (1957)
11. Sherif, M., Hovland, C.I.: Social judgment: Assimilation and contrast effects in communication and attitude change. Yale University Press, Oxford (1961)
12. Tajfel, H., Turner, J.C.: An integrative theory of intergroup conflict. The Social Psychology of Intergroup Relations, 33–47 (1979)
13. Axelrod, R.: The Dissemination of Culture: A Model with Local Convergence and Global Polarization. Journal of Conflict Resolution 41(2), 203–226 (1997)
14. Flache, A., Macy, M.W.: Local Convergence and Global Diversity: From Interpersonal to Social Influence. Journal of Conflict Resolution 55(6), 970–995 (2011)
15. Deffuant, G., Neau, D., Amblard, F., Weisbuch, G.: Mixing beliefs among interacting agents. Advances in Complex Systems 3(1-4), 87–98 (2000)
16. Huet, S., Deffuant, G., Jager, W.: Rejection mechanism in 2D bounded confidence provides more conformity. Advances in Complex Systems 11(4), 529–549 (2008)
17. Mäs, M., Flache, A., Kitts, J.: Cultural Integration and Differentiation in Groups and Organizations. In: Dignum, V., Dignum, F. (eds.) Perspectives on Culture and Agent-based Simulations, vol. 3, pp. 71–90. Springer International Publishing (2014)
18. Mäs, M., Flache, A., Helbing, D.: Individualization as driving force of clustering phenomena in humans. PLoS Comput. Biol. 6(10), e1000959 (2010)
19. Weidlich, W.: The statistical description of polarization phenomena in society. British. Journal of Mathematical and Statistical Psychology 24, 251–266 (1971)
20. Sznajd-Weron, K.J.S.: Opinion evolution in closed community. International Journal of Modern Physics C 11(6), 1157–1165
21. Galam, S.: Heterogeneous beliefs, segregation, and extremism in the making of public opinions. Physical Review E 71(4), 046123 (2005)
22. Castellano, C., Fortunato, S., Loreto, V.: Statistical physics of social dynamics. Reviews of Modern Physics 81, 591–646 (2009)
23. Gilbert, N.: Agent-Based Models. Sage Publishers, Thousand Oaks (2008)
24. Cioffi-Revilla, C.: Introduction to Computational Social Science: Principles and Applications. London and Heidelberg. Springer (2014)
25. Hegselmann, R., Krause, U.: Opinion dynamics and bounded confidence models, analysis, and simulation. Journal of Artificial Societies and Social Simulation 5(3) (2002), http://jasss.soc.surrey.ac.uk/5/3/2.html
26. Deffuant, G.: Comparing Extremism Propagation Patterns in Continuous Opinion Models. Journal of Artificial Societies and Social Simulation, 9(3) (2006)
27. Alstott, J., Bullmore, E., Plenz, D.: powerlaw: A Python Package for Analysis of Heavy-Tailed Distributions. PLoS ONE 9(4), e95816 (2014), doi:10.1371/journal.pone.0095816

28. Clauset, A., Shalizi, C.R., Newman, M.E.J.: Power-Law Distributions in Empirical Data. SIAM Review 51(4), 661–703 (2009)
29. Smith, M.A., Rainie, L., Shneiderman, B., Himelboim, I.: Mapping Twitter Topic Networks: From Polarized Crowds to Community Clusters. Pew Research Center (2014)
30. Kellstedt, P.: Race prejudice and Power Laws of Extremism. In: Cioffi-Revilla, C. (ed.) Power Laws and Non-Equilibrium Distributions of Complexity in the Social Sciences (2008)

Mapping the (R-)Evolution of Technological Fields – A Semantic Network Approach⋆

Roman Jurowetzki and Daniel S. Hain

Aalborg University, Department of Business and Management, IKE,
Fibigerstrde 11, 9220 Aalborg, Denmark
{roman,dsh}@business.aau.dk
http://www.ike.aau.dk

Abstract. The aim of this paper is to provide a framework and novel methodology geared towards mapping technological change in complex interdependent systems by using large amounts of unstructured data from various recent on- and offline sources. Combining techniques from the fields of natural language processing and network analysis, we are able to identify technological fields as overlapping communities of knowledge fragments. Over time persistence of these fragments allows to observe how these fields evolve into trajectories, which may change, split, merge and finally disappear. As empirical example we use the broad area of *Technological Singularity*, an umbrella term for different technologies ranging from neuroscience to machine learning and bioengineering, which are seen as main contributors to the development of artificial intelligence and human enhancement technologies. Using a socially enhanced search routine, we extract 1,398 documents for the years 2011-2013. Our analysis highlights the importance of generic interface that ease the recombination of technology to increase the pace of technological progress. While we can identify consistent technology fields in static document collections, more advanced ontology reconciliation is needed to be able to track a larger number of communities over time.

Keywords: Technological change, transition, technology forecasting, natural language processing, network analysis, overlapping community detection, dynamic community detection.

1 Introduction

Understanding the pattern of technological change is a crucial precondition to formulate meaningful long-term research and industry policy. Technological change usually happens along *technological trajectories* [1] focusing its pathway within a *scientific paradigm* [2]. Apart from defining the boundaries, a paradigm

⋆ We would like to thank Dan Mc Farland, Dan Jurafsky, Walter W. Powell, Chris Potts, all participants of the 2014 ISS Jena conference, the 2014 KID Nice workshop, and the 2014 Summer Term Stanford Network Forum for inspiration and feedback. All opinions, and errors, remain our own.

L.M. Aiello and D. McFarland (Eds.): SocInfo 2014, LNCS 8851, pp. 359–383, 2014.

often provides a set of generic *technology artifacts* which can be deployed along multiple trajectories [3]. Furthermore, recent trends towards modularization and the development of common interfaces have led to an increasing compatibility of technologies within and between paradigms. We argue that today we face an accelerating deterioration of burdens for technology (re-)combination through growing complementary of components [4,5]. In order to understand innovation activity in many modern technological fields, it therefore becomes pivotal to deploy conceptual frameworks, methods, and data geared towards the analysis of such dynamic and highly interdependent systems.

Common approaches to analyze technological change are yet limited to qualitative in-depth case studies [6,7], quantitative methods depending on data such as patents [8] or scientific publications [9], and more generic simulation models [10,11]. While undeniably useful, they either require massive effort to qualitatively analyze complex interaction patterns in technological space, or rely on quantitative data only available with non-negligible time delay, and only relevant for certain technology domains, often underestimating the context in which technology is used. During the last decade we have witnessed tremendous growth of freely available digital information, often in the form of unstructured text data from sources such as web-sites and blogs, written communication of communities in forums or via e-mail, and knowledge repositories (e.g. SSRN, Researchgate). The topicality and sheer amount of such data bear great opportunities for social science research in general, and particularly to timely analyze complex technological change, as we attempt to demonstrate in the following.

In this paper we present a framework and suggest a set of methods to map technological change by using large amounts of unstructured text data from various on- and offline sources. We conceptualize technological change as the reconfiguration of interaction patterns between *technology fragments*, and their clustering in space to *technological fields*, and in time to *technological trajectories*. To analyze such change, we propose the combination of techniques from the fields of natural language processing (NLP) and network analysis. We use the case of *technological singularity* to illustrate our approach graphically as well as with key measures derived from network analysis.

The remainder of the paper is structured as follows. Section 2 reviews and discusses literature and concepts of technological change, and provides a theoretical framework for our approach. In Section 3 we suggest a set of methods suitable to analyze such a framework, and illustrate it in Section 4 at the case of *singularity* technologies. Finally, Section 5 concludes, provides implications for theory, empirical research, and suggests applications for science and industry policy.

2 Conceptualization and Analysis of Technological Change

2.1 Conceptualization of Technological Change

The conceptualization of technological change has a long tradition in different academic communities. Generally, technology exists to fulfill or support some societal functions through direct application or indirectly through derived products. It is thus always embedded in and framed by a societal, political and organizational context, which co-evolves with it [12]. It is also understood as happening within broader *scientific paradigms* [2].

Scholars studying industrial dynamics further describe the development of technology as contextual to the evolution of industrial structures [1,13]. Technology is envisioned as a mean to problem solving in a particular context, which could usually be solved in various other ways using other technologies. *Technological trajectories* represent pathways spanning across the technological space delimited by the paradigm [1], focusing the problem solving process over time around one possible configuration of technologies. While this process usually unfolds gradually, sometimes significant technological discontinuities punctuate a trajectory [14]. Such disruptive change radically alters a trajectory's or even paradigm's internal logic, or completely replaces it in an act of Schumpeterian creative destruction [18]. Overall that suggests competition between substitutional trajectories. Yet, they can also be compatible and complementary to each other, since generic technological artifacts may feed the progress of multiple trajectories.

Drawing on work in theoretical biology [16], technological evolution can be conceived as a recombinatory process of novel and existing component technologies within complex adaptive systems [17]. Innovative recombinations can address fundamentally different problems from the ones that were initially targeted within the components' paradigms. This comes close to a Schumpeterian understanding, where the innovation process is envisioned as the recombination of existing resources in a novel way [18]. The result of such a development can also be envisioned as a complex system with a number of elements that collectively fulfill a single or various goals [19]. A main characteristic of such complex systems is a high degree of interdependence (or epistasis), meaning a functional sensitivity of a system to changes in constituent elements [17]. Thus, a change in one element will affect not only affect its own but also the functioning of epistatically related ones [20]. Since the complexity of the system increases with the number of elements and their degree of interdependence, in large epistatic systems one faces a *complexity catastrophe*, making it increasingly hard to find useful combinations [17].

A possible solution suggested to avoid the *complexity catastrophe* is to increase the systems modularity [4,5,21]. This approach aims at the development of standardized interfaces between more discrete elements to mediate interdependence [22], thus allowing to decrease the overall complexity while maintaining the number of possible recombinations. Modularity and common interfaces further ease the way to combine and recombine components stemming from different trajectories, perhaps even different paradigms. On a higher level, technological revolutions disrupting current techno-economic paradigms are usually accompanied by the emergence of such modules, which can be deployed in various contexts [14]. A recent and very obvious example for this development, the smartphone, is illustrated in Figure 1. The combination of voice and data communication with GPS, camera, compass and accelerometer technologies, bound together by a miniature touchscreen-computer, opened up for a uncountable number of not anticipated applications. Various standardized wireless connection technologies like bluetooth or WiFi allow for compatibility with many other external devises, thus increasing the functionality and re-purposing the phone.

We argue that today we are witnessing a rapid decline of the burdens to technology-combination through efficient modularization between components within artifacts such as the smartphone. Embracing this line of thought, we aim to develop a framework and methodology geared towards the analysis of evolving interdependent technology systems. Such a framework has to be able to capture the ongoing incremental adjustment of interaction pattern between its components (*technological evolution*) as well as disruptive changes fundamentally altering the systems logic (*technological revolution*).

2.2   Measurement and Analysis of Technological Change

Existing empirical research on technological change can broadly be divided in three fields. Work from scholars associated with the Science, Technology and Society (STS) tradition mainly relies on detailed ethnographic studies of the complex multidimensional setup around technological systems, and sheds light on the variety of factors that influence and shape its development [23,24,25].

A stream of more positivistic research in the fields of industrial economics and sciencometrics is primarily based on patent and scientific publication data as an approximation for technological development. Research so far mostly incorporates patent data as aggregated numbers to explain differences in scale [26], or in a network representation to explain structural differences [8,27] in the development of technologies across countries and industries. Patent data has also been used to study invention as a recombination process [17,28,29].[1]

Most recently, social scientists have also started to deploy methods from the fields of computational linguistic and NLP to advance empirical research on the development of science and technology [32,33,34,35]. In their essence, such linguistically informed methods are capable of identifying patterns of language usage in large bodies of text and communication. They range from simple measures of word co-occurrence across documents, corpora and over time [36], to complex linguistically informed probability model [37,38,39].

We perceive the latter as a fruitful way to analyze technological change, implicitly accounting for the socio-economic context in which it is embedded. Such an approach integrates the broad multidimensional perspective of qualitative researchers, that very importantly emphasizes the role of technology users, organizations and governments in innovation processes, with quantitative objectivity given by the machine learning based methodology.

---

[1] However, besides its merits and easy accessibility, there are widely recognized limits in the use of patent data [30,31] such as the high variation of importance across industries and countries, and over time and the long delay between the time research is conducted and the corresponding patent publication.

2.3   Technology Evolution as Structural Network Change

We conceptualize technology as a system of interdependent components [40] within their respective trajectories of development [1]. Representing such systems of interacting elements as networks has brought fresh perspectives and insights to the analysis of complex phenomena from the biological to the social sciences [41]. Embracing this approach, we attempt to analyze technological change as the ongoing structural reconfiguration of interaction between elements in a technology network, which allows us to deploy the rich set of network analysis.

On the lowest level of aggregation in a network representing a technological system, one finds what we call *technology fragments*. They represent atomic, non-reducible repositories of scientific/technological knowledge needed to fulfill certain narrow tasks. Scientific, technological and industrial applications such as machines, software and other devices (which we call *technological artifacts*) combine *technology fragments* in a functional relationship to produce some output. In our previous example, GPS devices, touchscreens and WiFi receivers represent *technology fragments*, which combined in a functional relationship can resemble the smartphone, a *technological artifact*. On a higher level, sets of complementary and substitutional artifacts form a *technological field* (which could be, let's say *mobile applications and devices*). Over time, such fields develop along *technological trajectories*, where accumulated sets of common configuration patterns reproduce over time and set the foundation for further combinations. Again, fragments and artifacts originating from one field might be reconfigured and redeployed in a different field to fulfill the same or even a different purpose. Furthermore, fragments as well as artifacts might not even mainly belong to one field, but be equally employable across multiple fields.

In summary, our conceptualization of technological change, and the suggested methods to analyze it, is based on the following assumptions:

Assumption 1: *Knowledge fragments are atomic, non-reducible repositories of scientific/technological knowledge*

Assumption 2: *Technology fragments can be arbitrary combined and recombined to resemble functional technological artifacts of varying quality*

Having clarified the elements (or edges) in such a network, one has to decide how to measure the functional relationships between them. In our case, identifying technology fragments in unstructured text data, we have to add the following assumption:

Assumption 3: *Co-location of technology fragments in documents imply a functional relationship between them*

3   Analyzing Technology Evolution: Dynamic Semantic Network Approach

After providing a conceptual framework to analyze technological change, in this section we suggest a set of methods to empirically study such changes. A illustration of the method pipeline is provided in Figure 5.

3.1 From Unstructured Text to Technology Fragments: Entity Extraction

First obvious choice to be made is which corpus of technology related text documents one wants to analyze. Such a corpus should optimally (i.) consist of technology related writings (ii.) ranging equally distributed over a time sufficient to observe technological change, and (iii.) not be biased towards particular technologies within the system. Examples for such data are scientific publications, patent descriptions, articles in industry journals, but also online sources such as collections of tech-blogs. In Section 4 we illustrate how to generate an online data corpus with socially enhanced web scraping techniques.

In a next step, it is necessary to convert the unstructured text documents to a machine readable representation.[2] For our means, the goal is to reduce each document to the contained technological concepts. Instead of using a probabilistic approach that stepwise excludes text-elements that are definitely not a technology, we try to detect mentioned technologies in the data. This task falls into the category of *named entity extraction*, which typically relies on tagged dictionaries and string-matching rules to identify the required concepts.

A number of applications related to this development target the identification of different concepts in unstructured text, among others technological and industrial terms. The advantage of these semantic web tools is that they are supported by large, centralized, constantly updated and optimized dictionaries and intelligent disambiguation functions. The result of a successful entity extraction returns a collection of documents that only contain the mentioned technology terms and their document appearance frequency. Referring to our conceptual framework in Section 2, the extracted technology term resemble the elements (nodes) in our technological system, which we label as *technology fragments*.

3.2 From Technology Fragments to a Network: Vector Space Modelling

After having defined the nodeset in our network of *technology fragments*, we have to create weighted edges between them, representing their technological relatedness and interaction. In a first step we construct a (hierarchical) 2-mode network between *technology fragments* and the corresponding documents they occur in. We weight the edges by the pairwise cosine similarity between the vectors of the *technology fragment* and document within a vector space, which we define by by training a Latent Semantic Indexing (LSI) model [42,43] on the

---

[2] Typically, this takes the format of a bag of words (BOW), a line-up of thematically relevant keywords, usually nouns and bi-gram noun phrases. The key assumption of this type of NLP applications is that statistically significant co-occurrence patterns of concepts across the corpus is indicative for actual association between them.

full corpus of documents.[3] Thus, our measure of edge weight indicates to which extent the term representing the *technology fragment* is semantically close to the entirety of other terms contained by the document (see. Section 5). To map technological change over time, we do this separated for every observation period.

While the entirety of *technology fragments* is stable over time, documents obviously experience a 100% turnover in population every observation period. To coerce a stable nodeset, we project the 2-mode to a weighted 1-mode network in technology space. Again, the underlying rationale is based on the assumption that co-occurrence in documents - at least on an aggregated level - also corresponds to a functional relationship between *technology fragments*. However, on a document level that will not always be true. While some documents may discuss technology in the realm of one particular *technological fields*, others might serve more as an overview on industry or research of a broader context, hence contain a collection of *technological fragments* from many otherwise distinct fields. Thus, we penalize documents containing more technology fragments in a similar spirit as the method used by [44], represented by the following equation [45]. Here w_{ij} represents the edge-weight between node i and j, and p the corresponding documents.

$$w_{ij} = \sum_p \frac{w_{i,p}}{N_p - 1} \tag{1}$$

We end up with a one-mode network of *technology fragments* connected by the pairwise projected semantic similarity values, associated with the corresponding period. Figure 3 illustrates these nodeset properties in dynamic networks.

Identifying Technological Fields: Overlapping Community Detection. We depict technological change as the structural reconfiguration of micro level interactions between *technology fragments*. When analyzing the structure, function, and dynamics of networks, it is extremely useful to identify sets of related nodes, known as communities, clusters, or partitions [46]. Such communities of closely connected technologies resemble what we call a *technological field*, a set of complementary or substitutional technologies following one *technological trajectory*, and clustering over time around a common objective. Therefore, we attempt to identify *technological fields* using a community detection algorithm of choice.[4]

---

[3] Before training the model, we apply TF-IDF weights to all terms within the documents. This appreciates the value of particularly important terms for the single document, while depreciating the value of generic terms that often occur across the corpus. Here we have chosen the established LSI algorithm for training the vector space model but other algorithms e.g. Latent Dirichlet allocation (LDA) or Random Projections would also be feasible to calculate pairwise cosines.

[4] An alternative approach would be to use to identify technological fields by the using topic modeling, an approach that lately started to gain traction in social science [35,34,37], create a two-mode network of terms and topics, and project it to an one-mode network of terms. However, for reasons described we here want to offer an alternative, where the topics are already identified using the powerful community detection methods offered by network analysis.

Early clustering and community detection algorithms, in network analysis and elsewhere, usually assumed that the membership of entities to one distinct group. However, depending on the meaning of edges and nodes, many real life networks show a high overlap of communities, where nodes at the overlap are associated with multiple communities. This especially tends to happen when relationship of different quality are projected in a one-mode network [47]. Ones' social interaction network for instance may consist of family members, work colleagues, members of the same karate club or other associations. The more diverse interests such a person has, the more different communities this person will be assigned into. In the same way, the more generic the nature of a *technology fragment* or artifact, the more technological fields will it have functional relationships with. Some *technological artifacts* (and the *technology fragments* resembling them) are that pervasive, they facilitate almost all other technologies in the way they work, such as by its time steam-power or nowadays semiconductors [14]. Embracing that line of thought, researchers recently stated to develop community detection algorithms able to cope with overlapping and nested community structures [48,49], which can be deployed to properly delimit interdependent *technological fields*.

Identifying Technological Trajectories: Dynamic Community Detection. *Technological fields* do not spontaneously appear and reassemble in a vacuum. They gradually change, grow or decline in an cumulative manner, following a historical *technological trajectory* which connects them over time. However, in times of disruptive technological change, former technology interaction pattern might completely reconfigure, particular new configurations might spin-off a main trajectory and so forth. Owing respect to the evolutionary nature of technology, we want to identify communities which are somewhat stable and thus to be found in multiple observation periods, but also allow *technological fields* to experience disruptive key-events in their life-cycle. Besides helping us linking changing communities over time, the identification of such effects in itself represents an interesting information. We consider the following significant events a community might experience during its evolution, also illustrated in Figure 4:

- Birth & Death: The first time a community C_i^t (which are the representation of a *technological field*) is observed and not matched with an already existing community C_j^{t-1}. This community, however, does not have to be stable over time. We in fact expect a substantial share of communities to only appear in on period but not sustain.
- Pause: Communities might be more stable than the reporting on them in the corpus. Thus, allowing them to pause for a period might smoothen birth & death dynamics.
- Merge: In case two communities develop substantial functional interdependence, the main interaction with the rest of the system only happens between them. Thus they merge and form a new community consisting of both

sets. Technically that happens when two or more different communities are matched with one dynamic community D_j in the previous period.

– Split: In the same manner, communities can also separate into independent disciplines. Technically a split occurs when one community C_i matches with two or more dynamic communities in the previous period.

We do so by applying a simple but effective heuristic threshold-based method allowing for many-to-many mappings between communities across different observation periods proposed by [50]. Here we compare an identified community C_i^t in observation period t with the set of dynamic communities in the previous period $\{C_1^t, \ldots, C_J^{t-1}\}$ by employing the widely adapted Jaccard coefficient J_{ij}^t, calculated as follows:

$$J_{ij}^t = sim(C_i^t, C_j^{t-1}) = \frac{|C_i^t \cap C_j^{t-1}|}{|C_i^t \cup C_j^{t-1}|} \qquad (2)$$

If the similarity exceeds the defined matching threshold $\theta \in [0,1]$, both communities are added to the dynamic community D_i. Using this has the advantage that is independent of the (static) community detection method of choice in the observation periods, hence represents a somewhat modular approach. It can also handle overlapping as well as (with some minor adjustments) weighted communities. A major advantage of this approach is the separation of static and dynamic community detection is the high flexibility in the choice of suitable algorithms.

4 Demonstration Case

In the following section we demonstrate the capabilities of our approach to deliver insightful results, and provide some illustrative examples of measures and graphical representations that can be used to gain further insights. We intended to find an empirical case of technological development that would combine a large number of components from traditionally disconnected *technological fields*. Additionally, the *technology field* in focus should be yet in a formative stage and have a potentially strong and broad social impact to generate enough attention and thus reporting texts online. We decided to explore the field of *singularity*. Rather then a clearly delineated *technological field*, singularity represents a future scenario and an umbrella term that summarizes a number of developments in areas as diverse as neuroscience and 3D printing. Based on the context of the technology under study and the characteristics of the corpus, we provide examples how to calibrate the techniques used in the different stages of or method pipeline.

4.1 Empirical Setting: The *Singularity* Case

Technological Singularity as a term has gained momentum since the publication of Ray Kurzweil's book in 2005 [51]. Observing various measures of technological progress over time, he argues that most technologies improved their performance exponentially and therefore it is only a matter of a few decades until we will have reached a point in history when artificial intelligence will supersede human intelligence. The most powerful technological advancement of the 21th century will happen when robotics, nanotechnology, genetic engineering and artificial intelligence reach a certain level of development and can be combined, what will have disruptive consequences for society, culture and the human nature.

Recently, *singularity* entered the European technology policy context, as a technological field within the Horizon 2020 programming. Since 2012, the Directorate General for Communications Networks, Content and Technology (DG CONNECT) is undertaking a foresight process to inform the ICT related programming of research to be financed under Horizon 2020, where *singularity* was identified as one of the 10 central technological fields. It is currently being examined closer to capture early signals and anticipate beneficial trends that should be supported within public research funding schemes.

4.2   Data Mining and Corpus Generation

Researchers, organizations and science journalists are increasingly using social media and the blogosphere to communicate findings and developments, far ahead of journal publication or conference proceedings. This makes microblogging platforms and in particular Twitter with over 200 million monthly active users (Feb. 2014) a valuable source of data. We now describe our data mining approach aiming at selecting relevant twitter updates by relevant users. Instead of using already available corpora to study technological change in *singularity*, such as patent description, scientific publications and industry journals, we choose to create an own out of a variety of online available technology relevant text documents, including publications, tech-blogs *et cetera*. Since *singularity* is a recent and very heterogeneous movement spanning various scientific, industries and tech-communities with distinct routines for communicating and publishing findings and progress, our final corpus therefore is supposed to be unbiassed towards a particular discipline.

To identify relevant documents, we employ a socially-enhanced search routine based on twitter tweets. Twitter's graph structure, built on followship links, is similar to citation networks in academic publications. This enables the construction of large directed graphs and allows applying network analysis methods, to identify central actors for a particular field or topic. For this study we constructed a large followship graph around the - somewhat arbitrarily selected - account *Singularity Hub*, which is an online news platform that actively reports on the topic. The initial *snowballed* network has 49,574 accounts. Using eigenvector centrality, we identify the most influential users and then manually reduce the number of nodes down to 34 twitter accounts that indicate an interest for the area in their profile.[5] Figure 6 shows a central fragment of the network.

---

[5] This selection is very restrictive but is likely to make the final corpus less noisy. Alternatively the manual reduction can be skipped and a corpus filtering built in, at a later stage.

Coloring represents communities, detected by the Louvain algorithm [53], merely for illustration. We can see that the red cluster seems to contain all the central organisations that are present on twitter and focused on singularity and transhumanism like the H+ movement, KurzweilAI, David Orban and more. The green cluster is mostly populated with users that are related to robotics and the violet to software architecture. An overview of the selected user accounts can be found in Table 1.

Micro-blogged tweets (status updates) by these actors often contain links to research papers, popular media articles or blog entries that the selected user considers as worth communicating. For each of these accounts we extract up to 3,200 status updates starting with the most recent, 63,000 in total. We discard all updates that do not carry a link. Relevant tweets were then identified using a vector space model powered semantic search. The text content behind the embedded links - outside of Twitter is then extracted and processed, and finally represents our document corpus for further analysis.

4.3   Identification of Technology Fragments: Entity Extraction

The documents in our corpus discuss technology from very different angles. Some talk about state-of-the art research in certain university labs, while others review the allocation of public research grants or venture capital investment strategies. When attempting to uncover functional relationships between technology fragments, it is crucial to avoid false positives caused by relationships that are non-technical in nature, such as *being funded by the same investor*, or *developed in the same country*. We rely on entity extraction when condensing documents to BOW representations. In the particular case we use OpenCalais, a free web service that performs entity identification across 39 different concepts within submitted text data. The great advantage of *cloudsourcing* in this case is given by the fact that the centralized machine learning algorithms of OpenCalais are trained on a very large amount of natural text and its dictionaries are constantly updated and optimized. An offline solution would hardly be able to compete in terms of performance and topicality.[6] When inspecting the results we find clear technology terms such as *dna profiling*, *robotic surgical systems*, *clinical genomics* or *regenerative stem cell technologies*, which come fairly close to how we understand technology fragments. These terms narrowly describe technology deployed for a fairly delimited task. However, we also find boarder technology terms such as *stem cells genomics*, which span across a somewhat larger field of applications and are likely to include some of the aforementioned terms, and on an even more generic level terms such as *biotechnology* or *robot*.While this clearly diverts from our theoretical framework, where we find on node level only functional interac-

---

[6] For an overview and performance evaluation of available systems see [52]. In addition, OpenCalais provides ontology reconciliation and disambiguation. Identified entities are in many cases enriched with metadata (e.g. profession for persons, ticker symbols for companies and geospatial coordinates for locations). Other detected entity types are not used in this analysis.

tion of atomic *technology fragments*, we do not consider that as worrisome for the analysis to come.

4.4   Network Generation, Technological Field and Trajectory Identification

For a very first inspection and illustration of the nodeset we create a simple static network of all documents connected by their similarity in terms of containing *technology fragments*, cluster them by applying the common Louvain algorithm [53], and plot them in Figure 7. For the three main communities detected we provide a tag-cloud, weighted by the fragments' TF-IDF scores. One can see at first glance that our *singularity* corpus very broadly consists of three fields, where the biggest is centered around robotics, and the two others around (stem) cell and brain research, or to be more interpretative: Robotics, biotechnology and neuroscience. Table 2 provides some key statistics on the networks, communities, and their development. While subject to some fluctuation, the networks seem to develop from many to less nodes and edges, and to less but denser communities. This might indicate *singularity* after an initial phase of experimentation to mature and establish more delimited fields and sub-disciplines, as life-cycle theories might suggest.

We now construct a set of two-mode networks between this nodes and the documents in our corpus,[7] containing only documents published in the corresponding observation period, which we choose to be half a year.[8] Finally, we project this structures on one-mode networks between technology fragments.

Now we identify *technological fields* with the link community detection algorithm proposed by [48], which is able to detect communities with highly pervasive overlap by clustering links between the nodes rather than the nodes themselves.[9] Each node here inherits all memberships of its links and can thus belong to multiple, overlapping communities (*technological fields*). By doing so, we owe respect to the overlapping and nested structure of technology, and are able to identify key *technological fragments* interacting with multiple distinct fields.We first run the community detection separated for every time step independently. We do not *a-priori* define a fixed amount of communities, but rather set the cutoff at the point where the average community density is optimized in every observation period.

Table 3 plots the network of *knowledge fragments* and their membership to *technological fields* for every timestep. Again, what can be seen is that *singularity*

---

[7] Vector space modeling is performed with the GENSIM package [54] within IPython, using LSI and a 400 dimensional model as suggested by [55].

[8] This choice has to be made according to the properties of the data to be analyzed, since best results can be achieved when the network structure shows some gradual change between the observation periods, but no radical turnover suggestion complete discontinuity. This corresponds roughly to a Jaccard index of the two networks between 0.2 and 0.8.

[9] We use the implementation of the link-community approach provided by [56] as package for the statistical environment R.

appears to develop from a broad area without clear boundaries and high inter-connectedness towards clearly delimited *technological fields*. However, we also find first hints that over time some very generic technologies such as *smartphones* and *artificial intelligence* appear to develop towards a very central position, where they serve as common interface between most other fields. While it seems unlikely that *smartphones* (as we understand them today) will be around for much longer then a decade, their centrality in the singularity discussion can be understood as the importance of mobile devices that enhance our by nature limited interaction range. In fact, *smartphones* became a rapidly adopted human enhancement device and currently a number of different wearable technologies are entering the mainstream markets. We also see the generic *artificial intelligence*, which is at the very core of the singularity debate, in a very central position as interface or generic technology between *technological.*

We now perform a threshold-based dynamic community detection[10], where we besides an immense turnover of briefly appearing and disseminating short-term trends indeed find identify a set of persistent *technological trajectories*. Table 4 illustrates the composition of some selected communities which proves to be somewhat stable over time.[11] The tag-cluster are a good way to visualize the interaction between the actual technologies, principal applications and challenges. The first cluster suggests for instance that an important area of application for biometric technologies in conjuncture with machine learning will be found within law enforcement. The second cluster addresses advancements in the area of augmented reality and connections to existent social network structures using primarily mobile devices.

5 Summary and Conclusion

The aim of this paper was to provide a framework and novel methodology geared towards mapping technological change in complex interdependent systems by using large amounts of unstructured data from various recent on- and offline sources. We combine techniques from the fields of NLP and network analysis. Our approach is based on the following steps:

- Using entity recognition techniques we identify technology related terms in the text document of our corpus, which resemble *technology fragments*.
- In a first step, using vector space modeling, we construct an undirected two-mode network between technology fragments and corpus documents for every observation period, where the edges are weighted by the pairwise cosine similarities between documents and terms.
- After projecting this network in technology space, we end up with an undirected one-mode network of technology fragments connected by their weighted co-occurrence in documents of the corresponding observation period.

---

[10] We use a C++ implementation provided by [50].
[11] For the sake of clarity, the technology fragments are weighted by their within-cluster centrality.

- To delimit *technological fields* in every observation period, we use overlapping community detection techniques, owing respect to the interdependent and nested nature of technology.
- To identify *technological trajectories*, we link *technological fields* between observation periods over time using

As empirical example we use the broad area of *Technological Singularity*, an umbrella term for different technologies ranging from neuroscience to machine learning and bioengineering which are seen as main contributors to the development of artificial intelligence and human enhancement technologies. We extract 1,398 relevant text documents all over the internet, using a social search routine that we built around the followship structure within the microblogging service twitter. Using entity recognition tools from the semantic web area, we reduce documents to technology-term representations and finally generate a semantic timestep network of technology fragments. Our community detection exercise identified many coherent technological fields within each community. Already the static clustering provides valuable insights in the emergence of new technological fields and applications for existing technologies. Overlapping community detection, allowed us also to identify certain *general* technologies that work as hubs between other technologies, stemming from a large number of different domains.

Yet, we find the results of the community-tracking over time unsatisfactory. The obstacle are *false negatives* that obstruct the identification of similar communities over time. Our language is full of synonyms, metaphors and unregulated terminology. The reader of this article has no difficulty comprehending that we use the terms *clusters* and *communities* interchangeably, a computer would not. While we are (yet) unable to *teach* the algorithm a deep understanding of ontology, we can try to normalize the terminology as far as possible. This future measure should increase the number of identical terms over time. Furthermore, there seem to be a trade-off between the thematic scope of a given corpus and the resolution of the analysis. Therefore, a broader corpus is most suitable for creating a broad-brush picture of technological change.

We believe a major advantage of our approach is that it conveys text data into a network representation suitable for a dynamic analysis of technology. It proves to be more flexible with respect to the corpus than other semantic or n-gram based methods in natural language processing. Furthermore, for subsequent quantitative analysis and graphical representation one can now draw from the large toolkit of powerful methods available for network analysis. The here performed dynamic community detection is one example, but other methods such as blockmodeling appear to be promising to gain further insights into the evolution of technology. Finally, networks are well established in many areas of social science and thus a representation of semantic features as networks is likely to help bridging the gap between scholars in computer and social science.

References

1. Dosi, G.: Technological paradigms and technological trajectories: a suggested interpretation of the determinants and directions of technical change. Research Policy 11(3), 147–162 (1982)
2. Kuhn, T.S.: The Structure of Scientific Revolutions. University of Chicago Press (1962)
3. Timothy, F.: Bresnahan and Manuel Trajtenberg. General purpose technologies engines of growth? Journal of econometrics 65(1), 83–108 (1995)
4. Baldwin, C.Y., Clark, K.B.: Design Rules: The power of modularity. MIT Press (2000)
5. Schilling, M.A.: Toward a general modular systems theory and its application to interfirm product modularity. Academy of Management Review (2000)
6. Davies, A.: Innovation in large technical systems: The case of telecommunications. Industrial and Corporate Change 5(4), 1143–1180 (1996)
7. Hekkert, M.P., Negro, S.O.: Functions of innovation systems as a framework to understand sustainable technological change: Empirical evidence for earlier claims. Technological Forecasting & Social Change 76(4), 584–594 (2009)
8. Verspagen, B.: Mapping technological trajectories as patent citation networks: A study on the history of fuel cell research. Advances in Complex Systems 10(01), 93–115 (2007)
9. Wagner, C.S., Leydesdorff, L.: Network structure, self-organization, and the growth of international collaboration in science. Research Policy 34(10), 1608–1618 (2005)
10. Dawid, H.: Agent-based models of innovation and technological change, vol. 2, ch. 25, pp. 1235–1272. Elsevier (2006)
11. Lopolito, A., Morone, P., Taylor, R.: Emerging innovation niches: An agent based model. Research Policy 42(6), 1225–1238 (2013)
12. Kaplan, S., Tripsas, M.: Thinking about technology: Applying a cognitive lens to technical change. Research Policy 37(5), 790–805 (2008)
13. Hain, D.S., Jurowetzki, R.: Incremental by design? on the role of incumbents in technology niches - an evolutionary network analysis. In: Conference Proceeding, 6th Academy of Innovation and Entrepreneurship Conference, Oxford, UK (2013)
14. Perez, C.: Technological revolutions and techno-economic paradigms. Cambridge Journal of Economics, bep051 (2009)
15. Schumpeter, J.A.: A Theory of Economic Development. Harvard University Press, Cambridge (1911)
16. Kauffman, S.A.: The Origins of Order. Self-Organization and Selection in Evolution. Oxford University Press (1993)
17. Fleming, L., Sorenson, O.: Technology as a complex adaptive system: evidence from patent data. Research Policy 30(7), 1019–1039 (2001)
18. Schumpeter, J.A.: Capitalism, Socialism and Democracy. Harper, New York (1942)
19. Simon, H.A.: The Sciences of the Artificial. MIT Press (1969)
20. Frenken, K.: A fitness landscape approach to technological complexity, modularity, and vertical disintegration. Structural Change and Economic Dynamics 17(3), 288–305 (2006)
21. Ethiraj, S.K., Levinthal, D.: Modularity and innovation in complex systems. Management Science (2004)
22. Langlois, R.N.: Modularity in technology and organization. Journal of Economic Behavior & Organization (2002)

23. Bijker, W.E., Hughes, T.P., Pinch, T., Douglas, D.G.: The Social Construction of Technological Systems. In: New Directions in the Sociology and History of Technology. MIT Press (2012)

24. Bijker, W.E.: Of Bicycles, Bakelites and Bulbs. In: Toward a Theory of Sociotechnical Change. The MIT Press (1997)

25. Hughes, T.P.: The Evolution of Large Technological Systems, pp. 51–82. The MIT Press (1987)

26. Pavitt, K.: R&D, patenting and innovative activities: a statistical exploration. Research Policy 11(1), 33–51 (1982)

27. Fontana, R., Nuvolari, A., Verspagen, B.: Mapping technological trajectories as patent citation networks. an application to data communication standards. Economics of Innovation and New Technology 18(4), 311–336 (2009)

28. Fleming, L., Sorenson, O.: Science as a map in technological search. Strategic Management Journal 25(89), 909–928 (2004)

29. von Wartburg, I., Teichert, T., Rost, K.: Inventive progress measured by multistage patent citation analysis. Research Policy 34(10), 1591–1607 (2005)

30. Griliches, Z.: Patent statistics as economic indicators: a survey. In: R&D and Productivity: The Econometric Evidence, pp. 287–343. University of Chicago Press (1998)

31. Pavitt, K.: Patent statistics as indicators of innovative activities: possibilities and problems. Scientometrics 7(1), 77–99 (1985)

32. Mohr, J.W., Bogdanov, P.: Introduction – topic models: What they are and why they matter. Poetics 41(6), 545–569 (2013)

33. Ramage, D., Rosen, E., Chuang, J., Manning, C.D., McFarland, D.A.: Topic modeling for the social sciences. In: NIPS 2009 Workshop on Applications for Topic Models: Text and Beyond, vol. 5 (2009)

34. McFarland, D.A., Ramage, D., Chuang, J., Heer, J., Manning, C.D., Jurafsky, D.: Differentiating language usage through topic models. Poetics 41(6), 607–625 (2013)

35. DiMaggio, P., Nag, M., Blei, D.: Exploiting affinities between topic modeling and the sociological perspective on culture: Application to newspaper coverage of us government arts funding. Poetics 41(6), 570–606 (2013)

36. Chen, S.-H., Huang, M.-H., Chen, D.-Z., Lin, S.-Z.: Technological Forecasting & Social Change 79(9), 1705–1719 (2012)

37. Hall, D., Jurafsky, D., Manning, C.D.: Studying the history of ideas using topic models. In: Proceedings of the Conference on Empirical Methods in Natural Language Processing, pp. 363–371. Association for Computational Linguistics (2008)

38. Ramage, D., Manning, C.D., McFarland, D.A.: Which universities lead and lag? toward university rankings based on scholarly output. In: Proc. of NIPS Workshop on Computational Social Science and the Wisdom of the Crowds (2010)

39. Nallapati, R., Shi, X., McFarland, D.A., Leskovec, J., Jurafsky, D.: Leadlag LDA: Estimating topic specific leads and lags of information outlets. In: ICWSM (2011)

40. Hughes, T.P.: The evolution of large technological systems. The social construction of technological systems: New directions in the sociology and history of technology, pp. 51–82 (1987)

41. Newman, M., Barabási, A.-L., Watts, D.J.: The Structure and Dynamics of Networks: Princeton University Press (2006)

42. Deerwester, S.: Improving information retrieval with latent semantic indexing. In: Proceedings of the 51st ASIS Annual Meeting (ASIS 1988), vol. 25 (1988)

43. Deerwester, S.C., Dumais, S.T., Landauer, T.K., Furnas, G.W., Harshman, R.A.: Indexing by latent semantic analysis. JASIS 41(6), 391–407 (1990)

44. Newman, M.: Scientific collaboration networks. ii. shortest paths, weighted networks, and centrality. Physical Review 64(1), 16132 (2001)
45. Opsahl, T.: Triadic closure in two-mode networks: Redefining the global and local clustering coefficients. Social Networks 35(2), 159–167 (2013)
46. Radicchi, F., Castellano, C., Cecconi, F., Loreto, V., Parisi, D.: Defining and identifying communities in networks. Proceedings of the National Academy of Sciences of the United States of America 101(9), 2658–2663 (2004)
47. Kivelä, M., Arenas, A., Barthelemy, M., Gleeson, J.P., Moreno, Y., Porter, M.A.: Multilayer networks. arXiv preprint arXiv:1309.7233 (2013)
48. Ahn, Y.-Y., Bagrow, J.P., Lehmann, S.: Link communities reveal multiscale complexity in networks. Nature 466(7307), 761–764 (2010)
49. Mucha, P.J., Richardson, T., Macon, K., Porter, M.A., Onnela, J.-P.: Community structure in time-dependent, multiscale, and multiplex networks. Science 328(5980), 876–878 (2010)
50. Greene, D., Doyle, D., Cunningham, P.: Tracking the evolution of communities in dynamic social networks. In: Proc. International Conference on Advances in Social Networks Analysis and Mining, ASONAM 2010 (2010)
51. Kurzweil, R.: The Singularity Is Near. When Humans Transcend Biology. Penguin (2005)
52. Rizzo, G., Troncy, R.: NERD: evaluating named entity recognition tools in the web of data. In: Workshop on Web Scale Knowledge Extraction, WEKEX 2011 (2011)
53. Vincent, D.: Blondel, Jean-Loup Guillaume, Renaud Lambiotte, and Etienne Lefebvre. Fast unfolding of communities in large networks. Journal of Statistical Mechanics: Theory and Experiment 2008(10), P10008 (2008)
54. Řehůřek, R., Sojka, P.: Software framework for topic modelling with large corpora. In: Proceedings of the LREC 2010 Workshop on New Challenges for NLP Frameworks (2010)
55. Bradford, R.B.: An empirical study of required dimensionality for large-scale latent semantic indexing applications. In: Proceeding of the 17th ACM Conference, p. 153. ACM Press, New York (2008)
56. Kalinka, A.T., Tomancak, P.: linkcomm: an r package for the generation, visualization, and analysis of link communities in networks of arbitrary size and type. Bioinformatics 27(14) (2011)

Appendix

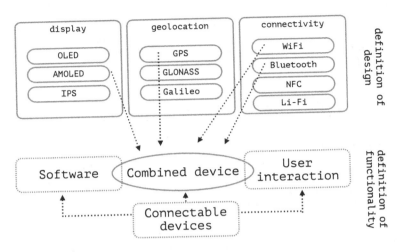

Fig. 1. Illustrative combination of technology components from different trajectories

Fig. 2. Example of pairwise semantic similarity between terms and documents

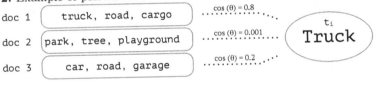

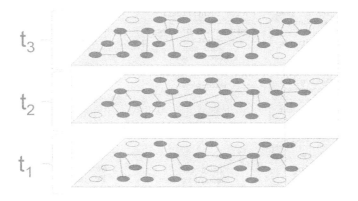

Fig. 3. Illustration of the development of a nodeset over time

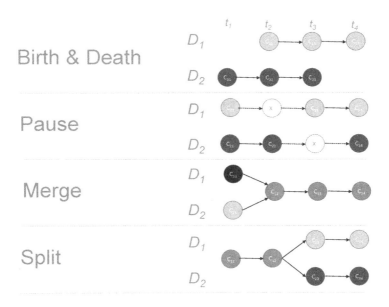

Fig. 4. Illustration of significant events in the evolution of communities, adopted from [50]

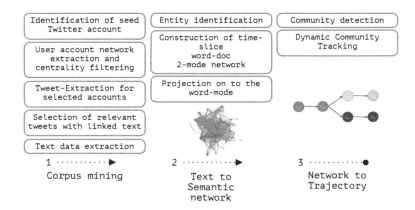

Fig. 5. Illustration of the method pipeline

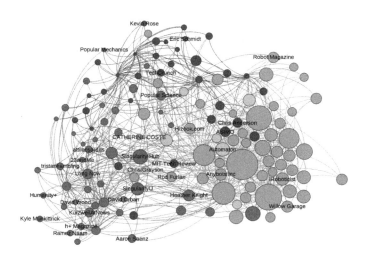

Fig. 6. The central fragment of the twitter account network with the finally selected profiles for text-extraction

Table 1. Overview over the "expert" Twitter-accounts that were used for the text extraction

Twitter-id	Name	Location	Description
121054992	CATHERINE COSTE	Genomic Entertainment	MIT certificate in Genomics. Genomic & Precision Medicine. UCSF. Blog. Ethics. Health & Death 2.0 - DTC Genomics
16870421	SingularityU	NASA Moffett Field, CA	Silicon Valley's leading experts on exponential technology. Follow @singularityhub @singularitylabs @suglobal @expotentialmed
60442721	Ramez Naam	Seattle	Author. Nexus / Crux / More Than Human / The Infinite Resource. Formerly a computer scientist at Microsoft. Interested in everything
18705065	Humanity+	Global	Humanity+ is dedicated to promoting understanding, interest and participation in fields of emerging innovation that can radically benefit the human condition
95661807	Kyle Munkittrick	Denver, NYC, San Fran	Bioethicist the unholy union of science, medicine, and philosophy. Blame no one but myself for what you find here.
16352993	Heather Knight		CMU Roboticist with a soft spot for interactive art & live robot performance. Founder @MarilynMonrobot, Director @robotfilmfest, Robo-Tech @RobotCombatSyFy!
28132585	Aaron Saenz		Writer for Singularity Hub, former Physics dude, Improv Comedian, Nomad
2443051	attilacordiac	Cambridge, UK	bioinformatician, EBI, regular Hadoop & R tinkerer, personal proteomics instigator, ex autochondrial-stem cell biologist driven by healthy lifespan extension
15243106	Singularity Hub	NASA Moffett Field, CA	News network covering science, technology & the future of humanity. Follow @singularityu — Become HUB Member http://t.co/XCGjd73k
16834443	Kurzweil AI News	California/Mass	KurzweilAI (http://t.co/KDf0Rg6D66p) is a newsletter/blog covering nano-bio-info-cogno-roenic breakthrough in accelerating intelligence
7445642	Chris Greyson	New York City / San Francisco	@Wearable / Advisor. http://t.co/ke23sJ3301 / Prior ECD. http://t.co/Q4pSuEbFu4 / Events: http://t.co/hqLtGO3G6o & http://t.co/GmiHcG38Ir
16931772	tristanhambling	New Zealand	Tracking future, tech, nano, bio, neuro, info stuff and anything new that scans past my event horizon. http://t.co/7aJFwAJkv7 also @fortuneseek
19004791	David Wood		Chair of London Futurists. Writer & consultant. PDA/smartphone pioneer. Symbian co-founder. Formerly at Psion and Accenture. Collaborative Transhumanist
15410587	Rod Furlan	USA	Artificial intelligence researcher, quant, Singularity University alum, Google Glass Explorer, serial autodidact, science lover & soon-to-be robot
23115748	h+ Magazine	USA	h+ Magazine covers technological, scientific, and cultural trends that are changing human beings in fundamental ways.
743915	David Orban	New York, NY	CEO, Dotsub / Advisor & Faculty, Singularity University. Analyzing and applying cycles of accelerating technological change. Flowing in wonderment.
19748200	Gizmag		I am a website about emerging technologies.
19722699	Popular Science		Science and technology news from the future! Tweets from @RossPastore
138222776	Neuralfuture		The future of life, humanity, and intelligence rests in the minds and hands of the innovators who we envision, guide, and build it.
594715367	Gianluca Robotics	Boston, MA, USA	Everything about consumer robotics, connected devices & IoT. Published by the first robotics investment company. Founder - @sdgrishin, feed editor - @Valery_Ko
66020845	Eric Topol	La Jolla, CA	Cardiologist, geneticist, digital medicine aficionado. Editor-in-Chief, Medscape, author of The Creative Destruction of Medicine
13808547	MIT Tech Review	Cambridge, MA	We identify important new technologies: deciphering their practical impact and revealing how they will change our lives.
10177570	Hizook.com	San Jose, USA	Robotics News for Academics & Professionals by Travis Deyle
44910688	Robot Magazine	Ridgefield, CT USA	The latest in hobby, science and consumer robotics.
16695266	ChiefRobot	Boston	Your daily dose of robots.
15164741	Robo-n-Wear		Clothing for humans, inspired by robots. Robot t-shirts, hats, polos and hoodies.
87468736	Eric Toro	Chicago, IL	Tweets about transhumanism, the singularity, AI, nanotech, biotech, robotics, life extension and human enhancement. All tweets and opinions are my own.
10351057	Willow Garage	Menlo Park, CA	Helping to revolutionize the world of personal robotics
67780032	Robert Oschler	Idaho	Artificial Intelligence and smart phone developer, currently focusing on speech recognition and natural language understanding applications and robotics.
18060713	robots_forever	Tokyo, Japan	Robot news, robotics research, combat and humanoid robot events, and other robot coverage from Japan.
81235022	Alexander Kruel	Germany	Transhumanist, atheist, vegetarian. Interested in math, programming, science fiction, singularity, philosophy, consciousness, the nature of reality.
22910080	Bob Spence Eyeborg	Toronto, Canada	We've built a wireless video camera eye. Tweets about privacy, cyborgs, prosthetics, eyepatches, Star Trek, The Bionic Man, and Augmented Reality
7786012	Transhumanists	New York, NY	Singularity, Transhumanism, Artificial Intelligence, Human Enhancement, Stem Cells, Nanotechnology, Renewable Energy
15784353	Sensium		Revolutionary body monitoring for healthcare: wireless, intelligent, continuous, low-cost
231102850	Popular Mechanics	New York City	The best in tech, science, aerospace, DIY and auto news. Customer Service: http://t.co/rYWTFWzg2R

Notes: Data extracted using the Twitter API in May 2014. Accounts can be freely accessed using http://twitter.com/intent/user?user_id=[insert here the twitter id]

Table 2. Network and community statistics over time

	2011, 2nd	2012, 1st	2012, 2nd	2013, 1st	2013, 2nd
N nodes	320	293	341	163	233
N edges	3,979	2,579	3,445	1,105	1,752
N communities	74	49	66	30	36
Max. community density	0.58	0.77	0.63	0.75	0.71
Max. nodes community	54	34	28	21	26

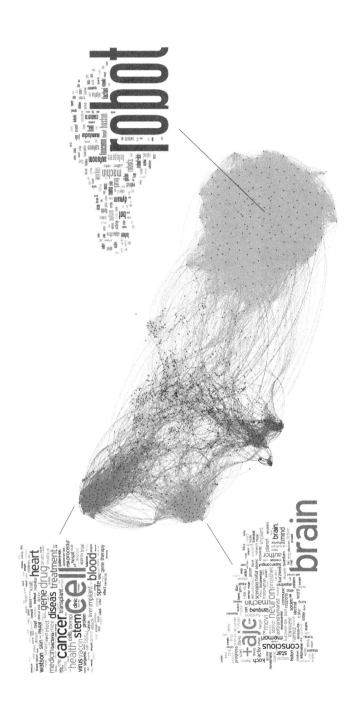

Fig. 7. Static Community Detection: Document similarity network of the whole corpus

Table 3. Network of Knowledge Fragments per Period after Overlapping Community Detection

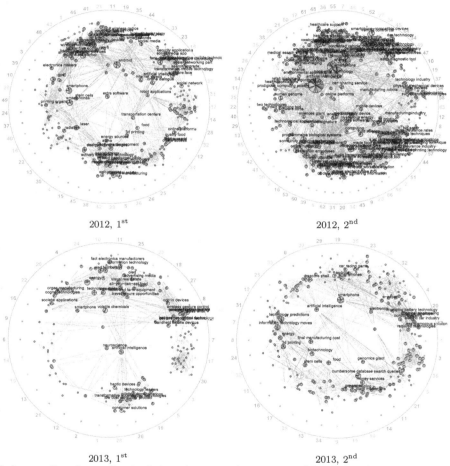

Nodes are aligned according to their main community, represented by the number outside the circle. Node size is scaled by number of communities the node belongs to. Multi-community membership is also indicated by multiple node color

Table 4. Exemplary identified technological fields and their knowledge fragments

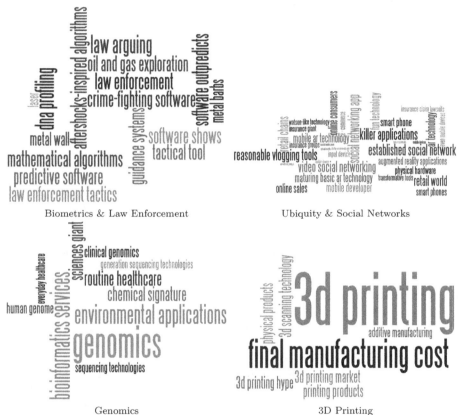

Biometrics & Law Enforcement

Ubiquity & Social Networks

Genomics

3D Printing

Nodes term representing the name of the technology fragment represented as tag-cloud. Size weighted by the nodes within community degree centrality.

Utilizing Microblog Data in a Topic Modelling Framework for Scientific Articles' Recommendation

Arjumand Younus[1,2], Muhammad Atif Qureshi[12], Pikakshi Manchanda[2], Colm O'Riordan[1], and Gabriella Pasi[2]

[1] Computational Intelligence Research Group, Information Technology, National University of Ireland, Galway, Ireland
[2] Information Retrieval Lab, Informatics, Systems and Communication, University of Milan Bicocca, Milan, Italy
{arjumand.younus,muhammad.qureshi,colm.oriordan}@nuigalway.ie,
{pikakshi.manchanda,pasi}@disco.unimib.it

Abstract. Researchers are actively turning to Twitter in an attempt to network with other researchers, and stay updated with respect to various scientific breakthroughs. Young and novice researchers have also found Twitter as a valuable source of information in terms of staying up-to-date with various developments in their field of research. In this paper, we present an approach to utilize this valuable information source within a topic modeling framework to suggest scientific articles of interest to novice researchers. The approach in addition to producing effective recommendations for scientific articles alleviates the cold-start problem and is a step towards elimination of the gap between Twitter and science.

1 Introduction

Social media services provide an important platform for people to express opinions, share ideas, receive updates relating to various topics of interest (e.g., science, sports, politics), and discover latest news. The famous microblog "Twitter" is one such social media service that has emerged as a significant medium of communication in the form of a social network where different resources are shared [14]. Researchers and scientists are actively turning to Twitter in order to engage in short scientific discussions, and spread scientific messages such as call for papers, research articles, and news about scientific breakthroughs [15]. As a result, young researchers are turning towards Twitter to connect with researchers within their field of interest in an attempt to stay up-to-date with the latest researches in a field [9].

Furthermore, young and novice researchers are faced with the enormous task of familiarizing themselves with the existing body of research literature in a given field. The complexity of this task increases due to information overload which is a consequence of an exponential growth in the rate of scientific publications every year [7]. For a novice researcher, it is challenging to find satisfactory results

L.M. Aiello and D. McFarland (Eds.): SocInfo 2014, LNCS 8851, pp. 384–395, 2014.

using the keyword search technique and therefore recent research literature has proposed recommendation systems for scientific articles to aid the researcher in organization of reading lists for a particular research field [6,7,17].

Most of the existing tools for scientific articles' recommendation utilize an existing pool of research articles to make further recommendations [3]; this paper however takes a different approach and utilizes Twitter activities of researchers to recommend scientific articles of interest. Given the modern tools of information access such as Twitter and in line with the proposal made by Letierce et al. [16], we aim to eliminate the boundary between experienced and young researchers by making use of the Twitter activities of young researchers to recommend to them scientific articles of their interest. Our system takes Twitter as a source to identify the research interests of a target user[1] via application of topic modeling over the tweets of the Twitter users that the target user is following, and on the list of titles of scientific articles. This step is followed by utilization of dominant topics discovered in the tweets to recommend scientific articles for the target user. The approach in addition to producing diverse recommendations is extensible in that it can also take into account other factors such as freshness of articles, popularity of venues, impact factor of authors etc. Experimental evaluations demonstrate the effectiveness of the approach as it outperforms a standard tf-idf based baseline.

The remainder of this paper is organized as follows. In Section 2 we review related work with a brief discussion on how we differ from existing works. In Section 3 we present the results of a survey conducted in order to analyze how and why early-stage researchers use Twitter. In Section 4 we explain the topic modeling recommendation framework in detail. In Section 5 we present the details of experimental evaluations with a description of the employed dataset and the recruited users who participated in our user-study. In Section 6 we conclude the paper with a discussion of future work.

2   Related Work

Our work touches various fields. In the following we review related work in scientific articles' recommendation along with works on analysis of researchers' activities on Twitter. Finally, we also provide an overview of recommendations systems built through utilization of Twitter data.

2.1   Recommending Scientific Articles

Existing academic search engines (such as Google Scholar, Microsoft Academic Search and CiteSeer) have failed to help researchers in the effective retrieval of scientific articles with the retrieved set of articles either being too large or too small [12]. Recommendation systems that generate a reading list of scientific articles have thereby emerged as a popular solution. The most popular approaches

---

[1] Here, target user refers to the researcher who wishes to get recommendation for scientific articles.

either use a set of papers as a query set or a corpus of papers relevant to a given area [20]. One of the earliest works by Woodruff et al. [23] uses a single paper for generation of a reading list through "spreading activation" over its text and citation data. El-Arini and Guestrin [7] use the notion of "influence" to capture the transfer of ideas as individual concepts among papers in the query set. Among the systems that utilize a large corpus of papers to extract core papers of a field, most utilize PageRank over the citation graph along with measures such as "download frequency", "citation count", and "impact factor" [5]. Finally, other techniques utilize collaborative filtering over research papers whereby the user-item ratings' matrix is obtained from the citation network between the papers [17] along with implicit behaviors extracted from a user's access logs [24]. Some works also use a hybrid approach that combines collaborative filtering with content-based filtering methods whereby content analysis is performed through probabilistic topic modeling [1,21]. We differ from the existing scholarly paper recommendation approaches in that we do not rely on an existing set of research papers and hence, the underlying citation network is not taken into consideration in our approach. Instead, we rely on scientific tweets that researchers post on Twitter.

2.2 Analysis of Researchers on Twitter

Recently, researchers have started investigating academic activities on Twitter. The earliest works use scientific tweets as a new measure for citation analysis where citation is defined as a tweet containing a URL to a peer-reviewed scientific article [8,19,15,22]. A more recent work by Hadgu and Jaschke [9] proposes a classification method to construct a directory of computer scientists on Twitter. Their approach starts from a seed set of Twitter accounts from which further Twitter accounts are derived and passed through a machine learning classifier that classifies the Twitter account into researcher or non-researcher. The work by Hadgu and Jaschke [9] can have potentially useful applications from the viewpoint of decreasing the gap between Twitter and science.

2.3 Twitter-Based Recommendation Systems

Over the past few years, recommendation systems technology has made significant progress with a number of recommendation systems built on top of Twitter data. As an example, one of the earliest recommendation systems by Phelan et al. (which we use as a baseline in our work) suggests a method that promotes news stories from a user's favorite RSS feeds based on Twitter activity through application of a content-based recommendation technique by mining terms from both the RSS feeds and the Twitter messages [18]. Other works mine data from Twitter to better suggest people to follow on Twitter [10,11] along with suggesting tweets of interest to a user [4]. To the best of our knowledge, none of the works that utilize Twitter for recommendations have focused on making use of scientists' Twitter data for recommendation of scientific articles; and our work is the first step in this direction.

3 Twitter Usage by Researchers

The underlying intuition behind the use of Twitter data for scholarly paper recommendations is common knowledge that scientists while attending conferences,
and/or while conducting experiments tweet about their experiences and these
tweets in turn can serve as a rich source for inferring research interests [9,16].
We set up an online survey to analyze the habits and motivations of early-stage
researchers with respect to their use of Twitter. We advertised the survey via
university mailing lists and social media services particularly targeting PhD students in early stages of their career and who had an active Twitter account. Our
main motivation was to observe the main motivations behind young researchers'
use of Twitter and whether or not they consider it as a valuable resource when
it comes to staying up-to-date about latest researches in a particular field. This
section presents details of the undertaken survey.

We received 280 responses distributed as follows: 65% were PhD students,
10% MSc students, 8% research assistants, 12% postdoctoral researchers, and
5% lecturers. The average number of years using Twitter among our respondents is 2.57 years. One outcome of this survey is that 93% of the respondents
use Twitter to stay up-to-date about latest research developments in their respective fields. Other uses of Twitter involved sharing knowledge about their
field of expertise and communication about their research projects; however,
these goals were shared by the senior researchers among our respondents with
the early-stage researchers mainly using it for learning about new researches
through the activity of following other researchers. Finally, 87% of our respondents follow approximately 50-100 researchers on Twitter with the average being
67.2 followed researchers per respondent and 69% consider it as highly beneficial
in terms of staying up-to-date with latest research in their fields..

Based on the findings of this initial survey of early-stage researchers we observe
that most early-stage researchers are turning to Twitter for discovering experts
with research interests similar to theirs and hence, their activities on an open
medium such as Twitter can be utilized towards the recommendation of scientific
articles. The following sections describe the proposed framework in detail.

4 Methodology

This section describes the proposed recommendation framework in detail. We
start with an overview of the processing pipeline and then explain the components in detail. The processing pipeline (see Figure 1) comprises the following
steps:

1. We follow a content-based filtering strategy in which we utilize tweets of
   Twitterers that a particular user follows (from this point on we refer to
   these followed Twitterers as followees) along with titles of scientific articles.
2. We apply a topic modeling algorithm over the tweets and titles simultaneously which is then used to rank all the followees of a user in order to produce
   lists of top-k researchers.

3. We then use the tweets of top-k researchers to discover the top-n topics from among which to generate the recommendations.
4. Finally, the tweets and paper titles corresponding to the top-n topics are utilized in a language modeling scoring function to generate the final set of recommended papers.

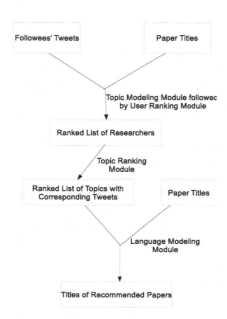

Fig. 1. Process Flow of our Recommendation Framework

4.1 Ranking Framework Using Twitter-LDA

The first step involves use of an LDA-based model in order to obtain the topics from tweets of all the followees of a target user and the paper titles of scientific articles. It is an unsupervised machine learning technique that discovers latent topics from a corpus. We use the model proposed by Zhou et al. namely Twitter-LDA [25] which is basically an author-topic model built upon following assumptions:

- There is a collection of 'K' topics in Twitter with each topic represented by a word distribution.
- Each user's interests are modeled through a distribution over topics.

- When writing a tweet, a user may choose to write a background word or he/she may choose a topic based on his topic distribution which may then lead to the choice of a word based on the word distribution of the chosen topic.

Twitter-LDA differs from the original LDA framework by Blei et al. [2] in that a single tweet is assigned a single topic instead of a distribution over topics. This is more suited to the task at hand as researchers' when tweeting about their research are very specific and focused, and mostly restricted to one topic. We apply the Twitter-LDA algorithm simultaneously on the followees' tweets and the paper titles with the number of topics set to 200^2. Our aim is to filter out and produce a ranking of those followees of a user who are involved in scientific research. To this aim, we utilize the intersection of topics found in both paper titles and followees' tweets. Each followee of a user is ranked as follows:

$$Rank_{followee} = (\sum_{t \in Topics_p} \frac{n(t, T_f)}{|T_f|}) * |Topics_p \cap Topics_f| \qquad (1)$$

where T_f denotes all tweets by a followee, $Topics_p$ denotes the set of topics defining the titles of scientific articles, $Topics_f$ denotes the set of topics defining the tweets of a followee and $n(t, T_f)$ the number of times a particular topic 't' from within $Topics_p$ occurs among the tweets of a followee. Based on the ranking scores of all followees of a particular user, we obtain top-k researchers followed by a target user. The next step involves using the topics from within tweets of these top-k researchers to find the top-n topics of interest to the target user so as to recommend scientific articles from within those topics.

The scoring framework for topics involves summing up scores for each topic and discovering the dominant topics. Note that we utilize the topics of the set $Topics_p$ from within the topics of top-k researchers and this helps avoid noisy topics in the recommendation process. We first determine a score for each topic using the following:

$$Score_{topic} = (\sum_{t \in Topics_p} \frac{n(t, T_r)}{|T_r|}) \qquad (2)$$

where T_r denotes all tweets by a followed researcher (determined using equation 1), and as before $Topics_p$ denotes the set of topics defining the titles of scientific articles. We use the topic score $Score_{topic}$ of each topic to compute a final score corresponding to each topic. We illustrate the process with the help of the example in Table 1.

---

[2] This number is determined empirically after determining the number which clearly distinguishes topics of tweets and paper titles.

Table 1. Example to Illustrate Ranking of Topics Related to Titles of Scientific Articles

Users	Topics					
	t_3	t_5	t_6	t_7	t_8	t_{10}
u_1	0.13	0.34	0.16	0	0.53	1.13
u_2	0	0.5	0	0.34	0.68	0.43
u_3	0.4	0.12	0.45	0.73	0	0
u_4	0	0.11	0.92	0.22	0	0.64
u_5	0.23	0	0	0.17	0.25	0.55
u_6	0	0.2	0.18	0	1.21	0
u_7	0	0.23	0.38	0.15	0.78	0
u_8	0.48	0	0.14	0.67	0	0.98
u_9	0.19	0	0	0.17	0.93	0
u_{10}	0	0.47	0	0	0.37	0.74

For the sake of understanding the example in Table 1, assume a total of 10 researchers (i.e., k of top-k researchers equalling 10) and 15 topics t_1-t_{15} with t_3, t_5,t_6,t_7, t_8 and t_{10} belonging to the set $Topics_p$. The scores for these topics are combined to produce a final ranking for the topics as shown in Table 2; and scientific articles corresponding to these topics are recommended in proportion to the contribution of each topic's score.

Table 2. Final Scores Assigned to Each Topic from within $Topics_p$

	Topics					
	t_3	t_5	t_6	t_7	t_8	t_{10}
Topic Scores	1.43	1.97	2.23	2.45	4.75	4.47

For the sake of continuing with this example we take 'n' to be 3 and hence, continue with top-3 topics from $Topics_p$ (i.e., t_8, t_{10} and t_7). Note that the incorporation of different topics at this stage enables the generated recommendations to be diverse.

4.2   Scoring Framework Using Language Models of Followed Scientific Experts

To the purpose of scoring each scientific article, we use a language modelling approach to compute the likelihood of generating an article a from a language model estimated from a user's Twitter followees as follows:

$$P(u)_{t_i}(a/T) = \prod_{a \in A} P(w \mid T)^{n(w,a)} \tag{3}$$

where w is a word in the title of articles corresponding to topic t_i (in our example t_i would be t_8, t_{10} and t_7 from among top-3 topics), $n(w,a)$ the term frequency of w in a, and u is the user for whom we want to generate the recommendations. Here, T is used to represent the uniform mixture of the Twitter model of researchers followed by a target user as follows:

$$P(w \mid T) = P(w \mid T_r) \qquad (4)$$

Here, T_r denotes the tweets by the researchers whom the user u follows corresponding to topic t_i. The Twitter model T_r can be estimated as:

$$P(w \mid T_r) = \frac{1}{|T_r|} \sum_{t \in T_r} P(w \mid t) \qquad (5)$$

The constituent language model for T_r are a uniform mixture of the language models of researchers' tweets' corresponding to topic t_i and employing Dirichlet prior smoothing:

$$P(w \mid t) = \frac{n(w,t) + \mu \dfrac{n(w, coll)}{|coll|}}{|t| + \mu}$$

where $n(w,.)$ denotes the frequency of word w in (.), *coll* is short for collection which refers to all tweets by top-k researchers, and |.| is the overall length of the tweet or the collection.

Note that equation 3 can be modified to include various factors such as freshness score of an academic article (i.e, a measure based upon year of publication), impact factor of venues and/or impact factor of authors. The content-based filtering strategy within our model enables it to be extensible and flexible in addition to being able to produce diverse recommendations. Moreover, the proposed model alleviates the cold-start problem commonly encountered in the recommendation systems' domain whereby user ratings for items to be recommended are not available; in this case however, the tweets by the followees of a target user serve as the starting point.

5   Experimental Evaluations

In this section we describe our experimental evaluations that demonstrate the effectiveness of of our proposed approach. We first describe the dataset of recruited users along with the dataset of scientific articles followed by details of experimental results.

5.1   Experimental Setup

Dataset: We recruited 64 active Twitter users with permission to use their Twitter data for the purpose of experimental evaluations. Using the Twitter API, we obtained the tweets of all their followees. Table 3 shows some basic statistics about the dataset. The titles of scientific articles are gathered by application of focused crawling to DBLP using the boilerpipe API [13]. A total of 50,252 titles were fetched from a record of various Computer Science conferences and journals from within diverse research fields such as databases, embedded systems, graphics, information retrieval, networks, operating systems, programming languages, software engineering, security, user interface, and social computing.

Table 3. Statistics about Employed Twitter Dataset

Average No. of Followees per User	237
Maximum No. of Followees	1022
Minimum No. of Followees	54
Average Tweets per Followee	508
Total Tweets in Collection	32,518

Parameters and Evaluation Measures: For the purpose of our experimental evaluations, we set 'k' described in Section 4.1 to 30, 60 and 90 respectively i.e., we use top-30, top-60 and top-90 researchers followed by a user for generating his/her list of scientific articles' recommendation. The number of topic 'n' of Section 4.1 is set to 15. As in standard information retrieval, top ranked documents are the most important since users often scan just the first ranks and hence, each user was asked to mark as relevant or irrelevant the top-20 articles recommended to him/her. We evaluated our recommender system using Mean Average Precision (MAP), Mean Reciprocal Rank (MRR) and Precision @ 10 (P@10)[3]. The content-based filtering strategy based on tf-idf by Phelan et al. is used as a baseline to compare the effectiveness of our recommendation model. Note that we modify Phelan et al.'s algorithm to include Twitter messages of the followees in order to ensure a fair comparison.

5.2   Experimental Results

We evaluate the performance of our proposed recommendation model using the relevance judgements obtained for the 64 users. Table 4 shows the experimental results i.e. MAP, MRR and P@10 values for our approach with the different parameter settings for 'k' and for the approach by Phelan et al.; we use student's t-test to verify the soundness of our evaluations and the results corresponding

---

[3] Note that we treat each user as a separate query.

to our model are statistically significant with $p < 0.05$. We report the results together across the judgements for all 64 users.

Our recommendation model built on top of a topic modeling framework is able to outperform the tf-idf baseline and this is due to terms introducing a significant amount of noise when recommending scientific articles. On the other hand, topics tend to be clean and better representative of research interests of a novice researcher. The model with parameter 'k' set to 60 outperforms all the other versions and intuitively this makes sense due to a limited amount of researchers the user is actually interested in (as the survey from Section 3 shows that an average of 67.2 researchers are followed by a particular user).

Table 4. Comparison of Retrieval Performance for our Proposed Personalization Model

Chosen	Measures		
Algo	MAP	MRR	$P@10$
top-30	0.461	0.667	0.512
top-60	0.651	0.878	0.681
top-90	0.511	0.728	0.643
Phelan et al.	0.384	0.528	0.496

6   Conclusions and Future Work

In this paper we demonstrated how Twitter can be used to eliminate the boundary between young and experienced researchers by taking advantage of tweets of those researchers that an early-stage researcher follows. We proposed a content-based recommendation model that makes use of a topic modeling algorithm specifically suited for short content such as tweets. The model is able to incorporate a diverse range of topics to produce the final recommendations and can be extended to include various factors such as an article's recency, a venue's or authors' impact factor.

To the best of our knowledge, this is the first work on scientific articles' recommendation that relies on a fully content-based strategy with other works making use of citation graphs for the collaborative filtering step. As future work, we aim to incorporate different measures from the Twitter graph to compute rankings for the followed researchers in addition to including various components from the citation graph.

References

1. Agarwal, D., Chen, B.-C.: flda: Matrix factorization through latent dirichlet allocation. In: Proceedings of the Third ACM International Conference on Web Search and Data Mining, WSDM 2010, pp. 91–100. ACM, New York (2010)
2. Blei, D.M., Ng, A.Y., Jordan, M.I.: Latent dirichlet allocation. The Journal of Machine Learning Research 3, 993–1022 (2003)

3. Bollacker, K.D., Lawrence, S., Giles, C.L.: Discovering relevant scientific literature on the web. IEEE Intelligent Systems and their Applications 15(2), 42–47 (2000)
4. Chen, K., Chen, T., Zheng, G., Jin, O., Yao, E., Yu, Y.: Collaborative personalized tweet recommendation. In: Proceedings of the 35th International ACM SIGIR Conference on Research and Development in Information Retrieval, SIGIR 2012, pp. 661–670. ACM, New York (2012)
5. Chen, P., Xie, H., Maslov, S., Redner, S.: Finding scientific gems with google's pagerank algorithm. Journal of Informetrics 1(1), 8–15 (2007)
6. Ekstrand, M.D., Kannan, P., Stemper, J.A., Butler, J.T., Konstan, J.A., Riedl, J.T.: Automatically building research reading lists. In: Proceedings of the Fourth ACM Conference on Recommender Systems, RecSys 2010, pp. 159–166. ACM, New York (2010)
7. El-Arini, K., Guestrin, C.: Beyond keyword search: Discovering relevant scientific literature. In: Proceedings of the 17th ACM SIGKDD International Conference on Knowledge Discovery and Data Mining, KDD 2011, pp. 439–447. ACM, New York (2011)
8. Eysenbach, G.: Can tweets predict citations? metrics of social impact based on twitter and correlation with traditional metrics of scientific impact. Journal of Medical Internet Research 13(4) (2011)
9. Hadgu, A.T., Jäschke, R.: Identifying and analyzing researchers on twitter. In: Proceedings of the 2014 ACM Conference on Web Science, WebSci 2014, pp. 23–32. ACM, New York (2014)
10. Hannon, J., Bennett, M., Smyth, B.: Recommending twitter users to follow using content and collaborative filtering approaches. In: Proceedings of the Fourth ACM Conference on Recommender Systems, RecSys 2010, pp. 199–206. ACM, New York (2010)
11. Hannon, J., McCarthy, K., Smyth, B.: Finding useful users on twitter: Twittomender the followee recommender. In: Clough, P., Foley, C., Gurrin, C., Jones, G.J.F., Kraaij, W., Lee, H., Mudoch, V. (eds.) ECIR 2011. LNCS, vol. 6611, pp. 784–787. Springer, Heidelberg (2011)
12. Huang, S., Wan, X.: AKMiner: Domain-specific knowledge graph mining from academic literatures. In: Lin, X., Manolopoulos, Y., Srivastava, D., Huang, G. (eds.) WISE 2013, Part II. LNCS, vol. 8181, pp. 241–255. Springer, Heidelberg (2013)
13. Kohlschütter, C., Fankhauser, P., Nejdl, W.: Boilerplate detection using shallow text features. In: Proceedings of the Third ACM International Conference on Web Search and Data Mining, pp. 441–450. ACM (2010)
14. Kwak, H., Lee, C., Park, H., Moon, S.: What is twitter, a social network or a news media? In: Proceedings of the 19th International Conference on World Wide Web, WWW 2010, pp. 591–600. ACM, New York (2010)
15. Letierce, J., Passant, A., Breslin, J., Decker, S.: Understanding how Twitter is used to widely spread Scientific Messages. In: Proceedings of the WebSci10: Extending the Frontiers of Society On-Line (March 2010)
16. Letierce, J., Passant, A., Breslin, J.G., Decker, S.: Using twitter during an academic conference: The #iswc2009 use-case. In: ICWSM (2010)
17. McNee, S.M., Albert, I., Cosley, D., Gopalkrishnan, P., Lam, S.K., Rashid, A.M., Konstan, J.A., Riedl, J.: On the recommending of citations for research papers. In: Proceedings of the 2002 ACM Conference on Computer Supported Cooperative Work, CSCW 2002, pp. 116–125. ACM, New York (2002)
18. Phelan, O., McCarthy, K., Smyth, B.: Using twitter to recommend real-time topical news. In: Proceedings of the Third ACM Conference on Recommender Systems, pp. 385–388 (2009)

19. Priem, J., Hemminger, B.H.: Scientometrics 2.0: New metrics of scholarly impact on the social web. First Monday 15(7) (2010)
20. Tang, Y.: The design and study of pedagogical paper recommendation (2008)
21. Wang, C., Blei, D.M.: Collaborative topic modeling for recommending scientific articles. In: Proceedings of the 17th ACM SIGKDD International Conference on Knowledge Discovery and Data Mining, KDD 2011, pp. 448–456. ACM, New York (2011)
22. Weller, K., Dröge, E., Puschmann, C.: Citation analysis in twitter: Approaches for defining and measuring information flows within tweets during scientific conferences. In: MSM, pp. 1–12 (2011)
23. Woodruff, A., Gossweiler, R., Pitkow, J., Chi, E.H., Card, S.K.: Enhancing a digital book with a reading recommender. In: Proceedings of the SIGCHI Conference on Human Factors in Computing Systems, CHI 2000, pp. 153–160. ACM, New York (2000)
24. Yang, C., Wei, B., Wu, J., Zhang, Y., Zhang, L.: Cares: A ranking-oriented cadal recommender system. In: Proceedings of the 9th ACM/IEEE-CS Joint Conference on Digital Libraries, JCDL 2009, pp. 203–212. ACM, New York (2009)
25. Zhao, W.X., Jiang, J., Weng, J., He, J., Lim, E.-P., Yan, H., Li, X.: Comparing twitter and traditional media using topic models. In: Proceedings of the 33rd European Conference on Advances in Information Retrieval, ECIR 2011, pp. 338–349. Springer, Heidelberg (2011)

Mining Mobile Phone Data to Investigate Urban Crime Theories at Scale

Martin Traunmueller[1], Giovanni Quattrone[2], and Licia Capra[1]

[1] ICRI Cities, Dept. of Computer Science, University College London
[2] Dept. of Computer Science, University College London
Gower Street, WC1E 6BT, London, UK
{martin.traunmueller.11,g.quattrone,l.capra}@ucl.ac.uk

Abstract. Prior work in architectural and urban studies suggests that there is a strong correlation between people dynamics and crime activities in an urban environment. These studies have been conducted primarily using qualitative evaluation methods, and as such are limited in terms of the geographic area they cover, the number of respondents they reach out to, and the temporal frequency with which they can be repeated. As cities are rapidly growing and evolving complex entities, complementary approaches that afford social scientists the ability to evaluate urban crime theories at scale are required. In this paper, we propose a new method whereby we mine telecommunication data and open crime data to quantitatively observe these theories. More precisely, we analyse footfall counts as recorded by telecommunication data, and extract metrics that act as proxies of urban crime theories. Using correlation analysis between such proxies and crime activity derived from open crime data records, we can reveal to what extent different theories of urban crime hold, and where. We apply this approach to the metropolitan area of London, UK and find significant correlations between crime and metrics derived from theories by Jacobs (e.g., population diversity) and by Felson and Clarke (e.g., ratio of young people). We conclude the paper with a discussion of the implications of this work on social science research practices.

Keywords: Urban crime, telecommunication data, open data, data mining.

1 Introduction

In modern society we are experiencing two phenomena: on one hand, there is a rapid population shift of people moving from rural areas into urban environments, with an annual growth of 60 million new city dwellers every year [29]. On the other hand, crime activities are on the rise (e.g., [5]), especially in densely populated areas [13]. Being able to understand and quantify the relationship between people presence and crime activity in an area has thus become an important concern, for both citizens, urban planners and city administrators.

The relationship between *people dynamics* and *crime* in urban environments has been researched extensively in architectural and urban studies over the last decades, with theories that sometimes appear to conflict with each other. Most influential theories lead back to the 1960's and 1970's: Jacobs [12] suggests that population diversity, activity

L.M. Aiello and D. McFarland (Eds.): SocInfo 2014, LNCS 8851, pp. 396–411, 2014.

and a high mix of functions lead to less crime for an area, whereas Newman [15] hypothesizes the opposite, supporting clear separation of public, semi-public and private areas towards urban safety. Each theory has been evaluated, and indeed supported, by means of qualitative research methods that enable in-depth investigations into the reasons behind certain phenomena. However, such methods are very expensive and time-consuming to run, so that studies are usually restricted to a rather small number of people (relative to the overall urban population) and constrained geographic areas (e.g., a neighbourhood); furthermore, they are almost never repeated over time, to observe potential changes. It becomes thus very difficult to collect sufficient evidence to explain under what conditions a certain theory holds.

In this paper we propose a new method to quantitatively investigate urban crime theories at scale, using open crime data records and anonymised mobile telecommunication data. From the former, we extract quantitative information about crime activity, as it happens across different urban areas of very fine spatial granularity. From the latter, we extract metrics that act as proxies for previously developed urban crime theories that link people presence in an area with crime. We then use correlation analysis between crime data and our defined metrics to validate urban crime theories at scale. We apply this method to data obtained for the city of London, UK, and find that, in this city and at the present time, Jacobs' theory of 'natural surveillance' [12] holds: we discover that age diversity, as well as the ratio of visitors in a given area, are significant and negatively correlated with crime activities; furthermore, Felson and Clarke theory [9] that links a higher presence of young people with higher crime is also confirmed. We believe the proposed method to be a powerful tool in the hands of social science researchers developing urban crime theories, as they can now complement qualitative investigations with quantitative ones: while the former afford them deep insights into the causality of certain phenomena, the latter afford them the ability to scale up findings in terms of population reach, geographical spread, and temporal evolution.

The remainder of the paper is structured as follows: we first provide a brief overview on background theories from architectural and criminological studies, and state-of-the-art follow-up research that has been grounded on them. We then present our method, in terms of the datasets we leverage, the pre-processing and data manipulation we have conducted, and the metrics we have extracted as proxies for urban crime theories. We discuss the results obtained when applying our method to data for the city of London, UK, and finally conclude by discussing implications, limitations and future steps.

2 Related Work

2.1 Background

Most well known architectural theories about the relationship between people dynamics, the urban environment and crime lead back to the studies of Jacobs [12] and Newman [15], with two different schools of thought. Jacobs [12] defines urban population as 'eyes on the street', a natural policy mechanism that supports urban safety through 'natural surveillance'. An open and mixed use environment supports this concept by enabling diversity and activity within the population using the area at different times. While Jacobs suggests that a high diversity among the population and a high ratio of visitors are contributing to an area's safety, Newman [15] argues the opposite. According to his theory, diversity and a high mix of people create the anonymity it needs for crime to take place. Newman suggests that a clear definition of public, semi-public and private space in a low dense and single use urban environment creates a 'defensible space' that is needed to support safety. Newman further argues that low population diversity, low visitor ratio and a high ratio of residents are contributing to an area's safety. Follow-up studies have tried to shed light onto these apparently conflicting theories. For instance, Felson and Clarke [9] have proposed the 'Routine Activity Theory', that studies people dynamics and crime in relation to specific points of interest; they have found that venues such as bars and pubs attract crime by pulling strangers into an area; the presence of middle aged women on the streets detracts crime instead.

These theories suggest different ways to design the built environment so to take advantage of the resulting social control of crime. But which one applies *where*, and also *when*? How do we know that theories developed in the '60s and '70s are still valid fifty years afterwards? To gain a deeper understanding of the context within which a certain theory holds, social science research needs a novel way to validate urban crime theories, that scales up in terms of the geographic urban area under exam, the population sample captured, and the frequency with which studies can be repeated.

2.2 Computational Science and Crime

In recent years, open data movements have made available large repositories of crime data to the public. These circumstances have been useful to start studying crime in a more systematic manner. Data mining has become a popular tool for crime research to detect crime patterns in an urban environment. Recorded crime data has been extensively mined to identify crime hotspots within a city [16,27,3,8], and can even be used for crime predictions [4]. These methods are capable of signaling where crime will happen; however, they do not shed light into possible *reasons* for incidents.

Recent architectural and urban design research has attempted to describe the relationship between the built environment and crime. Wolfe and Mennis [30] discuss the influence of green space in relation to crime, by using satellite images to detect green urban spaces and compare them to recorded crime data. Findings show clearly that well maintained green spaces contribute to less crime through an increased community activity and supervision, as also originally suggested by Jacobs. Hillier and Shabaz [18] investigate the relationship between street crime occurrences and the spatial layout of the street network for a London borough. Findings show an overall higher crime distribution along main roads compared to side roads, with the ratios changing throughout the day. These works show that there is a strong relationship between the built environment and location of crime. However, the findings above also point to the fact that there is a third and important dimension to the problem: people's dynamics. The very same built environment is appropriated and used by different people for different purposes and in different ways throughout the day. People dynamics thus need to be quantitatively explored in relation to crime too.

When it comes to analysing crime in relation to people, social and criminological research often uses census data. For instance, Tan and Haining [23] use spatial data of crime and census data to explore the impact of crime on population health for the city of Sheffield, UK. Song and Daqian [22] explored relationships between spatial patterns of property crime and socio-economic variables of a neighbourhood. Christens and Speer [6] use census data to explore the relationship between crime and population density, following Jacob's hypothesis that high population density would predict reduced violent crime; they found the hypothesis to be true for densely populated urban areas, but failed in suburban areas where population is less dense.

While shedding light into some important relationships between crime and demographics, census data is limited, in that it only offers a static image of the city (i.e., where people reside), without disclosing where people actually spend time throughout the day. Furthermore, census data is only collected every few years, so the information it provides may become quickly stale, especially for areas undergoing massive urbanization processes. According to Jacobs and Newman, it is these people dynamics that have great impact on the crime activities of a place which change steadily over time and space, so that we cannot use census data to analyse them.

People dynamics have started to be inferred from geo-located social networks, and used for different purposes. For instance, Prasetyo et. al. [17] use Twitter and Foursquare data to analyse the impact of major natural disasters on people; they do so for haze events in Singapore, and discuss how their approach can help both the private and public sector to better prepare themselves to similar future events. Wakamiya et. al. [26] use geo-located Twitter data to examine crowd interactions, from which social neighbourhood boundaries are defined, thus expanding upon the traditional concept of spatial, administratively-defined neighbourhoods. Discussing crime, Wang et. al. [28] use sentiment analysis to relate the content of Twitter messages to hit-and-run crime activity and demonstrate a high usability for crime prediction. Social media is a rich data source from which to derive information about people dynamics; however, it is also unrepresentative of the whole urban population, because of high bias in its adoption [2]. An alternative data source that can be used to mine people dynamics in urban areas, and that is subject to significantly lower bias than social media, is telecommunication data.

Telecommunication data has been recently used to understand the relationship between cities (and even whole countries) and socio-economic deprivation, both in the developed world [7] and in developing countries [20]. In relation to crime, recent work [1] uses a similar mobile phone data set as used in this paper in combination with census data to predict crime activity for urban areas of London. As results show the importance of variables extracted from the mobile phone data set predicting almost 70% of the cases when included, they underline the importance of people diversity in relation to crime activity in an area as described by Jacobs [12]. Focusing less on the predictive and more on the descriptive aspect, we believe the same data can be used to understand other established theories as well, as we will show next.

3   Method

In this section, we describe the method we propose to quantitatively explore previous architectural theories of urban crime. We start with a brief description of our datasets; we then present the pre-processing steps these datasets underwent, and finally elaborate on the metrics we extracted from them as proxies for urban crime theories.

Table 1. Record sample of mobile phone data, showing the number of people per area, per hour

Date	Time	Grid ID	Total	Home	Work	Visit	Male	Female	0–20	21–30	31–40	41–50	51–60	60+
10/12/2012	9:00:00	1122...	430	110	290	30	240	190	0	80	90	120	100	40
10/12/2012	10:00:00	2412...	910	210	160	540	520	390	0	180	180	260	170	120
10/12/2012	11:00:00	1092...	900	570	250	80	520	380	10	160	190	250	210	80
10/12/2012	12:00:00	2124...	690	80	120	490	410	280	10	120	150	190	140	80

Table 2. Record sample of open crime data, showing crime incidents, geo location and crime type

Crime ID	Month	Reported by	Lon	Lat	Location	LSOA Code	Crime Type
df0c4...	2012-12	Met Police	-0.219	51.568	near Clitterhouse Rd	E010...	Burglary
0f9a5...	2012-12	Met Police	-0.217	51.565	near Caney Mews	E010...	Burglary
62235...	2012-12	CoL Police	-0.221	51.570	near Claremont Way	E010...	Crim. damage & arson
194ed...	2012-12	CoL Police	-0.222	51.563	near Petrol Stn	E010...	Crim. damage & arson

3.1 Dataset Description

The method we propose requires access to two types of datasets: one providing information about people dynamics, and with information about crimes. For the purpose of this study, we chose datasets that cover the city of Greater London, UK. We did so as London represents a large and complex metropolitan city, composed of many different neighbourhoods, each with its own distinguishing characteristics in terms of built environment, demographics, and people dynamics. It thus represents a case where qualitative approaches to investigate urban crime theories would not scale, both because of the geographic span of the areas to study, and because of the time frequency with which one may wish to repeat these studies (e.g., to observe changes in relation to ongoing immigration processes [21]).

People dynamics. We use anonymised and aggregated data collected and made available by a mobile telecommunication provider in context of a data mining challenge with a 25% penetration in the UK. The dataset contains 12,150,116 footfall count entries for the Metropolitan Area of London for the course of 3 weeks in December 2012/January 2013. The geographic area is divided by the data provider itself into 23,164 grid cells of varying size: for the more densely populated areas within inner London, a grid size is about by 210×210 meters, while for the less densely areas of Greater London, the grid size increases to about 425×425 meters. For each cell, footfall counts are given on a per hour basis over the three week period, further broken down by gender (number of males/females), by type (number of residents, workers, visitors) and by age group. Table 1 shows a sample of our mobile phone dataset.

Crime data. We use open crime data records[1], which, for the area of Greater London, are made available by two authorities: the Metropolitan Police and the City of London Police. These records provide information about the reporting police district, the exact location (longitude and latitude) of the crime, the name and area code of the crime, and the crime type (which the UK police differentiates into 10 categories: i.e. burglary, drugs, robbery, shoplifting, etc.). Unfortunately, no timestamp is given of when the crime took place/was reported, and the only temporal information we have is the month

---

[1] Open–source crime data: `http://data.police.uk`. June, 2014

during which it took place. We thus collected crime data for the months of December 2012 and January 2013 (to temporally match our mobile phone data), and retrieved 83,526 recorded crimes in total. Table 2 shows a sample of our crime data set.

3.2 Data Pre-Processing

We first cleansed the telecommunication data, so to remove inconsistent entries (i.e., footfall count per area different from the sum of footfall counts broken down by gender, type or age). We further pruned grid cells that fell outside the Greater London area. This caused 1.8% of the raw telecommunication data to be removed.

In order to correlate people dynamics and crime data within an urban environment over time, we then needed to define a common spatio-temporal unit of analysis for both datasets. In terms of *spatial* unit of analysis, we operated at the level of grid cells defined by the telecomm operator. As mentioned before, these are rather fine-grained cells, varying from 210×210 meters for inner London, to 425×425 meters for outer London. As crime data is recorded in terms of latitude/longitude coordinates, the spatial association of crime data to grid cells was straightforward. For each grid cell, we can thus count the total number of crimes that took place there; we also break down such counter by crime type, distinguishing *street crime*, covering crime most likely happening on the streets (e.g., antisocial behavior, drugs, robbery and violent crime – a total of 47,238 entries), and *home crime*, including crime types happening most likely indoors (e.g., on burglary, criminal damage and arson, other theft and shoplifting – a total of 36,288 entries). In terms of *temporal* unit of analysis, we needed to align telecomm data, captured at hour-level unit of analysis, with crime data, captured at month-level unit of analysis. To do so, we computed average footfall counts per area per month; to reduce variance, we aggregated separately day-time hour slots (8AM-8PM) and night-time hour slots (8PM-8AM), as well as weekdays vs. weekends. For each grid area, we thus ended up with four footfall count averages. As subsequent correlation analysis results did not show significant differences across these four aggregation values, we will report results for the weekday/daytime case only. Having cleansed the data and defined a common spatial and temporal unit for analysis, we are now able to define the metrics we will use in our quantitative analysis.

3.3 Hypotheses and Metrics

Crime Count and Crime Activity. To begin with, we need to quantify crime per spatio-temporal unit of analysis. For each area i, we consider two complimentary metrics: crime count $CC(i)$, and crime activity $CA(i)$. The former simply counts the number of crimes that have taken place in area i; since most of the areas under study have comparable size, we may consider $CC(i)$ as a way of measuring crime normalized by area size. Areas have similar sizes, but not similar population density. To investigate possible differences caused by population density, we use $CA(i)$ to quantify crime normalized by population density instead; we can consider this metric as an indicator of the probability of being victim of a crime. We can compute crime activity $CA(i)$ by dividing the number of crimes in an area $CC(i)$ by the estimated population $P(i)$

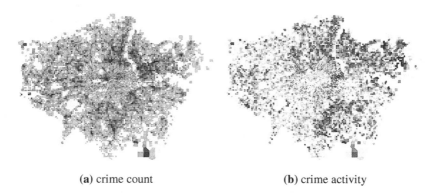

(a) crime count   (b) crime activity

Fig. 1. Choropleth maps showing crime count CC (left) and crime activity CA (right) all over Greater London for Dec 2012-2013, where the darker the shade of blue, the higher the crime rate in that area

present in area i. The number of crimes per area $CC(i)$ is ready available in our pre-processed crime dataset; as for the number of people present in the area, we considered all people present in area i in the 3 weeks covered by our phone call dataset. Since the crime dataset and telecommunication dataset covered different timespans (8 weeks for the former, 3 weeks for the latter), we multiplied by 3/8 so to have the average number of crimes per person in one week:

$$CA(i) = 3/8 \cdot \frac{CC(i)}{P(i)}$$

Figure 1 shows the spatial distribution of crime count and crime activity over Greater London (the darker the shade of blue, the higher the $CC(i)$ and $CA(i)$ values). As shown, crime count $CC(i)$ is found to be higher in the centre of London, with some other hotspots spread out all over the city (Figure 1a), whereas crime activity $CA(i)$ (that is, crime count normalised by people present in that area) is much higher outside inner London (Figure 1b). Having defined a metric that captures crime per spatio-temporal unit of analysis, we next define metrics that act as proxies for urban crime theories linking people dynamics with crime count and crime activity. We have a total of six metrics and associated hypotheses ($H1$ to $H6$).

H1 - Diversity of People. According to Jacobs, diversity of functions in an area supports the area's safety, as it attracts a greater diversity of people at different times that collectively act as 'eyes on the street'. Jacobs points out in her examples the importance of age diversity. Newman, on the contrary, suggests that high diversity of people in an area provides opportunities for crime to happen through anonymity. However, the two theories do not describe the term 'diversity' in further detail. From our telecommunication dataset, we are able to extract one metric of diversity, relative to age. For each area under exam, we have a footfall count breakdown relative to age in terms of these age groups: 0–20, 21–30, 31–40, 41–50, 51–60, 60+. We thus computed age diversity D_a as the Shannon-Wiener diversity index[2] over these counts. When correlating this metric with crime, according to Jacobs we would expect areas with higher age diversity to be safer than others, while following Newman's theory we would expect the opposite.

H2 - Ratio of Visitors. According to our reviewed theories, there are opposite opinions about the contribution towards crime of a high ratio of visitors for an area. Jacobs points out their importance for 'eyes on the streets', while Newman suggests that a high ratio of visitors actually brings crime to an area as a result of anonymity. To explore these apparently contrasting theories, we quantify the ratio of visitors R_v (relative to total footfall count) per area, and will then correlate these values with crime metrics. Following Jacobs, we would expect to have less crime where there are more visitors, whereas following Newman we would expect the opposite.

H3 - Ratio of Residents. A high number of residents in an area is strongly supported by Newman's territorial approach of 'defensible space' to reduce crime. Jacobs mentions residents as a less important factor for the 'natural surveillance' theory compared to shopkeepers, as residents provide less attention for street level activities. To validate Newman's theory, we compute the ratio of residents R_r compared to the overall population, and correlate them with crime metrics. According to Newman, we would expect a high ratio of residents in an area to correlate with less crime.

H4 - Ratio of Workers. Jacobs suggests that a high variety of functions in an area supports urban safety, pointing out the importance of shops in an area, as shop keepers and people who work in an area provide 'natural surveillance'. We will validate the statement by computing the ratio of workers R_w compared to the area's overall population for each area, and compute correlations with crime metrics. According to Jacobs' theory, we would expect to have less crime in areas with a higher ratio of workers.

H5 - Ratio of Female Population. Felson and Clarke suggest that a high ratio of women on the street is a positive sign towards urban safety, as they act as 'crime detractors'. To validate this, we will compute the ratio of female population R_f compared to the overall population for each area, and correlate the values with crime metrics. We would expect a lower crime activity in areas with a higher ratio of females according to the theory.

H6 - Ratio of Young People. According to Felson and Clarke, a higher ratio of young people leads to more criminal incidents in an area, as they show a higher aggression potential compared to elder people. We defined our young population group as those falling in the 0–20 and 21–30 age groups in our telecommunication dataset. We then compute the ratio of young (R_y) population relative to the area's overall population, and correlate it with the crime activity. In this case, the hypothesis is that areas with a higher ratio of young people also have higher crime rates.

---

[2] The Shannon diversity index is a measure that reflects how many different entries there are in a data set and the value is maximized when all entries are equally high [19].

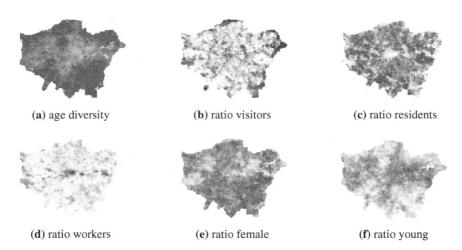

(a) age diversity   (b) ratio visitors   (c) ratio residents

(d) ratio workers   (e) ratio female   (f) ratio young

Fig. 2. Choropleth maps of our six metrics, where the darker the shade of blue, the higher the value of the metric

Summary of Metrics. Figure 2 illustrates the distributions of our six metrics across Greater London as choropleth maps. We observe that population's age diversity (Figure 2(a)) is generally low for Inner London, while it increases towards the edges. A high ratio of visitors is found in the centre of London (Figure 2(b)), which offers most points of interest as attractions and retail, and in some parts of the edges towards the north and the east. Ratios of residents (Figure 2(c)) and workers (Figure 2(d)) show a clear opposite picture between them: while workers concentrate in the central business districts, residents are found to be more widespread in less central boroughs. In Figure 2(e) we observe generally a higher female population ratio for the south of London, compared to the north. Finally, Figure 2(f) shows a higher concentration of young population in the centre of London spreading out towards the east, which is known to be popular among young people.

3.4   Correlation Analysis

Having defined metrics for crime count, crime activity and the six proxies relating to selected urban crime theories, the next step is to correlate these metrics. The major challenge of our approach was to manage the spatial autocorrelation present in our datasets. Spatial autocorrelation is rather common when studying spatial processes, whereby observations captured at close geographic proximity appear to be correlated with each other, either positively or negatively, more than observations of the same properties at further distance [14]. This is the direct quantitative demonstration of Tobler's First Law of Geography, which states that everything is related to everything else, but near things are more related than distant things [25]. Spatial autocorrelation violates the assumption that observations are independent; as such, common correlation analysis techniques that use Pearson, Spearman or Kendall coefficients to explore relationships

Table 3. Tjostheim Correlations r between crime metrics (crime count and crime activity) and individual variables; shown in bold are statistically significant results with p-value < 0.01

Hypothesis	Variable	crime count $CC(i)$			crime activity $CA(i)$		
		Total Crime	Street Crime	Home Crime	Total Crime	Street Crime	Home Crime
H1: diversity of people	D_a	**-0.27**	**-0.26**	**-0.23**	**-0.12**	**-0.14**	-0.10
H2: ratio of visitors	R_v	**-0.20**	**-0.20**	**-0.17**	**-0.28**	**-0.26**	**-0.23**
H3: ratio of residents	R_r	**0.17**	**0.19**	**0.14**	**0.27**	**0.26**	**0.21**
H4: ratio of workers	R_w	0.09	0.07	0.09	0.02	0.02	0.03
H5: ratio of females	R_f	-0.02	-0.02	-0.01	**0.16**	**0.14**	**0.16**
H6: ration of young	R_y	**0.31**	**0.31**	**0.25**	**0.13**	**0.17**	**0.10**

between variables cannot be applied. To address this issue, we will use the Tjostheim correlation index instead [24,11]; this index can be seen as an extension to Spearman and Kendall coefficients, so to explicitly account for spatial properties in our data. All results presented in the next section are thus to be interpreted as correlations r_t computed between crime count $CC(i)$, crime activity $CA(i)$ and the six metrics $H1 - H6$, using the Tjostheim correlation index.

4 Results

4.1 Correlation Results for Greater London

Table 3 presents the Tjostheim correlation coefficients between our two crime metrics ($CC(i)$ and $CA(i)$) and each variable introduced in the previous section. Note that the same correlation signs were found both when using crime count and crime activity, with only relatively small changes in actual correlation values. We interpret this as an indication of the robustness of our proposed metrics. The findings discussed below apply to both crime metrics used.

H1: Diversity of People. We find significant negative correlations between diversity of age and crime, both for total crime ($r_t = -0.27$ for CC and $r_t = -0.12$ for CA) and for street crime ($r_t = -0.26$ for CC and $r_t = -0.14$ for CA); for home crime, we found significant results only for the correlations with CC ($r_t = -0.23$) whereas for CA the p-value was found to be greater than 0.01 so the result is not statistically significant. These findings seem to support Jacob's theory of 'natural surveillance', where she linked different age groups in the same area to a variety of activities taking place in the same space, and this was further associated to less crime.

H2: Ratio of Visitors. We found a significant negative correlation between the ratios of visitors (R_v) of an area and crime. For total crime, we found $r_t = -0.20$ for CC and $r_t = -0.28$ for CA; for street crime, $r_t = -0.20$ and $r_t = -0.26$ respectively; and for home crime $r_t = -0.17$ and $r_t = -0.23$ (second row of Table 3). In all three cases, a higher ratio of visitors is linked to lower crime. These findings again support Jacobs' theory of 'eyes on the street', with consequent increase in the levels of safety of an area where visitors concentrate.

H3: Ratio of Residents. If we now focus on residents, we found a positive correlation between the ratio of residential population (R_r) in an area and crime. Newman's theory of 'defensible space' suggests that an increased ratio of residents is linked to urban safety, by clearly separating spaces for visitors from spaces for residents. However, our findings do not seem to support this. In fact, results show that a high ratio of residents is statistically correlated with crime (from a minimum of $r_t = 0.14$ for home crime correlated with crime count CC, to a maximum of $r_t = 0.26$ for street crime and crime activity CA (third row of Table 3).

H4: Ratio of Workers. Contrary to Newman, Jacobs suggests that residents are less involved with natural surveillance compared to, for example, shopkeepers, as they provide less attention to what is taking place around. Jacobs suggests to look at the relationship between the ratio of working people (R_w) in an area and crime instead. In particular, she posits that a high number of functions, especially shops, leads to increased safety as they attract people and support 'natural surveillance'. Unfortunately, our results do not help shed light into this controversy, as they are not statistically significant (fourth row of Table 3).

H5: Ratio of Female Population. A surprising result is found in the positive correlation between the female population (R_f) and crime activity CA in an area ($r_t = 0.16$ for total crime, $r_t = 0.14$ for street crime and $r_t = 0.16$ for home crime – fifth row of Table 3), though correlations with crime count CC were found not significant . This result shows the opposite of Felson and Clark's theory, suggesting that a higher ratio of female population in London is actually statistically correlated to a higher crime activity in an area. However, we should note a limitation of our metric in this case: in fact, R_f represents the overall ratio of female population for an area (residents, workers, or visiting), and not only the ratio of female population on the streets, so this result could have been affected by a relatively poor metric.

H6: Ratio of Younger Population. Finally, we have computed the ratio of young people (R_y) per area and we have correlated it with crime. Findings show a positive correlation between the younger population and crime (from a minimum of $r_t = 0.10$ for home crime and crime activity CA, to a maximum of $r_t = 0.31$ for total/street crime and crime count CC – last row of Table 3). This result would support Felson and Clarke's theory that a higher proportion of young population ratio is associated with more crime in an area.

4.2   Zooming in at Borough Level

We have shown how one may use our proposed methodology to quantitatively study the validity of certain urban crime theories at scale. However, one may wonder whether the chosen scale (that is, the whole metropolitan area of London) is appropriate for this type of investigations. As mentioned before, London is a very large and complex city, composed of many different neighbourhoods. Choosing the whole of London as a single context to study urban theories may thus hide the fact that, in practice, different

Table 4. Summary statistics of the Tjostheim correlations between total crime count CC and each individual variable on the 32 London boroughs. Stars indicate the percentage of Tjostheim correlations that are statistically significant in each quartile (p-values < 0.01): 0% ' ' 25% '*' 50% '**' 75% '***' 100%

Variable	Min		1st Qu.		Median		3rd Qu.		Max
D_a	-0.51	**	-0.27	***	-0.20	**	-0.12		0.23
R_v	-0.53	**	-0.30	***	-0.20	***	0.00	*	0.18
R_r	-0.16	**	-0.04	***	0.17	***	0.31	**	0.60
R_w	-0.28	***	-0.02	**	0.09	*	0.17	*	0.44
R_f	-0.28	*	-0.08	***	0.03	*	0.17	*	0.47
R_y	-0.18		0.18		0.24	***	0.40	**	0.54

Table 5. Summary statistics of the Tjostheim correlations between total crime activity CA and each individual variable on the 32 London boroughs. Stars indicate the percentage of Tjostheim correlations that are statistically significant in each quartile (p-values < 0.01): 0% ' ' 25% '*' 50% '**' 75% '***' 100%

Variable	Min		1st Qu.		Median		3rd Qu.		Max
D_a	-0.41	***	-0.19	***	-0.11		0.01	*	0.45
R_v	-0.57	***	-0.34	**	-0.27	***	-0.18	**	-0.03
R_r	-0.04	***	0.20	**	0.26	***	0.34	**	0.61
R_w	-0.32	***	-0.08		0.02	*	0.11	**	0.39
R_f	-0.18		0.02	*	0.15	***	0.25	**	0.47
R_y	-0.41	*	0.01		0.08	**	0.22	**	0.45

theories and correlations may hold in different London neighbourhoods. Indeed, theories by Jacobs and Newman had been previously investigated only at neighbourhood level, never at such a big geographic scale.

As our proposed methodology is not prescribed to a size of geographic area, we have repeated our analysis, this time separately considering the 32 administrative boroughs in which London is divided. We assigned grid cells to boroughs boundaries according to their centroids. Table 4 shows summary statistics of the correlations between crime count CC and each variable previously defined, as they vary across boroughs; Table 5 shows results obtained when using crime activity CA instead. By looking at these new results, and by comparing them with those in Table 3, we note that all the individual variables that were (positively or negatively) correlated to crime activity in the whole city of London, now show considerably higher (in positive or in negative) correlations in at least half of the 32 London boroughs. This indeed suggests that this smaller unit of analysis can be more appropriate to investigate the validity of urban crime theories. For those metrics for which we did not find significant statistical results when considering the whole of London, we now find significance in certain areas. For instance, our findings reveal that a quarter of London boroughs have a significant negative correlation between the ratio of working population (R_w), and both crime count CC ($-0.28 > r_w > -0.02$) and crime activity CA ($-0.32 > r_w > -0.08$), whereas for Greater London correlations of the same variable were found not to be significant (CA: $r_w = 0.02$, CC: $r_w = 0.09$). Interestingly, the results at borough level also show that, for another quarter of London boroughs, R_w is actually significantly and positively correlated with crime activity CA ($0.11 > r_w > 0.39$) and crime count CC ($0.17 > r_w > 0.44$) instead. These findings suggest that different, possibly conflicting theories may hold in different parts of the same metropolitan city; using our method, it is possible to investigate whether a theory holds at the full city scale or not. If not, the method also helps social science researchers identify the sub-areas that require further qualitative investigation.

5   Discussion, Limitations and Future Work

Summary. In this paper, we have presented a method to investigate architectural theories of urban crime and people dynamics in a quantitative way. The method requires access to two sources of information: crime data records and records about people presence in the built environment. From the former, we extracted two metrics of crime, crime count $CC(i)$ and crime activity $CA(i)$. From the latter, we extracted metrics that act as proxies for urban crime theories. Using correlation analysis, we have shown it is now possible to quantitatively investigate urban crime theories at large geographic scale and frequent intervals, at almost no cost.

Supported by the ongoing open data movement, an increasing amount of crime data for cities in different parts of the world is freely available and can be used for our purposes. Telecommunication data on the other hand is more difficult to access, but a variety of data mining challenges, such as the Data for Development challenge[3] and the Big Data Challenge,[4] show a clear trend of mobile phone providers towards making their data available to the public. This development suggests that the proposed methodology will become increasingly applicable in the next years.

Implications. The method we have proposed has both practical and theoretical implications. From a practical standpoint, tools can be built on top of it, to the benefit of different stakeholders, as citizens, administrators and city planners. To illustrate what such a tool would look like, we built an Ordinary Least Square (OLS) regression model for each of the 33 boroughs in Greater London separately, as well as for the whole of London. For each such regression model, we analysed the adjusted R^2 value, to understand the extent to which the built model was capable of 'explaining' crime variance. We found that, for a model that considers Greater London as a whole, the adjusted R^2 value is 0.12. However, when we build such model per borough, we are capable of reaching an adjusted R^2 between 0.20 and 0.30 for a quarter of the boroughs. We believe these results are quite promising, considering that we used a rather simple linear model, with just 'people dynamics' variables, as listed previously. A complete model of crime should also include other metrics, for instance, from census data for socio-economic factors, and from the built environment for the city's physical properties. Here we show that, even by just looking at metrics of people dynamics obtained from mobile phone data, we can gain a good insight into urban crime and we can explain up to 30% of its variance in the selected boroughs.

---

[3] D4D – Data for Development, by Orange: http://www.d4d.orange.com/en/home. June, 2014

[4] Big Data Challenge, by Telecom Italia: http://www.telecomitalia.com/tit/en/ bigdatachallenge.html. June, 2014

From a theoretical standpoint, the method offers social science researchers a new way to investigate past crime theories, as well as develop new ones. We have shown how to use the method to explore past theories for the city of London. The same method could be used for a multitude of cities around the world, so to advance knowledge in terms of the contexts within which past theories hold. The method can also be re-applied over time, on newly available data streams, to detect possible changes that call for social scientists to refine past theories or develop new ones. Even when looking at the single city of London in a single period, we have shown that some theories do not hold across all boroughs, thus calling for deeper qualitative investigations in selected areas. We foresee the proposed quantitative method to be used in conjunction with qualitative methods, during alternate phases of theory development and evaluation.

Limitations. Our work suffers from a number of limitations. First, the temporal unit of analysis used in the two datasets at hand was different (i.e., crime data was recorded on a monthly basis, while foot-counts were recorded on a hourly basis). This required a data-processing step that forces us to operate at the coarser level of granularity. This inevitably kept interesting questions unanswered. As previous studies suggest, different crime types follow different spatial and temporal patterns [10]; had we had access to crime timestamps, we would have been able to explore the relationship between people dynamics and crime in a more fine grained manner. Furthermore, our findings are based on mobile phone data collected by a single mobile phone provider. Being one of the major mobile phone providers in the UK with almost 25% market share in 2013, our dataset covers a high number and variety of people, but leaves a grey space for people using other providers or PayAsYouGo options that are excluded from the data. For those people covered from our dataset it stays unclear how the provider categorized them as resident, worker or visitor which could provide a more detailed insight. By including additional datasources, as for instance urban topology data, the ratio of workers could be discussed in more detail. Note that these limitations pertain the datasets used, and not the method proposed. As such, while actual results on the validity of the reviewed urban crime theories for the case study of Greater London would have to be revisited should more accurate and complete datasets become available, we believe the validity of the method withstands.

Future Work. Our future work spans two main directions: on one hand, we aim to expand the model, so to incorporate properties of people dynamics, the built environment, and census within a single framework. In so doing, we expect not only to predict crime activity with greater accuracy, but also to understand the dependencies between all such variables in relation to crime. On the other hand, we aim to apply the model to data from multiple cities in the world. In the last year, telecommunication data has been released both for cities in Europe (e.g., Milan) and in Africa (e.g., Dakar); we wish to apply the method presented in this paper in these very different settings, so to understand in what contexts certain theories hold, thus advancing knowledge in the area of urban crime.

References

1. Bogomolov, A., Lepri, B., Staiano, J., Oliver, N., Pianesi, F., Pentland, A.: Once upon a crime: Towards crime prediction from demographics and mobile data. In: ICMI (2014)
2. Boyd, D., Crawford, K.: Critical questions for big data. Information, Communication and Society 15(5), 662–679 (2012)
3. Chainey, S., Reid, S., Stuart, N.: When is a hotspot a hotspot? a procedure for creating statistically robust hotspot maps of crime. Innovations in GIS 9 Socio-economic Applications of Geographic Information Science (2002)
4. Chainey, S.P., Tompson, L., Uhlig, S.: The utility of hotspot mapping for predicting spatial patterns of crime. Security Journal 21(1-2), 4–28 (2008)
5. Chaplin, R., Flatley, J., Smith, K.: Home office statistical bulletin: Crime in england and wales 2010/11. Home Office Statistical Bulletin (2011)
6. Christens, B., Speer, P.W.: Predicting violent crime using urban and suburban densities. Behavior and Social Issues (14), 113–127 (2005)
7. Eagle, N., Macy, M.: Network diversity and economic development. Science (1029) (2010)
8. Eck, J., Chainey, S., Cameron, J., Leitner, M., Wilson, R.: Mapping crime: Understanding hot spots. Special Report NIJ (2005)
9. Felson, M., Clarke, R.: Opportunity Makes the Thief: Practical theory of crime prevention. Home Office (1998)
10. Felson, M., Poulsen, E.: Simple indicators of crime by time of day. International Journal of Forecasting (19), 595–601 (2003)
11. Hubert, L.J., Golledge, R.G.: Measuring association between spatially defined variables: Tjostheim's index and some extensions. Geographical Analysis (14), 273–278 (1982)
12. Jacobs, J.: The Death and Life of Great American Cities. Random House Inc. (1961)
13. Jansson, K.: British Crime Survey: Measuring crime for 25 years (2006)
14. Legendre, P.: Spatial autocorrelation: Trouble or new paradigm? Ecology 74(6), 1659–1673 (1993)
15. Newman, P.: Defensible Space: Crime Prevention Through Urban Design. Macmillian Pub. Co. (1972)
16. Paynich, R.: Identifying high crime areas. International Association of Crime Analysts (2) (2013)
17. Prasetyo, P.K., Gao, M., Lim, E.P., Scollon, C.N.: Social sensing for urban crisis management: The case of singapore haze. In: Proc of SocInfo 2013, pp. 478–491 (2013)
18. Sahbaz, O., Hiller, B.: The story of the crime: functional, temporal and spatial tendencies in street robbery. In: Proc of 6th International Space Syntax Symposium, Istanbul, pp. 4–14 (2007)
19. Shannon, C.E.: A mathematical theory of communication. The Bell System Technical Journal 27, 379–423, 623–656 (1948)
20. Clarke, C.S., Mashhadi, A., Capra, L.: Poverty on the cheap: estimating poverty maps using aggregated mobile communication. In: Proc of CHI 2014, pp. 511–520 (2014)
21. Snyder, M.: The impact of recent immigration on the london economy. Technical report, London School of Economics and Political Science (2007)
22. Song, W., Daqian, L.: Exploring spatial patterns of property crime risks in changchun, china. International Journal of Applied Geospatial Research 4(3), 80–100 (2013)
23. Tan, S.-Y., Haining, R.: An urban study of crime and health using an exploratory spatial data analysis approach. In: Gervasi, O., Taniar, D., Murgante, B., Laganà, A., Mun, Y., Gavrilova, M.L. (eds.) ICCSA 2009, Part I. LNCS, vol. 5592, pp. 269–284. Springer, Heidelberg (2009)
24. Tjostheim, D.: A measure of association for spatial variables. Biometrika (65,1), 109–114 (1978)

25. Tobler, W.R.: A computer movie simulating urban growth in the detroit region. Economic Geography 46, 234–240 (1970)

26. Wakamiya, S., Lee, R., Sumiya, K.: Social-urban neighborhood search based on crowd footprints network. In: Jatowt, A., Lim, E.-P., Ding, Y., Miura, A., Tezuka, T., Dias, G., Tanaka, K., Flanagin, A., Dai, B.T. (eds.) SocInfo 2013. LNCS, vol. 8238, pp. 429–442. Springer, Heidelberg (2013)

27. Wang, D., Ding, W., Lo, H., Stepinski, T., Salazar, J., Morabito, M.: Crime hotspot mapping using the crime related factors - a spatial data mining approach. Applied Intelligence 39(4), 772–781 (2006)

28. Wang, X., Gerber, M.S., Brown, D.E.: Automatic crime prediction using events extracted from twitter posts. In: Yang, S.J., Greenberg, A.M., Endsley, M. (eds.) SBP 2012. LNCS, vol. 7227, pp. 231–238. Springer, Heidelberg (2012)

29. U. H. WHO. Hidden cities: unmasking and overcoming health inequities in urban settings. WHO, Library Cataloguing-in-Publication Data (2010)

30. Wolfe, M.K., Mennis, J.: Does vegetation encourage or suppress urban crime? Evidence from Philadelphia, PA. Landscape and Urban Planning 108, 112–122 (2012)

Detecting Child Grooming Behaviour Patterns
on Social Media

Amparo Elizabeth Cano, Miriam Fernandez, and Harith Alani

Knowledge Media Institute, Open University, UK
{amparo.cano,m.fernandez,h.alani}@open.ac.uk

Abstract. Online paedophile activity in social media has become a major concern in society as Internet access is easily available to a broader younger population. One common form of online child exploitation is *child grooming*, where adults and minors exchange sexual text and media via social media platforms. Such behaviour involves a number of stages performed by a predator (adult) with the final goal of approaching a victim (minor) in person. This paper presents a study of such online grooming stages from a machine learning perspective. We propose to characterise such stages by a series of features covering sentiment polarity, content, and psycho-linguistic and discourse patterns. Our experiments with online chatroom conversations show good results in automatically classifying chatlines into various grooming stages. Such a deeper understanding and tracking of predatory behaviour is vital for building robust systems for detecting grooming conversations and potential predators on social media.

Keywords: children protection, online grooming, behavioural patterns.

1 Introduction

The online exposure of children to paedophiles is one of the fastest growing issues on social media. As of March 2014, the National Society for the Prevention of Cruelty to Children (NSPCC), reported that i) 12% of 11-16 year olds in the UK have received unwanted sexual messages; and ii) 8% of 11-16 year olds in the UK have received requests to send or respond to a sexual message [16]. The detection of children cyber-sexual-offenders is therefore a critical issue which needs to be addressed.

Children in their teens have started to use social media as their main means of communication [21]. Moreover a recent study of cognition, adolescents and mobile phones (SCAMP) has revealed that 70% of 11-12 year olds in the UK now own a mobile phone rising to 90% by age 14 [28]. While social media outlets (e.g., chat-rooms, images and video sharing sites, microblogs) serve as contact points for paedophile (predators) to potentially exploit children (victims), the automatic detection of children abuse on the Web is still an open question. A common attack from paedophiles is the so-called online child grooming, where adults engage with minors via social media outlets to eventually exchange sexually explicit content. Such grooming consists of building a trust-relationship with a minor, which finally leads into convincing a child to meet them in person [20].

Previous research on detecting cyberpaedophilia online, including the efforts of the first international sexual predator identification competition (PAN'12)[11], has focused

L.M. Aiello and D. McFarland (Eds.): SocInfo 2014, LNCS 8851, pp. 412–427, 2014.

on the automatic identification of predators in chat-room logs. However little has been done on understanding predators behaviour patterns at the various stages of online child grooming, which include Deceptive Trust Development, Grooming, and Seeking for Physical Approach (Section 2). Characterising such stages is a critical issue since most of the sexually abused children have been driven to voluntarily agree to physically approach the predator [36]. This suggests that understanding the different strategies a predator uses to manipulate children behaviour could help in educating children on how to react when expose to such situations.

Moreover the early detection of such stages could facilitate the detection of malicious conversations on the Web. We believe that a deeper characterisation of predator behaviour patterns in such stages could aid in the development of more robust surveillance systems which could potentially reduce the number of abused children. This paper advances the state of the art on predator detection by proposing a more fine-grained characterisation of predators' behaviour in each of the online child grooming stages [22]. The main contributions of this paper can be summarised as follows:

(1) We propose an approach to automatically identify grooming stages in an online conversation based on multiple features: i) lexical; ii) syntactical; iii) sentiment; iv) content; v) psycho-linguistic; and vi) discourse patterns.
(2) We generate classification models for each stage, using single and multiple features. Our findings demonstrate that the use of Label discourse pattern features alone can achieve on average a gain in precision (P) of 4.63% over lexical features. While the use of combined features in classifiers consistently boost performance in P with a gain of 7.6% in all grooming stages.
(3) We present a feature analysis to identify the most discriminative features that characterise each online grooming stage.

The rest of the paper is organised as follows: Section 2 introduces Olson's theory of luring communication which characterises predator's child grooming stages. Section 3 presents related work regarding detection of online predator-victim conversations as well as previous work in online child grooming. Section 4 presents the set of features selected to characterised the language used by predators. Section 5 introduces our methodology for characterising and identifying grooming stages. Results and discussion are presented in sections 6 and 7. Conclusions are presented in Section 8.

2   Online Child Grooming Stages

Child grooming is a premeditated behaviour intending to secure the trust of a minor as a first step towards future engagement in sexual conduct [20]. One of the psychological theories which explains the different child grooming stages in the physical world is Olson's theory of luring communication (LCT) [22]. Previous research has shown that such grooming stages resemble those used by predators in online child grooming [15][9]. According to LCT, once a predator has gained access to a child, the first stage is the *Deceptive Trust Development* which consists of building a trust relationship with the minor. In this first stage a predator exchanges personal information including age, likes, dislikes, former romances, etc. This stage allows the predator to build a common ground with the victim. In this way the predator gets information regarding the victim's support system. Once a trust relationship is established, the predator proceeds to the *Grooming stage*. In this stage the predator triggers the victim's sexual curiosity. This stage involves the use of sexual terms. In such a stage a predator is able to communicatively groom and entrap a child into online sexual conduct. Once the victim has been engaged in this stage, the so-called cycle of entrapment begins. In this cycle, the victim begins to entrust the predator. As the grooming process intensifies, the victim becomes isolated from friends and family, which promotes the predator-victim trust relationship.

In the final stage, the predator seeks to *Physically Approach* the minor. In this stage the predator requests information regarding, for example, the minor's and parent's schedules, and the minor's location. Table 1 presents extracts from the logs dataset provided by the Perverted Justice (PJ) foundation [14]. Here we can see how the different stages are represented in different sentences of the conversation. For example the sentence "I'm sorry your parents are at home all the time" indicates an intention of the predator to seek physical approach.

In the following section we present an overview of the different existing works targeting the detection of online predator-victim conversations as well as online child grooming.

Table 1. Conversation lines extracted from PJ conversations characterising the LCT child grooming stages

STAGE	PREDATOR	VICTIM
Deceptive Trust Development	where are you from?	Whats your asl?
Grooming	So do u masturbate?	not really that borin
Seek Physical Approach	Im sorry your parents home all the time	no

3   Related Work

Online grooooming detection has been widely researched in the past from both social [7] [32] and psychological perspectives [23][18] [35]. More recently the problem of predicting child-sex related solicitation conversations has started to be researched by applying data mining techniques. Simple text mining approaches have been applied to analyse paedophile activity in chat-rooms [24] [14] [15]. One of the major data sources for the automatic detection of paedophiles is the chat logs dataset provided by the Perverted Justice (PJ) foundation. In this foundation, adults volunteer to enter to chat rooms acting like minors. When a conversation involves sexual solicitation, the volunteers share the chat log with the foundation and authorities to prosecute the offenders. Those conversations that result in a predator's conviction are made available at this website.[1] Research involving the use of the PJ dataset for the detection of predators in chat-rooms includes the work of Pendar [24]. In his work he splits conversations into those of predators and those of pseudo-victims. He characterises this dataset by applying supervised (SVM) and non-parametric (kNN) classification models based on n-grams.

---

[1] Perverted Justice, http://www.perverted-justice.com

Kontostathis et al. [14] generate a tool which enables human annotators to tag conversation lines with child grooming stages. They consider the following four categories from Olson's theory of luring communication (LCT) [22]: Deceptive Trust Development, Grooming, Isolation, and Approach. Later on in [15] they apply a phrase-matching and rules-based approach to classify a sentence in a conversation as being related to grooming stages or not. Their results show that they can characterise non-grooming sentences with an accuracy of 75.13%. However, their work did not focus on finding out how accurately they can classify phrases to specific grooming stages.

Another study which focuses on child grooming stages is the one by Michalopoulos et al. [17]. They use a bag of words approach to characterise the stages proposed by [15], however, their goal is to detect a grooming attack rather than to characterise the particular stages within the grooming process. Their results are promising in the use of such stages as a discriminator of predator/non-predator behaviour in chat-room conversations. In [6], Escalante et al. propose a chain-based approach where the prediction of local classifiers are used as input to subsequent local classifiers with the aim of generating a predator-detection system. In their work, they use three classifiers which are applied in different segments of the conversations. Such classifiers are hypothesised to correspond to grooming stages. Based on such neural-network-based classifiers they generate a final classifier which characterises conversations as being from a predator or otherwise.

In [2], Bogdanova et al., approach the problem of discriminating cyber-sex conversations from child grooming conversations by characterising them using n-grams and high-level features. Such features include emotion, neurotism, and those proposed by Michalopoulos et al. [17]. In their task, emotion features appeared to be particularly helpful.

Our work differs from previous approaches in that, rather than characterising the predator-victim roles, we focus on characterising predators' behaviour in each of the child grooming stages. The study of grooming stages have been previously addressed by Gupta et al. [9]. They present an empirical analysis of chat-room conversations focusing on the six stages of online grooming introduced by O'Connell [19]. Their findings suggest that the relation-forming stage is more prominent than the sexual stage. However, while their study focuses on analysing online grooming stages, they do not provide an automatic classification of conversation lines into such stages. To provide such classification, our work introduces a novel set of features which pay particular attention on characterising the pyscho-linguistic and discourse patterns of the predator conversations. The complete set of features used in this work is presented in the following section.

4   Feature Engineering

In this work we use a collection of features which aim to characterise predator conversations in online grooming stages by profiling a predator based on the characterisation of: 1) bag of words (BoW); 2) syntactical; 3) sentiment polarity; 4) content; 5) psyhco-linguistic; and 6) discourse patterns.

The complete set of features is summarised in Table 2. As we can see in this table, the BoW patterns are represented using different sets of n-grams. To characterise the

Table 2. Description of features used for characterising patterns in predator conversation lines

Feature	Description
Bag of Words (BoW) Patterns	
N-grams	n-grams (n=1,2,3) BoW extracted from a sentence.
Syntactical Patterns	
Part-of-Speech tagging	POS tags extracted from a sentence.
Sentiment Patterns	
Sentiment Polarity	Indicates the average sentiment polarity of the terms contained in a sentence.
Content Patterns	
Complexity	Indicates the lexical complexity of a sentence. This is computed based on the cumulative entropy of the terms in a sentence (Section 4.1).
Readability	Computed following the Gunning fox index [8].
Length	Number of terms contained in a sentence.
Psycho-linghitic Patterns	
LIWC dimensions	62 dimensions caracterising psycho-linguistic patterns in English. Each dimension is composed of a collection of terms (Section 4.2).
Discourse Patterns	
Semantic Frames	Consists of a collection of over 10K words senses. This collection describes the lexical use of English in actual texts. A semantic frame can be understood as a description of a type of event, relation or entity and the participants in it (Section 4.3).

syntactical patterns we extract the part of speech (POS) tags of each sentence using the Stanford POS tagger [33]. Sentiment patterns are characterised by computing the sentiment polarity of the sentences. Since peadophiles are known to suffer from emotional instability and psychological problems, [18], we include the use of sentiment polarity as a feature which could describe those changes in a predator's discourse. To compute the sentiment polarity of a sentence we use Sentistrength.[2]

The features used to characterise content, psycho-linguistic and discourse patterns are a bit more complex and will therefore be explained in more detail in the following subsections.

4.1 Content Patterns

To derive content patterns we make use of a set of features which have been successfully used in the past for modelling engagement in social media [34][29]. These features include:

---

[2] Sentistrength http://sentistrength.wlv.ac.uk/

- **Complexity** captures the word diversity of a sentence. The complexity C of a sentence s is defined as:

$$C(s) = \frac{1}{|W|} \sum_{w=1}^{W} f_w(log|W| - log f_w)$$ (1)

where W is the total number of words in the sentence and f_w is the frequency of the word w in the sentence s.
- **Readability** gauges how hard a text is to parse by humans. The readability R of a sentence s is computed based on the Gunning Fox index [8] as follows:

$$R(s) = 0.4 \left(\frac{words}{sentence} + 100 * \left(\frac{complexwords}{words} \right) \right)$$ (2)

- **Length** indicates the number of words in a sentence.

4.2 Psycho-Linguistic Patterns

Previous work on authorship profiling [12] has shown that different groups of people writing about a particular genre use language differently. Such variations include the frequency in the use of certain words as well as the use of syntactic constructions. Authorship profiling based on such variations has been successfully used before for detecting personality features including for example neuroticism, and extraversion [12][30]. In this work we profile predator changes in the different grooming stages based on the variation of the use of different psycho-linguistic dimensions. Here we use the LIWC2007 dataset [26][25], which covers over 60 dimensions of language. These dimensions include style features like, for example, prepositions (e.g., for, beside), conjunctions (e.g., however, whereas), and cause (e.g., cuz, hence) as well as other type of dimensions relevant to psychological patterns like, for example: swearing (e.g., damn, bloody), affect (e.g.,agree, dislike), sexual(e.g., naked, porn). Each dimension is composed of a dictionary of terms. To compute the psycho-linguistic patterns appearing in a sentence we made use of the 62 dictionaries provided in LIWC [25]. To provide a representation of a sentence in these dictionaries, we propose the following approach:

LIWC

Let $LIWC_k$ be the vector representation of the k dictionary in LIWC. To calculate how close is a sentence s to this dictionary we compute the cosine similarity between the word-frequency vector representation of s and the vector $LIWC_k$. Therefore the representation of a sentence in LIWC, is a vector where each entry k corresponds to the cosine similarity of the sentence to the corresponding dictionary $LIWC_k$.

4.3 Discourse Patterns

Previous qualitative analysis [5] of PJ's predators transcripts revealed the frequent use of fixated discourse, showing the predator unwillingness to change a topic. Based on that, we believed that the use of features, which characterise the type of discourse in a conversation could be helpful to discriminate each online grooming stage. In this work we propose to make use of the the FrameNet semantic frames [1], which incorporate semantic generalisations of a discourse. A semantic frame is a description of context in which a word sense is used. These frames consists of over 1000 patterns used in English. Such patterns include: Intentionally_Act, Causality, Grant_Permission, and Emotion_Directed.

To obtain the semantic frames of the sentences produced by a predator in a conversation we apply SEMAFOR[4]. To understand this feature type consider the semantic frame extracted from the sentence "Your mom will let you stay home?, I'm happy" in Table 3. In this sentence two semantic frames (Grant_Permission, and Emotion_Directed) are detected and for each frame different semantic roles and labels can be extracted.

Table 3. Semantic frames parsed for two predator conversation sentences

| Sentence A: Your mom will let you stay home?, I'm happy | | | Sentence B: would you sleep with a guy like that | | |
FRAME	SEMANTICROLE	LABEL	FRAME	SEMANTICROLE	LABEL
Grant_Permission	Target	you	Capacity	Target	sleep
	Action	stay home		Theme	with a guy like that
	Grantee	you		Entity	you
	Grantor	your mom			
	Action	stay home			
Emotion_Directed	Target	happy	People	Target	a guy
	Experiencer	I			

From each parsed frame we generate three types of frame-semantic derived features. In this work we propose to use this information by incorporating them as features encoded in the following way:

Frame

The frame representation of a sentence is the bag of words (BoW) of frames parsed from the sentence. The frame feature representation for sentence A is therefore, {Grant_Permission, Emotion_Directed}.

Semantic Label

The Semantic Label representation of a sentence is the BoW of Labels extracted from the Semantic Frames parsed from the sentence. The Semantic Label feature representation for sentence A is therefore:
{you, stay home, your mom, happy, I}.

FRL

This feature combines Frames, Semantic Roles, and Labels extracted from the Semantic Frames parsed from a sentence. For the cases in which a Label is composed of two or more words we include the merged separated cases. Therefore the FRL feature representation of sentence A is: {Grant_Permission-Action-stay, -Grant_Permission-Action-home, Grant_Permission-Grantee-you, ..},where *Grant_Permission-Action-stay* is composed of the Frame *Grant_Permission*, the Semantic Role *Action* and the first part of the Label *stay home*.

In Section 5, we present how the set of features introduced in this section have been used to characterise and identify online child grooming stages.

5   Characterising and Identifying Child Grooming Stages

In this work we focus on the automatic identification of the three online grooming stages described in Section 2: *Trust Development, Grooming and Approach*. Since changes on predator's discourse are stage-dependent, we propose to characterise the language model used by predators per grooming stage. In this paper, we aim to understand which are the most discriminative features in each stage. To this end, we follow a binary classification approach. We trained three different classifiers, one per stage. Each classifier assigns a stage label to a conversation sentence.

Figure 1 presents a summary of the architecture used in our proposed framework. The first step consists of extracting predator lines from the PJ chat-log conversations, described in section 5.1. Each of these lines is then preprocessed as described in subsection 5.2. Each sentence is then represented into the feature space described in Section 4. To perform feature selection we followed an information gain approach. To build the classifiers we employed a supervised discriminative model (Support Vector Machine [3]) for our experiments.

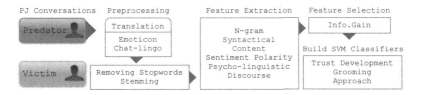

Fig. 1. Architecture for the characterisation and identification of child grooming stages

The following subsections describe the experimental set up used in this work including: i) the description of the selected dataset, Section 5.1, ii) the data preprocessing and feature extraction phases, Section 5.2 and, iii) the construction of the different classifiers to identify grooming stages, Section 5.3.

5.1   Dataset

In this work we make use of the dataset introduced by [15]. This dataset is based on chat conversation transcripts extracted from the PJ website. The provided dataset consists of 50 transcripts corresponding to conversations between convicted predators and volunteers who posed as minors. The length of these conversations varies from 83 to over 12K lines. During the annotation process each line produced by a predator was manually labelled by two trained analysts (Media and Comunication students). Only overlapping annotations were kept as final annotations. These annotations cover four labels: 1) Trust Development; 2) Grooming; 3) Seek for physical approach (Approach) and; 4) Other. The first three describing grooming stages presented in Section 2 and the latter describing the "Other" label for sentences belonging to none of the grooming stages.[3] General

---

[3] Criteria provided to the annotators during the labelling process is further explained in [15].

statistics of the number of sentences labelled for each stage are presented in Table 4. There were 10,871 sentences labelled as "Other". However, since we aim to classify the language model of grooming stages, we need to have a more balanced dataset to reduce potential bias in our experiments. Therefore we randomly picked a fixed set of 3,304 sentences (highest number of sentences per grooming stages) to represent the "Other" dataset.

Table 4. Statistics of the datasets used for generating the classifier of each grooming stage extracted from 50 predator-victim conversations

	Dataset			
	Trust Dev.	Grooming	Approach	Other
Sentences	1,225	3,304	2,700	3,304
After Processing	1,102	3,065	2,531	3,100

5.2   Data Preprocessing and Feature Extraction

One of the challenges of processing chat-room conversations is the appearance of non-standard English terms. It is common to find ill-formed words as well as chat and teen-age lingo. To overcome this issue we first generated a list of over 1,000 terms (including emoticons), which we then translated into standard English. Table 5 presents an extract of this list.

Table 5. Extract of the over 1K terms translated into standard English

CHAT-ROOM TERM	TERM-TRANSLATION	EMOTICON	EMO-TRANSLATION
ASLP	age, sex, location, picture	:'-(	I'm crying
AWGTHTGTTA	are we going to have to go through this again?	o /\o	High Five
BRB	be right back	@_@	I'm tired, trying to stay awake
CWOT	complete waste of time	('}{')	kiss

This first stage of preprocessing resulted in our base dataset. From the base dataset we computed syntactical, psycholinguistic, and frame features. Before computing n-grams, polarity, and content features we performed the following preprocessing: i) stop-words were removed and ii) remaining words were stemmed using Porter stemmer [27].

5.3   Generation and Assessment of Grooming Stage Classifiers

For each child grooming stage we built supervised stage classifiers using the independent feature types (i.e. n-gram, syntactical, polarity, content, psycholinguistic, and semantic frames) and the merged features (All). To generate binary classifiers for each stage, the Stage-labelled sentences (i.e., sentences labeled as belonging to Trust Development, or Grooming, or Approach stages- Section 5.1) were considered as the 'positive' set, while the sentences labelled as "Other" where considered as the 'negative' set.

To assess the classification impact of features in each of the stages, we use as a baseline the performance of a stage classifier using the unigram bag of words approach (1-gram). All the experiments reported in this paper where conducted using a 10 fold cross-validation 5 trial setting [31][13].

6   Results

In this study we report results for the performance of the supervised classifiers generated for the three online grooming stages. We also perform a feature analysis to identify the features that better characterise/discriminate the three child grooming stages.

6.1   Performance Analysis

Performance results are presented in Table 6. In all three stages the results obtained with the unigram baseline feature achieve a 100% recall while providing a precision of over 70%, and an F measure of over 80%. However although high recall values ensure good coverage of the stage-detected sentences, in this task we aim to also obtain high precision values in order to minimise the number of false positives.

When analysing the bigram and trigram features, we observe that in all three stages the use of n-gram feature representation did not improve upon the baseline in any of the performance metrics. The same trend follows for the syntactical features, which alone do not provide good classification performance. Moreover, the classification performance on all three stages drops particularly when using sentiment polarity and content features independently. This is surprising since we expected to find more verbose or complex patterns used by predators when trying to engage with minors, however this is not the case.

Figure 2 presents the distributions of such features per online grooming stage using box plots. Each box plot represents the distribution of positive and negative instances on the scale of values of a feature. For example in the case of sentiment polarity this scale goes from -1 (more negative) to 1 (more positive). The dark line within each green or red boxes represents the median, marking the mid-point of the data. We can see that in the *Trust Development* stage, the levels of complexity and length used within this stage (green box) and other-stage (red box) related conversations are very similar. While such levels slightly increase during the *Grooming* and *Approach* stages (i.e., sentences are longer and more complex).

Our results also show that sentiment polarity features alone are not good discriminators for characterising the online grooming stages. Based on Figure 2, sentiment levels are similar between positive and negative instances in the Trust Development and Grooming stages, while they present a slightly more negative polarity for the Approach stage.

Moving on to the psycho-linguistic features, we observe that, although such features alone do not improve upon the baseline, they do provide a more discriminative feature

Table 6. Presents results for the three stages in oline child grooming. The values highlighted in bold corresponds to the best results obtained in P, R, and F measure, while the light-shaded cells indicate the best feature which alone improve P upon the BoW baseline. Significance levels: p-value < 0.01.

	Child Grooming Stages											
	Trust Development			**Grooming**			**Approach**			**Average**		
	P	R	F1	P	R	F1	P	R	F1	P	R	F1
	N-gram Features											
$1-gram$	0.746	1.0	0.855	0.782	1.0	0.877	0.774	1.0	0.872	0.767	1.0	0.868
$2-gram$	0.629	1.0	0.772	0.663	1.0	0.798	0.654	1.0	0.791	0.649	1.0	0.787
$3-gram$	0.561	1.0	0.719	0.578	1.0	0.733	0.574	1.0	0.73	0.571	1.0	0.727
	Syntactic Features											
POS	0.653	0.344	0.451	0.584	0.559	0.571	0.628	0.671	0.649	0.621	0.525	0.557
	Sentiment Polarity Features											
$Polarity$	0.521	0.548	0.534	0.517	0.546	0.531	0.502	0.416	0.455	0.513	0.503	0.507
	Content Features											
$Readability$	0.565	0.315	0.405	0.513	0.293	0.373	0.584	0.595	0.590	0.554	0.401	0.456
$Complexity$	0.504	0.676	0.578	0.598	0.503	0.546	0.636	0.591	0.613	0.579	0.59	0.579
$Length$	0.512	0.417	0.460	0.614	0.187	0.287	0.693	0.288	0.407	0.606	0.297	0.385
	Psycho-Linguistic Features											
$LIWC$	0.662	0.719	0.689	0.724	0.619	0.668	0.666	0.668	0.667	0.684	0.669	0.675
	Discourse Features											
$Frame$	0.752	0.228	0.350	0.753	0.368	0.494	0.769	0.342	0.474	0.758	0.313	0.439
$Label$	0.778	0.306	0.439	0.850	0.414	0.557	0.813	0.400	0.536	0.814	0.373	0.511
FRL	0.755	0.235	0.358	0.751	0.377	0.502	0.742	0.365	0.490	0.749	0.326	0.45
	All Features											
All	**0.792**	0.823	0.807	**0.876**	0.888	**0.882**	**0.872**	0.887	**0.879**	**0.847**	0.866	0.856

space than those discussed so far. Based on combined feature selection [10] we obtained the top 5 most discriminative LIWC dictionaries of each stage. These top features are presented in Table 7. We see that the dictionaries characterising each stage reveal patterns highlighting the mindset of a predator on each stage.

Our results also show that discourse features are good discriminators in stage classification. In particular, for the *Trust Development* stage, all discourse features alone improve precision upon the baseline. Moreover the Label discourse feature consistently outperforms the baseline in precision for all three stages, providing an average boost in precision of 4.63% (t-test with $\alpha < 0.01$). Results for feature selection on the discourse features presented in Table 7 also provide an insight of the discourse patterns used in each stage, which will be further discussed in Section 6.2.

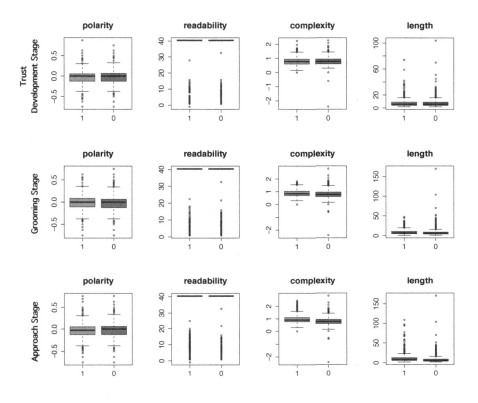

Fig. 2. Sentiment Polarity and Content Features distributions in the online grooming stages. From top to bottom, Trust Development, Grooming and Approach stages.

We finally trained classifiers combining all these features. Table 6 reports the best classification performance which where obtained by excluding bigrams and trigrams. We observed that although sentiment and content features alone are not good discriminators of the grooming stages they help in boosting performance when used with the rest of the features. Our results show that the combined-features classifiers do consistently outperform the baseline in precision on all three stages with an average boost of 8% (t-test with $\alpha < 0.01$) for the cost of a drop in recall of 13.3%. While the recall measure does not reach the one of the baseline on all stages, it does provide a good averaged recall of 86.6%. The combined-feature classifiers also improve upon the baseline in F-measure on the *Grooming* and *Approach* stages with an average boost of 0.6%.

6.2   Feature Analysis

In this paper we focus on the characterisation of three typical online grooming stages. Each of them presenting different variations in the use of language and therefore different complexity when being modelled in a classification system. Our results show that

Table 7. Top discriminative features for each online grooming stage

Feature	Child Grooming Stages		
	Trust Development	**Grooming**	**Approach**
LIWC Dictionaries	affect, assent, cog-Mech, future, home, insight, negate, ppron, see, tentant, you	assent, body, sex-ual,friends, death, filler, home, incl, sad, you	conj, discrep, funct, future, leisure, motion, prep, relativ, social, verbs
Frame Features	physical_artworks, similarity, coincidence, containers, desirabiity	observable_body_parts, activity_ongoing, cause_fluidic_motion, cause_to_be_wet, cloth-ing_parts	capability, arriving, come_together, stimu-lus_focus, visiting
FRL	act, coincidence, eval-uee, emotional_state, trust	manipulation, activity, agent, desiring, experi-encer	capability, event, goal, building, stimulus, vis-iting
Label	artifact, picture, send, act, trust	you, cock, pussy, body_part,sex	address, afternoon, beautiful, booted, call

the early stage, *Trust Development*, of the online grooming stages is more challenging when modelled using content, syntactical and sentiment features. However for all three stages the use of psycho-linguistic and discourse patterns appeared to be beneficial. Particularly the analysis of these two feature spaces facilitate the profiling of the predator discourse in each stage (see Table 7).

For the *Trust Development* stage, the top LIWC dictionaries in the psycho-linguistic profiling of the predator suggest the use of affect words (e.g, sweetheart, fun), assent (e.g.,absolutly, alright), cogMech (e.g, believe, secret) during the establisment of trust. For this stage the discourse pattern features FRL and Label, highlight the request and examination of media content (e.g., pictures). These features also suggest the relevance of emotional engagement in facilitating the building of trust relationships. For the *Grooming* stage the top psycho-linguistic features profiling predators reveal for example the use of body (e.g., naked, dick), sexual (e.g., condom, orgasm), and friends (e.g, sweetie, honey) related words. Such psycholinguistic patterns are similar to those highlighted by the discourse Label feature. Moreover the FRL and frame features characterise the context of the use of such words within this stage. Finally for the *Approach* stage, top psycho-linguistic features include conj (e.g.,also, then), discrep (e.g., hopefully, must) funct (e.g., immediatly, shall) words, while the discourse features suggest the use of stimulus frames as well as temporal (e.g., event) and locative-related frames (e.g., arriving, visiting) characterising the goal of a predator to achieve physical approach with a minor.

The sentiment polarity features studied in this paper do not appear to be discriminative of the stages. However top frames in each stage, including the *emotional_state*, *desiring*, and *stimulus_focus* frames, suggest that the use of more fine-grained emotions could be useful in characterising these stages.

7 Discussion

Previous work on the qualititative characterisation of online grooming stages in chat-room conversations [9] observed that in some cases the online grooming stages are not sequential. For example a predator could convince a child to meet in person during the *Trust Development* stage. Therefore it is possible for a conversation to move back and forth between stages indicating grooming obstacles or difficulties faced by the predator. In this work we focus on the categorisation of chat-room sentences into the typical online grooming stages. The classification of individual chat-lines enables the tracking of such stages at different points on the timeline of a conversation.

While in this work we did not focus on the appearance of such stages on a timeline, it could be possible to add temporal features to characterise such back-forth changes between stages within a conversation. Also the study of short vs. long conversations might need different tactics or yield new insights. Here we studied chat-lines of the merged conversations of our dataset, however we could study chat-lines at the level of independent predator conversations in order to generate multiple predator profiling. Moreover our study is based on those conversations which lead to convicted-paedophiles, however further studies could address differences between convicted and non-convicted peadophile conversations.

In chat-rooms it is common to find regular chat-conversations with sexual content between teens or between adults. These type of conversations pose serious challenges to systems which only focus on the predator-victim characterisation since in such systems the majority of features involves sexual content. The use of stages for the characterisation of predator conversations could potentially help systems in reducing the number of false positives when exposed to non-peadophile conversations with sexual content since predators' luring stages are not common in standard online sexual conversations [2].

One of the major policing concerns is to gather accurate evidence. Therefore providing systems with a low false positive rate is fundamental. While the proposed baseline offers a 100% recall, our experiments show that the proposed discourse patterns alone and the combined-merged classifiers can boost performance at the expense of a slight drop in recall reducing in this way false positive rates.

8 Conclusions

In this work we have presented a supervised approach for the automatic classification of online grooming stages. To the best of our knowledge this is the first study focusing on the automatic classification of such stages from the psycho-linguistic and discourse patterns perspective. Such features provide an insight of the mindset, and discourse patterns of predators in online grooming stages. Our experiments show that the discourse Label feature alone consistently outperforms our baseline in precision for all three stages. Moreover when using the combined-features classifiers our results show an improvement upon both precision and F-measure for both the *Grooming* and the *Approach* stages. Our results also show that the combined-features classifiers do consistently outperform the baseline in precision on all three stages with an average boost of 8% (t-test with $\alpha < 0.01$) for the cost of a drop in recall of 13.3%.

These results demonstrate the feasibility of the use of psycho-linguistic and discourse features for the automatic detection of online grooming stages. This opens new possibilities for adressing predator grooming behaviour online, where policing organisations can act in a preventive way by addressing grooming at early stages or in a reactive way by avoiding/intervining in the approach stage. We believe that the work in this paper has the potential to also open new possibilities into understanding the victim entrapment cycle.

Acknowledgments. This work was supported in part by the OU Policing Research Consortium in collaboration with Dorset Police, UK.

References

1. Baker, C.F., Fillmore, C.J., Lowe, J.B.: The berkeley framenet project. In: In Proc. of COLING/ACL (1998)
2. Bogdanova, D., Rosso, P., Solorio, T.: Exploring high-level features for detecting cyberpedophilia. Comput. Speech Lang. 28(1), 108–120 (2014)
3. Chang, C.-C., Lin, C.-J.: Libsvm: A library for support vector machines. ACM Trans. Intell. Syst. Technol. 2(3), 1–27 (2011)
4. Das, D., Schneider, N., Chen, D., Smith, N.A.: Semafor 1.0: A probabilistic frame-semantic parser. Technical report, Carnegie Mellon University Technical Report CMU-LTI-10-001 (2010)
5. Egan, V., Hoskinson, J., Shewan, D.: Perverted justice: a content analysis of the language used by offenders detected attempting to solicit children for sex. Antisocial Behavior: Causes, Correlations and Treatments 20(3), 273 (2011)
6. Escalante, H.J., Inaoe, L., Enrique, L., No, E., Villatoro-tello, E., Cuajimalpa, U., Juárez, A., Enrique, L., No, E., Villaseñor, L.: Sexual predator detection in chats with chained classifiers. In: Proc. of ACL (2013)
7. Webster, S., et al.: European online grooming project - final report. Technical report, European Comission Safer Internet Plus Programme (2012)
8. Gunning, R.: The technique of clear writing. McGraw-Hill, International Book (1952)
9. Gupta, A., Kumaraguru, P., Sureka, A.: Characterizing pedophile conversations on the internet using online grooming. CoRR (2012)
10. Hall, M.A.: Correlation-based Feature Subset Selection for Machine Learning. PhD thesis, University of Waikato, Hamilton, New Zealand (1998)
11. Inches, G., Crestani, F.: Overview of the international sexual predator identification competition at pan-2012. In: Forner, P., Karlgren, J., Womser-Hacker, C. (eds.) CLEF (2012)
12. Chambers, J.K., Trudgill, P., Schilling-Estes, N.: The Handbook Of Language Variation And Change. Blackwell, London (2004)
13. Kohavi, R.: A study of cross-validation and bootstrap for accuracy estimation and model selection. In: Proc. 14th IJCAI, vol. 2 (1995)
14. Kontostathis, A.: Chatcoder: Toward the tracking and categorization of internet predators. In: Proc. Text Mining Workshop 2009 held in conjunction with the Ninth SIAM International Conference on Data Mining. SPARKS, NV (2009)
15. Kontostathis, A., Edwards, L., Bayzick, J., Mcghee, I., Leatherman, A., Moore, K.: Comparison of rule-based to human analysis of chat logs. In: 1st International Workshop on Mining Social Media Programme, Conferencia de la Asociación Española Para La Inteligencia Artificial, 2009 (2010)

16. Lilley, C., Ball, R., Vernon, H.: The experiences of 11-16 year olds on social networking sites. Technical report, NSPCC (2014)
17. Michalopoulos, D., Mavridis, I.: Utilizing document classification for grooming attack recognition. In: Proceedings of the IEEE Symposium on Computers and Communications (2011)
18. Nijman, H., Merckelbach, H., Cima, M.: Performance intelligence, sexual offending and psychopathy. Journal of Sexual Aggression 15, 319–330 (2009)
19. O'Connell, R.: A typology of child cybersexploitation and online grooming practices. Technical report, Cyberspace Research Unit, University of Central Lancashier (2003)
20. Australian Institute of Criminology (AIC). Online child grooming laws. Technical report, High tech crime brief no. 17. Canberra: AIC (2008)
21. Australian Institute of Criminology (AIC). Children's use of mobile phones. Technical report, GSMA, NTT DOCOMO (2013)
22. Olson, L.N., Daggs, J.L., Ellevold, B.L., Rogers, T.K.K.: Entrapping the innocent: Toward a theory of child sexual predators luring communication. Communication Theory 17(3), 231–251 (2007)
23. Palmer, T., Stacey, L.: Just one click - sexual abuse of children and young people through the internet and mobile phone technology. Technical report. Barnardo's UK, Essex (2004)
24. Pendar, N.: Toward spotting the pedophile telling victim from predator in text chats. In: International Conference on Semantic Computing, ICSC 2007, pp. 235–241 (September 2007)
25. Pennebaker, J.W., Chung, C.K., Ireland, M., Gonzales, A., Booth, R.J.: The development and psychometric properties of liwc 2007. Technical report, Technical report, Austin, TX, LIWC.Net (2007)
26. Pennebaker, J.W., Francis, M.E., Booth, R.J.: Linguistic inquiry and word count (liwc): Liwc2001 manual. Technical report, Erlbaum Publishers (2001)
27. Porter, M.: An algorithm for suffix stripping. Program 14(3) (1980)
28. SCAMP Project. Study of cognition, adolescents and mobile phones (scamp). Technical report, Imperial College London (2014)
29. Rowe, M., Alani, H.: Mining and comparing engagement dynamics across multiple social media platforms. In: Proc. of ACM 2014 Web Science Conference, Bloomington, Indiana, USA, pp. 229–238 (2014)
30. Argamon, S., Koppel, M., Pennebaker, J., Schler, J.: Automatically profiling the author of an anonymous text. Communications of the ACM 52(2), 119–123 (2009)
31. Salzberg, S.L., Fayyad, U.: On comparing classifiers: Pitfalls to avoid and a recommended approach. Data Mining and Knowledge Discovery, 317–328 (1997)
32. Griffith, G., Roth, L.: Protecting children from online sexual predators. Technical report, NSW parliamentary library briefing paper no. 10/07 Sydney: NSW Parliamentary Library (2007)
33. Toutanova, K., Manning, C.D.: Enriching the knowledge sources used in a maximum entropy part-of-speech tagger. In: Proceedings of the Joint SIGDAT Conference on Empirical Methods in Natural Language Processing and Very Large Corpora (EMNLP/VLC-2000), pp. 63–70 (2000)
34. Wagner, C., Rowe, M., Strohmaier, M., Alani, H.: Ignorance isn't bliss: an empirical analysis of attention patterns in online communities. In: Proc. of 4th IEEE International Conference on Social Computing, Amsterdam, The Netherlands (2012)
35. Whittle, H., Hamilton-Giachritsis, C., Beech, A.: Victim's voices: The impact of online grooming and sexual abuse. Universal Journal of Psychology 1(2), 59–71 (2013)
36. Wolak, J., Mitchell, K., Finkelhor, D.: Online victimization of youth: Five years later. Technical report, Bulleting 07-06-025, National Center for Missing and Exploited Children, Alexadia, Alexandria, VA (2006)

Digital Rights and Freedoms:
A Framework for Surveying Users
and Analyzing Policies

Todd Davies*

Symbolic Systems Program and
Center for the Study of Language and Information
Stanford University
Stanford, California 94305-2150 USA
davies@stanford.edu
http://www.stanford.edu/~davies

Abstract. Interest has been revived in the creation of a "bill of rights" for Internet users. This paper analyzes users' rights into ten broad principles, as a basis for assessing what users regard as important and for comparing different multi-issue Internet policy proposals. Stability of the principles is demonstrated in an experimental survey, which also shows that freedoms of users to participate in the design and coding of platforms appear to be viewed as inessential relative to other rights. An analysis of users' rights frameworks that have emerged over the past twenty years also shows that such proposals tend to leave out freedoms related to software platforms, as opposed to user data or public networks. Evaluating policy frameworks in a comparative analysis based on prior principles may help people to see what is missing and what is important as the future of the Internet continues to be debated.

1 Introduction

In March of 2014, on the 25th anniversary of the proposal that led to the World Wide Web, its author Tim Berners-Lee launched an initiative called the Web We Want campaign, which calls for "a global movement to defend, claim, and change the future of the Web" [33]. The object of the campaign is an online "Magna Carta," "global constitution," or "bill of rights" for the Web and its users, which Berners-Lee argued was needed because "the web had come under increasing attack from governments and corporate influence and that new rules were needed to protect the 'open, neutral' system" [17].

---

* I would like to thank Laila Chima for assistance with the survey process, and Nathan Tindall for assistance with data analysis. I would also like to thank Jerome Feldman, Karl Fogel, Lauren Gelman, Rufo Guerreschi, Kaliya Hamlin, Dorothy Kidd, Geert Lovink, Mike Mintz, Brendan O'Connor, Doug Kensing, Steve Zeltzer, and the students who have taken Symsys 201 at Stanford for helpful discussions prior to the writing of this paper. Research costs were funded by the School of Humanities and Sciences at Stanford University.

L.M. Aiello and D. McFarland (Eds.): SocInfo 2014, LNCS 8851, pp. 428–443, 2014.

Although the weight of Berners-Lee's voice in calling for a users' "bill of rights" is a recent development, the idea of a comprehensive user rights framework has been floated by others previously (see section 3 below). With more limited scope, over the past three decades, many initiatives have emerged to promote particular rights, abilities, and influence for users over their online environments and data. Both codified and informal concepts such as Free Software [31], participatory design [16], Open Source software [25], Creative Commons [6] and free culture [18], data portability [8], and the DNT (Do Not Track) header [10] are attempts to establish and promote principles outside of public policy through which people can participate in the decisions that affect them as software users. Other concepts, such as the right to connect [2] and net neutrality [11] represent attempts to protect user rights and free access through public policy. This paper describes a broad set of principles guiding user freedom and participation, and relates these principles to past and ongoing initiatives introduced by others.

2 Rights, Freedoms, and Participation Principles

We can analyze users' rights with reference to ten principles, which might be present (or not) to differing degrees in a particular software environment or policy framework. The principles below outline a framework for *opinion assessment* and *comparative analysis* rather than being intended as a policy proposal. It is important to keep in mind this distinction for what follows. The principles and the concepts defined in relation to them below were derived empirically from users' rights policy proposals, but they are not meant to be exhaustive in any sense.[1]

2.1 User Data Freedoms

The first six of the principles (1-6) are amenable to adoption within a particular software platform or environment, which may be under either private or public ownership. Of these, the first three (1-3) pertain to the data generated by a given user, which are referred to in these descriptions as "their data."

Principle 1. *Privacy control.* The user is able to know and to control who else can access their data.

Some or all of the following concepts might appear in a privacy control policy.

a) *Originator-discretionary reading control.* The user who generates data is able to read and to determine who else can read their data, and under what circumstances, and cannot have this ability taken away. Generated data may be created by the user deliberately, e.g. by filling out a form online or by posting a photograph, or it may be created as a byproduct of the user's

---

[1] Principles and concepts that lack citations in this section are referenced in the documents analyzed in Table 3, section 4.1.

behavior, such as click stream data or cookies from the user's browser that are read and stored on a site which they use. (*Do Not Track* initiatives are attempts to provide users with a partial form of this type of control [10].)

b) *Data use transparency.* All policies and practices concerning the storage or transfer of a user's data are fully disclosed to the user prior to when the data are generated. This includes policies and practices of the software platform provider regarding the manner and length of time the user's data are stored.

c) *Usable privacy.* Access and control by a user of their data is practically feasible for the user. Access should be straightforward enough to be practical, and privacy settings should be as clear and easy to use as possible, including for novice users [14].

d) *Nonretention of data.* User data are not retained without the consent of the user.

Principle 2. *Data Portability.* The user is able to obtain their data and to transfer it to, or substitute data stored on, a compatible platform.

Some or all of the following concepts might appear in a Data Portability policy, as defined by the Data Portability Project [8].

a) *Free data access.* The user of a software platform is able to (a) download or copy all of their data, (b) download or copy all of the other data on the platform to which the user has access, and (c) know where their data are being stored, i.e. in what real world location or legal jurisdiction.

b) *Open formats.* The information necessary in order to read, interpret, and transfer data, i.e. application programming interfaces (APIs), data models, and data standards, are available to any user and are well documented.

c) *Platform independence.* The user is able to access data while using a software platform independently of whether those data are stored within the platform or outside it in a compatible platform. Principles put forward in Data Portability policies that elaborate on this concept include the ability of the user to (a) authenticate or log in under an existing identity on another platform, (b) use data stored on another platform, (c) update their data on another platform and have the updates reflected in the platform in current use, (d) update their data on other platforms automatically by undating them on the platform in use, (e) share data stored on the platform in use with other platforms, and (f) specify the location or jurisdiction of storage for their data within the platform [34].

d) *Free deletion.* The user can delete their account and all of their data, and these data will be removed or erased from storage in that platform consistent with the meaning of a transparently provided definition of deletion.

Principle 3. *Creative control.* The user is able to modify their data within the software platform being used, and to control who else can do so.

Some or all of the following concepts might appear in a creative control policy.

a) *Originator-discretionary editing control.* Subject to transparency requirements, the user who generates data is able to edit and to determine who else can edit their data, and under what circumstances, and cannot have this ability taken away. Transparency requirements such as visibly maintaining past versions and making their existence apparent to any user who can access a data item are safeguards against the abuse of editing, which could otherwise be used to alter the historical record.

b) *Authorial copyright support.* The creator of content holds any legally allowed copyright over their data, and a user has the ability to prevent others who have access to their generated data from copying it for access by a third party who lacks access to the original. (This definition reflects an adaptation of traditional copyright for digital content, applying the Fair Use exemption to copying for private viewing by a party who already has authorized access.)

c) *Reciprocal data sharing.* The user has the ability to permit people to copy (and possibly modify) their data for viewing by third parties subject to provisos such as the Attribution, Noncommercial, Share Alike, and No Derivatives requirements which can be imposed on the copying party in a *Creative Commons* license [6].

2.2   Software Platform Freedoms

Principles 4-6 pertain to the software platform in which users' data are created, edited, stored, and accessed. The descriptions of these principles distinguish different ways in which users may be able to participate in controlling, designing, and governing the operation of the software platform they use.

Principle 4. *Software freedom.* The user is able to modify code in the software platform being used, subject to rights of other users to control their own experience of the platform.

Some or all of the following concepts might appear in a software freedom policy.

a) *Open Source code.* The source code that operates the software platform can be legally read, copied, downloaded, and modified by any user. Source code includes all the code that is necessary to operate the platform and to serve data to the user. (Open Source software is defined in the Open Source Definition [25].)

b) *Reciprocal code openness.* The user and anyone else who modifies the platform's source code for use by others is legally bound to make their modified code available under licensing terms consistent with those under which the user legally accesses the source code. (This incorporates (a) the so-called "copyleft" provision of *Free Software* licenses such as the General Public License v.3, which require reciprocal sharing by anyone who distributes modified copies of the software to others in executable form [31], and sometimes also (b) the Affero clause in the Affero General Public License v.3, which requires that code modifications be reciprocally shared by anyone who executes their modified version of the source code in a networked environment (e.g. over the Web) for use by others [12].)

c) *User modifiable platform.* The user of a software platform has the ability to modify the code on the platform they are using, as long as doing so does not interfere with the rights of other users to experience the platform and interact with their data as they desire. (In its full form, this is a demanding provision that is not usually satisfied in practical platforms, though it is often fulfilled in limited ways, e.g. by permitting a user-selected, configurable interface. This concept is an extension of the ideas in [31] to networked platforms.)

Principles 1-4 are freedoms of individual users, which can be composed to define freedom for a community of users. Principles 5-6 are defined at the level of the group of users of a given software platform, which for each individual user means the freedom to participate in a collective process that determines the design and governance of a software and data environment. These last two freedoms allow for an especially large range of freedoms and participation mechanisms.

Principle 5. *Participatory design.* The design of the platform is produced by all of its users.

Some or all of the following concepts might appear in a participatory design policy [16,29].

a) *User-centered design.* The needs and desires of users are the primary or sole factor driving the design. Users' needs may be assessed in various ways, e.g. through ethnographic observation, surveys, one-on-one interviews, and focus groups, that focus on the problems and goals of users at a functional level.

b) *User input to design.* The users' preferences and beliefs about design choices are collected and influence the design of the platform. This type of input can include, for example, expressions of preference between different options that are presented by a designer.

c) *User-generated design.* Users participate in the creation of design solutions as actual partners in the design team, e.g. providing ideas through brainstorming with designers and/or other users, and helping to solve design problems creatively.

d) *Customizable design.* Users can individually or collectively redo or configure parts of the platform's design and this feature is itself part of the design.

Principle 6. *User self-governance.* The operation of the platform is governed by all of its users.

Some or all of the following concepts might appear in a user self-governance policy. Wikipedia self-governance implements all of these concepts in varying degrees [32].

a) *Participatory policy making.* Users are involved in creating and making decisions about the framework of rules and practices governing the platform they are using. (This can range from input on proposed policies, to voting, to full-fledged deliberative democracy online [9].)

b) *Participatory implementation.* Users are involved in executing and enforcing the policies that govern the platform. (This can include forms of participation such as monitoring one's own compliance with policies, notifying other users of policy violations, raising and discussing implementation questions in online forums, and serving in defined roles.)

c) *Participatory adjudication.* Users are involved in making judgments when human judgment (usually in a collective form) is a part of the platform's operation, e.g when a policy implementation question is in dispute. when content much be judged appropriate or not under defined criteria or procedures, or when the platform asks users for input in rendering a judgment or rating concerning user content.

2.3   Public Network Freedoms

Principles 7-10 generally require public policy adoption, such as legislation, executive orders, or international agreements.

Principle 7. *Universal network access.* Every person is legally and practically able, to the greatest extent possible, to access the Internet, and it is available everywhere in a form adequate for both retrieving and posting data.

Some or all of the following concepts might appear in a universal network access policy.

a) *Right to connect.* Internet access cannot be denied to a user or to a population of users wherever it is possible to provide access [2].

b) *Universal digital literacy.* Every person who possesses the intellectual ability to do so develops the skills to use the Internet as both a recipient and producer of information, to the maximal achievable for meaningful individual participation in a democracy.

c) *No- or low-cost service.* Cost is not a barrier to accessing the Internet.

d) *Omnipresent service.* The Internet is available everywhere and at all times.

e) *Accessibility.* Internet access is available to everyone in a way that matches their physical and mental abilities.

Principle 8. *Freedom of information.* Every person is legally and practically able to produce and receive information in the way that they want, to the maximal extent consistent with the rights of others.

Some or all of the following concepts might appear in a freedom of information policy.

a) *Right to privacy.* Private communications cannot be intercepted, monitored, or stored by governments or other entities without due process to establish a compelling public interest.

b) *Right to anonymous speech.* Everyone is able to both receive and produce public information without being required to identify themselves, either implicitly or explicitly.

c) *Freedom from censorship.* Free expression, without political restrictions, is protected both for producers and receivers of information.

d) *Open Access to all publicly funded data.* Government data and that which is produced through publicly funded research is available freely to everyone.

e) *Democratically controlled security.* Government security policies must be as transparent as possible to allow for them to be publicly debated, and those who oversee them must be accountable to everyone.

f) *Right to be forgotten.* Everyone is able to have information about them made inaccessible to others when these data are determined by established procedures to be either no longer relevant or unfairly stigmatizing to their subject(s) [27].

Principle 9. *Net neutrality.* All providers of Internet connections and services are legally and practically required to treat data equally as it is transmitted through the infrastructure they control.

Some or all of the following concepts might appear in a net neutrality policy. Disallowed forms of discrimination against data would include blocking data or charging fees in exchange for allowing it to be transmitted [11].

a) *Source neutrality.* Providers of network connections may not discriminate against data on the basis of its origin, e.g. another service provider or a particular social media platform.

b) *Format neutrality.* Providers of network connections may not discriminate against data on the basis of its format, e.g. MIME type, protocol, or port.

c) *Content neutrality.* Providers of network connections may not discriminate against data on the basis of its content, e.g. political expression with which the provider disagrees.

d) *End-user neutrality.* Providers of network connections may not discriminate against data on the basis of the end user's identity.

Principle 10. *Pluralistic open infrastructure.* Everyone has access to multiple independent but interoperating software platforms as options for their data.

Some or all of the following concepts might appear in a pluralistic open infrastructure policy.

a) *Multiplicity of platforms.* Policies ensure that all users have multiple software platforms to choose from as environments for their data.

b) *Decentralized control.* Software platforms are coordinated to interoperate in a way that is not controlled by any one government, authority, or interest.

c) *Transparent control.* Common infrastructure and standards are developed and documented in a way that is open and understandable to anyone.

3   A Survey of Internet Users

To illustrate the use of this analysis framework for surveying users, a demonstration survey was conducted using Amazon Mechanical Turk in the summer of 2014.

3.1 Participants and Method

A total of 780 survey takers completed a survey on the Qualtrics platform[24]. Each survey taker was shown a subset of the principles and concepts described in the framework of section 2, and asked to "rate [on a 0 to 10 scale, moved left or rigth from the midpoint] how important you think it is for the user of a software 'platform' (such as a website, app, operating system, or social network)" to have the particular right or freedom described in each statement they read.[2] The statements consisted of the unparenthesized and unbracketed portions of each principle and concept in section 2, with the title of each excluded. Participants were told: "These statements describe what *could* be true in some situations or hypothetically, not necessarily what *is* true now or in some particular situation."

The survey as designed assumed users were fluent in reading English, and able to understand digital concepts such as "data" and "software platform." Participants were recruited on the Amazon Mechanical Turk platform[1], with a link to the survey on Qualtrics. The survey was open only to U.S.-located respondents whose prior approval percentage by requestors on MTurk exceeded 98%. Participants were 39% female and 61% male. Respondents' reported age groups were 3% under 20, 41% 20-29, 33% 30-39, 12% 40-49, 8% 50-59, 3% 60-69, 0.4% 70-79, and 0% over 80. Thirteen percent reported being "very knowledgeable about digital rights and freedoms," while 72% reported being "somewhat knowledgeable" and 15% "not knowledgeable."

The participant pool, while not representative of the population of the United States as a whole (let alone the world), nonetheless represents a population of interest: relatively sophisticated users who could be expected to have heard of at least some of the concepts in our framework. The intent was both to test whether users would have consistent views of these statements, and to assess relative levels of support for the principles and concepts in the framework within the young-skewing demographic of high functioning Internet users. Although we will not do it here, the gender, age group, and knowledge data could be used to adjust for sampling bias relative to the general population of users in the United States. (A more complete demographic analysis is planned in a future paper.)

For this survey, participants randomly saw either a *broad* rating set consisting of random ordering of all ten of the primary principles in the framework (a *within*-group comparison of the ten principles), or a *narrow* random ordering of a subset of principles and concepts that included one of the primary principles and its associated concepts (a *between*-groups comparison of the ten principles). This allowed for cross-item comparative and correlational analysis for both the ten principles and for the concepts associated with each principle along with that principle itself. Random assignment of participants into the broad or narrow rating sets created an experiment for testing whether average importance ratings for the ten principles would remain stable across these two rating contexts.

---

[2] Survey materials and data for this study are published on the Harvard Dataverse Network at `doi:10.7910/DVN/27510`.

3.2 Survey Results

The mean importance ratings from the narrow sets for each principle and concept, together with sample sizes and standard deviations, are shown in Table 1. Standard errors ranged from 0.2 to 0.4 and are easily calculated from the table. As can be seen from Table 1, highly rated primary principles tend to have highly rated associated concepts, but there are occasional deviations within principle-concept groupings. The concepts associated with data portability, for example, ranged widely in support, from a 6.61 rating for open formats (2b) to an 8.96 rating for free deletion (2d). Pluralistic open infrastructure, as worded in the principle, drew less support (6.84) than any of its three associated concept statements, which ranged from 7.24 to 8.27. Every principle and concept was rated significantly above the midpoint (and starting point) of 5.0 in this survey, indicating that participants on average regarded each of them as at least somewhat important.

A full analysis of all of the principle-concept groups is beyond what we have room for in this paper, but Table 2 shows the basis for such an analysis of the ten primary principles. This table displays mean ratings first for the broad rating set – participants who rated all and only the primary principles – and compares them to the narrow set means. The aggregate means are simply the averages of the broad and narrow means. The correlation between the means of the within- and between-groups surveys is extremely high (.98), indicating that attitudes toward the principles are stable across these two different presentation contexts for this population. The most important primary principle in the eyes of participants was the statement that is labeled "privacy control" in section 2 (agg. mean 8.89), though again participants did not see the labels. Next highest were 7-universal access (8.49), 9-net neutrality (8.06), 8-freedom of information (7.94), and 2-data portability and 3-creative control (both 7.82). The remaining principles formed a less highly rated cluster: 10-pluralistic open infrastructure (6.69), 6-user self-governance (5.93), 4-software freedom (5.78), and 5-participatory design (5.30).

Table 2 also shows correlations between the importance ratings of pairs of principles for participants in the broad rating set condition: those who rated all ten of the main principles instead of just one. All of the significant correlations, and most of the nonsignificant ones, are positive, indicating a general disposition for individuals to be more or less favorable to digital rights and freedoms. Ratings for the lowest rated principles (4 and 5) were significantly correlated, but ratings between principle 4 or 5 and the other principles tended not to be significant. In the set of correlations involving just one of principles 4 and 5, only 3 out of 14 were significant, whereas in the remaining correlations, 25 out of 30 were significant. Consistent with their low overall average ratings, this indicates that principles 4 and 5 are evaluated differently by users compared to the rest of the principles ($p = .0001$ by a Fisher exact test).

3.3 Survey Lessons

The use of the framework in this survey has demonstrated that it is possible to obtain meaningful results about the relative importance that users attach to

Table 1. Importance Ratings of Principles and Concepts (Narrow Rating Sets)

Principle/Concept	Mean (0-10)	N	Std. Dev.
1–Privacy control	8.69	71	1.9
1a Originator-discretionary reading control	7.96	71	2.2
1b Data use transparency	8.06	71	2.2
1c Usable privacy	8.58	71	1.7
1d Nonretention of data	8.65	71	1.7
2–Data Portability	7.90	69	1.7
2a Free data access	7.74	69	2.0
2b Open formats	6.61	68	2.5
2c Platform independence	6.71	68	2.3
2d Free deletion	8.96	69	1.5
3–Creative control	7.77	65	2.4
3a Originator-discretionary editing control	7.36	66	2.3
3b Authorial copyright support	7.65	66	2.5
3c Reciprocal data sharing	6.64	65	2.5
4–Software freedom	6.01	73	2.7
4a Open Source code	5.85	71	2.7
4b Reciprocal code openness	5.52	72	2.6
4c User modifiable platform	6.63	74	2.6
5–Participatory design	5.48	77	2.4
5a User-centered design	7.08	75	2.1
5b User input to design	6.83	77	1.9
5c User-generated design	6.16	76	2.5
5d Customizable design	6.36	77	2.2
6–User self-governance	5.82	63	2.7
6a Participatory policy making	6.30	64	2.5
6b Participatory implementation	6.01	66	2.6
6c Participatory adjudication	†		
7–Universal network access	8.40	82	1.9
7a Right to connect	8.56	82	1.7
7b Universal digital literacy	7.45	82	2.2
7c No- or low-cost service	8.27	82	2.2
7d Omnipresent service	8.43	82	2.0
7e Accessibility	7.12	80	2.8
8–Freedom of information	8.01	74	1.7
8a Right to privacy	8.72	74	1.9
8b Right to anonymous speech	7.39	74	2.3
8c Freedom from censorship	8.46	74	1.8
8d Open Access to publicly funded data	8.20	74	2.0
8e Democratically controlled security	8.12	74	1.9
8f Right to be forgotten	7.59	74	2.3
9–Net neutrality	8.11	61	2.5
9a Source neutrality	8.39	61	2.2
9b Format neutrality	7.42	61	2.5
9c Content neutrality	8.56	61	2.2
9d End-user neutrality	8.52	61	2.2
10–Pluralistic open infrastructure	6.84	70	2.1
10a Multiplicity of platforms	7.24	70	2.0
10b Decentralized control	8.27	70	1.8
10c Transparent control	7.97	70	1.8

† Ratings for 6c were not meaningful: incorrect wording on survey.

Table 2. Comparing Importance Ratings of the Ten Principles

Principle Number	Broad Mean	Narrow Mean	Aggregate Mean	Correlations of Importance Ratings (Broad Set)								
				2	3	4	5	6	7	8	9	10
1	9.09	8.69	8.89	0.50‡	0.48‡	0.02	-0.06	0.24	0.40‡	0.66‡	0.15	0.06
2	7.74	7.90	7.82		0.35‡	0.08	0.05	0.31†	0.51‡	0.66‡	0.35‡	0.36‡
3	7.86	7.77	7.82			0.19	0.07	0.33†	0.22	0.48‡	0.37‡	0.13
4	5.55	6.01	5.78				0.36‡	0.22	0.14	0.18	0.05	0.31†
5	5.12	5.48	5.30					0.55‡	-0.02	0.18	0.28†	0.50‡
6	6.05	5.82	5.93						0.25†	0.37‡	0.49‡	0.43‡
7	8.58	8.40	8.49							0.48‡	0.26†	0.27†
8	7.86	8.01	7.94								0.36‡	0.42‡
9	8.02	8.11	8.06									0.43‡
10	6.55	6.84	6.69									

† denotes $p < .05$, and ‡ denotes $p < .005$.

different digital rights and freedoms. Meaningfulness in this case is demonstrated by the nearly perfect consistency between average ratings in two different contexts. In the narrow rating set (between-groups rating of principles), participants considered only one primary principle and several other concepts that were chosen for their close relationship to the primary principle. In the broad set (within-group rating of principles), participants saw all of the principles. These different contexts might have been thought to influence respondents differently. In terms of average ratings, that does not appear to happen in this population.

A second finding, which we can see in Table 2, is that while most of the principles tend to be significantly correlated with each other, indicating that people who tend to favor users' rights under one principle tend to favor them under other principles, there are exceptions to this pattern. The tendency of a user to favor privacy control, data portability, or universal network access (all of which are highly correlated with each other) is not predictive of a high rating for participatory design or software freedom. Indeed, in the narrow rating set, the principles fell into two groupings, and the lowest rated principles were those most associated with user participation in the software environment.

4   Users' Rights Frameworks

To illustrate the application of the principles to policy, we will analyze four policy frameworks that have been proposed over the past twenty years aimed at securing rights for users:[3] (1) *Rights and Responsibilities of Electronic Learners* (RREL, 1994). An early framework was developed as part of this project within the American Association for Higher Education (AAHE), after extensive input from the education community, and was described by American University computer science professor Frank W. Connolly [4].[4] (2) *A Bill of Rights for*

---

[3] For other users' rights framework proposals, see [7,13,23,26].

[4] An earlier paper laying out the motivations and a procedure for drafting such a document was published in 1990 by Connolly, Gilbert, and Lyman [5].

Users of the Social Web (BRUSW, 2007). Social media engineer Joseph Smarr and colleagues [30] delineated a set of "fundamental rights" to which "all users of the social web are entitled." (3) *Marco Civil da Internet* (MCdI, 2014). In recent years, Brazil has taken the lead in initiatives to define a "constitution of the Internet." In March and April, 2014, Brazil's two legislative chambers each passed the *Marco Civil da Internet* (Civil Rights Framework for the Internet). The priority placed on the *Marco Civil* followed a 2013 United Nations speech by the country's president, Dilma Rousseff, who "presented proposals for a civilian multilateral framework for the governance and use of the Internet, capable of ensuring such principles as freedom of expression, privacy of the individual and respect for human rights, as well as the construction of inclusive and non-discriminatory societies" [28,19].[5] (4) *NETmundial Draft Outcome Document* (NDOD, 2014). In its international role as a leader in recent Internet governance initiatives, Brazil was the host of the Global Multistakeholder Meeting on the Future of Internet Governance, also known as NETmundial, in April 2014. President Rousseff announced the meeting in October 2013, after revelations that the U.S. National Security Agency had monitored her phone calls and email messages [15], and the Draft Outcome Document was posted on the Web for open comment on April 14 [21].[6]

4.1   Comparison of Frameworks

An analysis based on the framework of section 2 of each of the four texts yields the results in Table 3, where a location reference means that the principle or concept is clearly and substantially present (explicitly or implied) in the text, in a positive way (meaning that the concept is affirmed as a right"; a blank entry means it is apparently not present; and a location reference followed by an asterisk "*" indicates ambiguity about whether the concept is present or not.

The table shows firstly that none of the frameworks covers all of the principles. But some are more comprehensive than others. While the RREL and MCdI frameworks span concepts in both the user data freedoms (principles 1-3) and public network freedoms (principles 7-10), the BRUSW and NDOD frameworks are more specialized. The BRUSW framework was put forward as a set of rights for social Web users, and is limited to user data freedoms. The NDOD framework, on the other hand, is a global Internet governance initiative that seeks

---

[5] Reportedly, Article 12 was struck from the draft version before final passage [20].

[6] The Draft Document was refined on April 24, 2014, into a "Multistakholder Statement," [22] but the draft document is used here because it is annotated with section references for easier analysis.

[7] Locations are coded as Article:Section as seen in [4].

[8] Locations are coded according to the inserted letters and roman numerals in the description of this framework above.

[9] The *Marco Civil* applies only to Brazil, so the public freedoms (Principles 7-10) must be understood in that light. Locations are coded as Article[:Section] as seen in [19].

[10] Locations are coded by paragraph number, as shown at
`http://document.netmundial.br/1-internet-governance-principles/`

Table 3. Analysis of Four Users' Rights Frameworks (see footnotes on previous page)

Principle/Concept	RREL[7]	BRUSW[8]	MCdI[9]	NDOD[10]
1–Privacy control				
1a Originator-discretionary reading control	I:3,IV:4	a,b	7:VII,8,10	
1b Data use transparency	I:3	b*	7	
1c Usable privacy				
1d Nonretention of data			15*,16*,17*	
2–Data Portability				
2a Free data access	I:5*	a*,c*		
2b Open formats		i		
2c Platform independence		c,i,ii,iii		
2d Free deletion	I:5*	b*	7:X	
3–Creative control				
3a Originator-discretionary editing control	I:3,I:5			
3b Authorial copyright support	I:5		20*	
3c Reciprocal data sharing				
4–Software freedom				
4a Open Source code				
4b Reciprocal code openness				
4c User modifiable platform				
5–Participatory design				
5a User-centered design				
5b User input to design				
5c User-generated design				
5d Customizable design				
6–User self-governance				
6a Participatory policy making				
6b Participatory implementation				
6c Participatory adjudication				
7–Universal network access				
7a Right to connect	I:1,IV:1		7:III	7,23
7b Universal digital literacy	I:2		7:XI,19:VIII,27	23
7c No- or low-cost service				23
7d Omnipresent service			7:IV	10,11
7e Accessibility			25	6,23
8–Freedom of information				
8a Right to privacy	I:3,IV:2		7,8,10,11	5
8b Right to anonymous speech	I:4			
8c Freedom from censorship	I:4			3
8d Open Access to publicly funded data				
8e Democratically controlled security			10:IV*	
8f Right to be forgotten				
9–Net neutrality				
9a Source neutrality			9	12*
9b Format neutrality			9	12*
9c Content neutrality			9	12*
9d End-user neutrality			9	12*
10–Pluralistic open infrastructure				
10a Multiplicity of platforms	III:1		19:VII+X	11
10b Decentralized control	III:3		19:I-VI	13,15,16,19-22,24
10c Transparent control				17,25

only to regulate at the international level. RREL and MCdI span two regions of the table for different reasons. RREL was an early and somewhat more vague attempt to establish principles that might apply either to public policy or to users of a specific platform. MCdI, on the other hand, is a draft law for a specific jurisdiction (Brazil) with authority to regulate software platforms that are subject to the country's laws, so that it may limit the freedom of platforms in the course of regulating at a national level.

None of the four frameworks analyzed above (or the additional ones referenced in footnote 3) include provisions that appear to enact what are herein called software platform freedoms (principles 4-6). It appears, from these data, that giving users power over their software environment, through software freedom, participatory design, and user self-governance, are not strong values among those who have constructed these frameworks. These were also the lowest rated three principles in the survey reported in section 3.

4.2   Benefits of an Analysis Framework

The principles and concepts of section 2 comprise an analysis framework, as opposed to the policy frameworks analyzed in Table 3. An analysis framework of this kind gives us the following types of leverage for understanding users' rights policies: (a) it allows for easier comparison across frameworks; (b) it allows us to see what is missing from a *particular* policy framework; (c) it facilitates further study of the dimensions that characterize users' rights, e.g. surveys of users and policy makers to determine strengths of priority for different freedoms; and (d) it allows us to see persistent gaps *across* policy frameworks, such as the apparent lack of attention to software platform freedoms (principles 4-6).

5   Conclusion

The current moment is one of revival for the idea of a "bill of rights" or "constitution" for users online. On one hand, some observers have expressed skepticism about the feasibility of this concept, particularly at the International level (e.g. [3]). On the other hand, the growing control as well as documented instances of misuse of power by governments appear to have fed this new level of interest on the part of figures such as Tim Berners-Lee and Dilma Rousseff. Whether this will translate into lasting change remains to be seen. But there remain many levels at which policies can be adopted, from particular software platforms and small online communities to the entire world.

It seems likely that discussions about the principles that govern Internet users will continue to pick up steam in the years ahead, and, if the history of other major shifts in civilization is a guide, the process will lag technological change considerably. If many people think carefully about the principles they want to govern their own use of the Internet, articulate those principles, and invoke them in discussing policy proposals, we may have a better chance of arriving at arrangements that satisfy most users and that meet the needs of contemporary societies.

References

1. Amazon Mechanical Turk, `http://www.mturk.com`
2. Cerf, V.: The Right to Connect and Internet Censorship. New Perspectives Quarterly 29(2), 18–23 (2012)
3. Clark, J.: Internet Users Bill of Rights: DOA. The Data Center Journal (March 13, 2014), `http://www.datacenterjournal.com/it/internet-users-bill-rights-doa/`
4. Connolly, F.W.: Who Are Electronic Learners? Why Should We Worry About Them? Change 26(2), 39–41 (1994)
5. Connolly, F., Gilbert, S.W., Lyman, P.: A Bill of Rights for Electronic Citizens. OTA, Washington, DC (1990)
6. Creative Commons, `http://creativecommons.org`
7. Curtis, R.: A Software User's Bill of Rights - Draft 1, `http://www.princeton.edu/~rcurtis/softrights.html`
8. Data Portability Project, `http://dataportability.org`
9. Davies, T., Gangadharan, S.P. (eds.): Online Deliberation: Design, Research, and Practice. CSLI Publications, Stanford (2009)
10. Do Not Track, `http://donottrack.us`
11. Economides, N.: "Net Neutrality," Non-Discrimination, and Digital Distribution of Content Through the Internet. I/S: A Journal of Law and Policy 4, 209–233 (2008)
12. GNU Affero General Public License. Version 3 (November 19, 2007), `http://www.gnu.org/licenses/agpl-3.0.html`
13. Karat, C.M.: The Computer User's Bill of Rights (1988), `http://theomandel.com/resources/users-bill-of-rights`
14. Karat, C.M., Brodie, C., Karat, J.: Usable Privacy and Security for Personal Information Management. Comm. ACM 49(1), 56–57 (2006)
15. Kelion, L.: Future of the Internet Debated at NETmundial in Brazil. BBC News (April 22, 2014), `http://www.bbc.com/news/technology-27108869`
16. Kensing, F., Blomberg, J.: Participatory Design: Issues and Concerns. CSCW 7, 165–185 (1998)
17. Kiss, J.: An Online Magna Carta: Berners-Lee Calls for Bill of Rights for Web. The Guardian (March 11, 2014)
18. Lessig, L.: Free Culture: The Nature and Future of Creativity. Penguin, New York (2005)
19. English Translation of the New Version of Brazil's Marco Civil, `http://infojustice.org/archives/31272`
20. Mari, A.: Brazil Passes Groundbreaking Internet Governance Bill. ZDNet (March 26, 2014), `http://www.zdnet.com/brazil-passes-groundbreaking-internet-governance-bill-7000027740`
21. NETmundial Draft Outcome Document (April 14, 2014), `http://document.netmundial.br/`
22. NETmundial Multistakeholder Statement (April 24, 2014), `http://netmundial.br/wp-content/uploads/2014/04/NETmundial-Multistakeholder-Document.pdf`
23. O'Reilly, D.: Time to Update the Software User's Bill of Rights (December 29, 2009), `http://www.cnet.com/news/time-to-update-the-software-users-bill-of-rights`

24. Qualtrics: Online Survey Software and Insight Platform,
    `http://www.qualtrics.com`
25. Parens, B.: The Open Source Definition. In: Di Bona, C., Ockman, S., Stone, M. (eds.) Open Sources: Voices From the Open Source Revolution, O'Reilly, Sebastopol (1999)
26. Patrianakos, B.: The Internet User's Bill of Rights (2014),
    `http://userbillofrights.org/`
27. Rosen, J.: The Right to Be Forgotten. Stanford Law Rev. Online. 64, 88 (2012)
28. Mrs, H.E.: Dilma Rousseff, President (September 24, 2013), `http://gadebate.un.org/68/brazil/`
29. Schuler, D., Namioka, A.: Participatory Design: Principles and Practices. Lawrence Erlbaum Associates (1993)
30. Smarr, J., Canter, M., Scoble, R., Arrington, M.: A Bill of Rights for Users of the Social Web (2007), `http://the.networkingur.us/post/2007/09/05/A-Bill-of-Rights-for-Users-of-the-Social-Web`
31. Stallman, R.: Free Software, Free Society: Selected Essays of Richard M. Stallman. Free Software Foundation, Cambridge, MA (2002)
32. Viégas, F.B., Wattenberg, M., McKeon, M.M.: The Hidden Order of Wikipedia. In: Schuler, D. (ed.) HCII 2007 and OCSC 2007. LNCS, vol. 4564, pp. 445–454. Springer, Heidelberg (2007)
33. Web We Want, `https://webwewant.org`
34. Your Portability Policy, `https://web.archive.org/web/20120126204316/http://portabilitypolicy.org/sample-policies.html`

Integrating Social Media Communications into the Rapid Assessment of Sudden Onset Disasters

Sarah Vieweg, Carlos Castillo, and Muhammad Imran

Qatar Computing Research Institute, Doha, Qatar
{svieweg,mimran}@qf.org.qa, chato@acm.org

Abstract. Recent research on automatic analysis of social media data during disasters has given insight into how to provide valuable and timely information to formal response agencies—and members of the public—in these safety-critical situations. For the most part, this work has followed a bottom-up approach in which data are analyzed first, and the target audience's needs are addressed later.

Here, we adopt a top-down approach in which the starting point are information needs. We focus on the aid agency tasked with coordinating humanitarian response within the United Nations: OCHA, the Office for the Coordination of Humanitarian Affairs. When disasters occur, OCHA must quickly make decisions based on the most complete picture of the situation they can obtain. They are responsible for organizing search and rescue operations, emergency food assistance, and similar tasks. Given that complete knowledge of any disaster event is not possible, they gather information from myriad available sources, including social media.

In this paper, we examine the rapid assessment procedures used by OCHA, and explain how they executed these procedures during the 2013 Typhoon Yolanda. In addition, we interview a small sample of OCHA employees, focusing on their uses and views of social media data. In addition, we show how state-of-the-art social media processing methods can be used to produce information in a format that takes into account what large international humanitarian organizations require to meet their constantly evolving needs.

Keywords: Crisis informatics, Microblogging, Humanitarian computing.

1 Introduction

The role of social media as a conduit for useful information during emergencies is increasingly acknowledged and accepted by formal response and humanitarian agencies [11]. We focus on the information gathering processes of a large, diverse, international humanitarian relief agency. We explain how the United Nations Office for the Coordination of Humanitarian Affairs (UN OCHA, or OCHA), views social media data, which are considered a legitmate source of information during the data-gathering process OCHA goes through when they respond to a crisis (though concerns exist) [27]. In addition, we discuss apprehensions some have about incorporating social media data in rapid assessment procedures. Informed by these observations, we perform both human and automatic analyses of tweets broadcast during Typhoon Yolanda, and provide the

L.M. Aiello and D. McFarland (Eds.): SocInfo 2014, LNCS 8851, pp. 444–461, 2014.

findings to OCHA. Each of these steps research leads to suggestions for how humanitarian agencies can use social media communications, and critically, on how methods to process social media data can effectively support these agencies.

1.1 Related Work

Much research on processing social media data during emergencies has focused on applying computational methods—such as Information Retrieval, Natural Language Processing, and/or Machine Learning—to the creation of systems for filtering, classifying, and summarizing messages (for a recent survey, see [12]).

In addition, an interdisciplinary line of work looks at how the end-users of these systems—including various formal response organizations and agencies—use social media during disasters to understand the situation, coordinate relief efforts, and manage information. For instance, several articles analyze how social media was used by response agencies during the 2010 Haiti earthquake. Starbird [25] studies how Twitter was used by hospitals to broadcast availability to care for victims. Sarcevic et al. [23] shows how it was used to report on the relief activities of various medical groups. Goggins et al. [9] analyzes an online discussion forum used by the US Navy to coordinate with NGOs during the same event, and finds that forum discussions correspond with "on the ground" activities during the earthquake.

Beyond the 2010 Haiti earthquake, Denef et al. [5] describe how two different police departments used Twitter in response to the 2011 riots that took place in England, and find that disparate adoption styles led to different images and relationships with the public. Cobb et al. [4] characterize the use of social media in disasters by "digital volunteers," and provide insight into how to best support the needs of geographically-dispersed volunteers when they respond to mass emergencies. Hughes [10] studies the usage of social media by Public Information Officers, who handle the public relations aspects of emergency response in the United States.

We contribute to this literature by showing how a large humanitarian organization can adapt its internal procedures—particularly those related to rapid assessment of an emergency situation—to incorporate social media communications. In particular, we focus on the information needs of OCHA, and give insight into the process of providing organizations with particular data that fits within their defined procedures and established information needs. By providing OCHA with Twitter data that contain information they specifically require to better oversee the management of dozens of organizations who perform myriad tasks related to disaster releif, the hope is that we ease the burden on those responsible for collecting and analyzing in the immediate post-impact period of a disaster. In addition, we show how existing methods can lead to more productive uses of social media data by stakeholders, and affected populations.

1.2 Background

We focus on Typhoon Yolanda (internationally known as Typhoon Haiyan), one of the strongest tropical cyclones ever recorded. The typhoon made landfall in the central Philippines on November 8, 2013, affecting over 14 million people, killing over 6,000, leaving over 4 million displaced, and causing billions of US dollars in property damage.

Though it was predicted and residents in some areas were able to take precautions, the typhoon brought about devastating effects.[1] Many national and international aid organizations were deployed to respond to Typhoon Yolanda, including OCHA.

According to the International Monetary Fund, the Philippines is an emerging market.[2] However, despite 24% of the population being classified as "poor" [2], telecommunications infrastructure is widespread. Nearly every Filipino adult has access to a mobile phone, and the cellular network covers almost the entire country.[3] Regarding Twitter use, the Philippines is ranked 10th in the world for number of Twitter accounts,[4] and English is widely spoken due to its former occupation by the United States. When Typhoon Yolanda struck the Philippines, the combination of widespread network access, high Twitter use, and English proficiency led to many located in the Philippines to tweet about the typhoon in English. In addition, outsiders located elsewhere tweeted about the situation, leading to millions of English-language tweets that were broadcast about the typhoon and its aftermath.

2   OCHA Procedures in Disaster

When a sudden-onset disaster happens and government capabilities are exceeded, OCHA is mobilized.[5] OCHA is tasked with quickly assessing the situation "on the ground" and coordinating response efforts. The UN uses a framework for division and organization of needs during humanitarian crises based on eleven different *clusters*, which are "groups of humanitarian organizations (UN and non-UN) working in the main sectors of humanitarian action ... [c]lusters provide a clear point of contact and are accountable for adequate and appropriate humanitarian assistance."[6] The clusters are: Logistics, Nutrition, Emergency Shelter, Camp Management and Coordination, Health, Protection, Food Security, Emergency Telecommunication, Early Recovery, Education, and Water, Sanitation and Hygiene (WASH).

We can think of the role of OCHA as overseeing the organization of humanitarian response in disasters. Their responsibilities include: assessing the situation, understanding the needs of the affected population and of responding organizations, deciding on priorities, obtaining access to affected areas (which may have political as well as logistical implications), ensuring sufficient funding and resources, consistently and clearly

---

[1] http://reliefweb.int/disaster/tc-2013-000139-phl

[2] http://www.imf.org/external/country/phil/index.htm?type=9988

[3] http://www.infoasaid.org/guide/philippines/telecommunications-overview

[4] http://business.inquirer.net/111607/telcos-report-record-number-of-customers
http://www.mediabistro.com/alltwitter/twitter-top-countries_b26726

[5] The UN differentiates between "slow-onset" and "sudden-onset" disasters. Sudden-onset disasters occur quickly and with little to no warning. Slow-onset disasters have a longer period of buildup, and extend for greater periods of time. [3].

[6] http://www.unocha.org/what-we-do/coordination-tools/cluster-coordination

communicating with the public, and monitoring progress.[7] Our goal in this research is to use OCHA's needs as the lens through which we examine Twitter communications broadcast during Typhoon Yolanda. In addition, we examine data compiled from interviews with OCHA employees and consider how information communicated via Twitter may (or may not) be used by aid organizations, and how advocates within the UN who strive to incorporate social media data into their assessments and decision-making process can further their cause.

In response to disasters, the international humanitarian community undertakes a series of actions called the "Humanitarian Program Cycle." OCHA manages the cycle while working with additional agencies, NGOs, and other technical bodies. One of the goals of the program cycle is to issue a MIRA (Multi Cluster/Sector Initial Rapid Assessment) report two weeks after disaster onset.

2.1 The MIRA Framework

The MIRA framework specifies how to quickly assess the needs of affected populations, and assign responsibilities to various response agencies soon after disaster impact.[8] The goal of MIRA is threefold: 1) to systematically collate and analyze secondary data to provide an accurate-as-possible understanding of the situation; 2) to perform a "community level assessment," where aid workers and volunteers interview and talk with affected populations to gather primary data, and; 3) to bring data together into a coherent picture that provides decision makers with a current report that incorporates information from the various clusters and allows them to have a common understanding of the situation. The MIRA process starts during the *immediate post-impact period* ([7, page 8]) of sudden-onset disasters.

The first step in the MIRA process is to produce a "Situation Analysis" report, written and released within the first 48 hours after impact. It starts with an overview that includes crisis severity, priority needs, and government capacity to respond. This is followed by the "humanitarian profile," which describes the population, estimated number of affected people, and casualties, among other details.

The Situation Analysis is comprised of "secondary data," and "primary data." Secondary data describe the population of concern under typical circumstances (i.e. poverty level), as well as data that have been collected from sources such as mainstream media and satellite imagery. The secondary data form an up-to-date picture of the situation that provides a common understanding among stakeholders. Primary data are collected on the ground in the immediate aftermath of the disaster. Teams of aid workers and volunteers conduct interviews with "key informants," i.e. those who are most likely to know the current state of the population, what are the most pressing needs, and how vulnerable populations are affected.[9] Due to the quick turnaround necessary to produce the Situation Analysis, and the potential for Twitter communications to provide data

---

[7] http://www.unocha.org/what-we-do/coordination/overview

[8] https://docs.unocha.org/sites/dms/Documents/mira_final_version2012.pdf

[9] OCHA recruits key informants knowing that they will obtain a purposive sample of data sources that are not necessarily representative of the affected population.

that is potentially useful as well as broadly representative of population needs in those first 48 hours of response, we focus on this aspect of the MIRA process.

2.2   Situation Analysis of Typhoon Yolanda

The Situational Analysis for Typhoon Yolanda [2] was released on November 10, 2013. It starts with a list of six priority needs, and goes on to provide a high-level overview of the current situation in hard-hit cities, explaining where people are without food and water, where electricity is not available, and what areas are inaccessible. Sources include the Philippines' government, Red Cross, Atmospheric, Geophysical and Astronomical Service Administration, and Department of Social Welfare and Development, in addition to existing demographic and census data.

This Situation Analysis also includes points such as "typical assistance needs," how impact "may" affect the local population, and how the typhoon "can" cause additional complications and problems. i.e. much of this report is based on previous knowledge of similar events, and describes what is *likely* to happen, and where supplies and services will *typically* be needed, but it does not include first-hand accounts of the situation. This is *not* a criticism of OCHA—it is not possible for any individual nor organization to grasp the situation on the ground so soon after a large-scale disaster that affects millions of people, across a large geographical area. OCHA's reliance on past experience and expected needs is necessary in the first days after disaster impact, when getting substantial data about the situation on the ground is so difficult.

Given the difficulty of assessing a disaster situation in such a short time period, OCHA is open to using new sources—including social media communications—to augment the information that they and partner organizations so desperately need in the first days of the immediate post-impact period. As these organizations work to assess needs and distribute aid, social media data can potentially provide evidence in greater numbers than what individuals and small teams are able to collect on their own.

2.3   Social Media Experiment in Typhoon Yolanda

OCHA attempted to gain information from Twitter communications during the immediate post-impact period of Typhoon Yolanda, to triangulate and/or augment information they had from other sources. The social media data OCHA employees had access to during the crisis were gathered and analyzed by MicroMappers.[10]

MicroMappers is a digital volunteer organization devoted to annotating and mapping tweets (and other data) produced during disasters. In the case of Typhoon Yolanda, OCHA contacted MicroMappers to see what information they could gain from Twitter communications. Starting on November 7, 2013 at 19:28 (GMT) engineers began to collect tweets that contained specific keywords that described the typhoon and/or relief efforts. Data collection continued until November 13, 2013 at approximately 12:00 (GMT). Details are provided in Section 4.1.

A set of tweets sent between the start of the collection and November 10, 2013 (at 07:00 GMT) were first sampled for quick response efforts on behalf of MicroMappers. These tweets were uploaded to the MicroMappers platform for volunteers to read

---

[10] http://www.micromappers.org/

and label based on the information they contained. A total of 3,678 tweets were labeled by volunteers. The categories volunteers used to label tweets at this early stage in the response efforts are detailed below. These categories were quickly identified during the typhoon response by MicroMappers volunteers working around the clock to label and organize information communicated via Twitter; they do not claim these categories were all-encompassing, nor that they were representative of all tweets about the typhoon. The three categories that appeared to be most salient in the immediate post-impact period, and which MicroMappers used to start their tweet-labeling procedure were:

- *Infrastructure Damage*: Information about destruction and/or damage of roads, bridges, buildings; disruptions to basis services, e.g. hospitals.
- *Community Needs*: Information about shelters, food, location of missing persons, water, and hygiene.
- *Humanitarian Support*: Information about deployment of aid, recovery services, and in-kind donations and contributions of goods and services.

In addition, volunteers produced a map including 600 of these tweets that were associated to a location based on geographical references contained in them. All the labeled tweets were then sent to OCHA for further analysis.

The Situation Analysis for Typhoon Yolanda mentions the Twitter data provided by MicroMappers, and includes approximately one page about social media data. This section on social media shows a general map containing data produced by MicroMappers volunteers. However, no specific information about tweet content is provided.

As expressed by the OCHA staff interviews we describe in the following section, OCHA's *hope* was that social media data could contribute to a better understanding of the situation on the ground in these three specific areas. However, at the time the Situation Analysis was published, they did not have sufficient data. Later analysis of the Typhoon Yolanda MIRA report and additional documents do not indicate that OCHA was able to garner additional information about these topics from social media communications. However, OCHA is hopeful that future disaster response efforts will successfully involve social media data.

3 Information Management in Practice

To better understand disaster response from OCHA's perspective, and how they perceive social media data in these situations, we interviewed OCHA employees about their experiences during the response to Typhoon Yolanda, and about the potential for social media data vis-à-vis response efforts going forward. OCHA staff described the procedures they follow, spoken candidly about the challenges they face when responding to disasters, and worked with us to formulate ideas on how to best use social media data when responding to disasters.

3.1 Information Management Roles

The OCHA employees tasked with "information management" during disaster situations are "information management officers" or "IMOs," as well as "humanitarian affairs officers," or "HAOs."

The duties of IMOs and HAOs in disaster response are to drive coordination. Different IMOs and HAOs have different skill sets, but as a group, they are tasked with gathering data, liaising with various cluster leaders, communicating with volunteers, updating databases and common data repositories, and producing a variety of documents. In the immediate aftermath of a disaster, they often experience "ad-hoc craziness" brought on by a need to complete myriad tasks in a short period of time [27]. Additionally, they answer requests from all manner of stakeholders, and are responsible for writing reports that provide up-to-the-minute information.

When IMOs and HAOs collect data in the field, they focus on eight themes that guide the MIRA framework: 1) drivers of the crisis and underlying factors; 2) scope of the crisis and humanitarian profile; 3) status of populations living in affected areas; 4) national capacities and response; 5) international capacities and response; 6) humanitarian access; 7) coverage and gaps; 8) strategic human priorities. The MIRA framework includes specific questions that coincide with each of these themes to guide OCHA employees, as well as others who may be working with them, to collect data.

3.2 Interviews with OCHA Staff

As yet, social media data are a somewhat amorphous source of information for OCHA. The population of OCHA staff who can speak to the use of social media data in disaster response is relatively small; thus, we were able to secure interviews in person, and via phone, with four OCHA staffers. While we recognize the small sample size, we nevertheless stress that—together with the documentation we analyzed—the insight and firsthand knowledge we gained by speaking to these interviewees provides a sufficient backdrop regarding the potential benefits and hindrances to using social media data, particularly in the large, multi-organization coordination efforts that OCHA undertakes.

Our first interview was with an OCHA staffer in New York, NY, United States, who we refer to as O1. O1 provided us with a big picture view of what OCHA staff are responsible for when they deploy in disaster situations, and also gave insight into the Typhoon Yolanda response effort. S/he laid out the initial background information regarding the role of OCHA, the types of information they need to collect and organize when assessing a situation, and how they usually perform the myriad tasks for which they are responsible. Our discussion with O1 provided us with the foundational information we needed to understand what OCHA does, and helped us frame questions and points for discussion for our subsequent interviews.

Our next interview was with an HAO based in Geneva, Switzerland (who we refer to as O2). His/her job in the first 48 hours after impact is to quickly compile as much information as possible about the area of impact and the current situation, organize it into a coherent narrative, and present it in a Situation Analysis.

For Typhoon Yolanda, in addition to the traditional sources that OCHA turns to, O2 and his/her colleagues were also open to seeing what information they could gain from Twitter data. They received the dataset of 3,678 labeled tweets from MicroMappers, in which each tweet was associated with one of three categories of information described in the previous section. O2 and colleagues looked at the content of these tweets; their impression was that around 200-300 of them provided what they considered relevant information. In addition, they found that reading and analyzing tweets was an interesting exercise, but it was very time consuming. During those initial hours of disaster response, so much work needs to happen so quickly that the OCHA employees who responded to Typhoon Yolanda are not sure the social media data augmented what they already knew. Overall, in this case, they felt "the time investment was too high." O2's experience speaks to the need for a way to process social media data that addresses and centers on their specific information requirements. In other words, O2 is implying that s/he needs to get Twitter data that are processed from a "top-down" perspective.

O2 was working from UN headquarters in Geneva during the days after Typhoon Yolanda made impact, and produced the Situation Analysis from there. Subsequently, O2 traveled to the Philippines after the Situation Analysis was released, and continued to work on the MIRA report, which is published two weeks after impact. During the time in the Philippines, s/he had access to the primary data that were collected from in-person interviews with key informants, and additional sources of local knowledge.

O2 observed that in comparing social media data to primary data, there seemed to be a considerable bias in the social media data toward those located in urban areas, with access to telecommunications networks.[11] However, despite the (well-understood and often inherent) bias of social media data, eyewitness accounts, first-hand knowledge and additional useful information captured via social media can still augment situational awareness, and OCHA is aware of this.

Subsequent discussions with HAOs and IMOs have provided further insight into the difficulty OCHA employees face when they perform a rapid needs assessment and write the Situation Analysis; getting a handle on the needs of a large population (i.e. millions of people) that has been severely affected by a disaster is a monumental task. IMOs, HAOs and other UN employees are ready and willing to use any viable source of information available to help them better understand and assess these situations. They *want* to use social media data; the question is how to provide them with these data in a timely, easily understandable format that they can use to triangulate and/or augment other sources within the immediate post-impact period—and which correlate to the specific information they are seeking.

Another interview with a OCHA employee (an IMO based in Geneva who we refer to as O3) revealed further difficulties with including social media data into disaster response procedures. Though O3 sees great potential in incorporating social media data in disaster response, s/he points out that using social media as a bona fide source of information in crises is a tough sell to UN management. O3 points to the notion among many at the UN is that social media data are more likely to contain "bad," "false," or "unverified" information persists. S/he also pointed out the problem of information expiration—information that is posted on social media sites often has a short period of time during which it is "true," or "actionable."

In further discussion, O3 also stressed that the role of UN agencies in disaster is not to respond to individual requests. If OCHA staff see a tweet about a trapped family, or where someone needs medical attention—regardless of whether the information is verified—they are not in a position to act upon such information. Rather, OCHA seeks a collective view of the situation, particularly with respect to the eleven clusters, and with respect to the location of various needs.

---

[11] This bias has been observed in other domains, particularly politics, see e.g. [8, 20].

In OCHA's view, social media data could contribute to this type of assessment by e.g. counting how many tweets are being sent (or not) from particular areas, how many tweets mention the need for food, water, or other supplies, and to locate tweets containing specific information about macro-level population needs, e.g. *"2,000 people in <village> are affected by the typhoon—all need shelter."* This assessment is provided by another interviewee—O4, an IMO also based in Geneva—who points to the potential for Twitter data to "complete the picture" when OCHA is trying to gain an overview of the situation and ascertain how to coordinate and activate the various cluster agencies.

Equipped with this understanding of how social media data can be of most use to OCHA, we show how current technologies can be used to develop reports of social media data that could be readily incorporated into OCHA processes, with a focus on humanitarian clusters and regional location of needs.

4   Data Analysis

Having spent time with OCHA staff who are open to using social media data, our next step was to perform an analysis of a separate dataset of tweets collected during Typhoon Yolanda (i.e. different from the datset that was labeled by MicroMappers.) Our express goal was to identify information that coincides with the UN humanitarian clusters. We then determined how Twitter data compares to and/or augments the information the IMOs and HAOs are typically able to collect within the first 48 hours of a disaster.

4.1   Data Collection and Pre-processing

To obtain a set of tweets sent during Typhoon Yolanda that were likely to include information about the event, we performed a keyword search using Twitter's Streaming API; keywords included: "YolandaPH," "Yolanda," "RescuePH," "TyphoonHaiyan," and many more that were identified during the typhoon by colleagues who were closely monitoring the Twitter stream as the event unfolded.[12] The keyword search resulted in a dataset of 2,302,569 tweets from November 7, 2013 19:28 (GMT) to November 13, 2013 12:00 (GMT), as shown in Table 1. Though many tweets about the typhoon were posted weeks after this time period, we stopped data collection on November 13 at 12:00 (GMT) because our OCHA interviewees stated that this six-day period would be of most interest to them. Further, we divided the dataset into two periods.

The first period represents the time frame OCHA considered for the Typhoon Yolanda Situation Analysis report. This period within our dataset consists of 1,173,850 tweets. In addition to the first set of tweets, we also consider tweets posted during the next 48 hours after the first period. The second period is from November 10, 2013 20:31 (GMT) to November 13, 2013 12:00 (GMT), and contains 1,128,719 tweets. We include tweets from this second time period in our analysis to determine: (i) to what extent the information posted on Twitter changes after the initial period, and (ii) to what extent those changes may affect OCHA's ability to gain situational awareness information that may

---

[12] More details on the data collection, including sampled keywords, are included in the data release, see URL at the end of this paper.

Table 1. Breakdown of tweets into two time periods

Period	Start (GMT)	End (GMT)	# of Tweets
First	Nov. 7, 2013 19:28	Nov. 10, 2013 20:30	1,173,850
Second	Nov. 10, 2013 20:31	Nov. 13, 2013 12:00	1,128,719
Total			2,302,569

be included in later reports (i.e. the MIRA report, and other reports that are generated after the Situation Analysis is released.) It was only after all tweets were collected and we had done some preliminary analysis that we spoke with OCHA employees.

4.2 Automatic Classification by Region

The Philippines are divided into seventeen different regions, or administrative divisions. The UN breaks down its analysis per region, as shown in Table A1 (in the Appendix). The information in this table is based on the most complete data provided by the UN, which they released on November 23, 2013. Regarding the UN-provided data, some regions have no data available; these regions were either unaffected, or no data were provided about them. Previous work points to social media activity increases in regions affected by a disaster [6, 24]. However, this is not always the case, as frequency of social media postings can increase in areas that are not strongly affected by a disaster.

We measure to what extent the number of tweets sent from particular regions correlate with amounts of damage or number of affected people. We classify tweets by region using two strategies. First, by *geolocation*, for tweets that include GPS coordinates, which are added by mobile clients when the user enables this functionality. Second, by *keyword*, i.e. we considered all tweets that mentioned the name of a region or the name of any municipality in that region. Table A1 shows the results. We note that while the activity on Twitter was in general more significant in regions heavily affected by the typhoon, the correlation is not perfect. For instance, there are more tweets from the National Capital Region and from CALABARZON, which were not among the most affected, than from the Bicol Region, were more than half a million people were affected. Though our results show that classifying tweets by region was not a reliable undertaking in this particular case, we maintain that it is a worthwhile exercise, as it can prove useful in some circumstances [21, 22, 24].

We also attempted to measure these correlations in relative terms, e.g. by expressing the affected people as a percentage of the population, and/or by expressing the number of tweets in proportion to the tweets "normally" present in each region (using a data sample from one month prior to the crisis). Results were similar, in terms of showing some correlation but not a perfect one. We did not expect that tweets would predict the number of affected people per region, for the same reason that they do not predict winners in political elections [8]. Again, we see an example that likely points to bias in the Twitter data; urban, affluent, tech-savvy people are more capable of posting to microblogging services than rural, poor populations. Knowing that these biases exist

Table 2. Initial classification task on a sample of tweets posted during Typhoon Yolanda

Category	Human Labeled		Automatically Labeled	
Informative	845	42%	1,109,480	48%
Not informative	613	31%	661,228	29%
Not related	542	27%	531,861	23%

and are likely to continue is critical for OCHA to take into account as they work to incorporate social media data into future response efforts.

4.3   Automatic Classification

We implement supervised machine learning to perform automatic classification of tweets. In this approach, a relatively small number of human-labeled items are used to train an automatic classification system (this is the "training set"). Then, this automatic classification system that has been trained on human-labeled data is used to classify the remaining tweets.

Automatic Classification of Informative Tweets. We filtered the datasets to identify messages that might contain useful information using the supervised classification approach of Random Forests. Tweets were first converted to binary feature vectors in which each word (unigram), or a sequence of two consecutive words (bigram), is a coordinate in the vector ([15, 28]). A random sample of 2,000 tweets was used as a training set. The choice of the learning approach, features types, and training set size was based the empirical evidence presented in [16]. We used crowdsourcing services from CrowdFlower, which provided us with workers who labeled tweets with the appropriate category. Workers were given the following instructions:[13]

> *Indicate if the item contains information that is useful for capturing and understanding the situation on the ground:*
> A. *Informative: contains useful information that helps you understand the situation.*
> B. *Not informative: refers to the crisis, but does not contain useful information that helps you understand the situation.*
> C. *Not related to this crisis.*

Two out of three workers' agreement was required to finalize a label.

Results are in Table 2. The resulting classifier has an AUC of 0.89, measured using 10-fold cross validation, which indicates fairly high classification accuracy. [14]

---

[13] CrowdFlower is an online crowdsourcing service that allows clients to upload tasks with instructions, which Crowdflower workers are then paid to complete: `http://crowdflower.com/`

[14] AUC is Area Under Receiver Operating Characteristic curve, 50% means a random classifier and 100% means a perfect classifier. We do not use accuracy, as it is misleading in this context.

Table 3. Classification of informative tweets posted during Typhoon Yolanda, according to the Humanitarian Clusters Framework. Up/down arrows indicate relative increase/decrease of 50% or more in period 2, proportional to the total number of tweets in each period.

Cluster	Human Labeled (period 1)	Automatically Labeled (period 1)	Automatically Labeled (period 2)
Food and nutrition	54	4,712	39,448 ↑
Camp and shelter	41	1,870	8,470 ↑
Education and child welfare	50	18,076	22,198 ↓
Telecommunication	90	8,002	5,899 ↓
Health	57	1,008	2,487
Logistics and transportation	51	2,290	3,259
Water, sanitation, and hygiene	31	1,210	82,568 ↑
Safety and security	87	7,884	4,970 ↓
Early recovery	216	14,602	46,388
None of the above	1,323	382,906	451,122 ↓
Total	2,000	442,560	666,809

Automatic Classification into the Cluster Framework. Next, we again turned to crowdsource workers to perform manual labeling, and used the output to train an automatic classifier.[15]

To reduce the amount of false positives—i.e. messages automatically classified as informative, but not containing useful information—we imposed the constraint that classification confidence must be higher than 0.8. The first data period yielded 270,781 tweets (23% of tweets during that period), from which 2,000 tweets were sampled uniformly at random and labeled according to the humanitarian clusters (the same constraint in the second data period yields 351,070 tweets: 53% of tweets during that period, which suggest an increase in informative content, consistent with Table 2).

Results of both the manual and automatic classification are shown in Table 3, where we also indicate whether there is an increase or decrease of 50% or more in the proportion of messages in each cluster. In the first time period (roughly the first 48 hours), we observe concerns focused on early recovery and education and child welfare. In the second time period, these concerns extend to topics related to shelter, food, nutrition, and water, sanitation and hygiene (WASH). At the same time, there are proportionally fewer tweets regarding telecommunications, and safety and security issues.

In general, Table 3 shows a significant increase of useful messages for many clusters between period 1 and period 2. It is also clear that the number of potentially useful tweets in each cluster is likely on the order of a few thousand, which are swimming in the midst of millions of tweets. This point is illustrated by the majority of tweets falling into the "None of the above" category, which is expected and has been shown in previous research [29].

---

[15] For this Crowdflower labeling task, we grouped "camp management" and "shelter" clusters together, and "food security" and "nutrition" clusters together for clarity, which gave us a total of 9 cluster categories from which workers could choose.

4.4   Drilling Down into Clusters: Topic Models

OCHA staff indicated their preference for being presented with aggregate information, as opposed to a list of individual tweets. In this section, we examine how information relevant to each cluster can be further categorized into useful themes. We employ topic modeling using Latent Dirichlet Allocation (LDA) [18]; a common method used to analyze datasets of thousands or millions of documents, and whose application to disaster-related tweets is described in [17].

Results of the topic modeling, including example tweets and representative words according to the LDA algorithm, are in Table A2. Due to a small number of items in the clusters, two themes were generated for most of them. However, some clusters e.g. "telecommunications, safety and security" resulted in only one theme because the majority of tweets in that cluster mention the same words/information.

Topic models allow us to quickly group thousands of tweets, and to understand the information they contain. In the future, this method can help OCHA staff gain a high-level picture of what type of information to expect from Twitter, and to decide which clusters or topics merit further examination and/or inclusion in the Situation Analysis.

Feedback from the UN. To find if we were on a helpful path regarding our post-hoc analysis of Typhoon Yolanda Twitter data, we asked OCHA staff to look at the information we present in Table A2. We provided a description of the data, and explained that though the data are from a past event, we were concerned with whether they could use this type of information in future events.

Feedback was positive and favorable. Regarding the information in Table A2, O4 said: "it could potentially give us an indicator as to what people are talking most about—and, by proxy, apply that to the most urgent needs." O4 goes on to say "There are two places in the early hours that I would want this: 1) To add to our internal "one-pager" that will be released in 24-36 hours of an emergency, and 2) the Situation Analysis: [it] would be used as a proxy for need." Another UN staffer, who works for a non-OCHA sector in disaster response, stated: "Generally yes this [information] is very useful, particularly for building situational awareness in the first 48 hours." One staffer (O1) did express concern that this level of analysis may be too general for some applications, saying that "the [topic] words seem to general." However, s/he went on to say that the table gives a general picture of severity, which is an advantage during those first hours of response. This validation from UN staff supports our continued work on collecting, labeling, organizing, and presenting Twitter data to aid humanitarian agencies with a focus on their specific needs as they perform quick response procedures.[16]

5   Discussion

Twitter is established as a place to communicate, gather and disperse information, and gain situational awareness during disasters. Furthermore, research suggests that there is abundant useful information broadcast on Twitter during mass emergencies [14, 21, 26, 29]. This has led many within OCHA to view Twitter communications as a way to triangulate what they know from other, more conventional, sources.

---

[16] We were unable to get feedback from all staff we interviewed earlier due to field deployments.

5.1 Obstacles

Emergency responders face technological and organizational barriers to the adoption of social media in their processes, including a growing need for institutional change [10]. OCHA has an overwhelming amount of work to do when tasked with assessing a crisis, identifying needs, and distributing reports that provide an overview of the situation. Social media communications are yet another item on the lengthy list of sources for them to consider when attempting to gain an accurate understanding of a crisis situation.

This is an obstacle noted by others: "Even when good data is available, it is not always used to inform decisions. There are a number of reasons for this, including data not being available in the right format, not widely dispersed, not easily accessible by users, not being transmitted through training and poor information management. Also, data may arrive too late to be able to influence decision-making in real time operations, or may not be valued by actors who are more focused on immediate action." [1]

Concerns about veracity of social media information were also voiced. These issues are not unlike those faced by Public Information Officers (PIOs) in the United States who also wrestle with knowing if they can trust information that is found on social media. [10]. However, regardless of the questions around "truth," and "trust," it is clear that social media data can be used to augment situational awareness. [26, 29].

5.2 Recommendations

Providing social media data to humanitarian organizations requires, first and foremost, an understanding of how those humanitarian organizations work. Organizations that have existed for decades will rarely re-invent themselves around a new technology. However, they can be guided toward making new tools and data an established feature of their processes. In this sense, OCHA staff cited the need to know what they are likely to find—and not find—on social media when they are in the midst of a response.

The next consideration is to present the information in a format that answers target users' questions. OCHA staff are supportive of incorporating social media in their processes, but they need data to be presented in a format that is easily consumable. This echoes concerns expressed by Public Information Officers interviewed by Hughes [10], who also note the complexity of social media as an information space. OCHA does not want to read thousands of tweets; they require a high-level snapshot that explains the Twittersphere, and which they can use to augment their assessment of the situation.

This research has shed light on the fact that providing the "big picture" of a crisis situation via an analytic view of tweets is helpful to OCHA, and potentially other aid agencies. While we do not deny the value of information found in individual tweets, organizations such as OCHA require a higher-level overview of the activities and behaviors that play out on Twitter in the immediate post-impact period of a disaster. Therefore, we suggest presenting results in *multiple levels*. For example, a higher level shows the number of tweets per geographical region, followed by the number of tweets per cluster, and the topics inside each cluster (the scheme we have followed in this paper).

Finally, it is important to have the right systems in place. Given the consensus among OCHA staff that social media data are particularly valuable during the early hours of a disaster, real-time acquisition and analysis of data is critical. This involves large amounts of time and effort on behalf of many people, so in addition to digital volunteer platforms such as MicroMappers—which employs humans as a sole source of information processing—we have pointed to other systems that perform real-time social media analysis using supervised machine learning, and which incorporate humans in the process when required [13, 19].

Data Release. The data we collected is available for research purposes at `http://crisislex.org/`.

Acknowledgments. The authors wish to thank Patrick Meier and the UN participants.

References

[1] Promoting innovation and evidence-based approaches to building resilience and responding to humanitarian crises. Tech. rep., Department for International Development, UK AID (2012)

[2] Situation Analysis Philippines, Typhoon Haiyan (Yolanda). Tech. rep., United Nations Office for the Coordination of Humanitarian Affairs (Nov 2013)

[3] Capelo, L., Chang, N., Verity, A.: Guidance for collaborating with volunteer and technical communities. Tech. rep., Digital Humanitarian Network (August 2012), `http://digitalhumanitarians.com/content/guidance-collaborating-volunteer-technical-communities`

[4] Cobb, C., McCarthy, T., Perkins, A., Bharadwaj, A., Comis, J., Do, B., Starbird, K.: Designing for the deluge: Understanding & supporting the distributed, collaborative work of crisis volunteers. In: Proc. of CSCW, ACM (2014)

[5] Denef, S., Bayerl, P.S., Kaptein, N.A.: Social media and the police: tweeting practices of British police forces during the August 2011 riots. In: Proc. of CHI, pp. 3471–3480. ACM (2013)

[6] Evnine, A., Gros, A., Hofleitner, A.: On Facebook when the earth shakes. Tech. rep., Facebook Data Science team (Aug 2014)

[7] Fischer, H.W.: Response to disaster: fact versus fiction & its perpetuation—the sociology of disaster. University Press of America (1998), `http://www.worldcat.org/isbn/0761811826`

[8] Gayo-Avello, D.: Don't turn social media into another "Literary Digest" poll. Commun. ACM 54(10), 121–128 (2011), `http://doi.acm.org/10.1145/2001269.2001297`

[9] Goggins, S., Mascaro, C., Mascaro, S.: Relief work after the 2010 Haiti earthquake: leadership in an online resource coordination network. In: Proceedings of the ACM 2012 Conference on Computer Supported Cooperative Work, pp. 57–66. ACM (2012)

[10] Hughes, A.L.: Participatory design for the social media needs of emergency public information officers. In: Proc. of ISCRAM (2014)

[11] Hughes, A.L., Peterson, S., Palen, L.: Social Media in Emergency Management. FEMA in Higher Education Program (2014)

[12] Imran, M., Castillo, C., Diaz, F., Vieweg, S.: Processing social media messages in mass emergency: A survey (August 2014), `http://arxiv.org/abs/1407.7071`

[13] Imran, M., Castillo, C., Lucas, J., Meier, P., Vieweg, S.: AIDR: Artificial intelligence for disaster response. In: Proc. of WWW (Companion), pp. 159–162. International World Wide Web Conferences Steering Committee (2014)

[14] Imran, M., Castillo, C., Lucas, J., Patrick, M., Rogstadius, J.: Coordinating human and machine intelligence to classify microblog communications in crises. In: Proc. of ISCRAM (2014)

[15] Imran, M., Elbassuoni, S., Castillo, C., Diaz, F., Meier, P.: Practical extraction of disaster-relevant information from social media. In: Proc. of Workshop on Social Media Data for Disaster Management, pp. 1021–1024 (2013)

[16] Imran, M., Elbassuoni, S.M., Castillo, C., Diaz, F., Meier, P.: Extracting information nuggets from disaster-related messages in social media. Proc. of ISCRAM, Baden-Baden, Germany (2013)

[17] Kireyev, K., Palen, L., Anderson, K.: Applications of topics models to analysis of disaster-related Twitter data. In: NIPS Workshop on Applications for Topic Models: Text and Beyond, vol. 1 (2009)

[18] Mark, G., Bagdouri, M., Palen, L., Martin, J., Al-Ani, B., Anderson, K.: Blogs as a collective war diary. In: Proc. of CSCW, pp. 37–46. ACM (2012)

[19] Melville, P., Chenthamarakshan, V., Lawrence, R.D., Powell, J., Mugisha, M., Sapra, S., Anandan, R., Assefa, S.: Amplifying the voice of youth in africa via text analytics. In: Proc. of KDD, pp. 1204–1212. ACM (2013)

[20] Mitchell, A., Hitlin, P.: Twitter reaction to events often at odds with overall public opinion. Tech. rep., PEW Research Center (March 2013), http://www.pewresearch.org/2013/03/04/twitter-reaction-to-events-often-at-odds-with-overall-public-opinion/

[21] Olteanu, A., Castillo, C., Diaz, F., Vieweg, S.: CrisisLex: A lexicon for collecting and filtering microblogged communications in crises. In: Proc. of ICWSM, pp. 376–385 (June 2014), http://chato.cl/papers/olteanu_castillo_diaz_vieweg_2014_crisis_lexicon_disasters_twitter.pdf

[22] Sakaki, T., Okazaki, M., Matsuo, Y.: Earthquake shakes Twitter users: real-time event detection by social sensors. In: Proc. of WWW, pp. 851–860. ACM, New York (2010), http://dx.doi.org/10.1145/1772690.1772777

[23] Sarcevic, A., Palen, L., White, J., Starbird, K., Bagdouri, M., Anderson, K.: Beacons of hope in decentralized coordination: learning from on-the-ground medical twitterers during the 2010 Haiti earthquake. In: Proc. of CSCW, pp. 47–56. ACM (2012)

[24] Shelton, T.: Visualizing the relational spaces of hurricane Sandy (August 2013), http://www.floatingsheep.org/2013/08/visualizing-relational-spaces-of.html

[25] Starbird, K.: Delivering patients to Sacré Coeur: collective intelligence in digital volunteer communities. In: Proc. of CHI, pp. 801–810. ACM (2013)

[26] Starbird, K., Palen, L., Hughes, A.L., Vieweg, S.: Chatter on the red: what hazards threat reveals about the social life of microblogged information. In: Proc. of CSCW, pp. 241–250. ACM, New York (2010), http://dx.doi.org/10.1145/1718918.1718965

[27] Verity, A., Mackinnon, K., Link, Y.: OCHA information management guidance for sudden onset emergencies. Tech. rep., UN OCHA (Feb 2014)

[28] Verma, S., Vieweg, S., Corvey, W.J., Palen, L., Martin, J.H., Palmer, M., Schram, A., Anderson, K.M.: Natural language processing to the rescue?: Extracting "situational awareness" tweets during mass emergency. In: Proc. of ICWSM (2011)

[29] Vieweg, S.: Situational Awareness in Mass Emergency: A Behavioral and Linguistic Analysis of Microblogged Communications. Ph.D. thesis, University of Colorado at Boulder (2012)

Appendix

In this appendix, we include details of people and houses affected by a disaster, compared with geo-located tweets by region (Table A1). We also include details of the results of topic modeling for each UN Cluster (Table A2).

Table A1. A chart showing the number of affected people and houses per region, compared with both the number of tweets geo-located in that region ("by geolocation"), as well as the number of tweets that contain a region or municipality name ("by keywords"). Source: United Nations Typhoon Yolanda data reports, November 2013.

	OCHA Information		Number of Tweets	
Region designation and name	Affected people	Affected houses	By geolocation	By keywords
I Ilocos Region	–	–	228	189
II Cagayan Valley	–	–	344	1,905
III Central Luzon	–	–	705	575
IV-A CALABARZON	27,076	840	2,034	2,524
IV-B MIMAROPA	425,903	33,499	150	1,339
V Bicol Region	695,526	12,129	1,372	1,214
VI Western Visayas	2,694,031	476,844	14,110	6,329
VII Central Visayas	5,180,982	101,789	19,075	7,938
VIII Eastern Visayas	4,156,612	504,526	1,110	19,224
IX Zamboanga Peninsula	–	–	25	165
X Northern Mindanao	19,592	20	381	2,174
XI Davao Region	5,175	40	847	1,217
XII SOCCSKSARGEN	–	–	74	39
XIII Caraga	45,063	549	198	660
ARMM Autonomous Region in Muslim Mindanao	–	–	175	442
CAR Cordillera Administrative Region	–	–	353	1,428
NCR National Capital Region	–	–	2,211	15,909

Table A2. Results of topic models with two topics per cluster. We include representative topic words generated by the topic model algorithm, and one example tweet per topic.

Cluster	Number of Tweets (period 1)	(period 2)	Topic Words	Example Tweet
Food and nutrition	2,340	17,559	food, need, please, goods, relief, help, volunteer	*Multi-climate ration packs or healthy army food. Folks need practical food specially the kids @sarah-meier @USArmy #yolandaPh #urgentneed*
	2,372	21,889	donate, food, wfp, families, water	*RT @radikalchick: Red Cross asks for help from police / military. their trucks w/ food and water for 25000 families are stopped in Tanauan*
Camp and shelter	846	3,447	homes, destroyed, areas, relief, moving, many	*Roxas says many homes in Leyte's coastal areas destroyed: They're like matchsticks that were flung inland & talagang sira*
	1,024	5,023	shelter, seek, millions, apart, super, rise, super	*Super typhoon Haiyan slams central Philippines millions seek shelter Read more: http://.../*
Education and child welfare	14,153	12,275	suspended, today, classes, work	*RT @AdamsonUni: Classes and work at all levels are suspended today Nov 8 in anticipation of Typhoon Yolanda. Stay safe Adamsonians. #wala*
	3,923	9,923	relief, kids, help, support, emergency	*Support UNICEF. emergency relief efforts for kids in the #Philippines. How to help:http://.../ #Haiyan http://.../*
Telecom-munications	8,002	5,899	satellite, call, image, mtsat, officials, countries	*MTSAT enhanced-IR satellite image of #YolandaPH as of 2:30 am 09 November 2013: http://.../ via @dost_pagasa RT @govph*
Health	542	1,030	medical, doctors, help, volunteer, charities, team	*MSF emergency & medical teams continue to closely monitor the #Typhoon #Haiyan situation and are ready to respond to needs*
	466	1,457	supplies, red cross, hospital, medical, send	*@KarloPuerto: Davao City 911 sends rescue and medical equipment and personnel to Tacloban City #YolandaPH*
Logistics and transport	1,138	1,649	goods, help, repack	*RT @DepEd_PH: DSWD needs volunteers to help repack relief goods. Call DSWD-NROC at 851-2681 to schedule your shift. #YolandaPH http://.../*
	1,153	1,609	roads, river, affected, debris	*Debris on roads in Tacloban is blocking delivery of aid from airport to victims of Typhoon #Haiyan in #Philippines http://.../*
Water, sanitation, and hygiene	613	34,825	water, clean, need, food, supply	*Heard from @ExtremeStorms who is still in Tacloban. Desperate need for drinking water. Need for military ship & supplies #haiyan #yolanda*
	596	47,743	donate, clean, water, millions, appeal	*No potable water supply power outage & impassable roads in Leyte. Immediate needs r clean water food & shelter-staff in OrmocMai #haiyan*
Safety and security	7,884	4,970	safe, dead, killed, ridiculous	*7000 kid's parents have been killed by the storm in the Philippines and #StayStrongJustin is trending... Ridiculous http://.../*
Early recovery	14,602	46,338	donate, relief, efforts, support, donations, goods	*Doing relief efforts now for #YolandaPH. Need free shipping line info. @indayevarona @juanxi @kwittiegirl*

Towards Happier Organisations: Understanding the Relationship between Communication and Productivity

Ailbhe N. Finnerty[1], Kyriaki Kalimeri[2], and Fabio Pianesi[1]

[1] FBK, via Sommarive 18, Povo, Trento, Italy
{finnerty,pianesi}@fbk.eu
[2] ISI Foundation, Via Alassio 11/c10126 Torino, Italy
{kalimeri}@ieee.org

Abstract. This work investigates in-depth the communication practices within a workplace to understand whether workers interact face to face or more indirectly with email. We analysed the interactions to understand how these changes affect our work (productivity, deadlines, interesting task) and our wellbeing (positive and negative affective states),by using a variety of data collection methods (sensors and surveys). Our analysis revealed that overall email was the most frequent medium of communication, but when taking into account just the communication within working hours (8am to 7pm), that face to face interactions were preffered. Correlation analysis revealed significant relationships between Affective States and Situational Factors while Longitudinal Analysis revealed an impact of communication features and measures of self reported Productivity and Creativity. These findings lead us to believe that different communication processes (synchronous and asynchronous) can impact Positive and Negative Affective States as well as how productive and creative you feel at work.

Keywords: Communication, Organisational Psychology, Multimodal Sensors, Growth Model.

1   Introduction

With constant developments in communication technology it has become important to examine the effects that this technology (smart phones, the virtual workplace), has on our everyday lives, at home and in the workplace. Within organizations, more and more workers are situated remotely from a designated office space, which is an increasing trend [4]. What is required is a way to enhance communication within an organization, due to the changing nature of its structure. A way to enhance communication is also necessary with the changes in how workers interact and collaborate and the effectiveness of working in teams [18]. Research into this particular area is important to determine whether technology helps or hinders our interactions with others. Within an organisation management needs to use an appropriate media when communicating with employees

L.M. Aiello and D. McFarland (Eds.): SocInfo 2014, LNCS 8851, pp. 462–477, 2014.

or to communicate work-related information throughout the organisation [12] in order to ensure worker productivity and satisfaction.

Previous research has focused on examining either email or face to face communication, but not on both within the same research population. What is necessary is to objectively examine the overall communication, face to face and email, as much of what has already been found is based on opinion and the attitudes of workers. This study will attempt to use a mixed methods approach utilising linear mixed models and growth models for a more complete understanding of how we communicate with others, also taking into consideration time varying phenomena. The type of communication available to us are the most common forms of interacting in a workplace, which are face to face interactions as well as internal company emails.

All media are not equally effective and although each type of media for communication has different characteristics, the reasons why managers choose one media over another are not clear, despite significant research. Four major theories have been developed to try to explain the reasons for different media choices for similar tasks [8], Media Richness Theory (MRT), Social presence theory, Social construction theory and Structuration theory [22]. We will focus on Media Richness Theory and a newer theory Media Synchronicity Theory which emerged from it [9]. These two theories, aim to explain how different media can have an impact, positive or negative, on communication, interactions and mood, due to their ability to create a shared context and convey the correct meaning, allowing for efficient communication practices, allowing us to better understand how to investigate communication.

Media Richness Theory (MRT) argues that productivity performance improves when team members use "richer" media for equivocal tasks [6]. This is central to the study as it is expected that using different media for communicating and collaborating, within and between groups, can have a positive or negative impact on the social interactions and mood of the participants. Media Synchronicity Theory (MST) develops upon MRT to focus on the capability of media to support synchronicity, such as, when individuals work together on the same activity at the same time i.e., having a common focus. The key to effective use of media is to match the media capabilities to the fundamental communication processes required to perform the task. Communication environments that support high immediacy of feedback and low parallelism encourage the synchronicity that is central to the convergence process, whereas communication environments that support low immediacy of feedback and high parallelism provide the low synchronicity that is central to the conveyance process. Because most work tasks require individuals to both convey information and converge on shared meanings, and media that excel at information conveyance are often not those that excel at convergence. Thus choosing one single medium for a task may prove less effective than choosing a medium or set of media for the task, which the group uses at different times in performing the task, depending on the current communication process (conveyance or convergence) [7].

MST can be applied to the data findings to understand why different media were used at different points of the study. This theoretical framework allows for better interpretation of the results of the study. By understanding that the way we communicate changes due to the availability of the media to convey the message in the most appropriate manner. This in turn allows us to give meaning to the changing patterns of communication and the effect that they had on the individuals over the course of the study. Many other theories have been used in the past and developed to accommodate the changes that are occurring within communication practices, however we believe that our choice of theoretical framework best matches the aims of the project.

2 Related Research

Previous research has focused on examining either email or face to face communication, but not on both with the same research population. It has been found that face to face interactions are of great importance for developing trust relationships in the workplace which is beneficial for relationships among workers, and increasing trust in the workplace has positive effects on weak relationships [10], [16]. Factors of trust need to be taken into account to be able to communicate effectively [26]. With virtual communication, certain issues can become misunderstood and come across as blunt without a context or shared working environment. The context of an interaction is as important as the message itself and when possible face to face communication is preferred even if electronic communication is available [12]. In a study of communication and training of electronic engineers, the workers felt that face to face communication develops a sense of community and allows small problems to be discussed and fixed rapidly [28].

Studies of productivity in the workplace show that using extensive digital networks can increase productivity by 7%; however, employees with the most cohesive face to face networks were the most productive with an increased productivity of 30 % [25]. In terms of working in group collaborations, email has been researched extensively and along with all its advantages, there are many disadvantages. High levels of emails can be stressful to try and manage. Due to this, emails are becoming less used and wiki's and blogs are becoming more commonplace for collaborating on group projects [17], and minimising email can actually improve communication. Features of electronic communication can be used to informally discuss aspects of working life with a colleague, but while it is a quicker form of communication it also is much less rich than face to face contact [3]. A study using sociometric badges, combining quantitative and qualitative data, found that an elevated level of face to face interactions preceded the launch of a new product, suggesting that this was the most effective form of communication in this period [19]. It could be that more face to face interaction at a certain stage was preceded by a successful outcome, while using email at the same time resulted in less productivity, but at an earlier point in the project it was a more common and efficient form of communication. This suggests that different methods of communication are more beneficial for different stages of a project and observing both email and Sociometric badge data can provide a less biased understanding of inter-team collaboration patterns [19].

Using sociometric badges along with survey data, as has been done in previous research projects e.g. [23]; can be useful for investigating groups as they collaborate. Using a mixed methods approach we can better evaluate how effective different methods of communication can be for the output of the project, as it records personal (mood, personality state) as well as social (social interactions, location) aspects of a person's working day. The reason for investigating the different methods of communication within and between the groups involved in this study is that the dynamics of research groups are constantly changing. Different groups should use different methods based on what is necessary to get the work done and in the method which works best for them, as the most effective means of reaching a satisfactory outcome. Understanding when face to face communication is more beneficial to a team, to produce better results, can help improve group interactions and collaborations. This is becoming a more important issue with the increase of the virtual workplace where teams are distributed and have to use alternative forms of communication, not just email but wiki's, group websites and shared files and document resources etc. By understanding the type of communication that leads to a more positive outcome can help to increase the effectiveness of media used by groups. A first step is to understand the impact of communication on the individual and applying the knowledge gained in studies like this to future studies on collaborative work.

3 Motivation of the Study

The aim of this project is to investigate communication in the workplace by analysing first what media is preferred (email vs. face to face) and second whether it has an effect on the pattern of interactions and mood of the participants in the study, such as leading to more positive and happier workers. We hypothesise that there will be a relationship between increased face to face interaction and positive mood, which can lead to a better understanding of how different media are used in organisations and their effectiveness in contributing to a positive working environment and positive outcomes of work projects. In this study to differentiate between communication mediums we will refer to two types of communicating with others; a) synchronous (immediate, happening in real time, e.g., face to face interactions) and b) asynchronous (delayed, when there is a gap between sending and receiving a message, e.g., emails). The specific research questions that this study aims at addressing are:

Question 1. What are the effects that technology has on friendships and formal relationships in the workplace when communicating face to face, or when using email as a primary medium of interacting.

Question 2. What are the effects that features of communication have on our social relationships and how communication can have a positive or negative impact on our well being, creativity and productivity.

Based on these research questions we form the following hypotheses:

Hypothesis 1. There will be more asynchronous (email) communication rather than synchronous (face to face), reflecting the changes in how we interact and communicate with each other.

Hypothesis 2. There will be positive affect associated with synchronous communication rather than asynchronous communication, such that more face to face and close interactions will lead to reports of positive affect.

4 Dataset

The Sociometric Badge Corpus [21] collected data from fifty four (Female=6) employees of a research centre over a six week period. The data collected by the Sociometric badges [24] consists of face to face interactions (infra-red) and social co-location (Bluetooth) as well as speech and bodily activity features. Electronic communication (email) amongst participants was registered in terms of email traffic (no content was saved to assure privacy) while all information regarding the identity of the subjects was fully anonymised. Using Experiencing Sampling methods [5] the participants filled in an online survey three times daily if they were present at work (11am, 2pm and 5pm), the questions related to affective states, personality, interactions (e.g. I was continuously interacting with those around me) and situations (e.g. I had a deadline, What I was doing was freely chosen by me). Furthermore, organisational information was collected regarding the collaboration on projects and social ties of the participants. The subjects were recruited on a voluntary basis to participate in the study and signed an informed consent form approved by the Ethical Committee of Ca' Foscari, University of Venice. The data were fully anonymised and participants were assigned an identification number for anonymity. Logs of data on electronic communication from social media (smartphones, personal email accounts, etc.) could not be recorded, due to privacy concerns over the content of the data.

4.1 Experience Sampling Data

Participants were asked to fill in a 6-items shortened version of the Positive and Negative Affect Schedule (PANAS) [29] three times a day (excluding week-ends). The items that comprise the PANAS are the most general dimensions that describe affective experience. They are the components of the structure of affect most often described by English language mood terms, and also make reference to the "basic" emotions of anger, disgust, fear, happiness, sadness and surprise [11]). Positive Affect and Negative Affect are the affective, emotional components of psychological or subjective well-being [27]. Following Fleeson [14], they were asked to respond to five situational items that described the interactional context. These items were: 1) During the last 30 minutes, how many other people were present around you? ("0, 1-3, 4-6, 7-9, 10 or more"); 2) I was continuously interacting with the other people around me, 3) What I was doing was freely chosen by me, 4) The deadline for what I was doing was very near and 5) What I was doing was extremely interesting to me. Two items regarding their self perceived Creativity and Productivity were also assessed. Due to the difficulty of measuring productivity in research [2] we simply asked the participants how Productive or Creative they felt in the past 30 minutes. We rely on the individual's subjective experience, however we found that the self reported values were within the normal range and no extreme within individual variation was found. All of the information from the participants relates to the 30 minutes before completing the survey. For full explanations of the dataset please see [21] and [20].

Friendship and Collaboration Social Network. In order to examine the communication by relationship the participants were divided into two groups; "friends" and "colleagues". The participants were asked to rate on a scale from 1- Strongly Disagree to 7- Strongly Agree their answer to the question "I consider this person a good friend of mine, someone I socialize with outside of work". Scores of 0 determined no relationship, scores of 1-3, "just colleagues", while neutral (4) and scores of 5-7 determined "friends". The relationships were then calculated as dyadic pairs. We collected a number of network measures that examined the participants relationships with each other. However we believe that the network on friendship was most applicable to our study on communication within the group.

4.2   Sociometric Badge Data

For this study, from the Sociometric Badges Corpus, we used the infra-red hits as a measure of face to face (synchronous) and email as a measure of electronic (asynchronous) communication. As well as the communication we used the data concerning, mood and context, which were recorded three times daily by means of Experience Sampling surveys. We examined the communication patterns of the participants as they went about their working day. The focus of the study is 1) in how they communicated with each other (face to face or by email). We were also interested 2) in whether their relationship with each other had an impact on how they interacted, such as having more face to face interactions with those who we consider friends. Then, 3) we wanted to understand whether the communication patterns had an impact on self reported measures of affect and 4) further if these changes depended on the context of the situation as defined by Fleeson [15].

Face-to-face Interaction - Infrared Sensor. The detection of another Infrared (IR) sensor can be used a good proxy for face-to-face interaction. For the IR sensor of one badge to be detected by the IR of another badge, the two individuals must have a direct line of sight and the receiving badge's IR must be within the transmitting badge's IR signal cone of height $h \leq 1$ meter and a radius of $r \leq h tan\theta$, where $\theta = \pm 15^o$ degrees. Infrared transmission rate (TR_{ir}) was set to 1Hz. The amount of F2F interaction is defined as the total number of IR detections per minute divided by the IR transmission rate.

Proximity - Bluetooth Sensor. Bluetooth (BT), and in particular the radio signal strength indicator (RSSI), can be used as a coarse indicator of proximity between devices, hence people. In particular, by analyzing our data we found that a BT hit with a RSSI value greater than, or equal to, -80 corresponded to a physical distance between the two sensors, hence the two subjects, of less than 3 meters ("strong signal"). Those BT hits can be taken as a good cue for small groups of people gathering at a conversational distance, as in meetings. We therefore distinguished between people being in *close* and in *intermediate* proximity, where the former corresponds to an RSSI range of $[-80, -60]$ (less than one meter, according to our data) and the latter to an RSSI range of $[-85, -80]$ (one to three meters).

While IR hits imply actual interaction between two people, the strict detection conditions (a direct line of sight and limited angles) mean that the device may fail to capture actual interaction in several situations such as group meetings (e.g., people sit around a big table) or when two interlocutors look at the same object (e.g., screen, whiteboard). In addition to IR, BT proximity can be used as a reliable method to sense face-to-face interaction with a low false negative rate. When using Bluetooth proximity data, the challenge is how to reduce its high false positive detection rate, which comes from its relatively long range compared to the face-to-face interaction. With these points in mind, we chose to combine both IR and BT data, for which we only keep BT hits with strong signal strength (high RSSI value). In order to reduce the false positive hits both from BT and IR, which were mostly due to the office arrangements. An office collocation map was created in the form of an adjacency matrix, based on the field knowledge regarding the institutes internal organisation.

Electronic Communication Data (E-Mail). The electronic communication from participant to participant, was registered; the emails with multiple recipients were treated as multiple one-to-one communications in order to be able to consider each exchange as similar to an one-to-one interaction in person. The emails between each pair of participants were totalled to have a measure of the strength of their (electronic) relationship.

5   Automatic Feature Extraction

In this work we move from the traditional static approaches in analysing communication patterns to concrete behavioral cues automatically extracted from wearable sensing devices. The sociometric and the e-mail data described previously provide the behavioral sequences that are aligned to the ground truth for affective states and situational factors. In the following paragraphs, we discuss the features we identified to represent those behavioral sequences, clustering them according to the sensor type they are based on. All the behavioral features, Infra-red hits, bluetooth and emails were then normalised in order to compare them to each other. All values representing the relationships between the pairs of participants were divided by the maximum value. This gave the communication a value between "0" and "1" for both infra-red and email, "1" being the maximum value, "0" being no interaction between the pair of participants.

Face-to-face Interaction - Infrared Sensor. For each subject and for each time window, we extracted: the number of people F2F interacting with the subject; the mean duration of the interactions; the number of friends the participant F2F interacted with; the amount of time spent with them; the overall level of the F2F interactions, computed as the fraction of friends over the total number of people the subject had F2F interacted with, the level of global formality of a given situation/window, computed as the fraction of collaborators who were present over the total number of present people.

Proximity - Bluetooth Sensor. For each time window and for each subject, we extracted: the number of people in close proximity based on the RSSI; the mean physical distance from other subjects. Besides measuring co-location and proximity between people, we also addressed spatial localization by means of 17 badges placed at fixed locations of common interest such as the organization's bar, cafeteria and meeting rooms. All Sociometric Badges, including base stations, broadcast their ID every five seconds using a 2.4 GHz transceiver (TR_{radio} = 12 transmissions per minute). Combining this information with the signal's strength, we extracted the amount of time spent at the canteen, at the the bar, and at meetings. Moreover, exploiting information subjects had provided in the initial survey about their acquaintances and friends, for each participant and for each window we extracted: the number of friends each participant interacted with; the amount of time spent with them; the level of global friendship of a given situation/window, computed as the fraction of friends who were present over the total number of present people. Similarly from the information subjects had provided regarding their collaboration with the other participants in terms of specific projects, for each participant and for each window we extracted: the number of collaborators each participant interacted with; the amount of time spent with them; the level of global formality of a given situation/window, computed as the fraction of collaborators who were present over the total number of people present.

Electronic Communication Data (E-Mail). For each subject and for each time window, the following features were extracted: the number of e-mails they received; the number of people they contacted; the consistency of the communication, defined as the average number of the emails sent per recipient; the standardized mean length of the sent e-mails measured by the number of characters used in the body of text; the mean number of recipients. Respectively, the same features were calculated for the emails received by each of the participants: the number of e-mails they received; the consistency of the communication, defined as the average number of the emails received per recipient; the mean length of the received e-mails measured by the number of characters used in the body of text; the mean number of senders.

6   Methodology

6.1   Linear Mixed Model Analysis

Multilevel models are fundamentally about modelling the non independence that occurs when the individual responses are affected by group membership which is further complicated with longitudinal analysis [13].

A linear mixed model can be represented as:

$$y = X\beta + Zu + \epsilon, \tag{1}$$

where, y is a vector of observations, with mean $E(y) = X\beta$, β is a vector of fixed effects, u is a vector of random effects with mean $E(u) = 0$ and variance-covariance matrix $var(u) = G$, ϵ is a vector of IID random error terms with mean $E(\epsilon) = 0$ and variance $var(\epsilon) = R$, X and Z are matrices of regressors relating the observations y to β and u, respectively.

We have a typical multilevel dataset with repeated measures Dependent Variable being the Communication features extracted from the data and the Independent Variables being the Affective States and the Situations. After comparing the communication types through Spearman correlations (Table 1), Linear Mixed Models analysis was used to investigate if the communication patterns over time had any effect on the self reported measures from the questionnaire data.

6.2   Growth Model Analysis

To further examine changes of communication and to compare it to Affective States the data was analysed as a Growth Model [1]. We used this measure to understand the relationship between the communication used by the participants, how it changed over the six week data collection period and what variables were associated with the fluctuations in patterns. We assume that over the course of the study there will be naturally occurring changes in the data and we are interested in whether these changes are as a result of external factors such as the context of the situation. We hypothesised that with changes in communication there would be associated changes in Affective States allowing us to draw conclusions on the effect of communication practices within the organisation. The type of analysis used is autoregressive correlation, which is a covariance structure used in multilevel models in which the relationship between scores changes in a systematic way. The notation AR("p") indicates an autoregressive model of order "p". The AR("p") model is defined as:

$$X_t = c + \sum_{i=1}^{p} \varphi_i X_{t-i} + \varepsilon_t , \tag{2}$$

where $\varphi_1, \ldots, \varphi_p$ are the parameters of the model, c is a constant, and ε_t is a white noise process with zero mean and constant variance σ_ε^2.

We used Time as a repeated measure for this study and was calculated as the number of the survey out of the total possible number of surveys (n=90) taken during the study. When the participants were absent from the workplace they were not able to take part in the surveys, which lead to some missing data.

7   Experimental Results and Discussion

7.1   Analysis of Variance (ANOVA)

Taking the total number of normalised hits and emails, and for both "friends" and "colleagues" groups we ran a series of Analysis of Variance (ANOVA) tests on the data to determine whether there were significant differences between the communication types and then communication types by relationship. An experimental design of 2 (Communication; Email, IR) x2 (Relationship; Colleagues, Friends) Analysis of Variance (ANOVA) was carried out on the data with the independent variable "Relationship" (Colleagues, Friends) and the dependent variable "communication" measured as a score of the interactions (email and IR) between the participants. The data considered all infra red hits and emails between the participants for the study.

The analysis revealed a main effect of Communication $F(51,1)=25.70$, $p <$ .001, with more email than infra-red (3.15 vs. 2.00) and a main effect of Relationship $F(51,1)=50.81$, $p < .001$, with more interactions between friends than colleagues (3.68 vs. 1.47). The analysis also revealed a significant interaction between Communication and Relationship $F(51,1)=6.31$, $p < .05$ Fig. 1. The interaction revealed that even though there was more email communication between the participants that the difference in the type of communication used between colleagues was very large, email was used as a form of communication much more than IR (2.26 vs. .65) while for friends the difference between the type of communication was much smaller difference indicating that friends interacted face to face nearly as much as they emailed (3.34 vs. 4.03).

As a second step we calculated the communication that was directly related to the Experience Sampling data, aggregated into thirty minute segments. This time frame was chosen as it was the time the participants were asked to consider when filling in the questionnaire (e.g. in the last thirty minutes "What I have been doing was freely chosen by me").

A second 2x2 ANOVA was carried out on the data. This analysis resulted in a significant main effect for Relationship $F(51,1)=67.65$, $p < .001$ with significantly more communication between friends (M = 2.56, SD = 0.23) than colleagues (M = 0.45, SD = 0.10). The analysis also revealed a significant interaction between Communication and Relationship $F(51,1)= 10.67$, $p < .05$ Fig. 2, where there was more communication between friends than colleagues, with a slightly greater value of face to face interaction than interaction via email for the friends group (3.0 vs. 2.17), while for colleagues it was found that there was a smaller proportion of face to face communication than email communication (0.13 vs. 0.77).

This difference in results could be due to the fact that infra-red data could only be measured during working hours, while the email data collected could have been taken at any time, during or outside of working hours. This could explain why email was marginally larger than infra red hits in our first ANOVA, while the second analysis supported our hypothesis that there would be more face to face interactions between friends than colleagues.

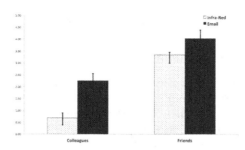

Fig. 1. Experimental design of 2 (Communication; Email, IR) x2 (Relationship; Colleagues, Friends) Analysis of Variance (ANOVA)

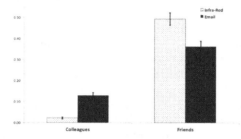

Fig. 2. Experimental design of 2 (Communication; Email, IR) x2 (Relationship; Colleagues, Friends) Analysis of Variance (ANOVA) on 30 minute segments

Table 1. Spearman Correlations between the communication, affective states and the situations. Note that * is significant at value $p < .000$ while all others are significant at $p < .05$

	HPA	HNA	LPA	LNA	Lonely	Discrete	Product	Create	Interact	FreeC	Interest
HPA			.697*			-.304	.778*	.838*	.416	.570*	.691*
HNA	.748*		-.476*	.577*	.878*					-.287	-.347
LPA				.755*	.720*						-.303
LNA						-.491*	.500*	.496*	.300	.482*	.490*
Lonely						.571*			-.369		
Discrete							-.319		-.305	-.373	-.392
Product								.770*	.427	.309	.582*
Create									.509*	.417	.633*
Interact										.355	.357
FreeC											.640*
Interest											

Using the data as formatted for the second ANOVA (in 30 minute segments prior to the survey) we ran a series of Spearman correlations on the communication data with the Affective States and Situations. We found that the communication had little impact on any of the variables ($sig > 0.05$). However we did find many significant correlations between the affective states and the situations (Table 1). While there were no significant correlations for communication (IR or Email) there were for the context of constantly Interacting with others, such as, High Positive Affect $rs[0.416]$, Productivity $rs[0.427]$ and Creativity $rs[0.509]$. We believe that this could be due to fluctuations in communication patterns over time, which could have accounted for a larger within subjects and between groups variance and the non significant results. We decided to continue our analysis using a method appropriate for longitudinal data analysis.

Taking the communication features as described in detail in Section 5 we ran an analysis using Linear Mixed Models in SPSS. This analysis expands the general linear model and allows the data to exhibit non consistent variability and adjusts for correlation due to repeated observations. This was done to examine the effect of communication patterns on the self reported questionnaire data and whether there were changing patterns over time.

7.2   Linear Mixed Model Analysis

We found that the number of infra red hits (synchronous interactions) between a larger number of friends explained variances in Productivity [$F(1, 317.80)$= -3.09, $p < 0.05$] and Creativity [$F(1, 1421.02)$= -3.03, $p < 0.05$]. While email sent and received (asynchronous interactions) between friends were related to the situational context, if you considered what you were doing as Interesting (email sent [$F(1, 265.45)$= 2.35]; email received $F(1, 2112.18)$= 3.283]) ($ps < .05$).

We did not find as many significant results for the Mixed Models analysis as expected. This could simply be due to the fact that the software used is not the most appropriate for the task. SPSS is not the best program for multilevel modelling [13] and more specialised software such as R and SAS are commonly used for this type of analysis. As the analysis here was inconclusive we decided to use growth models using R software to further examine the data as a time series. While we did not find the expected relationship between communication features and Affective States and did not support our second hypothesis, what was interesting were the significant relationships between the communication features and self reported measures of Productivity and Creativity. This could mean that the way that we interact with each other does affect us and that our interactions with those around us can make us feel more or less productive or creative.

7.3   Growth Model Analysis

It is assumed that the correlation between scores gets smaller over time and variances are assumed to be homogeneous. A benefit of using Growth Models is to be able to better understand the patterns of the communication and whether the variances over the course of the study can be accounted for by Time or the other variables present (Affective States, Situations).

For this, features extracted from infra-red and emails were used, as well as Bluetooth (a measure of co-location).

IR. The results found that interacting closely with more people face to face [$F(1, 3161)=2.78, p < 0.05$] was related to higher Positive Affect especially when your task was interesting [$F(1, 3161)=2.109, p < 0.05$], but also that higher number of face to face interactions with friends was negatively related to self reported measures of Productivity [$F(1, 3161)= -5.32, p < 0.001$] and Creativity [$F(3161)= -4.65, p < 0.001$].

BT. The same trend was found when considering co-location with others through BT signals, that being co located with friends [$F(1, 3157) =2.93, p < 0.05$] was related to higher reports of Positive Affect. It was also found that higher number of interactions, with more people in the canteen [$F(1, 3157= -4.99; F(1, 3158)= -3.95$] and when having coffee [$F(1, 3157= -6.11; F(1, 3158)= -3.10$] were found to negatively impact the reported levels of Productivity ($ps < .001$) and Creativity ($ps < .001$) respectively.

However for the BT features it was found that more time interacting [$F(1, 3157)= 2.53, p < 0.05$] and being co located with friends [$F(1, 3157)= 4.41, p < 0.001$] had a positive effect on Productivity but not Creativity, whereas being surrounded by more people [$F(1, 3158)= 3.92, p < 0.001$] regardless of relationship led to higher reports of Creativity.

EMAIL. Average email length was the only feature to have any impact on task status, if what you were doing was Interesting [$F(1, 3161)= 1.99, p < 0.05$] or if there was a Deadline [$F(1, 3162)= -1.98, p < 0.05$].

What we can draw from the Growth Model findings are that we found partial evidence to support our second hypothesis that more face to face interactions had an impact on feelings of positive affect and more importantly on self reported creativity and productivity, while email communication was only relevant when the participants were engaged in an interesting task or had a deadline to meet. Being co located with others differently impacted the productivity and creativity, in that when surrounded by those that are considered friends more productivity was reported but not creativity, while when around more people in general higher levels of creativity were reported. The findings here can have implications on future research on satisfaction and motivation in the workplace and also in studies attempting to understand how to improve productivity.

8   Conclusion

From the first analyses, our second ANOVA, evidence was found to support our first hypothesis, that there would be more asynchronous than synchronous communication between the participants. By analysing the data as a time series we found partial support for our second hypothesis, as over the course of the study different patterns emerged where the communication impacted the Affective States and self reported measure of the participant's Productivity and

Creativity. Interacting with "friends" face to face had an impact on Positive Affective state but a negative effect on self reported Productivity and Creativity, especially during lunch or coffee breaks. Co location just with friends found higher levels of self reported Productivity only, while being co located with more people in general was found to have a greater impact on self reported Creativity. Email was found to have minimal impact on any of the self reported variables.

Our findings were much more evident when it came to the positive effect of communicating and the link to productivity and creativity, which we believe can also be classified as positive states. We believe that by improving communication practices in the workplace can lead to more positive environments and boost worker morale, making it a happier place to be.

We focused on communication practices, mood and affective states, as this is an area that is changing with new working practices and means of communicating. This study extends the state of the art in the communication studies in organisational management, providing useful deep insights on the communication channels and attitudes of workers. These insights can be an important steppingstone for creating teams that are not only more productive, but more importantly engaged with their mission.

Future extension of this research includes not only development of predictive models that can accurately capture and explain the behavioural cues but also inclusion of other means of communications, such as virtual (online) communication, Skype, instant messaging as well as workers smartphones that are also becoming tools for communicating within the workplace. We aim not just to focus on improving productivity in the workplace, but on how it can be affected by our social interactions and our mood, exploiting simple broadly used productivity strategies and last but not least the effect of personality and individual characteristics on the communicational behaviours and preferences. The aim of future studies should focus on making this work generalisable to organisations and any collaborative situations.

Acknowledgment. Kyriaki Kalimeri acknowledges support from the Lagrange Project of the ISI Foundation funded by the CRT Foundation.

References

1. Bliese, P.: Multilevel Modeling in R (2.2)–A Brief Introduction to R, the multilevel package and the nlme package (2006)
2. Brown, M.G., Svenson, R.A.: Measuring R&D Productivity. Res. Manag. 41, 30–35 (1998)
3. Cameron, A.F., Webster, J.: Unintended consequences of emerging communication technologies: Instant messaging in the workplace. Comput. Human Behav. 21, 85–103 (2005)
4. Cascio, W.F.: Managing a virtual workplace. Acad. Manag. Exec. 14, 81–90 (2000)
5. Csikszentmihalyi, M., Larson, R.: Validity and reliability of the experience-sampling method. J. Nerv. Ment. Dis. 175, 526–536 (1987)

6. Daft, R.L., Lengel, R.H., Trevino, L.K.: Message equivocality, media selection, and manager performance: Implications for information systems. MIS Q., 355–366 (1987)
7. Dennis, A.R., Fuller, R.M., Valacich, J.S.: Media, tasks, and communication processes: A theory of media synchronicity. MIS Q. 32, 575–600 (2008)
8. Dennis, A.R., Kinney, S.T.: Testing media richness theory in the new media: The effects of cues, feedback, and task equivocality. Inf. Syst. Res. 9, 256–274 (1998)
9. Dennis, A.R., Valacich, J.S.: Rethinking media richness: Towards a theory of media synchronicity. In: Proceedings of the 32nd Annual Hawaii International Conference on Systems Sciences, HICSS-32, p. 10 (1999)
10. Dirks, K.T., Ferrin, D.L.: The Role of Trust in Organizational Settings. Organ. Sci. 12, 450–467 (2001)
11. Ekman, P.: Are there basic emotions (1992)
12. Ean, L.C.: Face-to-face versus computer-mediated communication: Exploring employees preference of effective employee communication channel (2010)
13. Field, A.: Discovering statistics using SPSS. Sage Publications (2009)
14. Fleeson, W.: Situation-based contingencies underlying trait-content manifestation in behavior (2007)
15. Fleeson, W.: Toward a Structure- and Process-Integrated View of Personality: Traits as Density Distributions of States. J. Pers. Soc. Psychol. 80, 1011–1027 (2001)
16. Jarvenpaa, S.L., Knoll, K., Leidner, D.E.: Is anybody out there? Antecedents of trust in global virtual teams. J. Manag. Inf. Syst., 29–64 (1998)
17. Johri, A.: Look ma, no email!: blogs and IRC as primary and preferred communication tools in a distributed firm. In: Proceedings of the ACM 2011 Conference on Computer Supported Cooperative Work, pp. 305–308 (2011)
18. Kim, T., Bian, L., Hinds, P., Pentland, A.: Encouraging cooperation using sociometric feedback. In: Proc. of the ACM Conference on Computer Supported Cooperative Work (2010)
19. Kim, T., McFee, E., Olguin, D.O., Waber, B., Pentland, A.: others: Sociometric badges: Using sensor technology to capture new forms of collaboration. J. Organ. Behav. 33, 412–427 (2012)
20. Kalimeri, K.: Traits, States and Situations: Automatic Prediction of Personality and Situations from Actual Behavior (2013)
21. Lepri, B., Staiano, J., Rigato, G., Kalimeri, K., Finnerty, A., Pianesi, F., Sebe, N., Pentland, A.: The SocioMetric Badges Corpus: A Multilevel Behavioral Dataset for Social Behavior in Complex Organizations. In: SOCIALCOM-PASSAT 2012: Proceedings of the 2012 ASE/IEEE International Conference on Social Computing and 2012 ASE/IEEE International Conference on Privacy, Security, Risk and Trust, pp. 623–628. IEEE Computer Society, Washington, DC (2012)
22. Moustafa, K.S.: Differences in the Use of Media Across Cultures. Encycl. Virtual Communities Technol. 131 (2006)
23. Olguin Olguin, D., Waber, B.N., Kim, T., Mohan, A., Ara, K., Pentland, A.: Sensible organizations: technology and methodology for automatically measuring organizational behavior. IEEE Trans. Syst. man Cybern. Part B Cybern. a Publ. IEEE Syst. Man Cybern. Soc. 39, 43–55 (2009)
24. Olgun, D.O., Gloor, P.A., Pentland, A.: (Sandy): Wearable Sensors for Pervasive Healthcare Management (2009)
25. Pentland, A.: How social networks network best. Harv. Bus. Rev. 87, 37 (2009)

26. Soto, F.: Virtual Organizations BET on People to Succeed! Human Factors in Virtual Organizations: Boundary-less Communication, Environment, and Trust. J. Strateg. Leadersh. 3, 24–35 (2011)
27. Terraciano, A., McCrae, R.R., Costa Jr, P.T.: Factorial and construct validity of the Italian Positive and Negative Affect Schedule (PANAS). Eur. J. Psychol. Assess. 19, 131 (2003)
28. Vest, D., Long, M., Thomas, L., Palmquist, M.E.: Relating communication training to workplace requirements: The perspective of new engineers. Prof. Commun. IEEE Trans. 38, 11–17 (1995)
29. Watson, D., Clark, L.A., Tellegen, A.: Development and validation of brief measures of positive and negative affect: The PANAS scales. J. Pers. Soc. Psychol. 54, 1063–1070 (1988)

Measuring Social and Spatial Relations in an Office Move

Louise Suckley and Stephen Dobson

Sheffield Hallam University, UK

Abstract. In this paper, we outline an investigation of the impact of an office move on the social relationships of staff and students in a university research department. Combining the techniques of Social Network Analysis to assess for changes in social relations and Space Syntax Analysis for measuring the spatial changes, we identify key changes in the social relations that can be defined by spatiality. A decline in the social connections taking place and a change in the structure of the social network, accompanied by significant changes in spatial connectivity suggests that the office locations are influencing the underlying complex social processes.

Keywords: Workspace, open plan, social network analysis, space syntax analysis, academic workspace.

1 Introduction

Applied case studies (Openshaw, 2013; Brennan et al, 2002; Cummings and Oldham, 1997; Dunbar, 1995) show there is a critical amount of interaction between workers: too much and there is not enough privacy and opportunities for reflection; too little and there is a reduction in innovation, which is largely accepted as a social process (Amabile *et al*, 1996). The workspace is considered to be a key determinant of social connectivity, from the accessibility of knowledge and stimulation (Csikszentmihalyi, 1996); linking layout with the exchange of information (Peponis et al, 2007); and engineering opportunities for serendipity (Sailer, 2011).

This research explores the social and spatial connectivity of research staff and students in a university research department that have undergone a change in their workspace. The extent, type and mode of interactions were measured on two occasions: prior to the relocation from traditional cellular office accommodation and following the relocation into an open plan workspace. The spatial visual connectivity of the different workspaces was also measured and adds to the conclusions that can be drawn from the changing social relations that emerged. Given the increase in visual and social connectivity of an open plan office configuration, it would be expected that the volume of interactions will increase.

This study contributes to the expanding field of research into the spatiality of organizational interaction and how the physical location of individuals influences the social processes at play within the workplace. Research has been undertaken in the field of facilities management, space management and environmental psychology into understanding the most conducive spaces to facilitate interaction, provide privacy and engineer serendipity (Martens, 2011; Parkin et al, 2011; Boutellier, 2008; Haynes, 2007).

L.M. Aiello and D. McFarland (Eds.): SocInfo 2014, LNCS 8851, pp. 478–492, 2014.
© Springer International Publishing Switzerland 2014

Of equal relevance is the field of organizational behavior which recognizes the social factors that influence interactions in the workplace. Job role, organizational structure, culture, individual preferences and historical contexts are all factors that should also be considered when interpreting social relations in the workplace (see Huczynski and Buchanan, 2007).

2   Background

2.1   Workspace Research

Research into the impact of workspace design has been undertaken from many theoretical perspectives. Classical organisation theorists such as Taylor (1911) regarded the workspace as being integral to management control and command in the pursuit of scientific management; Humanistic theorists such as Herzberg (1966) regarded the workspace as being a hygiene factor and therefore makes no contribution to worker motivation unless it was not delivered to the required level; Management theorists such as Drucker (1959) recognised the cultural elements of the workspace in their representation of organisational values and its capacity to enable cultural change; and Organisational ecology theorist such as Becker (2004) argue that the workspace can influence it occupiers and so can contribute to organisational effectiveness. All of these theorists consider the social aspects along with the spatial aspects and regard one to influence the other.

Further researchers have studied the use of workspace, its design and layout to facilitate social interaction. Peponis *et al* (2007) outlines two models of workspace design that links with the exchange of information and communication, these are the 'flow model' and the 'serendipitous communication' model. The 'flow model' suggests that communication is most effective if the office is designed around the required flow of information, such as placing people who need to communicate near each other. Proximity and communality is supported by others who suggest that *"Co-workers will more likely communicate with colleagues within their vicinity; face to face interaction declines rapidly after 30 meters"* (Allen, 1997). The 'serendipitous communication' model encourages facilitating opportunities for chance and informal interaction (e.g. the provision of a communal eating space, or seating around communal activity zones such as kitchens or print facilities). Fayard and Weeks (2007) suggested using communal focal points to foster informal serendipitous interaction which they developed into 'the water-cooler' effect. Further support for this approach is shown by Gladwell (2000) who draws on his experience of city design in New York: *"put all places where people tend to congregate - the public areas - in the center, so they can draw from as many disparate parts of the company as possible"*. Latour (2007) identified the value of what he termed 'actants' as having the power to influence the social reality of human actors. These are non-human, social artefacts that display mediatory or intermediary characteristics, and within the workspace this can be printers, photocopiers and kitchens.

Designing the workspace to facilitate interaction then moves into the debate of open plan offices vs. cellular office, which is a long-standing and often emotive debate for those affected. This debate is ever present in the field of academic workspaces in particular since the occupants require both collaboration and knowledge flow that could be delivered by open plan, but they also require space for concentrated work and reflection which cellular offices deliver. Individual enclosed cellular workspaces have had a long history in academia and have often been symbols of status and power (Becker and Sims, 2001). Since the expansion of UK university provision in the 1960s (Judt, 2005), workspaces have become open, accommodating open plan or combi layout arrangements. Van der Voort (2003) defines cellular and shared offices as housing 3 and 12 workstations respectively, which by definition suggests that the term open plan is for workspaces that house 13 work stations or more. Some support has been shown in the literature for multi-occupancy office environments in terms of the opportunity they engender for interaction. Dunbar (1995) found that discussions that scientists had with their lab colleagues in four world-leading research laboratories at US universities were critical to the interpretation of data that led to significant breakthroughs. Cummings and Oldham (1997) hold that interactions with colleagues are important for stimulating wider interests, boosting competitiveness and sharing knowledge and so employees should be in a populated environment.

Conversely there is evidence to demonstrate the negative influences of open plan working environments. Brennan et al (2002) found that they could be associated with a decreased level of motivation, productivity and work satisfaction; Brill et al (2001) found them to be responsible for increased noise, distraction and decreased psychological privacy. Parkin et al (2011) found that academics were more satisfied with combi-offices (small individual spaces surrounding a common shared space) than with the open plan design as they support both privacy and collaboration.

To support the flow and exchange of ideas through collaboration and the opportunity for privacy and concentration, a variety of workspaces should be included within an office environment (Duffy, 1997; Steele, 1998).

2.2   Social Network Analysis

An approach to mapping the social relationships that occupiers of a workspace have is Social Network Analysis (SNA). This is a method that has been used to understand the nature and characteristics of relational data (Scott, 1990) and through visualization it seeks: *"to describe patterns of relationships among actors, to analyze the structure of these patterns and discover what their effects are on people and organizations"* (Martinez et al 2003, p354). SNA offers a critical means for mapping multi-level, but often tacit, channels of collaboration and may be considered in terms of; general social practices, knowledge acquisition, knowledge management, and innovation - both within and between groups. This approach therefore extends social analysis beyond what might be gleaned from initial observations and discussion with participants, helping to further reveal, for example, barriers to communication or other such structural weaknesses; characteristics which may only become evident when formally modelling the structure of a workplace relationship network (Hanneman, 2001; Burt, 1992).

Social networks can be either sparse or dense dependent upon the number of connections between actors. Obstfeld (2005) argued that a more sparse network is valuable for the *generation* of good ideas, whereas a dense network is essential for the *promotion* of good ideas. Effective knowledge exchange is commonly regarded as a largely social process (Amabile, 1996; Csikszentmihalyi, 1996) and the generation and promotion of good ideas is one such feature. These interactions can be face to face or alternatively, to span greater spatiality, they can be through telephone or email communication. Interactions of a particular nature would tend to be sustained over time in order for them to become evident in the social network.

Openshaw (2013) investigated the social connectivity of knowledge workers in a pharmaceutical company to evaluate the impact of different workspaces on their work performance, as measured by relationships that are considered valuable to the role. He found that scientists who worked in a 'dense' work environment - an open plan environment populated by different groups of staff (scientists, administrators, project managers) had a level of connectivity that was more valuable to their performance than those that worked in more traditional, small cellular offices. Openshaw found that knowledge workers with large social networks performed better than those with small social networks since it provided access to different information that encouraged new thinking. A recent introduction by the case study company of a range of workspaces also impacted on the connectivity of the knowledge workers. There were quiet spaces available in the form of small offices and libraries; buzzy social spaces such as dense open plan seating; meeting spaces such as rooms, cafes, booths, breakout spaces; and war rooms/ project areas. Openshaw took a measure of connectivity before the change in office space and after and found that there followed a much greater integration of staff groups which led to better performance.

2.3   Space Syntax Analysis

Space Syntax Analysis (SSA) considers the spatiality of social relations, describing the social logic (Hillier and Hanson 1987) of spatial systems. The organization of space affects how it can be used, particularly with regard to how people move around and how they encounter other occupants (Hillier and Penn, 1991; Penn et al, 1999; Peponis, 1985). Particular layout patterns of offices and corridors will influence how people connect socially and using the technique provides an understanding of the spaces that could enhance or inhibit social exchange. Peponis et al (2007) use SSA as a method to help workplace design to support the socio-spatiality of communication and productivity.

Wineman et al (2014) used SSA along with SNA to explore the association between innovation and an organization's social and spatial structure. They undertook the spatial analysis to map the physical space and to calculate the mean distance between the occupants of the space as a measure of the likelihood for serendipitous interaction. SNA was used to capture the 'perceived' social network, and as a measure of reliability they also used location-tracking method to assess the network in 'real-time'. They found that spatiality with high levels of connectivity did provide opportunities for serendipitous interactions for individuals who come from disparate parts of the organization. They concluded that there are both spatial and social dimensions in the process of innovation.

3   Methodology

Measures of social and spatial connectivity were taken of the university engineering research department on two occasions. The first measures were taken in the final month of their occupation of a workspace they had occupied for 15 years. The second measures were taken after 4 months of occupying a temporary office space - an occupation that was scheduled for a total of 18 month during which time new combi- office space was being configured.

3.1   Sample

The wider department consisted of 63 academic research staff, 11 business administration staff and 78 research students (n=152), however only 40% of those were directly affected by the office move consisting of 26 academic research staff, 11 business administrators and 24 research students (n=61). The remaining 60% were located in other buildings across the university campus. The whole of the research department were included in the research in order to gain an understanding of the impact of the change in spatiality and the dispersion of spatiality on social relations.

The original workspace occupied by the 40% of the research department directly affected by the change in location, consisted of cellular offices accommodating 3-4 academics with small laboratories located in close proximity and spanning 2 floors. The business administration staff occupied an open plan layout that was situated outside the research centres' directors' individual offices (Fig. 1). The temporary office accommodation that they moved into was a single floor open plan layout with only 3 cellular offices that were occupied by the research centre director and two research groups dealing with confidential data. The remaining staff were in the open plan space arranged into desks of 5-6 according to the specific research group with tall storage facilities dividing the office space (Fig. 2).

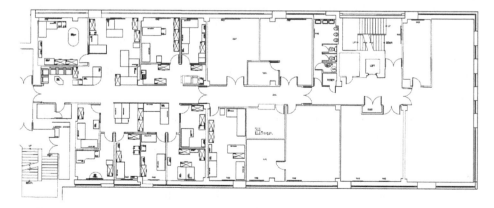

Fig. 1. Previous workspace

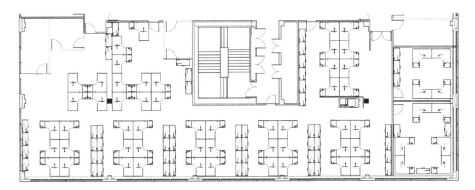

Fig. 2. Temporary workspace

3.2   Social Network Survey

All members of the research department were asked to complete a web-based so-ciometric survey. This survey collected data on the nature of the relationships be-tween the different research departments; participating groups (research academics, business development and post-graduate research students); and the most frequent mode of communication for a variety of interaction purposes. This was gathered by asking respondents to *"indicate the people with whom you have formal or informal working relationships which you consider influence or impact on you"* with regard to:

- completing everyday work processes
- developing new ideas
- discussing social topics
- making improvements to everyday working practices
- seeking expert advice
- finding out what's going on
- making decisions

Respondents made their own judgement in interpreting which relationships they felt influenced or impacted on them and so it was open to a range of time frames, office locations and cultural references. This approach was considered to have higher validity. The survey was completed over a 2 week period, one month prior to the re-search department moving out of their cellular office space; and after occupying the open plan office space for 4 months. The results were compiled as a case by case adjacency matrix (see Scott, 2000) and processed through the SNA software package yEd (yWorks[1]).

The social network variable used in this study to model interactions was connectiv-ity which is a measure of the number of instances an individual is acknowledged for each interaction type outlined above.

---

[1] http://www.yworks.com/en/products_yed_about.html

3.3   Spatial Layout

Measured floor plans of each of the office spaces were input into the software Syntax 2D to analyze the space for its grid depth and connectivity.

Depth provides a measure of the most private/ complex space to navigate to from a visual perspective and at the other end of the spectrum Connectivity describes the areas with the greatest number of connecting spaces. These are the areas where most activity is likely to take place and they tend to also be the most integrated and easiest places to find. A specific calculation of the degree of depth or connectivity is given for each point on the floor plan.

4   Findings/ Discussion

A total of 60 (39%) members of university engineering research department completed the pre-move survey and 71 (46%) members completed the post-move survey. In each of these samples, 60% (pre-move n=36; post-move n=42) were directly affected by the office move. However only 27 respondents directly affected by the office move completed both the pre and post move surveys, which makes assessing for an exact change in connectivity difficult given the small sample size. Other analysis undertaken on the average level of changes in spatial and social connectivity measures taken prior to and following the office move, has revealed a number of key findings.

4.1   Decreasing Interactions

There was a significant reduction (<0.05) in the number of interactions taking place between members of the research department since re-locating to the temporary office space. According to the results of the SNA, the number of connections that individuals are making has declined for all of the social interactions included, this is particular prevalent in connecting for completing everyday work processes and seeking expert advice (Table 1).

Table 1. Average number of connections before and after the change in workspace

Reasons for interaction	Average Pre-move connections	Average Post - move connections	% differ-ence
completing everyday work processes	18.8	9.4	-0.50
developing new ideas	8.8	8.7	-0.01
discussing social topics	7.4	6.7	-0.10
making improvements to everyday working practices	10.0	6.7	-0.33

seeking expert advice	12.7	7.7	-0.39
finding out what's going on	8.0	6.8	-0.14
making decisions	10.2	7.4	-0.27

The reduction in the number of interactions is surprising given that there was an increase in density of occupation, spatial proximity and visual connectivity, it was expected that interaction levels would increase. As found by Dunbar (1995) and Cummings & Oldham (1997) the visual and spatial proximity of open plan spaces should facilitate interaction among occupants, but this is not supported in the academic research department.

It should be noted that the SNA takes a cross-sectional measurement of social connectivity at a particular time so the changes in the number of interactions could be due to the nature and volume of the interactions taking place within an individual's frame of reference in that particular period. It is impossible to ascertain what people judge to be 'formal or informal working relations' or how they have interpreted the different types of interactions, all are subjective. However these have been objectified through the responses given by individuals in the survey.

The decline in connections could be due to:

- less of a need to work with others in these areas during this time;
- a clearer understanding of who to go to for these types of interactions due to the open-plan nature of the space from over-hearing conversations or observing others;
- a clearer awareness of availability for individuals for connecting on these issues;
- inaccessibility of individuals due to their new office location e.g. wanting to discuss confidential issues or having the confidence to communicate when can be overheard;
- fear of disturbing others in the open plan accommodation by making connections;
- the need to complete more immediate individually-focused tasks due to the disturbances created from the office move.

As an example, one individual that saw a significant decline in their interaction levels was the research department's business development manager. Previously this extroverted individual was located in an office on their own, so they were highly mobile around the workspace to maintain connections and facilitate knowledge exchange that was essential to their role. In the open plan space however, their extroverted nature, high mobility and regular conversations, were not conducive with the needs for quiet and concentration of those that surrounded him. Consequently, the manager reduced the conversations they were having and stifled his mobility, thus resulting in lower levels of interaction. This example and the others that are noted from the results are of concern, particularly when knowledge exchange, creativity and innovation are widely accepted as social processes (Amabile, 1996; Csikszentmihalyi, 1996; Wineman, 2014).

4.2   Change in the Network Structure

There was a change in social relations following the change in spatiality. Individuals that exhibited high levels of connectivity before the move to the temporary office accommodation (such as the example of the business development manager given above) were no longer interacting at the same level in the post-move stage.

Figures 3 and 4 show the structure of the social network for seeking expert advice prior to the change in workspace and after the move into the temporary workspace. Each square on the figures represents an individual and the size of the square reflects the number of times the individual was cited by others as someone they go to for expert advice.

There is a clear difference in the structure of the network with much fewer individuals dominating the network whilst occupying the temporary workspace (post move). Two of the six individuals that were key sources of expert advice before the workspace change maintain their status in the social network for this type of interaction. The remaining individuals and others that were cited as a key source of expert advice when located in the previous workspace are now no longer as prominent. If this expert advice is no longer being sought in this research department, there could be implications for the quality of the work that is being produced in terms of, for example, the laboratory work that is undertaken, funding applications that are made or the journal articles written. However this result would also suggest that there is a more inclusive network for expert advice since more individuals are being consulted for their expertise. Larger social networks were considered by Openshaw (2013) to be positively associated with productivity in knowledge workers, so this change in network could be beneficial.

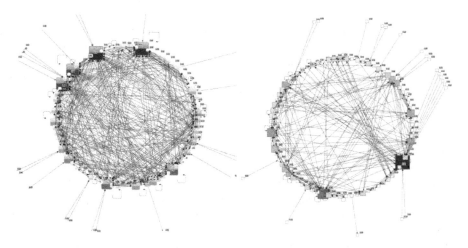

Fig. 3. SNA connectivity for seeking expert advice - Pre move

Fig. 4. SNA connectivity for seeking expert advice - Post move

It is again recognized that SNA takes a cross-sectional measure of social interaction at one particular point in time. This result and the changes found in the social networks for the other discussion topics could be the result of the timeframe of reference that respondents make in their judgment. However the changes in the network structure could also be the result of the changing workspace where the individuals that previously dominated the network for expert advice are feeling inhibited by the open plan temporary workspace or it has become more clear from the increased visual, spatial and auditory perspective who has the required expertise.

4.3   Changes in Spatiality

There was a great difference in the level of spatial connectivity between the two workspaces occupied by the engineering research department. The overall results of the SSA for each of the offices are shown in Figures 5 and 6. The hotter colours display a higher level of connectivity and are where most activity and chance meetings are likely to take place; the colder colours display lower levels of visual connectivity (and subsequently greater depth) which are the most private spaces.

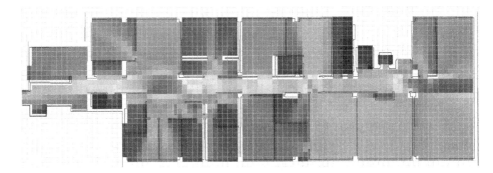

Fig. 5. SSA Connectivity Pre-move

There were relatively low levels of connectivity (indicated in blue) in the original workspace occupied by the academic research department which signals there was a very high level of privacy. The lab spaces (towards the right of Fig. 5) tended to have mid-levels of connectivity and the 'natural meeting points' (indicated by the hotter colours) tended to be the open plan reception area and central corridor space.

In comparison, the temporary workspace has a much larger area of spatial connectivity (Fig. 6). The open plan area in red represents a significant space for high levels of proximal connectivity.

By comparing the connectivity measure pre-move to that post move for each individual it is possible to establish who has seen the biggest change in terms of their workspace syntax (Fig. 7). Those on the left of the figure have seen the biggest decrease in privacy, and those on the right have experienced the biggest increase in privacy between the previous and the current spatial arrangements.

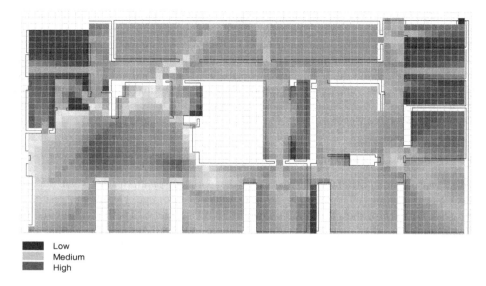

Low
Medium
High

Fig. 6. SSA Connectivity Post-move

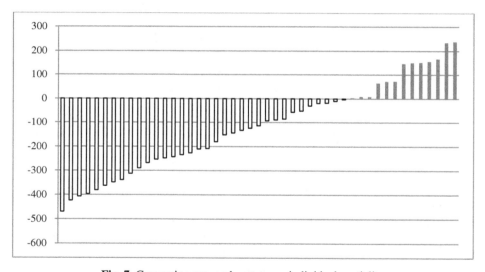

Fig. 7. Comparing pre- and post-move individual spatiality

Correlational analysis established that there was no correlation between the changes that people experienced in the level of spatial connectivity and those experienced in social connectivity. For example a reduction in social connectivity was not accompanied by a reduction in spatial connectivity. Statistically therefore the association of changes have not been found, although the small sample size makes any significance testing of this conclusion difficult. Anecdotal evidence however suggested that those who occupied the more spatially connected spaces in the temporary workspace compared to a relatively private desk space in the previous workspace did in fact feel more socially connected.

5   Conclusions

The change in the workspace that had been experienced by the university research department in this case study had a profound impact on both the spatial and social relations.

Contrary to previous research which suggests that open plan office space supports interactions and collaborative working (Dunbar, 1995; Cummings and Oldham, 1997) the relocation of this research department actually and surprisingly inhibited their interactions. This may be explained by the improvement in visual connectivity that the open plan accommodation affords which allows the occupiers to be more efficient in their interactions, given that they can now see the availability of individuals and so only interact as necessary. Experience of the workspace however suggests that the decline in interaction level is more likely to be the result of a fear of disturbing others, the lack of confidentiality and the enforced requirement for concentration. A set of office protocols drawn up by the senior managers of the research department at the start of their occupation of the temporary open plan accommodation, play a vital role in the social relations that exist. The protocols were used as a tool to reassure people and placate their fears of working in an open plan environment. They covered the use of phones, desk space, general office, kitchen and meeting rooms and were very authoritative about things that MUST be done, for example try not to shout when speaking on your mobile, try not to conduct meetings at your desk, keep your voice low and do not stand about chatting close to other people's workspace. They have been continually reinforced throughout the 4 months of occupation of the temporary workspace and have resulted in a creation of very low auditory levels across the open plan accommodation, described by the occupiers as 'library quiet'.

These office protocols have also had a significant effect on the structure of the network. The more extraverted individuals that featured as key sources of interaction with regard to expert advice, discussing new ideas and making decisions, when located in the previous cellular workspace, are interacting far less. Their need for high levels of interaction to suit their personality and their work roles, are not conducive with the dominant need for concentration in the research department. This has resulted in them retracting from the network and adopting coping mechanisms such as wearing headphones, making phone calls in the corridor or working from home. This is to the detriment of the research department given that knowledge exchange is a key part of the role.

The change in the spatiality of the office accommodation has also impacted on the social relations. The open plan configuration of the temporary workspace has particularly influenced the auditory levels in the workspace. Rather than the high noise levels that would be expected from a workspace of this layout (Brennan, 2002; Brill, 2001) with all the interactions that is allows, it in fact affords extremely low noise levels. Research academics located in the area with a high level of visual connectivity in the temporary workspace are more demanding of the need for quiet than others located elsewhere in the space. As a source of tension, they are located close to the business administration staff who require the high level of visual connectivity that this area of the workspace offers. Had these measures been taken prior to the allocation of individuals to workspace and used to inform workspace allocation, then this contradiction would have been identified and the tension could have been avoided. The combination of measures of spatiality with social relations therefore makes a valuable contribution to practice, as the right type of space can be found for the types of social interactions that are needed by a work role. This will help to reduce the negative affects that can accompany organisational change, such as work stress and reduced productivity, and can be used as a source of reassurance for those involved in the change.

There are limitations to this research that would need addressing through further studies before widespread conclusions can be drawn. As a case study, this two time series approach to measuring changes in spatial and social relations would need repeating elsewhere to establish further reliability. A larger sample size would also be beneficial to addressing reliability, as the current study is severely limited in the conclusions that can be drawn by its sample size. A further limitation is the inequality in the time period that each type of workspace had been occupied. The comparison of social relations that emerged from a space occupied for 15 years, with those occupied for 4 months, is somewhat uneven. Further data should be gathered from the research department over time to establish how social relations develop over time during their remaining occupation of the temporary workspace and into the new combi-workspace.

Nevertheless this study is of theoretical significance in the combination of measures that has been adopted, contributing to the cross disciplinary body of knowledge that considers workspaces and draws particular attention to the profound impact that organisational culture has on social relations and use of space. This supports the views of management theorists, such as Drucker (1959) who recognise the capacity of the workspace in enabling cultural change, which is not always for the better as has been found to be the case here.

References

Allen, T.: Managing the flow of technology: Technology transfer and the dissemination of technological information within the R&D organization. MIT Press, Cambridge (1997)

Amabile, T.M., Conti, R., Coon, H., Lazenby, J., Herron, M.: Assessing the work environment for creativity. The Academy of Management Journal 39, 1154–1184 (1996)

Becker, F.: Offices at Work: Uncommon workspace Strategies that add value and improve performance. Jossey-Bass, San Francisco (2004)

Becker, F., Steele, F.: Workplace by Design: mapping the higher-performance workspace. Jossey-Bass, San Francisco (1995)

Boutellier, R., Ullman, F., Schreiber, J., Naef, R.: Impact of office layout on com-munication in a science-driven business. R&D Management 38(4), 372–391 (2008)

Brennan, A., Chugh, J.S., Kline, T.: Traditional versus Open office design: A longitudinal field study. Environment and Behaviour 34(3), 279–289 (2002)

Brill, M., Weidemann, S., Olson, J., Keable, E.: Disproving widespread myths about workplace design. Kimball International, Buffalo (2000)

Burt, R.: Structural Holes. Harvard University Press, Cambridge (1992)

Csikszentmihalyi, M.: Creativity: Flow and the psychology of discovery and invention. Harper Perennial (1997)

Cummings, A., Oldham, G.R.: Enhancing creativity: managing work contexts for the high potential employee. California Management Review 40, 22–38 (1997)

Drucker, P.: Landmarks of tomorrow. Harper, New York (1959)

Duffy, F.: The New Office. Conran Octopus, London (1997)

Dunbar, K.: How scientists really reason: scientific reasoning in real-world labora-tories. In: Sternberg, R.J., Davidson, J. (eds.) Mechanisms of Insight. MIT Press, Cambridge (1995)

Fayard, A.L., Weeks, J.: Photocopiers and water-coolers: The affordances of in-formal interac-tion. Organization Studies 28, 605–634 (2007)

Gladwell, M.: Designs for Working. The New Yorker, 61-70 (December 11 (2000)

Hanneman, R.A.: Introduction to social network methods. Riverside, University of California (2001)

Haynes, B.P.: The impact of the behavioural environment on office productivity. Journal of Facilities Management 5(3), 158–171 (2007)

Herzberg, F.: Work and the Nature of Man. World Publishing Company, New York (1966)

Hillier, B., Hanson, J.: The Social Logic of Space. Cambridge University Press, Cambridge (1984)

Hillier, B., Penn, A.: Visible Colleges Structure and randomness in the place of discovery. Science in Context 4(1), 23–49 (1991)

Huczynski, A., Buchanan, D.: Organizational behaviour: an introductory text, 6th edn. Pearson Education Ltd. (2007)

Judt, T.: Postwar: A history of Europe since 1945. Walter Heinmann, London (2005)

Latour, B.: Reassembling the Social. Oxford University Press, Oxford (2007)

Martens, Y.: Creative workplace: instrumental and symbolic support for creativity. Facili-ties 29(1/2), 63–79 (2011)

Martinez, A., Dimitriadis, Y., Rubia, B., Gomez, E., de la Fuente, P.: Combining qualitative evaluation and social network analysis for the study of classroom social interactions. Computers and Education 41, 353–368 (2003)

Obstfeld, D.: Social networks, the Tertius Iungens orientation, and involvement in innovation. Administrative Science Quarterly 50, 100–130 (2005)

Openshaw, R.: Places for Innovation: A way to measure space's effect on knowl-edge worker productivity. Workplace Week Convention, London, November 8-12 (2013)

Parkin, J., Austin, S., Pinder, J., Baguely, T., Allenby, S.: Balancing collaboration and privacy in academic workspaces. Facilities 29(1/2), 31–49 (2011)

Penn, A., Desyllas, J., Vaughan, L.: The Space of Innovation: interaction and communication in the work environment. Environment and Planning B: Planning and Design 26, 193–218 (1999)

Peponis, J.: The spatial culture of factories. Human Relations 38, 357–390 (1985)

Peponis, J., Bafna, S., Bajaj, R., Bromberg, J., Congdon, C., Rashid, M., Warmels, S., Zhang, Y., Zimring, C.: Designing Space to Support Knowledge Work. Environment and Beha-viour 39(6), 815–840 (2007)

Sailor, K.: Creativity as social and spatial process. Facilities 29(1/2), 6–18 (2011)

Scott, J.: Social Network Analysis. Thousand Oaks, Sage (1990)

Steele, F.: Workspace Privacy: A changing equation (1998),
http://www.steelcase.com

Taylor, F.W.: Scientific Management: Comprising Shop Management, The Princi-ples of Scientific Management and Testimony before the Special House Committee. Harper & Brothers (1911)

Van der Voordt, D.J.M.: Costs and benefits of innovative workplace design. TU Delft Centre for People and Buildings (2003)

Wineman, J., Yongha, H., Kabo, F., Owen-Smith, J., Davis, G.F.: Spatial layout, social struc-ture, and innovation in organizations. Environment and Planning B: Plan-ning and De-sign 14, 1–14 (2014)

Determining Team Hierarchy from Broadcast Communications

Anup K. Kalia[1], Norbou Buchler[2], Diane Ungvarsky[2], Ramesh Govindan[3],
and Munindar P. Singh[1]

[1] North Carolina State University, Raleigh, NC 27695, USA
[2] US Army Research Lab, Aberdeen Proving Ground, MD 21005, USA
[3] University of Southern California, Los Angeles, CA 90089, USA

Abstract. Broadcast chat messages among team members in an organization can be used to evaluate team coordination and performance. Intuitively, a well-coordinated team should reflect the team hierarchy, which would indicate that team members assigned with particular roles are performing their jobs effectively. Existing approaches to identify hierarchy are limited to data from where graphs can be extracted easily. We contribute a novel approach that takes as input broadcast messages, extracts communication patterns—as well as semantic, communication, and social features—and outputs an organizational hierarchy. We evaluate our approach using a dataset of broadcast chat communications from a large-scale Army exercise for which ground truth is available. We further validate our approach on the Enron corpus of corporate email.

1 Introduction

In an organization, a team is a purposeful social system created to get work done. Therefore, it is important to understand and characterize the degree to which team members coordinate with each other. In most organizations, a team hierarchy exists among the team members wherein a higher ranking team member sets high-level goals, and guides or motivates lower ranking team members, who are expected to carry out such commands. Although team members have clearly delineated roles, it is important to evaluate whether they are performing their jobs well or whether the team needs restructuring. One important factor for evaluating team performance is communication between team members. Eaton [4] provides insight that communication is essential for team members to build their inter-personal relationships which indirectly enhance team performance. Leonard and Frankel [13] describe that for effective teamwork communication is important because it creates predictability and agreement between team members. Resick et al. [17] suggest that information elaboration is important in evolving teams to maintain team performance. Our premise is that we can determine such indicators of organizational effectiveness and team member performance from members' communications, such as chats and emails, which provide an account of actual behavior while being unobtrusive.

L.M. Aiello and D. McFarland (Eds.): SocInfo 2014, LNCS 8851, pp. 493–507, 2014.

Several works have identified team hierarchies from graphs extracted from online social networks such as Twitter, Flickr, Prison, and Wikivote [8,14,15] and text such as emails and short message service (SMS) communications [7,18,21]. Gupte et al. [8] and Enys et al. [14,15] provide hierarchical measures called social *agony* and *global reaching centrality* (GRC), respectively, to extract hierarchies from online social networks. Rowe et al. [18] extract an undirected graph from Enron emails [5,10] based on the number of emails exchanged between Enron employees whereas Wang et al. [21] compute hierarchy from Enron emails as well as from call and SMS data. Gilbert [7] emphasized analyzing text content to extract phrases that indicate hierarchy. The above works apply when social graphs can be extracted, such as from online social networks and directed messages (emails and SMS). However, these approaches do not apply for broadcast messages, where the receiver is not clear.

Our approach takes in broadcast messages recorded from a multiparty event and produces a team hierarchy among the participants. The basis of our approach is to identify communication patterns from messages that indicate a possible team hierarchy. Broadly, we identify three patterns: *directive, question,* and *informative*. We select these patterns based on the existing literature [7,16] and the fact that they occur frequently in broadcast messages. The overall approach approximates Gilbert [7]. Whereas his approach identifies communication content that indicates power and hierarchy, we additionally compute the ranks and validate our approach versus ground truth. Also, Gilbert's approach is domain-dependent, whereas our approach is domain-independent and applies to broadcast as well as directed communications.

We analyze *semantic, communication,* and *social* features that can be extracted from messages to compute hierarchy. Semantic features include *responses* to communication patterns and *emotions* expressed in responses features extracted from text content. Communication features include the *average response time delay* and *messages* sent features. Social features include the *degree centrality* and *betweenness centrality* features. We hypothesize that semantic features, which capture the meaning of interactions, are better indicators of hierarchy than social features, which merely capture network statistics.

To identify the patterns, we select two chat rooms from a military exercise dataset. We use one chat room to refine our methods to identify patterns and test our method on the second chat room, obtaining an F-measure of 83% for identifying the patterns. From the patterns identified, we collect the features described above. Using these features we determine participants' ranks computed via hierarchical clustering. We evaluate our results against actual known ranks. In addition, we evaluate the generalizability of our approach to directed communications, as in Enron email corpus. In directed communications, emails exchanged between senders and receivers provide good indicators of hierarchy.

We find that for the chat corpus the accuracy in identifying ranks using the *informative* pattern is significantly higher than for the *directive* and *question* pattern. Additionally, we find that semantic features along with communication features are better indicators of hierarchy than social features. For Enron, we obtain similar results regarding the identification of patterns though we find that social features are better indicators of hierarchy than semantic features, possibly because compared to the military dataset, the Enron corpus is much larger with more participants and messages. And it may be that in such a large corporate organization, the roles, responsibilities, and influence need to be ascertained socially. Also, compared to participants in Enron, participants in military communication networks have well-defined functional roles and prescribed work flows that lead to more structured communication and hence, semantic features may perform better than social features.

2   Communication Patterns in Broadcast Messages

Broadcast messages are sent by participants in a group and hence, everyone in a group can see and respond to messages. Before we infer a hierarchy from broadcast messages it is important to understand what each message means. For example, a message can indicate different illocutions [1] such as directives and commissives. Based on the literature [7,16] and our preliminary analysis, i.e., manually finding the distributions of meanings of the messages in the military dataset, we hypothesize that hierarchical information can be extracted from messages via three communication patterns: directive, question, and informative. A *directive* is an order or request; a *question* is an inquiry; an *informative* is a report. *Directives* and *questions* correlate with the sender having a higher rank than the receiver; *informatives* the reverse.

An important challenge in dealing with broadcast messages is that the recipient of a message is not clear. To tackle the challenge, we define a *window* $\mathcal{W}$ consisting of two consecutive messages where we assume that the second message $\mathcal{W}^{next}$ is a response to the first message $\mathcal{W}^{curr}$. The two messages must occur in the same chat room and have different senders. A window $\mathcal{W}$ is instantiated as a *directive, question,* or *informative* pattern if, respectively, $\mathcal{W}^{curr}$ is a *directive, question,* or *informative* and correspondingly $\mathcal{W}^{next}$ is an acknowledgment, response, or acknowledgment. Table 1 provides examples of these patterns from military data.

Table 1. Examples of communication patterns from military chat data

Window	Sender	Messages	Pattern
$\mathcal{W}_a^{curr}$	8_6i_256_s3	Cos, send all reports up to BN over this net	*Directive*
$\mathcal{W}_a^{next}$	8_6i_256_b_cdr	rgr	
$\mathcal{W}_b^{curr}$	8_6i_256_s3	B, whats your status on personnel?	*Question*
$\mathcal{W}_b^{next}$	8_6i_256_b_cdr	no casualties	
$\mathcal{W}_b^{curr}$	8_6i_256_b_cdr	have been engaging with SAF and MTRs with no effect	*Informative*
$\mathcal{W}_b^{next}$	8_6i_256_cdr	ack, keep me posted	

3   Process

Figure 1 shows the process we follow. In the process, we separately consider the *directive, question, informative* patterns as well as the combination of *directive* and *question* patterns to compute ranks. Next, we evaluate the accuracy of the ranks computed based on different patterns.

As an illustration, consider computing ranks using *directive* patterns. For each participant P in chat messages we extract the following features. First, we extract *directive* patterns $\mathcal{W}$ where $\mathcal{W}^{curr}$ indicates a directive message and P is the sender of $\mathcal{W}^{next}$. From the patterns, we assume that P responds to $\mathcal{W}^{curr}$ and hence, we calculate the total number of such responses to directives for P.

Second, we determine whether $\mathcal{W}^{next}$ indicates a positive, neutral, or negative emotion. We extract emotions because we hypothesize that they can be indicators of hierarchy. For example, P may be a team leader and may display positive emotions to motivate subordinates or P may be a subordinate and may express emotions with respect to outcome of his or her actions. We include responses to patterns and emotions within semantic features.

Third, based on the patterns $\mathcal{W}$ we find the *average response time delay*, i.e., the average of the time lags between $\mathcal{W}^{curr}$ and $\mathcal{W}^{next}$ extracted for P. Fourth, we find the *number of messages* that P broadcasts. We include the average response time delay and number of messages broadcast as communication features.

From the patterns $\mathcal{W}$ we create a graph that contains directed edges from responders (P) to respondees. Using the graph, we compute social features for P, i.e., P's *degree centrality* and *betweenness centrality* [2,6]. We aggregate all features—semantic, communication, and social—for P. We repeat the feature extraction for all participants P^*. Finally, based on P^*s' features we compute hierarchical ranks for each P. We evaluate computed ranks against the ground truth of actual ranks. We carry out the above process for the informative and question patterns.

Prior works [8,15,18] focus primarily on social and communication features to compute ranks whereas we include semantic features based on the intuition that semantic features, being based on the message content, can reveal important hierarchical information. Below, we discuss the extraction of features in detail.

3.1   Extracting Semantic Features

To extract semantic features for each participant, first, we identify patterns $\mathcal{W}$. To identify patterns, we create a rule-based approach using training data and evaluate it on a test data. Both training and test data consist of broadcast messages labeled *directive, question*, or *informative*. To support our rules, for each dataset, we build a domain-specific lexicon of action verbs that includes words occurring frequently in the data.

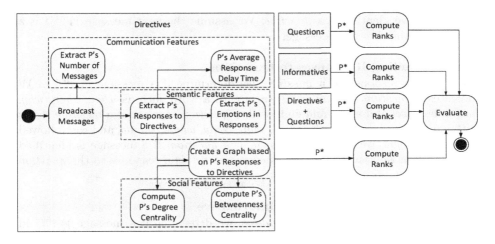

Fig. 1. Process followed to compute ranks with respect to directive, question, and informative patterns, and directives and questions combined for participants (P*) from broadcast messages

3.1.1 Extracting Responses to Directives

To extract a response to a *directive*, we determine if a message $\mathcal{W}^{curr}$ in $\mathcal{W}$ indicates a *directive*. To do so, we parse a message $\mathcal{W}^{curr}$ using the Stanford Natural Language Parser [9] and extract a parse tree. Figure 2 represents a parse tree for a sample message "Cos, send all reports up to BN over this net." In the parse tree, first, we look for a verb phrase (VP) indicated by the shading in Figure 2. Then, in the VP we look for an action verb (VB). If the action verb matches a verb in our domain-specific lexicon, we extract the rest, i.e., noun (NP) and prepositional phrase (PP), as shown in Figure 2. Hence, the words extracted from the example message are "send all reports up to BN over this

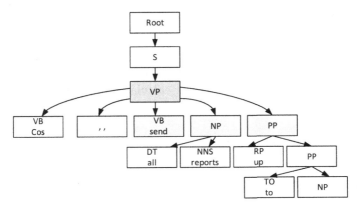

Fig. 2. Parse tree derived from "Cos, send all reports up to BN over this net" where Cos is the Chief of Staff position and BN is the Battalion

net" which we identify as a *directive*. We assume the next message $\mathcal{W}^{next}$ is a response to the *directive* message.

3.1.2 Extracting Responses to Questions

To extract a response to a *question*, we determine if a message $\mathcal{W}^{curr}$ in $\mathcal{W}$ indicates a *question*. If a message starts with a word such as *what, when, why, has, how, have*, and so on and ends with a *question mark* or if a message starts with a modal verb (MD) such as *will, shall, could, would, should*, and *can* followed by the word *you*, we mark the message as a *question*. If a message is identified as a *question*, we assume the next message $\mathcal{W}^{next}$ is a response to the *question* regardless of its grammar or content.

3.1.3 Extracting Responses to Informative

To extract a response to an *informative*, we determine if a message $\mathcal{W}^{curr}$ in $\mathcal{W}$ indicates an *informative*. If a message begins with the following *rgr, Roger, ack, yes, yup, yep, okay, ok, thanks*, and so on we tag the message as the informative. Although some of the words (e.g., *Roger* and *ack*) are domain-specific, other words (*thanks, yes*, and *okay*) are domain independent. Such generic words make this pattern domain-independent. The next message $\mathcal{W}^{next}$ we assume is a response to the *informative* message.

For each participant, we calculate the count of all $\mathcal{W}^{next}$ or responses extracted for each pattern.

3.1.4 Extracting Emotions in Responses

For each communication pattern $\mathcal{W}$, we determine if the response message $\mathcal{W}^{next}$ indicates an emotion, which could be positive, neutral, or negative. We use the Stanford Sentiment Parser [20], which computes the emotion corresponding to a message. For each participant, we compute the sums of the emotion polarities identified from response messages.

3.2   Extracting Communication Features

For each participant we extract two communication features. One, the number of messages sent by the participant and second, the *average response time delay* for a participant based on the messages that indicate *responses* to a pattern. The number of messages is a network statistic calculated independently of *responses* to patterns.

3.3   Extracting Social Features

To extract social features, we create a graph represented as an adjacency matrix $\mathcal{A}_{ij}$. In the matrix i and j represent the participants. An edge ij in $\mathcal{A}$ exists from the sender (responder) of $\mathcal{W}^{next}$ toward the sender (respondee) of $\mathcal{W}^{curr}$, if $\mathcal{W}^{next}$ indicates a response to a pattern, i.e., directive, question, or informative. If an edge ij exists, we mark $\mathcal{A}_{i,j} = 1$ else we mark $\mathcal{A}_{i,j} = 0$. We also mark $\mathcal{A}_{i,j}$

$= 0$ if i equals j because we assume a sender does not respond to itself. We mark $\mathcal{A}_{i,j} = 1$ irrespective of one or more responses between i and j. From $\mathcal{A}_{i,j}$ we can construct a directed graph $G(V, E)$ where V represents the participants and E represents the directed edge between the participants.

Using the directed graph $G(V, E)$ extracted from a pattern, we compute the social features of *degree centrality* and *betweenness centrality*. We consider these social features for two reasons. One, they have been used in the literature to interpret Rowe et al.'s [18] hierarchy. Two, we consider chatrooms that contain more intrateam messages than interteam messages, possibly, because we assume graphs derived from intrateam messages may be strongly connected than graphs derived from interteam messages. Our assumption is based on the notion that a chatroom mapping is not one-to-one direct and in general, people subscribe to chatrooms. In that sense the degree distribution is shared widely (observed) by all.

- **Degree centrality** is defined as the degree of a node or the number of edges directed to a node. The *degree centrality* $dc(v_j)$ of a node v_j equals the number of edges ij directed to v_j, i.e., $\sum_i a_{ij}$ [2].
- **Betweenness centrality**, defined as the number of shortest paths passing through a node, is a measure of how important a node is. The *betweenness centrality* of a node v_j is calculated as $\sum_i \sum_k \frac{\delta_{ijk}}{\delta_{ik}}$ where δ_{ijk} is the number of shortest paths between i and k that include j and δ_{ik} is number of shortest paths between i and k [2,6].

3.4 Computing Ranks

We compute ranks based on features extracted for participants. We adopt hierarchical clustering for two reasons. First, it being an unsupervised technique can be applied to datasets of any size. This is useful because we don't need to create a model from a large dataset and then use the model to produce predictions for a new dataset. Second, we want to infer a hierarchy among team members. The method helps cluster employees with similar rankings.

To compute ranks, we normalize all features extracted for each participant to the interval [0,100]. We construct a feature vector for each participant and use the *Euclidean distance* between them as a basis for hierarchical clustering. We plan to evaluate other distance metrics in future. We adopt the *single link* algorithm [19], which is a simple and popular technique. Figure 3 shows an example of a hierarchical cluster as a *single link dendrogram*. In Figure 3, d_1, ..., d_5 represent distances between the clusters. We assume that participants in the same cluster have the same rank. Next we provide rules to estimate rank orders between participants in clusters. We derive these rules by checking the consistency in rank outputs by applying the rules on multiple datasets.

Rank Rule 1. *For the directive and question patterns, increasing distance between clusters from bottom to top indicates decreasing rank.*

Rank Rule 2. *For the informative pattern, increasing distance between clusters from bottom to top indicates increasing rank.*

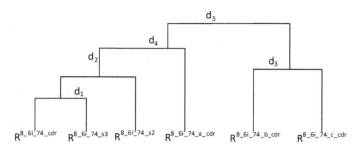

Fig. 3. An example of a single link dendrogram with distance d between clusters, applied to estimate rank R (bottom row)

4   Evaluation

We evaluate our approach primarily on our military broadcast chat dataset and secondarily on the Enron (directed) email dataset. The evaluation has two steps. First, we evaluate our methods to extract communication patterns, as described in Section 3.1. Second, we evaluate our estimation of ranks based on the patterns, as described in Section 3.4.

To evaluate the extraction of patterns we use the following metrics: precision, recall, and F-measure. Precision is given by $\frac{true\_positive}{true\_positive+false\_positive}$, recall by $\frac{true\_positive}{true\_positive+false\_negative}$, and F-measure by $\frac{2 \times precision \times recall}{precision+recall}$. The mean absolute error (MAE) of a rank prediction is $\frac{\sum_i^N |predicted\_rank_i - actual\_rank_i|}{N}$. The accuracy of a rank prediction is $\frac{N-MAE}{N}$, where N is the highest rank.

4.1   Data Description

4.1.1 Military
The military dataset was provided by the Mission Command Battle Lab at Fort Leavenworth, Kansas, and the US Army Research Laboratory, Maryland, from an Army simulation experiment (SIMEX). The dataset contains 20 chat rooms, on average, with 42 participants each and 6,998 messages. From the dataset, we consider the following chat rooms: Infantry Brigade Combat Team (IBCT), USMC Maneuver Brigade (MEB), Cavalry (CAV), and Commander (CDR) to evaluate our results. MEB has 546 messages and 50 participants, CAV has 481 messages and 48 participants, CDR has 409 messages and 37 messages, and IBCT Intel has 1027 messages and 64 participants. We consider these chat rooms because, first, they have more messages than the mean number of messages and, second, they have more intrateam messages than interteam messages.

The dataset includes the participants' actual ranks. (Rank 1 is the highest.) Table 2 shows the ranks of a few participants who sent more than one broadcast message and belong to a particular military team.

Some participant IDs in the dataset have OCR errors. For example, the ID 8_6i_256_s3 has spurious variants 8_61_256_s3 and 8_6i_256_53 in which i is substituted by 1 and s by 5, respectively. Such errors make it difficult to identify the IDs automatically. To handle such spurious IDs, we select participant IDs with the highest number of messages. For example, if 8_6i_256_s3, 8_61_256_s3, and 8_6i_256_53 have sent 25, 34, and 10 messages respectively, then for our evaluation we consider 8_61_256_s3 with 34 messages.

Table 2. Ranks of participants selected from the chat rooms

Rank MEB		CAV	CDR	IBCT Intel
1.	2meb_cdr	8_6i_74_cdr	8_6i_256_cdr	8_6i_s2
2.	2meb_s2	8_6i_74_s3	8_6i_256_s3	8_6i_156_s2
3.	2meb_s3	8_6i_74_s2	8_6i_256_s6	8_6i_256_s2
4.	2meb_fso	8_6i_74_fso	8_6i_256_fso	8_6i_256_s3
5.	2meb_mech_bn_cdr	8_6i_74_a_cdr	8_6i_256_alo	8_6i_35_s2
6.	2meb_mech_bn_s3	8_6i_74_b_cdr	8_6i_256_a_cdr	–
7.	2meb_mech2_bn_cdr	8_6i_74_c_cdr	8_6i_256_b_cdr	–
8.	2meb_helo_sqdn_cdr	8_6i_74_jtac	8_6i_256_c_cdr	–
9.	–	–	8_6i_256_wpn_cdr	–

4.1.2 Enron

In the Enron email dataset [5,10], we arbitrarily consider 62 employees who have sent 38,863 emails with a total of 360,708 email sentences. Prior to the evaluation, we obtain the actual ranks of these 62 employees [7]. The distribution of ranks from 0 to 6 is as follows: 8%, 2%, 29%, 11%, 6%, 36%, and 8%.

4.2   Results

We describe the results of our evaluation for extracting patterns and computing ranks on both the military chat dataset and the Enron dataset.

4.2.1 Extracting Communication Patterns

We created the rule-based approach given in Section 3.1 using CDR (training data) and evaluated it on CAV (test data). Figure 4(a) shows distributions of the communication patterns in these datasets. Notice the high frequency of the *informative* pattern. Two raters (both graduate students in Computer Science) labeled the data with the various patterns. Their inter-rater agreement (*kappa* score [3]) was 0.76, which is fairly high. We arbitrarily selected one of the rater's assigned labels as the ground truth, because we cannot take the average. There are advanced approaches that use Bayesian techniques to estimate a ground truth probability for each classification [12], but this is beyond the current scope and means of the paper.

Based on the training data, we constructed our rules, as described in Section 3.1, and evaluated them on the test data. For the training and test data, we found that the F-measures are respectively 0.71 and 0.64 (for the *directive* pattern), 0.83 and 0.91 (the *question* pattern), 0.95 for each (for the *informative* pattern), and 0.84 and 0.83 (overall). Considering the F-measure to identify different *patterns* as 0.83, we predicted the patterns for the dataset MEB and IBCT Intel.

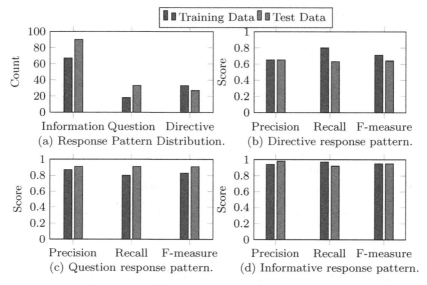

Fig. 4. Panel A: Distribution of response patterns. Panels B, C, D: F-measure scores for the response patterns (highest for the informative response pattern in Panel D).

4.2.1 Computing Ranks via Different Patterns

We used the hierarchical clustering approach described in Section 3.4 to compute ranks. Specifically, we considered eight features F_1 to F_8 extracted for each pattern. F_1 represents the counts of *responses* to patterns, i.e., either *directive*, *question*, or *informative*; F_2, F_3, and F_4 represent the number of negative, neutral, and positive *emotions*, respectively; F_5 represents the *average response time delay*; F_6 represents the number of *messages* sent; F_7 represents the *degree centrality*; and F_8 represents the *betweenness centrality*. Since the *directive* and *question* patterns have the same relationship, we combined them into the *directive+question* pattern with the assumption that it would yield improved results over treating them separately.

Using the clustering method, we calculated the percentage accuracies for the four datasets MEB, CAV, CDR, and IBCT Intel for the four patterns respectively. From the mean absolute errors (MAE) we computed the percentage accuracy based on the highest rank N considered for the evaluation. Figure 5 describes the overall result. In each panel, the x-axis shows the patterns, i.e., *directive, question, informative* and *directive+question* and the y-axis shows the percentage accuracy. From the result, we observed that the percentage accuracy for *informative* is the highest for all the datasets (73.4%, 76.5%, 69.5%, 68%), which suggests that the *informative* pattern is a better indicator of hierarchy than other patterns. In addition, we performed one-tailed t-test to check if the accuracy for *informative* is significantly higher than for *directive, question,* and *directive+question* at the significant level of 5%. We find that the accuracy for *informative* is indeed significantly higher than *directive* (p=0.03) and *directive+question* (p=0.002), but not significantly so for question (p=0.06).

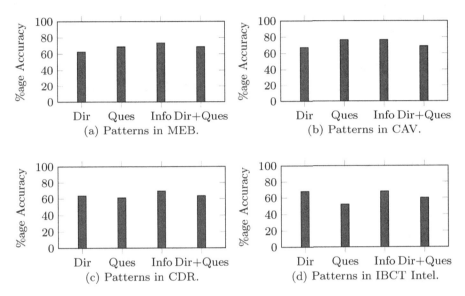

Fig. 5. Percentage accuracy of computing hierarchy using different response patterns directive (Dir), question (Ques), informative (Info), and directive+informative (Dir+Info) via different datasets

4.2.2 Evaluating Features

We compared MAEs obtained using only the semantic features with those obtained using only the social features. For the comparison, we performed one-tailed t-tests on the MAEs obtained from the four chat rooms for all patterns.

Table 3 summarizes these hypotheses and the results obtained.

In the table, we have stated hypotheses that compare the mean (μ) of the MAEs obtained using features F_1 to F_8 for the patterns. When the null hypothesis is rejected we accepted the alternative hypothesis, i.e., one mean is significantly less than the other. Among the features, F_1 to F_4 represent the semantic features, F_5 and F_6 represent the communication features, and F_7 and F_8 represent the social features. We found that the MAEs obtained based on features F_1 to F_4 were not significantly lower than the MAEs obtained based on features F_7 and F_8. Recall that F_5 is *average response time delay* and F_6 is *number of messages*. We found that the MAEs obtained based on features F_1 to F_5 or obtained based on features F_1 to F_6 were significantly lower than those obtained based on F_7 and F_8. When we added F_5 and F_6 to features F_7 and F_8, the MAEs obtained were not significantly lower than the MAEs obtained considering features F_1 through F_4. Similarly, when we added F_6 to features F_7 and F_8, the MAEs obtained were not significantly lower than the MAEs obtained using features F_1 through F_5. The foregoing suggests that the semantic features are better indicators of hierarchy than the social features.

Table 3. Statistically comparing semantic features with social features (sem-semantic, comm-communication, soc-social, avg-average, resp-response, del-delay, msg-messages, hyp-hypotheses, rej-rejected)

# Alt. Hypotheses	Null hyp. p-val	Null hyp. rej. at 5%?
1. **sem.** $(\mu_{F_1 to F_4}) <$ **soc.** $(\mu_{F_7 to F_8})$	0.23	no
2. **sem. & avg. resp. time del.** $(\mu_{F_1 to F_5}) <$ **soc.** $(\mu_{F_7 to F_8})$	0.04	yes
3. **sem. & comm.** $(\mu_{F_1 to F_6}) <$ **soc.** $(\mu_{F_7 to F_8})$	0.00	yes
4. **soc. & comm.** $(\mu_{F_5 to F_8}) <$ **sem.** $(\mu_{F_1 to F_4})$	0.08	no
5. **soc. & no. of msg** $(\mu_{F_6 to F_8}) <$ **sem.** $(\mu_{F_1 to F_4})$	0.13	no

4.2.3 The Enron Dataset

We evaluated our approach on the Enron email dataset [5,10] as well. A major challenge we faced is to create conversation threads based on a subject or a topic. Whereas in the military dataset we considered the counts of the response messages to *directive*, *question*, and *informative* messages, for the Enron dataset, we considered the counts of *directive* and *question* messages sent by an employee. We considered a message whose subject begins with "RE:" as an *informative* because it indicates that the message responds to a prior message. To identify patterns we used the rules described in Section 3.1. Once the messages were identified, we computed ranks using the rules provided in Section 3.4. The features we considered to compute ranks were F_1, F_6, F_7, and F_8. We did not consider features F_2 to F_5 (*emotions* and *average response time delay*) because we could not create conversation threads. We constructed F_7 and F_8 based on the *number of messages* exchanged between employees.

We found that ranks computed using the *informative* pattern have higher accuracy (75%) than the directive (74.4%) and question (70.1%) patterns. This results coheres with our finding over the military data. However, unlike the military data, the accuracy from social features (72%) was slightly higher than for the semantic and communication features (71%). We also found that adding semantic features to social and communication features (75%) slightly improved the accuracy over considering only social and communication features (74%). Therefore, along with social and communication features semantic features were important in predicting hierarchy.

5   Discussion and Future Work

We provide a novel approach to computing team hierarchy from broadcast messages. To compute the hierarchy, first, we identify three patterns via text mining obtaining F-measures of 80%, 95%, and 60% respectively for *question*, *informative*, and *directive* patterns, and 83% overall. Second, once we identify the patterns, we extract disparate features: semantic, communication, and social. Third, using the features we compute ranks using the hierarchical clustering method. We find that the *informative* pattern is a better indicator of hierarchy than the other patterns, thus validating our approach. We find that semantic features added with communication features (i.e., the network statistics) are better indicators of ranks than using social features alone. We obtain similar results regarding the usage of patterns to infer hierarchy on the Enron dataset. We also find that semantic features added to social and communication features improve accuracy in predicting hierarchy. However, social features in Enron are better indicators of hierarchy than semantic features. This could be because the Enron dataset is much larger than the military dataset: on average, Enron participants sent more messages than military participants.

Although we consider only two datasets, our study provides some hints as to the differences in how people use chat communications versus email, at least in work-related settings. Email communications would tend to respect predefined organizational relationships (who writes to whom) and thus social features are predictive of hierarchy. In contrast, broadcast communications at the level of connectivity do not respect any predefined relationships. Thus their semantic features are better predictive of hierarchy. In the military setting, the ranks of the participants are well defined. We conjecture that, in settings where ranks are not predefined, such as in collaborations between peers as in open source software development or nascent political movements, broadcast communications would be a way for true hierarchies to emerge.

This work estimates intrateam hierarchy. In future work, we will consider interteam hierarchy. Also, we hope to extend our work on the estimation of a hierarchy to the estimation of team cohesion, trust, and performance. We plan to improve our domain-specific military lexicon to further improve performance. We expect that our results would be stronger on larger datasets where participants communicate more frequently with each other.

6   Related Work

There has been a small amount of research on inferring hierarchy from communications. Nishihara and Sunayama [16] compute hierarchy by two measures: based on request actions communicated by a speaker and the number of sentences sent by a speaker. In contrast, instead of identifying requests, we identify patterns such as *directive, question,* and *informative.* Moreover, Nishihara and Sunayama do not incorporate features such as *emotions, average response time delay,* or centrality features that can provide important clues to identify hierarchy. Also, they evaluate their work on directed messages but not on broadcast messages.

Gilbert [7] identifies words and phrases from Enron emails [10,5] that indicate team hierarchy. This work is limited to finding such words and phrases rather than computing a hierarchy. Also, Gilbert's approach is domain-dependent because it requires words and phrases related to hierarchy. Preparing such lexicons for new datasets can be cumbersome. In contrast, we provide ways to identify patterns that generalizes to different datasets. Also the lexicon we prepare is easy to extract as the verbs are extracted based on their frequencies.

Rowe et al. [18] compute team hierarchy by extracting an undirected graph based on emails exchanged between senders and receivers. They consider centrality measures to compute hierarchy and do not focus on analyzing the content of emails. Hence, Rowe et al.'s contribution does not handle broadcast messages. In contrast, we emphasize understanding the content of the messages to identify the patterns and consider broadcast messages. In addition, we find that patterns and emotions extracted from messages are better indicators of hierarchy than are centrality measures.

Krafft et al. [11] propose a probabilistic model to visualize topic-specific subnetworks in email datasets. In specific, they associate an author-recipient edge (or an email) with different subtopics using K-dimensional topic-specific communication patterns. In our work we take a similar approach where we extract different communication patterns and features from emails and broadcast messages for participants to infer their hierarchy.

Acknowledgment. This work was primarily supported by the Army Research Laboratory in its Network Sciences Collaborative Technology Alliance (NS-CTA) under Cooperative Agreement Number W911NF-09-2-0053. Kalia and Singh were partially supported by the NCSU Laboratory for Analytic Sciences.

References

1. Austin, J.L.: How to Do Things with Words. Clarendon Press, Oxford (1962)
2. Borgatti, S.P., Everett, M.G.: A graph-theoretic perspective on centrality. Social Networks 28(4), 466–484 (2005)
3. Cohen, J.: A coefficient of agreement for nominal scales. Educational and Psychological Measurement 20(1), 37–46 (1960)
4. Eaton, J.W.: Social processes of professional teamwork. American Sociological Review 16(5), 707–713 (1951)

5. Fiore, A., Heer, J.: UC Berkeley Enron email analysis (2004), http://bailando.sims.berkeley.edu/enron_email.html
6. Freeman, L.C.: Centrality in social networks conceptual clarification. Social Networks 1(3), 215–239 (1979)
7. Gilbert, E.: Phrases that signal workplace hierarchy. In: Proceedings of the ACM Conference on Computer Supported Cooperative Work, pp. 1037–1046 (2012)
8. Gupte, M., Shankar, P., Li, J., Muthukrishnan, S., Iftode, L.: Finding hierarchy in directed online social networks. In: Proceedings of the 20th International Conference on World Wide Web, pp. 557–566 (2011)
9. Klein, D., Manning, C.D.: Accurate unlexicalized parsing. In: Proceedings of the 41st Meeting of the Association for Computational Linguistics, Stroudsburg, pp. 423–430 (2003)
10. Klimt, B., Yang, Y.: The enron corpus: A new dataset for email classification research. In: Boulicaut, J.-F., Esposito, F., Giannotti, F., Pedreschi, D. (eds.) ECML 2004. LNCS (LNAI), vol. 3201, pp. 217–226. Springer, Heidelberg (2004)
11. Krafft, P., Moore, J., Desmarais, B.A., Wallach, H.M.: Topic-partitioned multinetwork embeddings. In: Proceedings of 25th Annual Conference on Neural Information Processing Systems 2012, pp. 2807–2815. Curran Associates, Inc., Lake Tahoe (2012)
12. Lehner, P.: Testing the accuracy of automated classification systems using only expert ratings that are less accurate than the system. In: The MITRE Corporation, pp. 1–28 (2014)
13. Leonard, M.W., Frankel, A.S.: Role of effective teamwork and communication in delivering safe, high-quality care. Mount Sinai Journal of Medicine 78(6), 820–826 (2011)
14. Mones, E.: Hierarchy in directed random networks. Physical Review E 87(2), 022817 (2013)
15. Mones, E., Vicsek, L., Vicsek, T.: Hierarchy measure for complex networks. PLoS ONE 7(3), e33799 (2012)
16. Nishihara, Y., Sunayama, W.: Estimation of friendship and hierarchy from conversation records. Information Sciences 179(11), 1592–1598 (2009)
17. Resick, C.J., Murase, T., Randall, K.R., DeChurch, L.A.: Information elaboration and team performance: Examining the psychological origins and environmental contingencies. Organizational Behavior and Human Decision Processes 124(2), 165–176 (2014)
18. Rowe, R., Creamer, G., Hershkop, S., Stolfo, S.J.: Automated social hierarchy detection through email network analysis. In: Proceedings of the 9th WebKDD and 1st SNA-KDD 2007 Workshop on Web Mining and Social Network Analysis, San Jose, pp. 109–117 (2007)
19. Sibson, R.: SLINK: An optimally efficient algorithm for the single-link cluster method. The Computer Journal 16(1), 30–34 (1972)
20. Socher, R., Perelygin, A., Jean, Y., Wu, J.C., Manning, C.D., Ng, A.Y., Potts, C.: Recursive deep models for semantic compositionality over a sentiment treebank. In: Proceedings of the Conference on Empirical Methods in Natural Language Processing, Seattle (2013)
21. Wang, Y., Iliofotou, M., Faloutsos, M., Wu, B.: Analyzing communication interaction networks (cins) in enterprises and inferring hierarchies. Computer Networks 57(10), 2147–2158 (2013)

Cultural Attributes and Their Influence on Consumption Patterns in Popular Music

Noah Askin[1] and Michael Mauskapf[2]

[1]INSEAD, Fontainebleau, France
noah.askin@insead.edu
[2]Kellogg School of Management, Northwestern University, Chicago, IL
m-mauskapf@kellogg.northwestern.edu

Abstract. In this paper we leverage recent developments in the way scholars access, collect, and analyze data to reexamine consumption dynamics in popular music. Using web-based tools to construct a dataset that distills songs' musical content into a handful of discrete attributes, we test whether and how these attributes affect a song's position on the *Billboard Hot 100* charts. Our analysis suggests that attributes matter, beyond the effect of artist, label, and genre affiliation. We also find evidence that the relational patterns formed between attributes—what we call cultural networks—crowds songs that are too similar to their neighbors, adversely affecting their movement up the charts. These results suggest that culture possesses its own sphere of influence that is partially independent of the actors who produce and consume it.

Keywords: culture, consumption, networks, attributes, music.

1   Introduction

Over the last two decades, a small but growing group of scholars has worked to import and develop new techniques to understand better the antecedents, consequences, and qualities of culture (e.g., DiMaggio 1994; Mohr 1998; Weber 2005; Lizardo 2006; Lena and Peterson 2008). These developments have been bolstered by methodological advancements in "big data" and computer science, which increasingly influence the work of sociologists and organization and management scholars. Despite this bridge between the social and computational sciences, however, issues of empirical measurement—how to operationalize conceptual variables and processes, determine appropriate indicators for them, and specify discrete units to model and test—remain central for scholars interested in the study culture (Mohr and Ghaziani 2014).

In this paper, we engage with these issues by leveraging recent advances in the way we access, collect, and analyze data to reexamine cultural consumption and evaluation in popular music. To do this, we have constructed a unique dataset describing over 25,000 songs that appeared on the *Billboard Hot 100* charts between 1958 and 2013. Published weekly by *Billboard Magazine*, these charts contain information on and rankings of the most popular songs in public circulation. The analyses conducted in this paper test whether chart performance is a function of certain musical characteristics,

L.M. Aiello and D. McFarland (Eds.): SocInfo 2014, LNCS 8851, pp. 508–530, 2014.
© Springer International Publishing Switzerland 2014

measured here using a suite of algorithmically-determined attributes that represent the sonic fingerprint of each track (e.g., loudness, tempo, valence, etc.).[1]

Introducing a new means of measuring cultural content in the domain of popular music allows us to investigate two questions that have important implications for the study of cultural production and consumption outcomes. First, we ask to what extent aesthetic attributes affect consumers' evaluations of cultural products. Extant literature on this topic emphasizes the role producers, distributors, critics, and consumers play in determining what kinds of culture is created and how it is received (e.g., Peterson 1990). In other words, factors such as artist familiarity, payola, label size, and peer recommendation are the primary determinants of whether a song is considered a hit or a flop (Dowd 1992; Salganik, Dodds, and Watts 2006; Rossman 2012). While some of the work in this area has hinted that there may be other forces at play, no one has provided compelling evidence to explain how cultural content—the attributes of the cultural products themselves—shapes this process. We propose here that, although people are ultimately responsible for the production and evaluation of new cultural forms and practices, attributes play a significant role in determining if and how cultural products (e.g., songs) are consumed.

Yet even if we agree that cultural attributes play a predictive role in the evaluation process, the way in which such influence asserts itself remains unspecified. To address this puzzle, we ask whether certain patterns of attribute similarity affect the position and performance of cultural products in the competitive marketplace. Specifically, we propose that the relational structures formed between bundles of sonic attributes may affect when and where new songs appear on the charts. In other words, a song's position within its cultural network—the system of relationships defined by shared attributes between songs, and the cultural fabric within which a song is embedded—endogenously influences how that song is evaluated by consumers.

After providing a brief review of relevant work in cultural sociology and network science, we introduce our conception of cultural networks and generate a series of models designed to test whether such relationships exist, and to what effect. Although these analyses are exploratory in nature, they provide preliminary evidence that attributes matter, both independently and in the way they structure songs' relationships to each other. By recognizing the possibility that culture has its own sphere of influence that is partially independent of the actors who produce and consume it, we rethink some of the basic mechanisms associated with cultural production and consumption. Employing endogenous explanations to understand how songs are evaluated suggests that the engine of production and consumption can arise from cultural structures themselves, generating new insights into the dynamics of a multi-billion dollar industry while reinforcing the notion that large-scale quantitative comparison can effectively extend scientific research on culture.

2   Culture, Measurement, and a Move toward Relationality

The heart of culture research, both in the humanities and social sciences, lies in its contextual richness and interpretive complexity. Much of the research in this area has

---

[1]   More information about these attributes, and how they are constructed, can be found in the Data & Methods section of this paper and in the Appendix.

employed qualitative methodologies, which are ideal for generating detailed descriptions of and theory about culture, meaning, and its relationship to the social world. The resulting research landscape has caused some to argue that cultural sociology is, in fact, "methodologically impoverished" (DiMaggio, Nag, and Blei 2013). However, recent advancements in data collection, analysis, and visualization techniques offer scholars an opportunity to shirk this condemnation (Bail 2014). A growing body of themed conferences and special issues devoted to new means of studying culture suggests that researchers are beginning to take this opportunity seriously (e.g., Mohr and Ghaziani 2014).

One area of research that has been particularly fruitful in this regard is the study of social networks. While relationality has become central to our understanding of how social systems operate (Emirbayer and Goodwin 1994), it has not played a significant role in the study of culture until recently. In the context of Broadway musicals, Uzzi and Spiro (2005) find that when collaborations between artists and producers display small world properties, their cultural productions are more likely to achieve critical and commercial success. In the context of popular music, Lena and Peterson (2008) argue that genres emerge as a form of symbolic classification that helps to organize artists' work and shape audience consumption patterns. And Phillips's (2011) work on jazz explicitly ties cultural reproduction (in this case, re-recordings) to the geographic location of its genesis. Songs from "disconnected" locations were more likely to be re-recorded, in spite of the fact that innovation in the form of original music was less likely to come from disconnected cities.

These and other studies (e.g., Dowd and Pinheiro 2013) highlight the means through which cultural practices have been shaped and determined: via geographic, interpersonal, and/or professional collaboration networks. While these are sensible applications of relational analysis to the study of culture, there are other ways in which these two areas might be integrated. Rather than simply being embedded in a static cultural environment, social networks can be altered by a dynamic set of tastes and practices, recasting culture and social structure as mutually constitutive (Lizardo 2006; Vaisey and Lizardo 2010). At the intersection of culture and cognition, Amir Goldberg (2011) has developed a network-inspired method—relational class analysis (RCA)—to identify groups of individuals that share distinctive ways of understanding cultural categories. RCA's explicit emphasis on the patterns of relationships between individuals' attitudes, rather than the attitudes themselves, reaffirms "the relation" as an important unit of analysis for cultural studies and social scientific inquiry more broadly.

2.1   Cultural Networks: What Are They, and Why Might They Matter?

Although all of the studies cited above are interested in cultural phenomena of one sort or another, most of them employ social characteristics as their primary explanatory variable, and those that employ cultural attributes are interested in the effects of culture on social structure. In this paper, we posit that the relational patterns between cultural attributes themselves—what we call "cultural networks"—represent an important and as of yet unstudied phenomena that might enhance our understanding of various production and consumption processes.

The idea of a cultural network is not new, but existing considerations of the concept are neither fully formulated nor well understood. Mohr (1998) proposes that cultural networks can be conceived of in two ways: via similarities in attributes (e.g., aesthetic or ideological characteristics), or through actors' cognitive judgments and categorizations (e.g., consumer evaluations and consumption habits). Nearly all extant research that invokes this idea emphasizes the second approach, explaining the emergence and transmission of cultural practices through the social relationships of producers and consumers. Rather than conceptualizing cultural networks as patterns of cultural choice that position a person as a bridge between cultural worlds, we define them simply as the system of relationships between some bundle of attributes, practices, or ideas. In the domain of popular music, cultural networks contain songs that are tied to each other when they display some degree of attribute overlap or similarity.

To be clear, this view of culture is explicitly structural in nature, wherein relational patterns are defined by cultural content. These relations are connected to but conceptually distinct from the networks of actors that produce, distribute, and consume culture. Importantly, this distinction has implications for the way we understand the dynamics and consequences of cultural production, consumption, and evaluation. For example, whether a cultural product garners popular appeal or is deemed worthy of praise is often understood to be a function of producer-level characteristics. Cultural attributes and their resonance with different audiences may also affect how products are received and evaluated (e.g., Schudson 1989), but what about the performance consequences of cultural networks?

Like their social counterparts, cultural networks consist of structural signatures that generate opportunities differentially, rendering certain products more likely to perform well depending on their relative position within the broader cultural milieu. Structural holes theory argues that the presence of unfilled gaps between mutually connected actors will incite brokerage and lead to new and better performance outcomes (Burt 1992; Zaheer and Soda 2009). Cultural holes operate in much the same way, representing an opportunity space between two cultural products that have yet to be connected. Lizardo (2014) finds evidence for the generative effects of cultural holes, showing how omnivorous consumers can exploit previously unconnected cultural practices to shape taste. And in the case of academic paper citations, Vilhena and colleagues (2014) argue that communicative efficiency in academic discourse is a function of scholars' ability to bridge cultural holes and communicate across fields. They find that the more distinct and indecipherable a field's scientific jargon, the less likely scientists are to leverage insights (e.g., cite previously published papers) from outside their home discipline.

Like science, music represents an ideal setting in which to test the consequences of certain network properties, due in part to its reliance on an internally consistent grammar (e.g., discrete combinations of pitch, harmony, and rhythm). At the level of the genre, Lena and Pachucki (2013) have developed a new measurement technique to explore the association between status, sampling, and popularity in rap music. At the level of the individual composition, Salganik and colleagues (2006) simulated a musical marketplace to show that popular appeal is determined by both social influence and artistic quality. Measuring artistic quality objectively requires a comprehensive understanding of a song's form and attributes. Due to the specialized skills needed to identify, categorize, and evaluate such attributes reliably, research on this topic is practically nonexistent. The work that has been done employs musicological techniques to construct a system of comparable musical codes that may be more or less present in a particular musical work (Cerulo 1988; La Rue 2001a, 2001b). Yet even if social scientists learned these techniques, or collaborated more often with musicologists, it would be extremely difficult to apply and automate such complex codes at scale.

Lucky for us, these difficulties have been partially attenuated by the rise of big data, machine learning, and new computational methods. Developed first by computer scientists and then adopted by mainstream social science, these technologies have begun to filter into the toolkits of cultural sociologists. While most of the work in this area has employed topic modeling to crunch large amounts of text (e.g., Mohr et al. 2013), there are opportunities for applications in other domains as well. Advances in the field of music information retrieval (MIR) and machine learning have made possible new research prospects that were previously considered impractical or impossible.

3   Data and Methodology

Using new web-based data that distills the high dimensionality of songs' musical content into a handful of discrete attributes, we look at how these attributes and the implicit relational structures they form shape audience evaluations. Our primary data come from the weekly *Billboard Hot 100* charts, which we have reconstructed from their inception on August 4, 1958 through May 11, 2013. While the eponymous magazine originally published the charts, the data we use comes from an online repository of *Billboard* charts known as "The Whitburn Project." Joel Whitburn collected and published anthologies of the *Billboard* charts (Whitburn 1986, 1991) and, beginning in 1998, a dedicated fan base started to collect, digitize, and add to the information contained in those guides. This augmented existing chart data, adding metadata and additional details about the songs and albums on the various *Billboard* charts. Our timeframe includes 25,762 songs for which we were able to obtain complete data.

3.1   Independent Variables and Controls

The *Hot 100* chart was initially created as a means for music industry insiders and observers to gain insight into the most popular music in the United States. While the algorithm used to create the charts has changed over the years—starting with a combination of radio airplay and a survey of record stores across the country, and gradually evolving to include actual unit sales (see Anand and Peterson 2000 for implications of this change), digital sales, and even streams (Billboard.com 2007, 2012)—they remain the industry standard. As such, they have been used extensively in social science research on popular music (Alexander 1996; Scott 1999; Anand and Peterson 2000; Dowd 2004; Peterson 2005; Lena 2006; Lena and Pachucki 2013; Mol et al. 2013), and are widely defended for their reliability as indicators of popular taste (e.g., Eastman and Pettijohn II 2014). Although the *Hot 100* is ostensibly *Billboard's* "all-genre" chart, its aim is to measure broad-based popularity across the United States in a given week. The scope of these analyses is thus limited to the most popular songs and genres in the U.S. over the past 55 years.

While thorough, the Whitburn data requires augmentation in order to capture more fully the multifaceted social and compositional elements of songs and artists. First, we added genre designations. Although genres are often-changing and potentially contentious (Lena and Pachucki 2013), they provide a system by which producers, consumers, and critics of music structure their listening patterns and tastes (Bourdieu 1983, 1984; Frith 1996; Hesmondhalgh 1999; Lena 2012), and therefore impact the way that music is perceived and its attributes evaluated. Our genre data for the *Hot 100* come from discogs.com, a crowd-sourced but community-monitored site containing extensive artist, album, and track data. Unlike many music data sources, Discogs does not limit genre assignments to just artist or album, nor does it cap the number of potential genres for a given artist or song at one. For many of the singles in our data, we were able to obtain multiple track-level genre categorizations. In our analyses, dummy variables for each genre were created, and songs were assigned a '1' for each genre assignment they received, and a '0' otherwise. While every attempt was made to provide genre assignments at the song level, data limitations often required using an artist or band's genre assignment for all of their songs.[2]

Given our interest in cultural product attributes, we endeavored to collect more detailed information for each song in our dataset. For these data, we turned to The Echo Nest, an online music intelligence provider that offers access to much of their data via a suite of application programming interfaces (APIs). This organization represents the current gold standard in music information retrieval (MIR). Using web crawling and audio encoding technology, the Echo Nest has collected—and continuously updates—information on over 30 million songs and over 2 million artists. Their data contains objective and derived qualities of audio, text analyses based on artist appearances in articles and blog posts, and qualitative information about artists. While this data has been used to conduct research in computer learning (e.g., Shalit, Weinshall, and Chechik 2013), it has not yet been explored by social scientists (see Serrà et al. 2012 for a notable exception).

We are using the Echo Nest API to collect the following for each song in our dataset: audio characteristics such as tempo, loudness (decibel level), and key, as well as some of the company's own creations like "danceability," and "acousticness" (see the Appendix for a detailed explanation of all attributes). Together with the Discogs genre categorizations and a dummy variable reflecting whether the song was released on a major or independent label, nine of the Echo Nest's audio attributes serve as our initial set of independent variables and controls. Table 1 provides the descriptive statistics for these variables.

3.2   Dependent Variables

Our use of the *Billboard* charts provides us with real-world performance outcomes, which can otherwise be difficult to find for cultural markets, especially music. Unlike movie box-office results or television show ratings, music sales are often tightly guarded by the content owners, leaving songs' diffusion across radio stations

---

[2]   We were able to collect track or album-level genre categorizations for 96% of the songs in our data, while 2.3% of the genre data information is at the artist level. We were unable to collect genre data for 1.7% of our songs.

(Rossman 2012) or their chart position as the most reliable performance outcome (see Bradlow and Fader (2001) and Giles (2007) for differing approaches to modeling *Billboard* chart performance, and Dertouzos (2008) for an examination of the link between *Billboard* chart position and sales). In their examination of fads in baby naming, Berger and Le Mens (2009) use both peak popularity and longevity as key variables in the measurement of cultural diffusion processes; we use them here as our dependent variables.

To measure songs' maximum popularity, we reverse coded peak chart position by subtracting from 101 each song's peak position in our *Billboard* data. Number one songs are therefore coded as 100, and positive coefficients on the independent variables indicate a positive relationship with performance. For additional peak performance analyses, we also created dummy variables for any entry that becomes a number one song or reaches the top ten. The second performance measure we use is the number of weeks on the chart, a measure of sustained popularity. Though these two outcome variables are related to one another in our data set (i.e., songs that reach a higher peak chart position are likely to remain on the charts longer, $R \approx .74$), we believe it is important to look at both peak performance and its longevity, even if the longest-lived songs in our data only made it to 76 weeks.

4   Estimation and Results

Our first suite of models was designed to test the direct relationship between audio attributes and chart performance. We conducted three different analyses. First, we ran pooled, cross-sectional OLS regressions on all of the songs in our data set as a means of demonstrating between-song differences related to being aligned with a particular genre or having certain values of audio attributes. The second set of analyses adds artist-level fixed effects to control for time-invariant artist- and band-level traits and effects. These include the intrinsic talents of individual bands and artists; the levels of institutional (e.g., marketing and PR) support that each artist receives; and "superstar" effects (Krueger 2005; Elberse and Oberholzer-Gee 2006), which operate much like Matthew Effects (Merton 1968; Bothner et al. 2010) and aid certain artists due to their previous songs' success. The third set of analyses employs logistic regression to examine the between-song odds of reaching the number one or a top ten position on the charts.

4.1   Results

Table 2 presents estimates for six models demonstrating the relationship between genre assignments, audio attributes, and chart performance. Model 1 includes 14 genre categories and the major label affiliation dummy ("pop," the fifteenth genre, acts as the reference category, as does indie label affiliation). Though these only account for a small portion of the variance in chart performance ($R^2 = .027$), some expected patterns emerge. First, we see that many of the niche genre categories simply are not as successful as major categories: songs from the blues, Folk/World/Country, and jazz genres all fare worse on the charts than the more mainstream genres of rock and pop. Moreover, songs from major labels tend to generally do better than their indie peers.

Table 1. Select Correlations and Descriptive Statistics

	[1]	[2]	[3]	[4]	[5]	[6]	[7]	[8]	[9]	[10]	[11]	[12]	[13]	[14]
[1] Loudness (normed)	1													
[2] Tempo (normed)	-0.004	1												
[3] Energy	0.712	0.148	1											
[4] Speechiness	0.008	-0.050	0.033	1										
[5] Acousticness	-0.470	-0.033	-0.669	0.046	1									
[6] Mode	-0.027	-0.013	-0.056	-0.079	0.085	1								
[7] Danceability	0.055	0.106	0.114	0.203	-0.276	-0.126	1							
[8] Valence	0.236	0.157	0.447	0.192	-0.309	-0.103	0.565	1						
[9] Liveness	0.025	-0.008	0.137	0.061	-0.009	0.005	-0.238	-0.02	1					
[10] Major Label Dummy	0	0.004	0.010	-0.006	-0.034	0.014	-0.020	-0.09	0.017	1				
[11] Song Length (seconds)	0.068	-0.004	0.055	0.107	-0.200	-0.092	0.132	-0.08	-0.019	0.180	1			
[12] Week on Charts	0.050	-0.005	0.023	0.037	-0.083	-0.019	0.075	-0.03	-0.009	0.062	0.199	1		
[13] Cosine Similarity (all songs)	0.231	0.075	0.303	-0.092	-0.301	0.684	0.222	0.28	-0.147	-0.015	-0.091	-0.018	1	
[14] Cosine Similarity (songs in overlapping genres)	0.210	0.071	0.280	-0.055	-0.281	0.525	0.184	0.25	-0.115	-0.009	-0.060	-0.006	0.801	1
Mean	0.3231	0.4369	0.5854	0.0714	0.2628	0.7287	0.6031	0.5806	0.2421	0.6971	215.6366	8.7240	0.8088	0.7955
Standard Deviation	0.1530	0.0908	0.2184	0.0883	0.2931	0.4446	0.1517	0.2486	0.2244	0.4595	48.7111	6.9040	0.0566	0.0667
Min	0.0019	0.1622	0.0026	0.0218	0.0000	0	0	0	0.0132	0	62	1	0.4650	0.1924
Max	0.8876	1	0.9999	0.9587	0.9998	1	0.9902	0.9995	1	1	570	76	0.9112	0.9168

Table 2. Models Predicting *Billboard Hot 100* Peak Chart Position (Inverted) & Chart Longevity

		(1) Peak (Inverted)	(2) Peak (Inverted)	(3) Peak (Inverted)	(4) Weeks on Charts	(5) Weeks on Charts	(6) Weeks on Charts
	Blues	-9.206***	-2.149	-1.930	-2.179***	-1.694***	-0.773**
		(1.431)	(2.075)	(2.082)	(0.342)	(0.339)	(0.390)
	Brass & Military	4.592	11.18	17.54**	-1.484	1.107	1.150
		(10.11)	(8.713)	(8.716)	(2.415)	(2.343)	(2.759)
	Children's	5.118	0.0626	0.335	-3.283**	-2.574*	-2.296
		(5.758)	(6.607)	(6.515)	(1.376)	(1.336)	(1.614)
	Classical	3.363	11.60	11.82	0.521	1.317	2.764*
		(5.375)	(8.998)	(9.019)	(1.284)	(1.267)	(1.626)
	Electronic	5.950***	1.420	1.183	3.416***	1.885***	0.550**
		(0.605)	(0.995)	(0.997)	(0.145)	(0.153)	(0.275)
Discogs.com Genre Dummies	Folk, World & Country	-4.890***	-3.406**	-3.192**	0.923***	0.620***	-0.847***
		(0.722)	(1.370)	(1.370)	(0.172)	(0.173)	(0.315)
	Funk / Soul	-0.472	-0.198	-0.215	-0.415***	-0.564***	-0.165
(Pop music is		(0.532)	(1.039)	(1.040)	(0.127)	(0.130)	(0.243)
reference catgeory)	Hip Hop	4.934***	1.127	0.965	4.625***	2.485***	0.351
		(0.644)	(1.530)	(1.530)	(0.154)	(0.178)	(0.457)
	Jazz	-6.251***	-2.182	-2.121	-1.954***	-1.014***	-0.587*
		(1.076)	(1.835)	(1.837)	(0.257)	(0.261)	(0.336)
	Latin	-3.153	-1.444	-1.269	0.821*	0.621	0.427
		(2.007)	(3.405)	(3.406)	(0.480)	(0.474)	(0.842)
	Non-Music	-2.490	4.213	4.357	-1.487*	-0.971	2.063
		(3.549)	(6.664)	(6.578)	(0.848)	(0.917)	(1.720)
	Reggae	-2.116	3.411	3.369	1.638***	0.972	-0.0733
		(2.603)	(5.070)	(5.035)	(0.622)	(0.619)	(1.359)
	Rock	5.769***	2.403***	2.543***	0.941***	0.875***	0.102
		(0.493)	(0.861)	(0.862)	(0.118)	(0.121)	(0.201)
	Stage & Screen	1.165	4.451**	4.516**	-1.826***	-1.549***	0.803**
		(1.369)	(1.809)	(1.803)	(0.327)	(0.325)	(0.357)
	Major Label Dummy	3.951***	-0.0932	-0.346	1.197***	0.352***	-0.348*
		(0.409)	(0.966)	(0.967)	(0.0977)	(0.0998)	(0.209)
	Loudness (Normed)		-0.440	-0.721		-0.118	-0.890
			(2.462)	(2.462)		(0.451)	(0.615)
	Tempo (Normed)		6.999***	6.600**		0.498	0.749
			(2.635)	(2.633)		(0.517)	(0.620)
	Energy		-8.554***	-7.681***		-1.436***	-2.184***
			(2.206)	(2.209)		(0.418)	(0.515)
	Speechiness		2.574	2.776		-1.285**	0.416
			(3.628)	(3.635)		(0.617)	(0.932)
	Acousticness		-2.700**	-2.271*		-0.882***	-0.498*
			(1.197)	(1.197)		(0.227)	(0.272)
	Mode		0.892*	0.912*		0.229**	0.249**
			(0.526)	(0.526)		(0.104)	(0.124)
	Danceability		8.423***	8.538***		5.171***	3.198***
			(2.308)	(2.308)		(0.427)	(0.546)
	Valence		-2.783**	-2.032		-2.080***	-0.403
			(1.410)	(1.413)		(0.267)	(0.334)
	Liveness		6.619***	6.577***		1.268***	1.490***
			(1.077)	(1.078)		(0.216)	(0.248)
	Song Length (seconds)			0.124***		0.108***	0.0544***
				(0.0332)		(0.00514)	(0.00686)
	Song Length^2			-0.000190***		-0.000167***	-9.25e-05***
				(6.83e-05)		(1.12e-05)	(1.49e-05)
	Constant	49.66***	52.62***	34.99***	9.118***	-5.465***	3.754***
		(0.519)	(2.163)	(4.512)	(0.124)	(0.692)	(0.939)
	Fixed Effects for Artist	N	Y	Y	N	N	Y
	Observations	25,762	24,392	24,377	25,762	24,377	24,377
	R-squared	0.026	0.349	0.351	0.080	0.143	0.430

Standard errors in parentheses
*** p<0.01, ** p<0.05, * p<0.1

Of interest, however, are the positive coefficients for Electronic and Hip-Hop, two genres that do not enter the musical mainstream until the 1980s but have in recent years fared relatively well on the charts. A closer examination of the data reveals that many Hip-Hop, Funk/Soul, and Pop songs are actually categorized secondarily as Electronic songs, which could account for this effect.

In model 2, we enter both the audio attributes and artist-specific fixed effects.[3] Though left unreported due to space considerations, the coefficients for the artist dummies indicate that there are some artists who fare better on the charts than others. When controlling for audio attributes, several genres no longer predict chart performance significantly. Coefficients for tempo, danceability, and liveness are all positive, while those for valence (emotional scale), energy, and acousticness are negative. The negative coefficient for our energy variable is perhaps the most surprising, and likely reflects the fact that hard rock songs and remixes tend to score highly on energy but do not necessarily take the charts by storm. Model 3 replicates model 2, with the addition of linear and quadratic song length variables. Lasting a little bit longer generally benefits songs, but only up to a point: once over five minutes long, song length adversely affects peak performance.

Next we move from regressions on peak performance to chart longevity. Slightly different patterns emerge in model 4: while some niche genres tend to spend less time on the charts than do mainstream genres like rock and pop, we find evidence of some of the more robust niche genres benefitting from their faithful audience. For example, Folk/World/Country, though a hybrid category, is predominantly comprised of Country songs in our data, especially after the 1960s. Country music, though not necessarily mainstream (Peterson 1997; Eastman and Pettijohn II 2014), has a strong fan base and is clearly delineated from other genres on the chart. The combination of differentiation and a large, dedicated audience, similar to Hip-Hop and Electronic music, may explain why songs from these genres last longer on the charts. Songs released on major labels again benefit when compared to their indie label peers.

We believe that chart longevity has less to do with a particular artist and more to do with the "catchiness" of a given song, so we wanted to explore the full effects of these variables on our duration outcome before adding artist-level fixed effects. In model 5, the genre effects stay the same, but indicators of more upbeat songs—songs written in major key (mode = 1) and highly "danceable" songs—predict sustained success on the charts. Conversely, higher energy (again, often hard rock songs and remix tracks), "acousticness," and "speechiness" all contribute to shorter shelf lives. As was the case with peak performance, songs fare better as they get longer, up until just over five minutes.

To control for individual artists' tendencies to experience extended success on the chart, the final model in table 2 adds artist-level fixed effects to model 5. Although many of the coefficients are directionally similar to those in model 5, two differences stand out. First, the switch in sign from positive to negative for the Folk/World/Country genre dummy suggests that, when an artist or band releases a song in a genre with which they have not been previously associated, that song is not likely to last as long on the charts. Perhaps this is the result of audience alienation, or an inability to appeal to the core fan base of the Folk/World/Country genre category. A similar switching of signs occurs for the major label dummy, suggesting that, as an

---

[3] We normalized the two audio attribute variables that were not already on a 0-1 scale—loudness and tempo—in order to make them more directly comparable. Tempo was normalized by dividing each song's tempo by the max tempo in the panel, while loudness, which is computed in decibels, was converted with the following equation:

$$\text{norm\_loud}_i = 10\text{^}(dB_i/20) \tag{1}$$

artist moves from an independent to a major label, they do not necessarily benefit from added resources and visibility. This could be the result of a "sell out" effect, or it could simply be that major labels do not provide the kind of grassroots promotional support expected of independent labels.

While the first set of models reveals relationships between songs' genres, label affiliations, and audio attributes and chart performance, the second set attempts to connect these variables with more concrete performance outcomes. In Table 3, we present two cross-sectional logistic regression models that estimate a song's chance of reaching the top of the charts. Model 7 presents the odds ratios of our full set of variables on whether a song reaches the number one position on the *Hot 100*. We again find evidence that songs categorized in niche genres tend to suffer from a performance discount. One interesting exception appears to be Stage & Screen, which is likely a function of popular movie theme songs' success. The impact of audio attributes on songs' chances of reaching the top spot remain in line with what we found previously: "danceability" is king when it comes to the *Hot 100*.

Our final model in set of analyses (model 8), examines the odds of reaching the top 10. Here, the mainstream genres again perform well, although the difference between models 7 and 8 suggests that Hip-Hop songs experience a ceiling effect. Songs in this genre generally do quite well, often reaching the top ten but rarely attaining the top spot on the charts. This effect may be due to audience listening habits, broad-based popularity (or lack thereof), and/or the construction of the chart itself, but the data suggests a performance ceiling exists.

5   Relational Crowding Analyses

Before testing for the effects of cultural network properties on chart performance, we thought it important to establish face validity for the genres and audio attributes that undergird our operationalization of a relational network in music. Although the scope of the results presented in tables 2 and 3 is limited, these findings provide evidence of attributes' role in determining cultural consumption and evaluation patterns. We move now to the creation of two new relational measures.

5.1   New Independent Variables

After a review of the relevant literature, we found that a common and relatively straightforward means of examining the effects of networks on performance outcomes was to look at crowding (Podolny, Stuart, and Hannan 1996; Bothner, Kang, and Stuart 2007). As with many social phenomena (e.g., Phillips and Zuckerman 2001; Bothner, Kim, and Smith 2011), we expect an inverted U-shaped result from attribute crowding among songs: put succinctly, songs that are either stacked on top of one another or isolated will see their future chart prospects suffer, while those spaced comfortably proximate to other songs—signaling a reference group for listeners while displaying some degree of optimal distinctiveness—will be more likely to ascend the charts.

Table 3. Cross-Sectional Logistic Regression Models
Predicting *Billboard Hot 100* Achievement Levels

		(7) Predicting #1	(8) Predicting Top 10
Discogs.com Genre Dummies (Pop music is reference category)	Blues	0.296*** (0.134)	0.476*** (0.0801)
	Brass & Military		2.102 (1.714)
	Children's	0.896 (0.920)	0.403 (0.298)
	Classical	0.916 (0.946)	2.059* (0.875)
	Electronic	1.598*** (0.148)	1.500*** (0.0777)
	Folk, World & Country	0.227*** (0.0479)	0.344*** (0.0305)
	Funk / Soul	0.825** (0.0771)	0.959 (0.0465)
	Hip Hop	0.832 (0.0981)	1.201*** (0.0752)
	Jazz	0.386*** (0.110)	0.769** (0.0852)
	Latin	0.816 (0.282)	0.840 (0.153)
	Non-Music	0.421 (0.430)	0.875 (0.346)
	Reggae	1.256 (0.467)	1.153 (0.250)
	Rock	1.047 (0.0898)	1.417*** (0.0640)
	Stage & Screen	1.450* (0.293)	1.090 (0.131)
	Major Label Dummy	1.186** (0.0885)	1.072* (0.0403)
Echo Nest Audio Attributes	Loudness (normed)	1.298 (0.418)	1.254 (0.209)
	Tempo (Normed)	0.989 (0.374)	1.092 (0.213)
	Energy	0.323*** (0.0959)	0.428*** (0.0661)
	Speechiness	0.628 (0.277)	0.486*** (0.116)
	Acousticness	1.090 (0.174)	0.901 (0.0761)
	Mode	1.106 (0.0817)	1.039 (0.0398)
	Danceability	2.280*** (0.700)	2.424*** (0.386)
	Valence	1.118 (0.212)	0.877 (0.0864)
	Liveness	1.585*** (0.234)	1.640*** (0.127)
	Song Length (seconds)	1.011*** (0.00376)	1.006*** (0.00192)
	Song Length^2	1.000* (7.78e-06)	1.000* (4.07e-06)
	Constant	0.00662*** (0.00341)	0.0663*** (0.0173)
	Observations	24,368	24,377

Coefficients report odds ratios; standard errors in parentheses
*** p<0.01, ** p<0.05, * p<0.1

To create our crowding measures, we followed earlier work concerned with proximity, which uses cosine similarity to measure the distance between two distinct entities (Evans 2010; Aral and Alstyne 2011). First, we transformed the data into a long panel, giving each of the nearly 280,000 song-weeks in our dataset its own row. Next, we created similarity measures by taking the eleven audio attributes provided by The Echo Nest, normalizing them across a 0-1 scale, and then collapsing them into a single vector for each song, V_i. We took each chart separately and calculated the cosine distance between each song's vector of attributes, using the following equation:

$$dist_{ij} = \cos(\mathbf{v}_i, \mathbf{v}_j) \tag{2}$$

The resulting square matrix $\mathbf{A}$ has dimensions matching the number of songs on each week's charts, with cell A_{ij} representing the similarity between song i and song j. As every song is perfectly similar to itself (i.e., has a cosine similarity of 1), we removed $\mathbf{A}$'s diagonal from all calculations; leaving it in would be equivalent to adding a constant to our variable. Finally, we took the average of each row in $\mathbf{A}$ to give songs' a chart specific cosine similarity value, *all_cosine_sim*. If we map the space of attribute similarity across songs using nodes and ties, weekly "cultural networks" are formed (see Figure 1). These networks represent a dynamic system of relationships between songs that display some degree of attribute overlap, and consist of clusters that not surprisingly approximate stylistic designations such as genre.

Our second similarity variable was calculated using a similar procedure, but considers crowding within specific genres. We believe this is necessary for two reasons. First, as listener groups often focus their attention on particular genres of music (e.g., "I listen to Pop and Hip-Hop, but never listen to any Country music"), the impact of crowding may be differentiated within these categories. Second, given the reductive nature of the audio attributes in our data, we wanted to increase the likelihood that songs that *appear* to be similar to one another in our data are interpreted as such when listened to. For example, one can imagine a situation where a reggae song and a country song sound nothing alike, yet when reduced to a few attributes, seem to be quite similar. The creation of two variables—one that compares all songs on the same chart to each other, and one that only compares those that have overlapping genres—reflects our interest in exploring the possibility that two songs from distant genres may actually be quite similar, but may also be quite distinct.

5.2   New Dependent Variables

Our second set of models also requires new outcome variables, as we are more interested in measuring the effect of crowding on subsequent performance on the charts, rather than static chart position. Because attribute similarity is measured for each song pair in a given week, we wanted to capture the immediate impact of crowding on a song's future chart performance. Thus, a positive sign for our similarity variables would imply that greater crowding leads to higher chart position the following week.

We also generated a second new outcome variable: the weekly change in chart position. Using inverted position values, we subtracted last week's position from each song's current position to measure its ascent up (or descent down) the charts. We consider this

Fig. 1. Cultural Network of *Billboard Hot 100* Chart – August 15, 1964[4]

an important step, in part because songs that are near the bottom of the charts have more room for improvement, while those at the top of the charts are less likely to experience gains. While we could have included lagged chart position in our models to control for past performance, the resulting standard errors when running fixed-effects analyses generate concerns (Nickell 1981), and findings are usually more robust when using change scores as a dependent variable (Morgan and Winship 2007).

5.3   Crowding Analysis

We now turn to the results presented in Table 4, which examine the impact of crowding on subsequent chart performance. All of these models we estimated include linear

---

[4] The presence of a tie connotes a *within-genre* cosine similarity above the mean value for that variable (~.79 across the entire dataset). For ease of demonstration, connections only occur between songs that share at least one genre assignment (each song can have up to three). Nodes are colored by primary genre.

and quadratic control variables for the number of weeks a song has already been on the charts. We do this in light of the fact that the average "life span" of a song is just over 11 weeks, and songs generally follow a parabolic trajectory through the charts. Most songs initially improve in chart position, but begin to decline after about 6 weeks. It is rare for songs to move up substantially after that point in time, though it does happen on occasion. Songs that last an unusually long time on the charts usually climb relatively slowly.

Beginning with model 9, we employ *all_cosine_sim* as our independent variable of interest in a pooled, cross-sectional analysis predicting a song's position on the subsequent week's *Hot 100* chart. The coefficient here is strongly negative, indicating that in a given week, songs that are more crowded by similar others are also more likely to suffer a performance disadvantage. In model 10, we add the quadratic form of the independent variable as well as song-level fixed effects. This serves to remove time-invariant characteristics of the song—its genre assignments, audio attributes, performing artist, and label, among others—and allows us to explore how shifting relation dynamics around a focal song impact its subsequent performance. The results show some support for the use of a quadratic term, although the point at which similarity begins to help is at the very top of the *all_cosine_sim* range, comprising less than .03% of all songs. In both cross-sectional and within-song analyses, high levels of crowding appear for the most part to be harmful to a song's future chart performance.

Beginning with model 11, we limit our explanatory variable of interest to within-genre attribute similarity (*genre_cosine_sim*). When analyzed in cross-section, we initially find evidence for a *beneficial* effect of crowding. That is, songs that are similar to other songs on the chart at the same time appear to perform better, perhaps moving up in tandem as certain genres ebb and flow in popularity. This story remains unchanged when adding the quadratic term (model 12), although the benefits of similarity crowding are less significant. Beginning just below the mean value of *genre_cosine_sim* ($\approx$.78), songs' future performance is harmed as they become more like the other songs in their genres. These results support our expectation of a curvilinear effect of attribute similarity, at least in cross-sectional analysis.

This changes once we add song-level fixed effects in model 13. Here, we again find evidence that, as similar songs within overlapping genres encroach upon a focal song, they adversely affect its chart performance. This raises an interesting question around why, in cross-section, genre-based attribute similarity seems helpful, but within-song changes in similarity over time render crowding hurtful. One potential answer lies in the specific distribution of crowding within a chart: if, as we previously demonstrated, a handful of mainstream genres tend to crowd among the top 10 or top 20 chart positions—where it is more difficult to improve rank from week-to-week—then this result should not be particularly surprising. For example, if similar songs from Rock and Pop (which often appear simultaneously as genre assignments for the same song) find themselves regularly in the upper segment of the charts, they will likely "compete" for the top few positions that remain. Here, the effects of attribute crowding would show up as harmful.

To disentangle this further, we estimate models that predict change in rank. Model 15 includes only the linear form of *genre_cosine_sim* and fixed effects for songs. We again find evidence for crowding's role in discounting future performance outcomes, indicating that the previous effect in model 13 was likely not due only to crowding at

Table 4. Cross-sectional and Fixed Effects Models Predicting (Inverted) Chart Position & Δ (Inverted) Chart Position on Billboard Hot 100, 1958-2013

	(9) Chart Position	(10) Chart Position	(11) Chart Position	(12) Chart Position	(13) Chart Position	(14) Chart Position	(15) ΔChart Position	(16) ΔChart Position
All-Cosine Similarity	-5.521*** (0.940)	-162.5** (80.88)						
All-Cosine Similarity^2		90.09* (50.38)						
Genre-Cosine Similarity			4.891*** (0.801)	82.08*** (8.340)	-6.291** (2.655)	-15.36 (14.95)	-3.514*** (1.165)	12.17* (6.559)
Genre-Cosine Similarity^2				-52.13*** (5.607)		6.466 (10.49)		-11.18** (4.601)
Weeks on Chart	2.189*** (0.0178)	0.839*** (0.0141)	2.183*** (0.0178)	2.181*** (0.0178)	0.833*** (0.0141)	0.833*** (0.0141)	-2.084*** (0.00620)	-2.084*** (0.00620)
Weeks on Chart^2	-0.0456*** (0.000555)	-0.0414*** (0.000452)	-0.0455*** (0.000556)	-0.0455*** (0.000556)	-0.0413*** (0.000453)	-0.0413*** (0.000453)	0.0378*** (0.000199)	0.0378*** (0.000199)
Constant	45.67*** (0.769)	124.4*** (32.46)	37.51*** (0.647)	9.345*** (3.098)	57.37*** (2.114)	60.47*** (5.448)	16.56*** (0.927)	11.21*** (2.390)
Fixed Effects for Songs	No	Yes	No	No	Yes	Yes	Yes	Yes
Observations	253,642	253,642	250,979	250,979	250,979	250,979	250,979	250,979
R-squared	0.071	0.613	0.071	0.071	0.611	0.611	0.449	0.408
Number of sid		23,134			22,851	22,851	22,851	22,851

Standard errors in parentheses; all variables lagged one week. Outcome variables created by subtracting song's chart position from 101.

*** $p<0.01$, ** $p<0.05$, * $p<0.1$

the top of the charts. Instead, this appears to be a real effect of genre-based crowding. Additional support is provided by model 16, which adds the quadratic term for within-genre attribute similarity. While these results appear to suggest that crowding helps a song up to a point, the similarity value at which the effect becomes negative is actually quite low (0.544) given the distribution of our data. This suggests that only true isolates—those songs that are more than three standard deviations below the mean for genre-based similarity—benefit from crowding. When compared to other songs that appear on the same chart and are categorized within the same genres, these isolates are helped by "bridging" songs: those songs that are not too distant from the rest of their genre-peers, but distinct enough so as to provide a bridge to the isolate. These "bridges" or brokers may help to cushion audiences' evaluations of more unusual sounding songs. In most instances, however, crowding appears to harm songs' subsequent performance on the *Billboard Hot 100*.

6   Discussion and Conclusions

"Culture" represents one of the most well-trodden topics in the humanities and social sciences, but until recently our understanding of how culture really works, and to what effect, has been relatively circumscribed. This need not be the case. Although the findings presented above are exploratory in nature, they provide preliminary evidence that sonic attributes matter, both independently and in the way they structure songs' relationships to each other. Controlling for artist familiarity and genre preferences, consumer assessments of popular music is shaped in part by the content of the songs themselves, perhaps suggesting that listeners are more discerning than we sometimes give them credit for (Salganik, Dodds, and Watts 2006). Indicators of more upbeat songs, including higher tempo and danceability, predict both a higher peak position and a longer shelf life on the charts. Our empirical proxy for the effect of cultural network configurations—crowding caused by attribute similarity—also has a significant effect on how songs are evaluated. Songs that are musically too similar to their neighbors suffer a performance discount, both within and across genres; indeed, only those songs that are especially unique benefit from being crowded by similar others. These results support the argument that, to become a hit, a song must achieve some degree of optimal distinctiveness.

We believe that the ideas presented in this paper make several contributions. First, we import methods traditionally associated with big data science and network analysis to enhance our understanding of large-scale cultural dynamics. While these tools necessarily simplify the intrinsic high-dimensionality of culture, they also empower us to learn new things that might otherwise remain unknown. Although many new cultural measurement tools originate from advances in computer science and other disciplines, social scientists must critically develop and apply them appropriately and thoughtfully (Bail 2014). Other scholars have mapped meaning structures (Carley 1994; Mohr 1994, 1998), charted diffusion patterns (Rossman 2012; Long and So 2013), and conceptualized the link between culture and action (Swidler 1986; Weber 2005), but there has been no attempt to theorize and measure the role cultural content plays in the emergence and diffusion of new production and consumption patterns. We collected and analyzed machine-learned data on discrete musical attributes to extend our collective understanding of how consumers evaluate cultural products, generating new insight into the world of popular music.

Second, we develop and test the effects of cultural networks, which provide a new means of understanding the fabric within which fields of cultural production are enmeshed. We argue that the system of relations between attributes is theoretically and analytically distinct from networks of cultural producers and consumers. In so doing, we raise the possibility that cultural content asserts its own partially independent influence over evaluation outcomes through crowding and other mechanisms. This conceptualization of culture is dynamic, and pushes network scholars to theorize new ways in which mapping techniques might be used to describe different kinds of relationships. While existing research on networks focuses largely on interpersonal ties, substantive relationships exist between all sorts of actors, objects, and ideas. Redefining what constitutes a node and an edge might help scholars rethink how inanimate actors assert influence or agency, thereby addressing a critical issue in social theory more broadly (Berger and Luckmann 1966).

We also recognize several important limitations of our study. Although the data we use to measure cultural attributes is relatively comprehensive and sophisticated, it represents a significant distillation of a song's musical complexity. Reducing such a high dimensional object into eleven fixed attributes inevitably simplifies its cultural fingerprint and alters its relationships with other like-objects. Our data also does not allow us to account for listeners' interpretations of attributes or lyric similarity between songs. In the future we expect to use additional data to conduct comparative analyses that match songs which appeared in the *Billboard Hot 100* charts with songs that did not, allowing us to define the limits of our findings to learn more about the effects of cultural attributes on performance outcomes. We also hope to conduct more dynamic analyses to understand better the nature and implications of specific cultural structures that appear in our dataset. Carving the chart into distinct segments, toying with different time lags, and mapping the social life of individual songs via their chart trajectory should provide additional insight into the dynamism of cultural networks.

Finally, although we provide robust evidence for how musical attributes affect songs' performance on the charts, our explanation of evaluation outcomes is limited to characteristics of the production environment. The analyses presented in this paper do not account for the external consumption environment, making it difficult to identify the mechanisms by which audiences shape the composition of the charts. It is also unclear how contingent our findings are on the domain of cultural production being studied (e.g., popular music) or the specifics of the historical and social situation. Collecting additional data and employing other methods (e.g., instrumental variables, cluster analyses, binary random classifiers) will allow us to better account for the complex and likely interdependent nature of these dynamics.

Appendix: Echo Nest Audio Attributes[5]

Attribute	Scale	Definition
Acousticness	0 to 1	Represents the likelihood a recording was created by solely acoustic means such as voice and acoustic instruments, as opposed to electronically such as with synthesized, amplified, or effected instruments.
Danceability	0 to 1	Describes how suitable a track is for dancing. The combination of musical elements that best characterize danceability include tempo, rhythm stability, beat strength, and overall regularity.
Duration*	Seconds	The length of a track.
Energy	0 to 1 (continuous)	A perceptual measure of intensity throughout the track. Typical energetic tracks feel fast, loud, and noisy. Perceptual features contributing to this attribute include dynamic range, perceived loudness, timbre, onset rate, and general entropy.
Key*	1 to 12 (integers only)	The estimated overall key for a track. The key identifies the tonic triad, the chord, major or minor, which represents the final point of rest of a piece.
Liveness	0 to 1	Detects the presence of an audience in the recording. The more confident that the track is live, the closer to 1 the attribute value.
Loudness	Decibels (dB)	The average loudness of a track. Loudness is the quality of a sound that is the primary psychological correlate of physical strength (amplitude).
Mode	0 or 1	Indicates the modality (major or minor) of a track.
Speechiness	0 to 1	Detects the presence of spoken words in a track. The more exclusively speech-like the recording (e.g. talk show, audio book, poetry), the closer to 1 the attribute value.
Tempo	Beats per minute (BPM)	The overall estimated tempo of a track. In musical terminology, tempo is the speed or pace of a given piece and derives directly from the average beat duration.
Time Signature*	Beats per measure	An estimated overall time signature of a track. The time signature (meter) is a notational convention to specify how many beats are in each bar (or measure).
Valence	0 to 1	Describes the musical positiveness conveyed by a track. Tracks with high valence sound more positive (e.g., happy, cheerful, euphoric), while tracks with low valence sound more negative (e.g. sad, depressed, angry). This attribute in combination with energy is a strong indicator of mood.

---

[5] Attributes marked with a * were used to construct cosine similarities between songs, but they do not appear as stand-alone variables in our models because there is no clear interpretation of what a unit increase (or decrease) in these attributes connotes.

References

1. Alexander, P.J.: Entropy and Popular Culture: Product Diversity in the Popular Music Recording Industry. American Sociological Review 61(1), 171–174 (1996)
2. Anand, N., Peterson, R.A.: When Market Information Constitutes Fields: Sensemaking of Markets in the Commercial Music Industry. Organization Science 11(3), 270–284 (2000)
3. Aral, S., Van Alstyne, M.: The Diversity-Bandwidth Trade-Off. American Journal of Sociology 117(1), 90–171 (2011)
4. Bail, C.A.: The Cultural Environment: Measuring Culture with Big Data. Theory and Society 43, 465–482 (2014)
5. Berger, J., Le Mens, G.: How Adoption Speed Affects the Abandonment of Cultural Tastes. Proceedings of the National Academy of Sciences 106(20), 8146–8150 (2009)
6. Berger, P., Luckmann, T.: The Social Construction of Reality: A Treatise in the Sociology of Knowledge. Doubleday & Company, Inc., Garden City (1966)
7. Billboard.com, Billboard Hot 100 To Include Digital Streams Billboard (2007)
8. Billboard.com. 2012. Hot 100 Impacted by New On-Demand Songs Chart Billboard
9. Bothner, M.S., Haynes, R., Lee, W., Smith, E.B.: When Do Matthew Effects Occur? The Journal of Mathematical Sociology 34(2), 80–114 (2010)
10. Bothner, M.S., Kang, J.-H., Stuart, T.E.: Competitive Crowding and Risk Taking in a Tournament: Evidence from NASCAR Racing. Administrative Science Quarterly 52(2), 208–247 (2007)
11. Bothner, M.S., Kim, Y.-K., Smith, E.B.: How Does Status Affect Performance? Status as an Asset vs. Status as a Liability in the PGA and NASCAR. Organization Science 23(2), 416–433 (2011)
12. Bourdieu, P.: The Field of Cultural Production, or: The Economic World Reversed. Poetics 12(4–5), 311–356 (1983)
13. Bourdieu, P.: Introduction. In Distinction: A Social Critique of the Judgment of Taste, pp. 1–7. Harvard University Press, Cambridge (1984)
14. Bourdieu, P.: The Rules of Art: Genesis and Structure of the Literary Field. Stanford University Press, Stanford (1996)
15. Bradlow, E.T., Fader, P.S.: A Bayesian Lifetime Model for the 'Hot 100' Billboard Songs. Journal of the American Statistical Association 96(454), 368–381 (2001)
16. Burt, R.: Structural holes: The social structure of competition. Harvard University Press, Cambridge (1992)
17. Burt, R.: Structural Holes and Good Ideas. American Journal of Sociology 110, 349–399 (2004)
18. Carley, K.M.: Extracting Culture through Textual Analysis. Poetics 22, 291–312 (1994)
19. Cerulo, K.: Analyzing Cultural Products: A New Method of Measurement. Social Science Research 17, 317–352 (1988)
20. Dertouzos, J.N.: Radio Airplay and the Record Industry: An Economic Analysis. National Association of Broadcasters, USA (2008)
21. DiMaggio, P.: Introduction: Meaning and measurement in the sociology of culture. Poetics 2(4), 263–371 (1994)
22. DiMaggio, P.: Cultural Networks. In: Scott, J., Carrington, P.J. (eds.) The Sage Handbook of Social Network Analysis, pp. 286–298. Sage, London (2011)
23. DiMaggio, P., Nag, M., Blei., D.: Exploiting Affinities between Topic Modeling and the Sociological Perspective on Culture: Application to Newspaper Coverage of U.S. Government Arts Funding. Poetics 41, 570–606 (2013)

24. Dowd, T.J.: The Musical Structure and Social Context of Number One Songs, 1955 to 1988. An Exploratory Analysis. In: Wuthnow, R. (ed.) Vocabularies of Public Life: Empirical Essays in Symbolic Structure, pp. 130–157. Routledge, London (1992)

25. Dowd, T.J.: Concentration and Diversity Revisited: Production Logics and the U.S. Mainstream Recording Market, 1940-1990. Social Forces 82(4), 1411–1455 (2004)

26. Dowd, T.J., Pinheiro, D.L.: The Ties among the Notes: The Social Capital of Jazz Musicians in Three Metropolitan Areas. Work & Occupations 40(4), 431–464 (2013)

27. Eastman, J.T., Pettijohn II, T.F.: Gone Country: An Investigation of Billboard Country Songs of the Year across Social and Economic Conditions in the United States. Psychology of Popular Media Culture (2014)

28. Elberse, A., Oberholzer-Gee, F.: Superstars and Underdogs: An Examination of the Long Tail Phenomenon in Video Sales. Division of Research, Harvard Business School (2006)

29. Emirbayer, M., Goodwin, J.: Network analysis, culture, and the problem of agency. American Journal of Sociology 99, 1411–1454 (1994)

30. Evans, J.A.: Industry Induces Academic Science to Know Less about More. American Journal of Sociology 116(2), 389–452 (2010)

31. Frith, S.: Music and Identity. In: Hall, S., Du Gay, P. (eds.) Questions of Cultural Identity, pp. 108–125. Sage, London (1996)

32. Giles, D.E.: Survival of the Hippest: Life at the Top of the Hot 100. Applied Economics 39(15), 1877–1887 (2007)

33. Goldberg, A.: Mapping Shared Understandings Using Relational Class Analysis: The Case of the Cultural Omnivore Reexamined. American Journal of Sociology 116(5), 1397–1436 (2011)

34. Hannan, M.T.: Partiality of Memberships in Categories and Audiences. Annual Review of Sociology 36(1), 159–181 (2010)

35. Hesmondhalgh, D.: Indie: The Institutional Politics and Aesthetics of a Popular Music Genre. Cultural Studies 13(1), 34–61 (1999)

36. Krueger, A.B.: The Economics of Real Superstars: The Market for Rock Concerts in the Material World. Journal of Labor Economics 23(1), 1–30 (2005)

37. La Rue, J.: Significant and Coincidental Resemblance between Classical Themes. The Journal of Musicology 18(2), 268–294 (2001a)

38. La Rue, J.: Fundamental Considerations in Style Analysis. The Journal of Musicology 18(2), 295–312 (2001b)

39. Lena, J.C.: Social Context and Musical Content of Rap Music, 1979-1995. Social Forces 85(1), 479–495 (2006)

40. Lena, J.C.: Banding Together: How Communities Create Genres in Popular Music. Princeton University Press, Princeton (2012)

41. Lena, J.C., Peterson, R.A.: Classification as Culture: Types and Trajectories of Music Genres. American Sociological Review 73(5), 697–718 (2008)

42. Lena, J.C., Pachucki, M.C.: The Sincerest Form of Flattery: Innovation, Repetition, and Status in an Art Movement. Poetics 41, 236–264 (2013)

43. Lizardo, O.: How cultural tastes shape personal networks. American Sociological Review 71, 778–807 (2006)

44. Lizardo, O.: Cultural Correlates of Ego-Network Closure. Sociological Perspectives 54(3), 479–487

45. Lizardo, O.: Omnivorousness as the Bridging of Cultural Holes: A Measurement Strategy. Theory and Society 43, 395–419 (2014)

46. Long, H., So, R.: Network Science and Literary History. Leonardo 46(3), 274 (2013)

47. Merton, R.K.: The Matthew Effect in Science: The Reward and Communication Systems of Science Are Considered. Science 159(3810), 56 (1968)
48. Mohr, J.W.: Soldiers, Mothers, Tramps, and Others: Discourse Roles in the 1907 New York City Charity Directory. Poetics 22, 327–357 (1994)
49. Mohr, J.W.: Measuring Meaning Structures. Annual Review of Sociology 24, 345–370 (1998)
50. Mohr, J.W., Wagner-Pacifici, R., Breiger, R.L., Bogdanov, P.: Graphing the Grammar Motives in National Security Strategies: Cultural Interpretation, Automated Text Analysis and the Drama of Global Politics. Poetics 41, 670–700 (2013)
51. Mohr, J.W., Ghaziani, A.: Problems and Prospects of Measurement in the Study of Culture. Theory and Society 43, 225–246 (2014)
52. Mol, J.M., Chiu, M.M., Paleo, I.O., Wijnberg, N.M.: Calling Out Around the World, Are You Ready for a Brand-New Beat? Genre Formation in Popular Music. Working Paper (2013)
53. Morgan, S.L., Winship, C.: Counterfactuals and Causal Inference: Methods and Principles for Social Research. Cambridge University Press, Cambridge (2007)
54. Nickell, S.: Biases in Dynamic Models with Fixed Effects. Econometrica: Journal of the Econometric Society, 1417–1426 (1981)
55. Pachucki, M.A., Breiger, R.L.: Cultural Holes: Beyond Relationality in Social Networks and Culture. Annual Review of Sociology 36, 205–224 (2010)
56. Padgett, J.F., Powell, W.W.: The Problem of Emergence. In: Padgett, J.F., Powell, W.W. (eds.) The Emergence of Organizations and Markets, pp. 1–29. Princeton University Press, Princeton (2012)
57. Peterson, R.A.: Why 1955?: Explaining the Advent of Rock Music. Popular Music 9(1), 97–116 (1990)
58. Peterson, R.A.: Creating Country Music: Fabricating Authenticity. University of Chicago Press, Chicago (1997)
59. Peterson, R.A.: In Search of Authenticity. Journal of Management Studies 42(5), 1083–1098 (2005)
60. Phillips, D.J.: Jazz and the Disconnected: City Structural Disconnectedness and the Emergence of a Jazz Canon, 1897–1933. American Journal of Sociology 117(2), 420–483 (2011)
61. Phillips, D.J., Zuckerman, E.W.: Middle-Status Conformity: Theoretical Restatement and Empirical Demonstration in Two Markets. American Journal of Sociology 107(2), 379–429 (2001)
62. Podolny, J.M., Stuart, T.E., Hannan, M.T.: Networks, Knowledge, and Niches: Competition in the Worldwide Semiconductor Industry. American Journal of Sociology 102(3), 659–689 (1996)
63. Rossman, G.: Climbing the Charts: What Radio Airplay Tells Us about the Diffusion of Innovation. Princeton University Press, Princeton (2012)
64. Salganik, M.J., Dodds, P.S., Watts, D.J.: Experimental Study of Inequality and Unpredictability in an Artificial Cultural Market. Science 311, 854–856 (2006)
65. Schudson, M.: How Culture Works: Perspectives from Media Studies on the Efficacy of Symbols. Theory and Society 18, 153–180 (1989)
66. Scott, A.J.: The US Recorded Music Industry: On the Relations between Organization, Location, and Creativity in the Cultural Economy. Environment and Planning A 31(11), 1965–1984 (1999)
67. Serrà, J., Corral, Á., Boguñá, M., Haro, M., Arcos, J.L.: Measuring the Evolution of Contemporary Western Popular Music. Scientific Reports 2 (2012)

68. Shalit, U., Weinshall, D., Chechik, G.: Modeling Musical Influence with Topic Models. In: Proceedings of the 30th International Conference on Machine Learning (ICML-13), pp. 244–252 (2013)
69. Swidler, A.: Culture in Action: Symbols and Strategies. American Sociological Review 51, 273–286 (1986)
70. Uzzi, B., Spiro, J.: Collaboration and Creativity: The Small World Problem. American Journal of Sociology 111, 447–504 (2005)
71. Vilhena, D.A., Foster, J.G., Rosvall, M., West, J.D., Evans, J., Bergstrom, C.T.: Finding Cultural Holes: How Structure and Culture Diverge in Networks of Scholarly Communication. Sociological Science 1, 221–238 (2014)
72. Weber, K.A.: Toolkit for Analyzing Corporate Cultural Toolkits. Poetics 33, 227–252 (2005)
73. Whitburn, J.: Joel Whitburn's Pop Memories, 1890-1954: The History of American Popular Music: Compiled from America's Popular Music Charts 1890-1954. Record Research (1986)
74. Whitburn, J.: Joel Whitburn's Top Pop Singles, 1955-1990: Compiled from Billboard's Pop Singles Charts, 1955-1990. Record Research (1991)
75. Zaheer, A., Soda, G.: Network Evolution: The Origins of Structural Holes. Administrative Science Quarterly 54(1), 1–31 (2009)

Migration of Professionals to the U.S.

Evidence from LinkedIn Data

Bogdan State[1,2], Mario Rodriguez[1], Dirk Helbing[3], and Emilio Zagheni[4]

[1] LinkedIn Corporation
[2] Stanford University
[3] ETH Zürich
[4] University of Washington

Abstract. We investigate trends in the international migration of professional workers by analyzing a dataset of millions of geolocated career histories provided by LinkedIn, the largest online platform for professionals. The new dataset confirms that the United States is, in absolute terms, the top destination for international migrants. However, we observe a decrease, from 2000 to 2012, in the percentage of professional migrants, worldwide, who have the United States as their country of destination. The pattern holds for persons with Bachelor's, Master's, and PhD degrees alike, and for individuals with degrees from highly-ranked worldwide universities. Our analysis also reveals the growth of Asia as a major professional migration destination during the past twelve years. Although we see a decline in the share of employment-based migrants going to the United States, our results show a recent rebound in the percentage of international students who choose the United States as their destination.

The United States is in the middle of a fierce debate over an immigration reform that would, among others, increase the number of temporary visas for skilled workers, boost the number of visas available to foreign students who earn advanced degrees in STEM disciplines (science, technology, engineering and mathematics), and create new visas awarded on the basis of a scoring system intended to favor "merit" [11].

The United States has always been a country of immigration, a top destination for scientists [6, 16] and, more broadly, for holders of a doctorate degree [2]. It has been found that "individuals making exceptional contributions to science and engineering (S&E) in the United States are disproportionately drawn from the foreign born" [9] and that the US has largely benefited from talent educated abroad [9]. Most of the public discussion around immigration reform has focused on the potential consequences of the immigration bill for employment and wages of United States citizens. Less attention has been paid, however, to the position of the United States in the context of recent changes in the composition and destinations of highly skilled migrants around the world.

The past decades have seen a general increase of worldwide migration [1, 19], including a jump in the migration of professionals [10]. In turn, employment-based migration to the United States has been governed by a complicated system of visa regulations, which in some cases (e.g. the H1-B visa) include absolute caps on the number of

L.M. Aiello and D. McFarland (Eds.): SocInfo 2014, LNCS 8851, pp. 531–543, 2014.
© Springer International Publishing Switzerland 2014

individuals admitted to the country.[1] The combination of these two processes leads us to expect the emergence of other destinations for professional migrations, as has been observed at the turn of the century [16].

There is a large body of literature, mainly in the disciplines of sociology, demography, economics, and geography, about international migration, and, more specifically, highly-skilled migration. It is beyond the scope of this article to discuss theories of migration and the rich and healthy debate about them (for an overview see, for instance, [4, 8, 12, 17]). With this article we emphasize an outstanding problem in migration research: the lack of timely, consistent and comparative data sources about international migrants. We address the issue by proposing an analysis based on new and innovative data from LinkedIn, the largest online platform for professionals. More specifically, we investigate recent trends in the composition of international students and highly-educated migrants in the US. We hope that presenting new empirical findings in an interdisciplinary context will contribute to improvements in our theoretical understanding of migration dynamics.

New Data for the Analysis of Migration Patterns

Monitoring international flows of migrants is key to designing effective policies. However, migration data tend to be coarse-grained, inconsistent across countries, expensive to gather, and available only with a considerable delay [5, 20]. The increasing availability of geolocated data from online sources or cellphone call records has opened new opportunities to identify migrants and to follow them, in an anonymous way, over time. Cellphone data have been used mainly to evaluate patterns and regularities of internal mobility for a country (e.g., [3, 7]). IP address geolocation has been used to evaluate internal mobility [14]. Analogously, recent trends in international flows of migrants have been estimated by tracking the locations, inferred from IP addresses, of users who repeatedly login into Yahoo! services [18, 21]. More recently, geolocated Twitter 'tweets' have proven useful to monitor trends in short-term international mobility [22].

The relevance of new digital records for migration studies can be evaluated along three main dimensions: i) scope, ii) time series length, and iii) accuracy of geolocation. Most data sources rarely excel in all the three dimensions. For instance, cellphone call detail records are quite accurate in terms of geolocation, but often available only for single countries or small geographic regions. IP geolocated logins to websites are not constrained by country borders, but have low granularity within a country. Geolocated Twitter data provide precise estimates of geographic coordinates and the scope is global. However, the time series are relatively short and little demographic information can be extracted from Twitter profiles.

We analyzed recent trends in international migration of highly skilled workers using a dataset of unprecedented detail, extracted from LinkedIn, the social networking website for professionals. LinkedIn counts over 200 million members in more than 200 countries and territories [13]. People typically use their LinkedIn profiles to post their

---

[1] The American Community Survey documents a flat trend in the number of college-educated individuals who migrated to the United States during the period 2000-2010.

employment and educational history. When aggregated and anonymized, that information provides the most comprehensive and up to date picture of international flows of highly skilled migrants.

Trends in Highly Skilled Migration to the US

We tracked the proportion of migrants whose destination was the United States, out of all migrants observed during a particular calendar year, for the period 1990-2012. Figure 1 shows the fraction of world migrants who moved to the United States, over time. The trends are broken down by level of education and by sector of employment (STEM vs. non-STEM). In our sample of LinkedIn users we observed a slight increase of the conditional probability of migrating to the United States during the 1990s, followed by a downward trend after the year 2000. The trend that we observed suggests

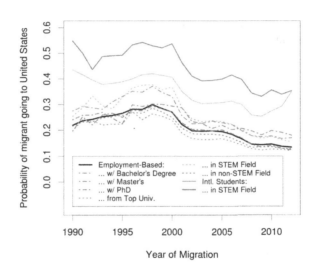

Fig. 1. Conditional Probability of Migration to United States by Year, 1990-2012

that a smaller fraction of highly skilled migrants seeking employment have made their way to the United States as the first decade of the 21st century progressed. The patterns that we observed could be related to both increasing opportunities outside the United States or a reduction of the demand in the United States. For instance, during the first decade of the 21st century, the United States experienced two major economic crises: the collapse of the "dot-com bubble" during 1999-2001, and the financial crisis of 2008. These crises adversely affected opportunities for immigrants in the United States. The nature of our dataset has allowed us to assess the decline in migration likelihoods by educational attainment at the time of migration. As Figure 1 shows, 33% of professional migrants with Bachelors' degrees achieved by the time of migration were likely to reach the US in the year 2000, compared to 17% in 2012. Analogous figures are 27% in 2000 and 12% in 2012 for migrants with Master's degrees, 29% (2000) and 18% (2012) for migrants with PhDs.

The current policy debate has centered around the availability of temporary and permanent visas for highly-skilled migrants in STEM fields. To address this area of interest we classified individuals according to their broad occupational field. A downward trend is observed in STEM as well as non-STEM fields, although the overall decrease in the probability of migrating to the US was higher in STEM (22 percentage points, from 37% to 15%) as compared to non-STEM fields (12 percentage points, from 25% to 13%). Our findings suggest that, in addition to short-term crises, such as the "dot-com bubble", there are long-term structural changes in the global system of employment-based, highly-skilled migration. The United States continues to occupy a central place in the global migration system. However, its dominant position is no longer indisputable. Figure 2 shows that, while the U.S. became a less prominent destination for professional migrations during the 2000s, Europe and Canada also saw a decrease in their share[2] of the world's professional migration flows – albeit a gentler one – while Australia and Oceania, Africa and Latin America increased their proportional intake.[3] The most prominent increase was recorded for Asian countries, which attracted, in our sample, a cumulative 25% of the world's professional migrants in 2012, compared to only 10% in the year 2000. The observed decline of the United States as a professional

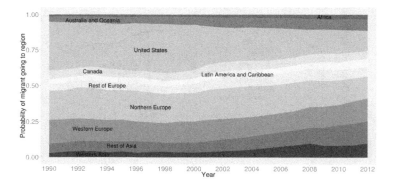

Fig. 2. Distribution of Migration Flows, by year and region of destination, 1990-2012

migration destination may be a reflection of increased competition for highly skilled migrants from other countries, of declining demand for highly skilled migrants in the United States, of an increased worldwide supply of highly skilled migrants, or of inefficiencies created by current US migration laws. While the mechanism is most likely a

---

[2] Europe attracted 40.8% of the world's professional migrants in 2000, and 37.8% in 2012, while Canada attracted 6.2% of the flow in 2000 and 5.5% in 2012.

[3] Africa increased from 1.3% in 2000 to 3.3% in 2012, Australia and Oceania from 5.7% in 2000 to 7.9% in 2012, Latin America and the Caribbean increased from 3.7% in 2000 to 5.7% in 2012.

multi-factorial one, the overall conclusion seems to suggest the possibility of a fundamental change in the international migration patterns of professionals.

Robustness of the Results

Although our dataset allows an otherwise-unattainable glimpse into the global system of highly skilled migration, there are a number of limitations that we would like to acknowledge and discuss. First, we do not know the citizenship status of individuals in our sample. As a result, our dataset does not directly distinguish between the return migration of US expatriates and in-migration of foreign persons. However, this is expected to be a minor factor, as relatively few American professionals migrate outside of the United States, and fewer return to their country of origin.[4] Another relatively minor source of uncertainty in our data concerns cases of circular migrations, back and forth from the United States, of foreign persons, which are expected to be rare events. Indeed, 92% of migration events in our dataset were due to individuals who generated only one migration event.

LinkedIn users are not a representative sample of the entire population of highly-skilled migrants. As a result our estimates may be biased. A potential problem of our data is the mechanism through which individual migrants are selected into the sample. We thus verified the robustness of our main result, the downward trend in fraction of migrants to the United States, with further analyses. Since LinkedIn is a United States company, those individuals who joined earlier were more likely to be located in the United States at the time of their registration, and thus more immigrants to the United States are expected to be included in the early sample of our data. However, we checked that the size of this potential source of bias is small and does not affect our results. In order to control for unobserved users' characteristics associated with the choice of registering with LinkedIn, we divided our dataset into ten separate subsets, one for each annual cohort[5] of new LinkedIn users since 2004. For all of the ten cohorts we found a statistically significant downward trend in migrants' likelihood to move to the United States after the year 2000.[6]

As a further test of the validity of the results, we compared predictions derived from our model against the American Community Survey (ACS) (http://www.census.gov/acs), using a dataset provided by the IPUMS project (https://usa.ipums.org/usa/sda/). To our knowledge, the ACS – a survey continuously run by the US Census Bureau – represents one of the most authoritative data sources available to estimate migrations to the United States. We compared the

---

[4] This consideration is even more likely to hold for graduates of non-US top global universities.

[5] A cohort of users comprises all those individuals who joined LinkedIn during the same calendar year. Regardless of when a user joins, we observe events both before and after their joining of LinkedIn, from the user's professional history as reported on their LinkedIn profile.

[6] Statistical significance was established using a logistic regression where the year of migration and the year of user registration were dummy-coded. The ratio between the cohort-specific likelihoods of migrating to the United States in 2012 and 2000 ranged between 0.47 and 0.72. The similar ratio against the year 1999 ranged between 0.47 and 0.62. There was no monotonic relation between user cohort and decrease of likelihood of migrating to US.

yearly rate of change in the US in-migration rate estimated from our data and from the ACS, for the period 2001 to 2010. ACS and LinkedIn estimates were computed for individuals who had at least a Bachelor's degree at the time of migration. Pearson's ρ between the two time-series is 0.70, whereas Spearman's rank-correlation coefficient is 0.83. The time-series are plotted together in Figure 3. The plot shows the two time-series tracking each other quite closely until 2005 (Pearson's $\rho = 0.96$, Spearman's rank-correlation coefficient 0.9). After 2005, estimates based on LinkedIn data give a higher immigration rate. It is possible that ACS underestimates professional migration, due to underreporting. Alternatively, our approach based on LinkedIn data may tend to overestimate professional migration to the US during the late 2000s. This observation further strengthens our main result. If estimates of migration rates from our LinkedIn dataset tend to overestimate recent migration of professionals to the US (i.e., if the population of LinkedIn users is more mobile than the overall population of highly-skilled professionals), then the downward trend in conditional probabilities of professional migration to the United States may be even steeper than what we expect. In other words, in spite of the fact that LinkedIn data may overestimate recent migration of highly-skilled individuals to the United States, in our sample professional migrants appear less likely to go to the United States in the second half of the last decade than in the first.

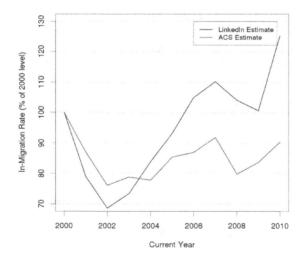

Fig. 3. U.S. In-Migration rate, computed from LI and ACS data

An additional potential confounding factor in our data concerns the definition of a "highly-skilled migrant". A skeptical argument would be that the quality of university degrees might have been diluted by increases in the number of higher education institutions worldwide. By this token, the United States is receiving the same share of the "truly" highly-skilled migrants in the world, but the (likely) increasing number of university graduates is hiding this fact. We falsified this hypothesis by computing the conditional migration probability to the United States for a subset of individuals in our sample: those whose latest degree at the time of migration came from one of the top-500 worldwide universities, as listed in the Quacquarelli-Symonds (QS) ranking (2013).[7] Once more we observed the same overall pattern of decreasing probabilities of migration to the United States: in our sample, 24% of migrants who were graduates of the top 500 universities worldwide went to the United States in the year 2000, but only 12% did so during 2012.

Discussion

Highly-skilled migration is an important demographic phenomenon with relevant consequences, for instance in terms of human capital formation, a central issue in the study of economic development. Despite the importance of highly-skilled migrations for a number of disciplines and for policy making, it is extremely difficult to find reliable data on the flows of highly-skilled migrants. This is due to a number of factors. There is no uniform international definition of migration, and even migration data sources that provide time-series data caution against assuming either within- or between-country consistency in the measurement of migrations. In some cases the data sources are so indirect as to render them useless in a comparison against our dataset. For instance, data for the United States in the OECD international migration database come from the Department of Homeland Security count of new permanent residencies, though a great number of migration episodes to the United States start out with a "non-immigrant" visa status (e.g., the H1-B, F-1 visa, etc.).

For this article, our aim is to measure highly-skilled – rather than overall – migration flows. There is even less consistent data available for this task, and to our knowledge no large-scale survey of the world's professional migration flows has currently been compiled. The boundaries of the concept of a "highly-skilled" migrant are relatively porous, rendering its measurement difficult with traditional demographic instruments. We believe that complementing existing data sources with social media data may improve our understanding of migration patterns. LinkedIn, with a website interface in 20 languages and an aggressive strategy emphasizing growth outside of the United States, provides innovative data to investigate population processes for highly-skilled professionals.

In this article, we showed that LinkedIn data provide important insights about recent trends in migrations of highly-skilled migrants to the United States. At the same time, the sample of LinkedIn users is a convenience sample. It is a large and interesting sample, but not representative of the entire population of highly-skilled migrants. We provided analyses that support the robustness of our results. Nonetheless, there is a tradeoff between generating new information from social media, and the statistical confidence in the results. Whenever large datasets exist for calibration of estimates from

---

[7] There were 406 non-US universities in the Quacquarelli-Symonds top 500. We only included non-US universities because individuals who have attended US schools and are currently abroad are by definition return migrants to the United States, whereas we are primarily interested in first-time migrants.

social media data, then our uncertainty about the outcomes is low. In those situations, the novelty of the results is also low. Whenever little traditional data exist for calibration, social media may provide more novel information, but with higher uncertainty. The challenge for social scientists and computer scientists is to incorporate existing data sources, from official statistics to social media data, into a unified framework.

The rise of very large datasets has the potential to reshape both science and policy in innumerable ways, as long as appropriate methods will be developed to make inference from unstructured data. Traditional measurement methods have not been enough to generate timely estimates consistent across countries. We believe that the use of social media data in this area will be very fruitful, especially in combination with existing data sources. Measuring migrations is a relatively well-defined problem. Thus it will be possible to evaluate the predictive power of models that incorporate social media data. Our article is intended to provide a first step towards the study of highly-skilled migrations using social media data. As such, we hope to stimulate the discussion about the use of social media data to improve our understanding of population processes. We believe that social scientists will not only benefit from new and large data sets, but also increasingly contribute to the emerging field of Web science by developing new and innovative methods.

References

1. Anich, R., Appave, G., Aghazarm, C., Lacko, F., Kigouk, A.: International Migration Trends, World Migration Report Series, ch. 2. IOM (February 2011)
2. Auriol, L.: Careers of Doctorate Holders: Employment and Mobility Patterns, OECD Science, Technology and Industry Working Papers, OECD Publishing (April 2010)
3. Blumenstock, J.E.: Inferring Patterns of Internal Migration from Mobile Phone Call Records: Evidence from Rwanda. Information Technology for Development 18(2), 107–125 (2012)
4. Czaika, M., de Haas, H.: The Globalization of Migration: Has the World Become more Migratory? International Migration Review 48(2), 283–323 (2014)
5. De Beer, J., Raymer, J., Van Der Erf, R., Van Wissen, L.: Overcoming the Problems of Inconsistent Migration Data: A New Method Applied to Flows in Europe. European Journal of Population 26, 459–481 (2010)
6. Franzoni, C., Scellato, G., Stephan, P.: Foreign-born Scientists: Mobility Patterns for 16 Countries. Nature Biotechnology 30, 1250–1253 (2012)
7. Gonzalez, M.C., Hidalgo, C.A., Barabasi, A.-L.: Understanding Individual Human Mobility Patterns. Nature 453(7196), 779–782 (2008)
8. Hatton, T.J., Williamson, J.G.: What Fundamentals Drive World Migration? National Bureau of Economic Research (2002)
9. Levin, S.G., Stephan, P.E.: Are the Foreign Born a Source of Strength for U.S. Science? Science 285(5431), 1213–1214 (1999)
10. Lowell, L.: Highly-Skilled Migration. World Migration Report Series, ch. 2. IOM (April 2008)
11. Malakoff, D.: Visa Reform Advances in Senate as House Offers STEM Ideas. Science 340(6136), 1027 (2013)
12. Massey, D.S., Arango, J., Hugo, G., Kouaouci, A., Pellegrino, A., Taylor, J.E.: Theories of International Migration: A Review and Appraisal. In: Population and Development Review, pp. 431–466 (1993)

13. Nishar, D.: 200 Million Members!, LinkedIn Blog (January 9, 2013),
    `http://blog.linkedin.com/2013/01/09/linkedin-200-million/`
14. Pitsillidis, A., Xie, Y., Yu, F., Abadi, M., Voelker, G.M., Savage, S.: How to Tell an Airport
    from a Home: Techniques and Applications. In: Proceedings of the 9th ACM SIGCOMM
    Workshop on Hot Topics in Networks, vol. 13, pp. 1–6 (2010)
15. Quacquarelli Symonds Ltd., World University Rankings (August 1, 2013),
    `http://www.topuniversities.com/qs-world-university-rankings`
16. Shachar, A.: Race for Talent: Highly Skilled Migrants and Competitive Immigration
    Regimes. The. NYUL Rev. 81, 148 (2006)
17. Stalker, P.: Workers Without Frontiers: The Impact of Globalization on International Migra-
    tion. International Labor Organization (2000)
18. State, B., Weber, I., Zagheni, E.: Studying International Mobility through IP Geolocation.
    Proceedings of Web Search and Data Mining, 265–274 (2013)
19. United Nations Population Division/DESA. Trends in International Migrant Stock: The 2013
    Revision (2013)
20. Van Noorden, R.: Science on the Move. Nature 490, 326–329 (2012)
21. Zagheni, E., Weber, I.: You are where you E-mail: Using E-mail Data to Estimate Interna-
    tional Migration Rates. In: Proceedings of Web Science, pp. 348–358 (2012)
22. Zagheni, E., Garimella, V.R.K., Weber, I., State, B.: Inferring International and Internal Mi-
    gration Patterns from Twitter Data. In: Proceedings of WWW, pp. 439–444 (2014)

Appendix

Extracting Information from LinkedIn Profiles

From the initial population of over 200 million LinkedIn users worldwide, we extracted the subset of inter-country migration events related to changes in individuals' places of employment, for migrations lasting at least one calendar year between 1990 and 2012. We measured migrations by examining country-level locations associated with positions held by individuals across their careers, as listed in their LinkedIn profiles. Part of the geolocated positions are standardized data, where the user selects the position's location from a drop-down menu. We inferred the remaining positions' location by combining various sources of information: free-text entered by the user (addresses), IP geo-location, location of the company associated with the position, colleagues' locations, and the location associated with the next and previous positions in the individual's profile. To combine the various sources of information, we used a Naive Bayes classifier trained on the standardized location data. The decision threshold that we chose achieved 99% precision and 54% recall against a held-out dataset.

We represented each individual's career as an ordered tuple $(p_{i,1}; p_{i,2}; \ldots p_{i,k})$, where $p_{i,j}$ denotes the j-th position held by individual i, with the order determined by each position's start date. We projected each person's tuple of geolocated positions into month-level observations that specify their location during a particular month. In cases where location information is missing from a person's career for a period of less than or exactly twelve months, we interpolated the location with respect to the nearest (in time) non-missing observation. Where two non-missing and different observations are equally close (e.g. location A six months before and location B six months later), we selected an imputation at random from the two possibilities. We then inferred the place of residence for each user, at regular intervals of time (i.e., during the month of January) over the course of several years.

We define a migration event by querying the location of each individual at the beginning of every calendar year. If the individual's estimated place of residence is in a different country, compared to the beginning of the previous year, we assume that a migration event has occurred during the past calendar year. For the purposes of this article, immigration rates are defined as the ratio $N_{\to C}^{(y)}/N_C^{(y)}$ between the number of individuals who moved to country C during year y ($N_{\to C}^{(y)}$), and the number of individuals who were observed in country C at the end of year y ($N_C^{(y)}$).

We mapped employment-based positions to their Standard Occupational Classification (SOC) code. From each position we extracted the job title as reported by the user. Job titles were then mapped through an internal algorithm to a number of standardized titles, which in turn were mapped by human coders to their Standard Occupational Classification code. Positions were considered to be STEM if their SOM code was either 15-1000 (Computer Occupations), 15-2000 (Mathematical Science Occupations) 17-1000 (Architecture and Engineering Occupations), 19-0000 (Life, Physical and Social Science Occupations), and 25-1000 (Postsecondary Teachers). The decision to include all Postsecondary Teachers in the STEM field is motivated by the great deal of overlap between academia and STEM fields.

Table 1. Probability that Migration Destination is U.S. (cf. Figure 1)

Year		Employment-Based Migration							Education-Based	STEM
	Overall	Degree Prior to Migration				Top	STEM Field		Overall	
		Bac.	Mst.	PhD	School	Yes	No		Field	
1	1990	0.22	0.27	0.20	0.24	0.20	0.25	0.22	0.44	0.55
2	1991	0.24	0.29	0.25	0.26	0.24	0.29	0.23	0.42	0.50
3	1992	0.24	0.29	0.22	0.26	0.24	0.33	0.23	0.40	0.44
4	1993	0.25	0.28	0.26	0.27	0.22	0.30	0.25	0.38	0.49
5	1994	0.26	0.31	0.27	0.25	0.22	0.28	0.25	0.38	0.49
6	1995	0.27	0.33	0.25	0.23	0.22	0.33	0.26	0.39	0.49
7	1996	0.28	0.35	0.28	0.30	0.27	0.36	0.27	0.40	0.53
8	1997	0.28	0.35	0.28	0.29	0.24	0.37	0.27	0.42	0.54
9	1998	0.30	0.37	0.29	0.30	0.26	0.38	0.29	0.42	0.53
10	1999	0.28	0.35	0.28	0.30	0.25	0.36	0.27	0.41	0.52
11	2000	0.27	0.33	0.27	0.29	0.24	0.37	0.25	0.41	0.54
12	2001	0.22	0.26	0.23	0.29	0.19	0.29	0.21	0.35	0.46
13	2002	0.20	0.23	0.20	0.23	0.17	0.24	0.19	0.32	0.41
14	2003	0.20	0.23	0.20	0.23	0.16	0.23	0.18	0.30	0.39
15	2004	0.20	0.24	0.20	0.21	0.16	0.24	0.18	0.31	0.39
16	2005	0.19	0.23	0.20	0.22	0.17	0.23	0.18	0.31	0.40
17	2006	0.18	0.22	0.19	0.20	0.16	0.22	0.17	0.31	0.42
18	2007	0.17	0.20	0.17	0.21	0.14	0.20	0.15	0.30	0.40
19	2008	0.15	0.18	0.14	0.20	0.13	0.18	0.13	0.26	0.34
20	2009	0.14	0.17	0.13	0.18	0.12	0.17	0.13	0.25	0.33
21	2010	0.15	0.18	0.14	0.20	0.13	0.18	0.14	0.27	0.36
22	2011	0.14	0.17	0.13	0.19	0.13	0.16	0.13	0.29	0.34
23	2012	0.13	0.17	0.12	0.18	0.12	0.15	0.13	0.35	0.35

Notes: Employment-based migration: migrant (first) obtains job in destination country. Education-based migration: migrant (first) pursues educational program in destination country. If migrant pursues both employment and education upon arriving in destination country, migration event is assumed to be education-based. Prior degree must have been received during the previous year. "Top schools" are all non-US schools in the top 500 universities in the Quacquarelli-Symonds ranking. For STEM field identification, see main text.

Table 2. Definition of Regions used in Figure 2

Region	Countries
Africa	Algeria; Angola; Benin; Botswana; Burkina Faso; Burundi; Cameroon; Cape Verde; Central African Republic; Chad; Comoros; Congo, Republic Of; Congo, The Democratic Republic Of; Cote D'ivoire; Djibouti; Egypt; Equatorial Guinea; Eritrea; Ethiopia; Gabon; Gambia; Ghana; Guinea; Guinea-bissau; Kenya; Lesotho; Liberia; Libyan Arab Jamahiriya; Madagascar; Malawi; Mali; Mauritania; Mauritius; Mayotte; Morocco; Mozambique; Niger; Nigeria; Reunion; Rwanda; Saint Helena; Sao Tome And Principe; Senegal; Seychelles; Sierra Leone; Somalia; South Africa; Sudan; Swaziland; Tanzania, United Republic Of; Togo; Tunisia; Uganda; Zambia; Zimbabwe
Australia and Oceania	American Samoa; Australia; Cook Islands; Fiji; French Polynesia; Guam; Kiribati; Marshall Islands; Micronesia, Federated States Of; Nauru; New Caledonia; New Zealand; Northern Mariana Islands; Palau; Papua New Guinea; Samoa; Solomon Islands; Tonga; Tuvalu; Vanuatu
Canada	Canada
Latin America and Caribbean	Anguilla; Antigua And Barbuda; Argentina; Aruba; Bahamas; Barbados; Belize; Bermuda; Bolivia, Plurinational State Of; Brazil; Cayman Islands; Chile; Colombia; Costa Rica; Cuba; Dominica; Dominican Republic; Ecuador; El Salvador; Falkland Islands (Malvinas); French Guiana; Greenland; Grenada; Guadeloupe; Guatemala; Guyana; Haiti; Honduras; Jamaica; Martinique; Mexico; Montserrat; Netherlands Antilles; Nicaragua; Panama; Paraguay; Peru; Puerto Rico; Saint Kitts And Nevis; Saint Lucia; Saint Pierre And Miquelon; Saint Vincent And The Grenadines; Suriname; Trinidad And Tobago; Turks And Caicos Islands; Uruguay; Venezuela, Bolivarian Republic Of; Virgin Islands, British; Virgin Islands, U.s.
Northern Europe	Aland Islands; Denmark; Estonia; Faroe Islands; Finland; Guernsey; Iceland; Ireland; Isle Of Man; Jersey; Latvia; Lithuania; Norway; Svalbard And Jan Mayen; Sweden; United Kingdom
Rest of Asia	Afghanistan; Bangladesh; Bhutan; Brunei Darussalam; Cambodia; China; Hong Kong; India; Indonesia; Iran, Islamic Republic Of; Japan; Kazakhstan; Korea, Democratic People's Republic Of; Korea, Republic Of; Kyrgyzstan; Lao People's Democratic Republic; Macao; Malaysia; Maldives; Mongolia; Myanmar; Nepal; Pakistan; Philippines; Singapore; Sri Lanka; Tajikistan; Thailand; Timor-leste; Turkmenistan; Uzbekistan; Vietnam
Rest of Europe	Albania; Andorra; Belarus; Bosnia And Herzegovina; Bulgaria; Croatia; Czech Republic; Gibraltar; Greece; Holy See (vatican City State); Hungary; Italy; Macedonia, The Former Yugoslav Republic Of; Malta; Moldova, Republic Of; Montenegro; Poland; Portugal; Romania; Russian Federation; San Marino; Serbia; Slovakia; Slovenia; Spain; Ukraine
United States	United States
Western Asia	Armenia; Azerbaijan; Bahrain; Cyprus; Georgia; Iraq; Israel; Jordan; Kuwait; Lebanon; Oman; Palestinian Territory, Occupied; Qatar; Saudi Arabia; Syrian Arab Republic; Turkey; United Arab Emirates; Yemen
Western Europe	Austria; Belgium; France; Germany; Liechtenstein; Luxembourg; Monaco; Netherlands; Switzerland

Table 3. Distribution of World Migrations (cf. Figure 2)

	Afr.	Aus.	Can.	L. Am.	N.Eur.	R.of Asia	R.of Eur.	U.S.	W. Asia	W. Eur.	Total
1990	0.01	0.04	0.07	0.03	0.20	0.06	0.05	0.34	0.03	0.17	1.00
1991	0.01	0.04	0.06	0.04	0.20	0.06	0.05	0.33	0.04	0.16	1.00
1992	0.01	0.04	0.06	0.03	0.21	0.07	0.05	0.32	0.04	0.16	1.00
1993	0.01	0.04	0.06	0.04	0.22	0.07	0.05	0.32	0.04	0.15	1.00
1994	0.01	0.05	0.06	0.04	0.21	0.08	0.05	0.32	0.03	0.15	1.00
1995	0.02	0.05	0.06	0.04	0.20	0.07	0.05	0.32	0.04	0.15	1.00
1996	0.01	0.05	0.05	0.04	0.21	0.08	0.05	0.33	0.03	0.14	1.00
1997	0.01	0.05	0.06	0.04	0.21	0.07	0.05	0.34	0.04	0.14	1.00
1998	0.01	0.05	0.06	0.03	0.20	0.06	0.06	0.35	0.04	0.14	1.00
1999	0.02	0.05	0.06	0.03	0.20	0.07	0.06	0.33	0.03	0.15	1.00
2000	0.01	0.06	0.06	0.04	0.20	0.07	0.06	0.32	0.03	0.15	1.00
2001	0.02	0.06	0.07	0.04	0.21	0.08	0.07	0.27	0.04	0.15	1.00
2002	0.02	0.07	0.07	0.04	0.21	0.08	0.08	0.24	0.04	0.14	1.00
2003	0.02	0.07	0.07	0.04	0.21	0.09	0.08	0.24	0.05	0.14	1.00
2004	0.02	0.07	0.06	0.04	0.21	0.09	0.08	0.23	0.05	0.13	1.00
2005	0.02	0.07	0.06	0.04	0.20	0.10	0.08	0.23	0.06	0.13	1.00
2006	0.02	0.07	0.06	0.04	0.20	0.10	0.08	0.22	0.08	0.13	1.00
2007	0.02	0.08	0.06	0.04	0.20	0.10	0.08	0.20	0.08	0.13	1.00
2008	0.03	0.08	0.06	0.04	0.19	0.11	0.08	0.17	0.10	0.14	1.00
2009	0.03	0.07	0.06	0.05	0.18	0.13	0.08	0.17	0.08	0.15	1.00
2010	0.03	0.08	0.06	0.05	0.17	0.14	0.07	0.17	0.08	0.14	1.00
2011	0.03	0.08	0.06	0.06	0.16	0.15	0.07	0.16	0.09	0.15	1.00
2012	0.03	0.08	0.05	0.06	0.16	0.16	0.07	0.14	0.10	0.16	1.00

Note: Table reflects all observed migrations, whether employment- or education-based.

U.S. Religious Landscape on Twitter

Lu Chen[1,*], Ingmar Weber[2], and Adam Okulicz-Kozaryn[3]

[1] Kno.e.sis Center, Wright State University, Dayton, OH, USA
chen@knoesis.org
[2] Qatar Computing Research Institute, Doha, Qatar
iweber@qf.org.qa
[3] Rutgers-Camden, Camden, NJ, USA
adam.okulicz.kozaryn@gmail.com

Abstract. Religiosity is a powerful force shaping human societies, affecting domains as diverse as economic growth or the ability to cope with illness. As more religious leaders and organizations as well as believers start using social networking sites (e.g., Twitter, Facebook), online activities become important extensions to traditional religious rituals and practices. However, there has been lack of research on religiosity in online social networks. This paper takes a step toward the understanding of several important aspects of religiosity on Twitter, based on the analysis of more than 250k U.S. users who self-declared their religions/belief, including *Atheism, Buddhism, Christianity, Hinduism, Islam*, and *Judaism*. Specifically, (i) we examine the correlation of geographic distribution of religious people between Twitter and offline surveys. (ii) We analyze users' tweets and networks to identify discriminative features of each religious group, and explore supervised methods to identify believers of different religions. (iii) We study the linkage preference of different religious groups, and observe a strong preference of Twitter users connecting to others sharing the same religion.

1 Introduction

Religiosity is a powerful force shaping human societies, and it is persistent – 94% of Americans believe in God and this percentage has stayed steady over decades [30]. It is important to study and understand religion because it affects multiple domains, ranging from economic growth [1], organizational functioning [10] to the ability to better cope with illness [3]. A key feature of any belief system such as religion is replication – in order to survive and grow, religions must replicate themselves both vertically (to new generations) and horizontally (to new adherents). The Internet already facilitates such replication. Traditional religions are likely to adapt to the societal and historic circumstances and take advantage of social media. Many churches and religious leaders are already using social networking sites (e.g., Twitter, Facebook) to connect with their believers.

---

* This work was done while the first author was an intern at Qatar Computing Research Institute.

L.M. Aiello and D. McFarland (Eds.): SocInfo 2014, LNCS 8851, pp. 544–560, 2014.
© Springer International Publishing Switzerland 2014

While social networking and social media become important means of religious practices, our understanding of religiosity in social media and networking sites remains very limited. In this paper, we take a step to bridge this gap by studying the phenomenon of religion for more than 250k U.S. Twitter users, including their tweets and network information.

Twitter, because of its global reach and the relative ease of collecting data, is becoming a great treasure trove of information for computer and social scientists. Researchers have studied various problems using Twitter data, such as mood rhythms [14], happiness [12], electoral prediction [7], or food poisoning [8]. However, studies that explore the phenomenon of religion in social networking sites are still rare so far. To date, the most relevant study investigates the relationship between religion and happiness on Twitter [29]. It examines the difference between Christians and Atheists concerning the use of positive and negative emotion words in their tweets, whereas our work focuses on the religiosity of Twitter users across five major religions and Atheism. One recent study [26] addresses the prediction of users' religious affiliation (i.e., Christian or Muslim) using their microblogging data, which focuses on building the classification model but not studying the phenomenon.

We collected U.S. Twitter users who self-reported their religions as *Atheism*, *Buddhism*, *Christianity*, *Hinduism*, *Islam*, or *Judaism* in their free-text self-description, and further collected their tweets and friends/followers. Our dataset comprises 250,840 U.S. Twitter users, the full lists of their friends/followers, and 96,902,499 tweets. In particular, we explore the following research questions in this paper:

1. *How does the religion statistics on Twitter correlate with that in the offline surveys?* Our correlation analysis shows that: (1) There is a moderate correlation between survey results and Twitter data regarding the distribution of religious believers of a given denomination across U.S. states, e.g., the macro-average Spearman's rank correlation of all the denominations is $\rho = .65$. (2) Similarly, the fraction of religious people of any belief within a given U.S. state in surveys matches well with that of Twitter users referencing any religion in their profiles with a Pearson Correlation of $r = .79$ ($p < .0001$).

2. *Whether or not do various religious groups differ in terms of their content and network? Can we build a classifier to accurately identify believers of different religions?* Specifically, (1) By looking at discriminative features for each religion, we show that users of a particular religion differ in what they discuss or whom they follow compared to random baseline users. (2) We build two classifiers that detect religious users from a random set of users based on either their tweets' content or the users they follow, and we find that the network "following" features are more robust than tweet content features, independent of the religion considered.

3. *Does the in-group linkage preference exist in any particular religious denomination?* Our main findings include: (1) We find strong evidence of same-religion linkage preference that users of a particular denomination would have an increased likelihood to follow, be-followed-by, mention or retweet

other users of the same religion. For example, our results show that following someone of the same religion is 646 times as likely as following someone of a different religion based on a macro-average of six denominations. (2) We show that "the pope is not a scaled up bishop" in that hugely popular religious figures on Twitter have a higher-than-expected share of followers without religious references in their profiles.

Our findings may not only improve the understanding of religiosity in social media but, as the Internet is becoming a medium for religious replication, also have implications for religion *per se*.

2   Related Work

Religion shapes human society and history, and defines a person in many ways. It has such a large influence on people that it can be used as a measure of culture [18]. Fundamentally, religiosity satisfies "the need to belong", that is, people who are religious and live in religious societies, feel that they are part of that society [27,28]. Religiosity does predict multiple outcomes such as economic growth, happiness, trust, and cooperation [1,3,27,28,31,15].

In the past few years, religion has been the subject of some Social Informatics research, particularly examining the role of Internet-based technologies in religious practice [38,36,16]. For example, Wyche et al. [38] explore how American Christian ministers have adopted technologies such as the World Wide Web and email to support the spiritual formation and communicate with their laity. In another study [36], researchers discuss the design and evaluation of a mobile phone application that prompts Muslims to their five daily prayer times. There is a study of "church" (and "beer") mentions on Twitter, which corroborates our results showing more religiosity in South Eastern states[1]. In addition, by examining how religious people use various technologies (e.g., home automation technology, information and communications technology) for their religious practices, and whether that is different from their secular counterparts, implications can be gained to guide the future design of technologies for religious users [34,35,37]. Another line of research in this context investigates the process of "spiritualising of Internet" – how religious users and organizations shape and frame the Web space to meet their specific needs of religious rituals and practices [4,2]. It is also suggested by some researchers that studying religion on the Internet provides a microcosm for understanding Internet trends and implications [5].

Some other studies have focused on online religious communities. For example, McKenna and West [22] conduct a survey study of the online religious forums where believers interact with others who share the common faith [22]. Lieberman and Winzelberg [21] examine religious expressions within online support

---

[1] http://www.floatingsheep.org/2012/07/church-or-beer-americans-on-twitter.html

groups on women with breast cancer. It is reported that the same self and social benefits (e.g., social support, emotional well-being) found to be associated with the involvement in traditional religious organizations can also be gained by participation in online religious communities.

While much research effort has been made to understand religious use of Internet technologies, we know very little about religiosity in online social networks. On the other hand, there is recently an explosion of studies on Twitter [14,12,23,8,20,11,19], yet we do not know much specifically about religiosity on Twitter. Wagner et al. [32] develop classifiers to detect Twitter users from different categories, including category *religious*; Nguyen and Lim [26] build classifiers to identify Christian and Muslim users using their Twitter data, but neither of the two studies addresses the analysis of the phenomenon of religion on Twitter. [29] appears to be the most relevant study, which focuses on exploring the relationship between religion and happiness via examining the different use of words (e.g., sentiment words, words related to thinking styles) in tweets between Christians and Atheists. Our present work differs both in scope and purpose.

3   Data

Identifying religiosity on Twitter is non-trivial as users can belong to a particular religious group without making this affiliation public on Twitter. In this section we describe how we collect data, with a general focus on precision rather than recall, and how we validate the collected data. Concerning the selection of religions we decided to limit our analysis to the world's main religions, concretely, *Buddhism, Christianity, Hinduism, Islam*, and *Judaism*. We also included data for *Atheism*, and an *"undeclared"* baseline set of users. We focused our data collection on the U.S. as this allowed us to obtain various statistics about the "ground truth" distribution of religions across U.S. states.

The advantage of Twitter is that data are captured unobtrusively (free from potential bias of survey or experimental setting). However, Twitter has its own biases and the issues of representativeness need to be taken into account when interpreting the results. For example, according to a study[2] published in 2012, Twitter users are predominantly young (74% fall between 15 to 25 years of age). It is reported in another study [24] in 2011 that Twitter users are more likely to be males living in more populous counties, and hence sparsely populated areas are underrepresented; and race/ethnicity is biased depending on the region.

3.1   Data Collection and Geolocation

To obtain a list of users who are most likely believers of the six denominations of interest, we search Twitter user bios via Followerwonk[3] with a list of

---

[2] http://www.beevolve.com/twitter-statistics/
[3] https://followerwonk.com/bio

Table 1. Description of the dataset

User group	Atheist	Buddhist	Christian	Hindu	Jew	Muslim	Undeclared
# of users	7,765	2,847	202,563	204	6,077	6,040	25,344
Mean # of tweets per user	3976.8	2595.7	1981	2271.5	2095.7	3826.5	1837.3
Mean # of tweets per user per day	3.3	2	1.8	1.9	1.8	4.2	1.9
Stdev of # of tweets per user per day	8.3	5.8	5.3	4.8	6	9.2	5.8
Median of # of friends	179	144	151	119	163	166	114
Mean # of friends per user	442.8	452.3	370	277.2	399.6	344.1	295.5
Stdev of # of friends per user	659	1825.4	2179.6	470.1	882.9	243.1	991.6
Median of # of followers	79	77	77	74	112	104	52
Mean # of followers per user	707.5	628.9	418.2	308.2	665.2	467.9	400
Stdev of # of followers per user	23987.4	4873.6	6834.6	889.8	5063.2	2855	6691.6

keywords[4]. From Followerwonk, we obtain these users' screen names, with which we collect more information of these users through Twitter API, including their self-declared locations, descriptions, follower counts, etc.

In addition, we collect another group of Twitter users who do not report any of the above mentioned religions/beliefs in their bios. Specifically, we generate random numbers as Twitter user IDs [5], collect these users' profiles via Twitter API, and remove the users who appear in any of the user collections of the six denominations from this set. We label this user group as *Undeclared*.

We then identify users from the United States using users' self-declared locations. We build an algorithm to map location strings to U.S. cities and states. The algorithm considers only the locations that mention the country as the U.S. or do not mention any country at all, and uses a set of rules to reduce incorrect mappings. For example, "IN" may refer to the U.S. state "Indiana" or be a part of a location phrase, e.g, "IN YOUR HEART". To avoid mapping the latter one to "Indiana", the algorithm considers only the ones where the token "IN" is in uppercase, and mention either the country U.S. or a city name. If a city name is mentioned without specifying a state, and there are more than one states that have a city named that, the algorithm maps it to the city and state which has the largest population.

We keep only the users whose location string is mapped to one of the 51 U.S. states (including the federal district Washington, D.C.), the language is specified as "en", the self-description bio is not empty[6], and tweet count is greater than 10. Overall, this dataset contains 250,840 users from seven user groups. Using Twitter API, we also obtain the collection of tweets (up to 3,200 of a user's most recent tweets as by the API restrictions), and the list of friends and followers of these users. Table 1 provides an overview of the dataset. If we measure the

---

[4] We realize that this keyword list is not complete (e.g. Mormons self-identify as Christians) of these denominations, and leave it for the future research to explore an extended list. Our current focus is on precision, with a potential loss in recall.

[5] We registered a new Twitter account and obtained its ID, then we generated random numbers ranging from 1 to that ID, i.e., 2329304719. Note that Twitter IDs are assigned in ascending order of the time of account creation.

[6] This only happened for the undeclared users as the other users were found by searching in their bio. We removed such users with a empty bio as they were likely to have a very different activity pattern than users providing information about themselves.

Table 2. Example user bios. Example 1-5 are true positive, and 6-9 are false positive.

1	Animal lover.Foodie.Model.**Buddhist**.
2	**Atheist**, Doctor Who fan, the left side of politics, annoyed by happy-horseshit & pseudo-spiritual people
3	**ISLAM** 100%
4	a little bit cute,a loving sis,a good follower of **jesus**,.,.. a friendly one..
5	**Christian**, Wife of @coach_shawn10, Mother of 3 beautiful daughters, Sports Fan, AKA. I'm blessed and highly favored!
6	Worked with The **Hindu** Business Line & Dow Jones News-wires. Tracking/Trading Stock market for over 15 years.
7	PhD in Worthless Information. Surprisingly not **Jewish** or Amish. We Are! Let's Go Buffalo!
8	my boss is a **Jewish** Carpenter
9	**JESUS**! I get paid to go to football games. Social life? What is that? Follow @username for all things Sports. I think I'm funny, I'm probably wrong.

active level of users in terms of the number of tweets, friends and followers, on average, *Atheists* appear to be more active than religious users, while the *Undeclared* group generally appears to be less active than other groups. Among the five religious groups, *Muslim* users have more tweets, both *Muslim* and *Jew* users tend to have more friends and followers, compared with other religions.

It is important to note that only the Twitter users who publicly declare their religion/belief in their bios are included in our data collection, while vast majority of believers may not disclose their religion in their Twitter bios and thus not included. This may lead to bias toward users who are very religious or inclined to share such information.

3.2 Data Validation

Mentioning a religion-specific keyword (e.g., "Jesus") in the bio may not necessarily indicate the user's religious belief. Table 2 shows example user bios including both true positives (religion/belief is correctly identified) and false positives (religion/belief is not correctly identified). To evaluate the quality of our data collection, we randomly selected 200 users from each user group, and manually checked their bios and religion labels. The precision of religion identification is represented as $\frac{\#true\ positive}{\#total}$. Overall, macro-averaged precision across all the groups is 0.91, which shows that our way of identifying religiosity is quite precise. The identification of Jewish users is found to be the least accurate (0.78), because it contains the largest fraction of false positives (mostly indicating opposition and hatred) as illustrated by Examples 7 and 8 in Table 2[7]. Sadly, "digital hate" seems to be on the rise [6].

We also evaluate the geolocation results of the same data sample. The authors manually identified U.S. states from location strings of users in the sample. Among all the 1,400 users, 329 users' locations were mapped to U.S. states by the authors. The algorithm identified 298 U.S. users and mapped their locations to states, among which 289 were consistent with the manual mapping. The algorithm achieved a precision of $\frac{289}{298} = 0.97$ and a recall of $\frac{289}{329} = 0.88$.

---

[7] We chose not to show offensive profile examples here. Disturbing examples can, however, be easily found using `http://followerwonk.com/bio/`

4 Correlation Analysis of Religion Statistics between Twitter Data and Surveys

In this section, we explore how religion statistics we observed in our Twitter dataset correlate with that in offline surveys.

Pew Research U.S. Religious Landscape Survey[8] By counting the Twitter users of each denominations for each state, we get estimates of the religious composition in each of the 51 states. Pew Research Religious Landscape Survey also provides the religious composition by U.S. states, which covers nine categories including Buddhist, Christian, Hindu, Jew, Muslim, Unaffiliated, Other World Religion, Other Faiths, and Don't know/refused. The Unaffiliated category includes Atheist, Agnostic, and Nothing-in-particular. Since our data collection does not include categories such as Other World Religion, Other Faiths, or Don't know/refused, and our group of Atheist does not include Nothing-in-particular category, we remove these categories that are not included in our data collection and recalculate the composition among the remaining ones.

The per-value correlation across all the religions and states is $r > .995$, but since Christians are dominant in every state, it's easy to get a high correlation by just guessing Christian $= 100\%$ in every state. So we also conduct the correlation analysis of each religion across the 51 states. The Pearson's r on Christian and Jew are $r = .73$ and $r = .77$ ($p < .0001$ in both cases), respectively. The Spearman's rank correlation on Christian, Jew and Buddhist are $\rho = .77$, $\rho = .79$ and $\rho = .75$ ($p < .0001$ in all three cases). But the correlations on Muslim and Hindu are only at $.15 < r < .30$ ($.03 < p < .3$) and $.48 < \rho < .50$ ($p < .0004$).

The proportions by denomination in our Twitter sample from Table 1 can also be compared with the actual proportions – for instance according to Pew[9] there are about twice as many Jews as Buddhists in the U.S., and our sample shows the same proportions; there are about 2 times more Buddhists than Hindus; yet our sample has 10 times more Buddhists than Hindus.

The most plausible reason for non-perfect fits, especially for the geographic distribution of Muslims and Hindus in the U.S., is simply that the Twitter population is a biased selection of the general population as explained in Section 3. The sample size is another potential reason. Especially for small U.S. states we have only few non-Christian users in our set. Finally there are most likely also religion-specific differences in terms of the inclination to publicly state one's religious affiliation in a Twitter profile.

Gallup U.S. Religiousness Survey[10] Gallup's survey measures religiousness based on respondents' self-reported importance of religion in their daily lives and their attendance at religious services [25]. The survey provides the proportions

---

[8] http://religions.pewforum.org/
[9] http://religions.pewforum.org/reports
[10] http://www.gallup.com/poll/125066/State-States.aspx

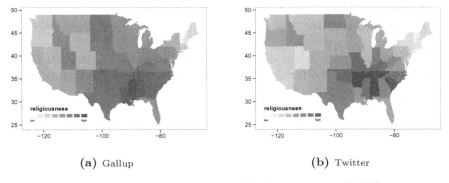

(a) Gallup   (b) Twitter

Fig. 1. Map of State Variations of Religiousness in the U.S.

of *very religious*, *moderately religious* and *non-religious* residents in each U.S. state. We get the percentage of religious residents by adding the very religious and moderately religious proportions together.

We count the number of religious Twitter users (including Buddhist, Christian, Hindu, Jew, and Muslim) in each state, which is $N^R(s)$, where s can be any of the 51 U.S. states, e.g., $N^R(Ohio)$. By adding them together we get the total number of religious users in the U.S., i.e., $N^R(all)$. Then the fraction of religious users of state s is $\frac{N^R(s)}{N^R(all)}$. In a similar way, we can get the fraction of undeclared users of state s as $\frac{N^U(s)}{N^U(all)}$, where $N^U(s)$ is the number of undeclared users in state s, and $N^U(all)$ is the total number of undeclared U.S. users. Note that we do not differentiate users on Twitter according to degrees of religiosity for this study as, we believe, users that explicitly state their religious affiliation online are likely to be comparatively more religious.

Then we measure the religiousness of state s as $\frac{N^R(s)}{N^R(all)} / \frac{N^U(s)}{N^U(all)}$. The higher the score, the more religious the state as it has a larger-than-expected number of Twitter users with a self-stated religious affiliation. Correlating this religiousness score per state against the Gallup survey shows a respectable fit of Pearson's $r = .79$ ($p < .0001$). Figure 1 shows state variations of religiousness by both the survey data and Twitter data. They agree on 11 of the top 15 most religious states (e.g., Alabama, Mississippi, and South Carolina) and 11 of the top 15 least religious states (e.g, Vermont, New Hampshire, and Massachusetts). However, Utah is the second most religious state according to Gallup survey, but is one of the least religious states according to our data collection. The main reason might be that Mormonism (the dominant religion in Utah) is underrepresented in our dataset as we did not scan for related terms in the users' profiles.

In addition, the Pearson's r between the number of undeclared users per state and the population of those states is .986 ($p < .0001$), which suggests a good level of representativeness in terms of the number of Twitter users.

5   Identification of Believers of Various Religious Denominations

In this section, we explore the discriminative features of a religion that differentiate its believers from others, and build classifiers to identify religious Twitter users of various denominations.

By exploring the features that are effective for identifying Twitter users of a certain religious denomination, we would gain insight on the important aspects of a religion. For example, the comparison of tweet content based features and network based features in a classifier would show whether it is more about "the company you keep" or "what you say online" that tells you apart from others of a different religious belief. In addition, by looking at how easy/difficult it is for a classifier to recognize believers of a particular religion, we could see which religions are "most religious" in that they differ most from "normal" behavior on Twitter. This is not just a classification question but also a societal question: religions that could be told easily by with whom you mingle (network) are probably more segregated, and possibly intolerant towards other groups – in general religiosity and prejudice correlates [15]. Again, this has broader societal implications because these linkage or group preferences are likely to be present in the real world as well – for instance, real world traits and behaviors such as tolerance, prejudice, and openness to experience are likely to be correlated with our findings. For example, differences in hashtag usage between Islamists and Seculars in Egypt has been found to indicate "polarization" in society [33].

5.1   Discriminative Features

What Do They Tweet? We first study the discriminative words in tweets that differentiate the users of one particular religious group from others by chi-square test. Specifically, we get the words from the tweet collection, and keep only the ones that appear in no less than 100 tweets. Each user group is represented by a vector of words extracted from its tweet collection, in which the words are weighted by the frequency of how many users of that group used them in their tweets (including retweets and mentions). Then a chi-square test is applied to the vector of each religious group (i.e., *Atheist, Buddhist, Christian, Hindu, Jewish, Muslim*) against the vector of the *Undeclared* user group. The top 15 words that are most positively associated with each group are displayed in Figure 2. The font size of a word in the figure is determined by its chi-square score.

These discriminative words are largely religion-specific, which may refer to religious images, beliefs, experiences, practices and societies of that religion. For example, the top 20 discriminative words of Christianity cover images (e.g., *jesus, god, christ, lord*), beliefs (e.g., *bible, gospel, psalm, faith, sin, spirit*, etc.), practices (e.g., *pray, worship, praise*), and societies (e.g., *church, pastor*). On the other hand, Atheists show apparent preferences for topics about science (e.g., *science, evolution, evidence*), religion (e.g, *religion, christians, bible*) and politics (e.g., *republicans, gop, rights, abortion, equality*).

Fig. 2. The top 15 most discriminative words of each denomination based on a chi-square test

Fig. 3. The top 15 most frequent words for each denomination

Generally, the most interesting observations relate to non-religious terms appearing as discriminative features. This includes "evidence" for Atheist[11], or "bjp", referring to Bharatiya Janata Party[12], for Hindu. In a sense, if our observations were to hold in a broader context, it could be seen as good for society that followers of religious groups differ most in references to religious practice and concepts, rather than in every day aspects such as music, food or other interests. This leaves more opportunities for shared experiences and culture.

Whereas Figure 2 shows discriminative terms, those terms are not necessarily the most frequently used ones. Figure 3 shows tag clouds that display terms according to their actual within-group frequencies. As one can see, there are lots of commonalities and terms such as "love", "life", "people" and "happy" that are commonly used by believers of all religions. This illustrates that the differences in content are not as big as Figure 2 might seem to imply.

Whom Do They Follow? We apply essentially the same methodology to study how religious people are distinguished by whom they follow on Twitter. We represent each user group by a vector of their friends, where each entry (of the vector) represents a friend being followed by the users in that group. Similar to weighting ngrams by how many users use them in the previous section, the friends in the vector are weighted by how many users from that group follow them. We then apply chi-square test to the vector of each religious group against the vector of the *Undeclared* user group. Figure 4 displays the top 15 Twitter accounts (i.e., friends' screen names) that are most positively associated with each group. The font size of an account in the figure is determined by its chi-square score.

As before, we found that the most discriminative Twitter accounts of a particular denomination are specific to that religion. E.g., *IslamicThinking*,

---

[11] This is in line with recent work examining the relationship between religion and happiness on Twitter which also found Atheists to be more "analytical" [29]. Atheists are overrepresented among scientists, including top scientists (members of the Academy of Sciences) [9].

[12] It is one of the two major parties in India, which won the Indian general election in 2014.

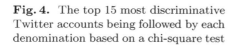

Fig. 4. The top 15 most discriminative Twitter accounts being followed by each denomination based on a chi-square test

Fig. 5. The top 15 Twitter accounts being followed by most users of each denomination

MuslimMatters, *YasirQadhi*[13], *ImamSuhaibWebb*[14], and *icna*[15] are the top 5 Twitter accounts followed by Muslims which are assigned the highest chi-square scores. The top 5 Twitter accounts that characterize Atheists all belong to atheistical or irreligious celebrities, including *RichardDawkins*, *neiltyson*, *rickygervais*, *billmaher* and *SamHarrisOrg*. This may have broader societal implications because these linkage or group preferences are likely to be present in the real world as well – for instance, real world traits and behaviors such as tolerance, prejudice, and openness to experience are likely to be correlated with our findings [15].

An analysis of the frequently followed users (see Figure 5) continues to show differences though and only few accounts are followed frequently by different religions. In a sense, people differ more in whom they follow rather than what they tweet about. Exceptions exist though and, for example, @BarackObama would be frequently followed by followers of most of the religions we considered.

5.2   Religion Classification

We then build classifiers to identify religious users of each denomination based on their tweet content and friend network. Specifically, we first extract a set of unigrams and bigrams (denoted as S) which appear in no less than 100 tweets in our tweet collection. We represent each user as a vector of unigrams and bigrams (in S) extracted from their tweets, where each entry of the vector refers to the frequency of that ngram in the user's tweets. The users are labeled by their denominations. We build a gold standard dataset for training and evaluating the binary classification of each denomination against the *Undeclared* user group. The different sizes of the datasets affect the classification performance, e.g., the classification of Christian benefits from larger dataset. To be able to compare the performance for different denominations, we downsample the datasets of all the denominations to the same size of the Hindu dataset, the smallest one. We balance each dataset to contain the same number of positive and negative instances. For each religious group, we train the SVM classifiers using LIBLINEAR

---

[13] The Twitter account of Yasir Qadhi, who is an Islamic theologian and scholar.

[14] The Twitter account of Suhaib Webb, who is the imam of the Islamic Society of Boston Cultural Center.

[15] The Twitter account of Islamic Circle of North America.

Table 3. The performance of tweet-based and friend-based religiosity classification of Twitter users

	Atheist	Buddhist	Christian	Hindu	Jew	Muslim	Macro-average
Tweet-based							
Precision	0.747	0.6657	0.7193	0.6653	0.6977	0.7248	0.7033
Recall	0.7869	0.7388	0.7285	0.6529	0.7526	0.6529	0.7188
F1	0.7658	0.6993	0.7231	0.6588	0.7241	0.6868	0.7097
Friend-based							
Precision	0.7726	0.733	0.7681	0.7201	0.7676	0.7992	0.7601
Recall	0.8557	0.8488	0.7285	0.7148	0.7595	0.8351	0.7904
F1	0.8117	0.7864	0.7477	0.7169	0.7635	0.8167	0.7738

[13], and apply 10-fold cross validation to its dataset. Similarly, we also represent each user as a vector of their friends, where each entry of the vector refers to whether the user follows a user X (1 - if the user follows X, and 0 - otherwise.) For each denomination, we build the gold standard dataset, balance it, train the SVM classifiers, and estimate the performances by 10-fold cross validation.

Table 3 reports the results. The tweet-based classification achieves a macro-average F1 score of 0.7097, and the friend-based classification achieves a macro-average F1 score of 0.7738. It demonstrates the effectiveness of content features and network features in classifying Twitter users' religiosity, and network features appear to be superior to content features. According to the F1 score, the difficulty level of recognizing a user from a specific religious group based on their *tweet content* is (from easiest to hardest): Atheist < Jew < Christian < Buddhist < Muslim < Hindu, while the difficulty level of recognizing a user from a specific religious group based on their *friend network* is (from easiest to hardest): Muslim < Atheist < Buddhist < Jew < Christian < Hindu.

6   Linkage Preference

In this section, we focus on exploring ingroup and outgroup relations. We construct four directed networks based on religious users following (friend), being-followed-by (follower), mentioning, and retweeting others, respectively. Following and being-followed-by relations are extracted from users' friend and follower lists, respectively. Mention and retweet relations are extracted from tweets, i.e., whether user A retweeted at least one tweet from user B, and whether user A mentioned user B in at least one of his/her tweets, respectively. Here we do not separate reply from mention. If a tweet addresses a specific user by including "@" followed by the user's screen name and it is not a retweet (e.g., marked with "RT"), we call it a mention.

For each user in our dataset, we count the numbers of all his/her connections (i.e., friends, followers, retweets, or mentions) and the connections with each religious group. Then we calculate the proportions of his/her ingroup (same-religion) connections and the connections to users from other groups. We get the average proportions of ingroup and outgroup connections for each group by adding that proportions of all the users in the group together and dividing by the number of users. The raw proportion may not reflect the linkage preference since

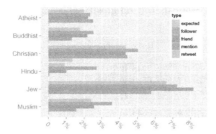
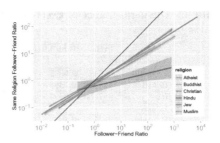

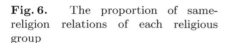

Fig. 6. The proportion of same-religion relations of each religious group

Fig. 7. Follower-friend ratio vs. same-religion follower-friend ratio. Linear smoothing is applied.

it is affected by the number of users in a group. The connections to Christians may always account for the biggest proportion because there are much more Christians than others in the dataset and even random linkage would give the illusion of preferring connection to Christians. So in addition to the raw proportion, we also estimate the expected proportion of connections to a specific user group by the fraction of users of a certain religion in a random user sample.

To be specific, in Section 3.1 we describe how we generate random numbers as Twitter user IDs, and collect user profiles from Twitter by these IDs. From all the valid U.S. user profiles collected in this way, we identify the users included in any religious denominations from our sample, and get the proportion of users of each denomination as the expectation of how likely a Twitter user connects with a user from a certain group. The expected proportions of connections are 0.0466% (Atheist), 0.0259% (Buddhist), 1.3358% (Christian), 0.0013% (Hindu), 0.0207% (Jew) and 0.0414% (Muslim). Note that these proportions are low as the vast majority of Twitter users do not explicitly state a religious affiliation in their profile. We then use the relative difference of the proportion to its expected value to represent the linkage preference. For example, Christian-Christian following accounts for 4.33% of all followings of a Christian user in average, and its relative difference compared to the expected value is $\frac{4.33\% - 1.3358\%}{1.3358\%} = 2.2$. These values are often referred to as "lift" in statistics.

We observe a preference for religious users to connect to others that share the similar belief to them, e.g., religious users are much more likely to follow other users of the same religion than of a different religion. For example, the same-religion followings of Hindu account for a proportion of 0.99% and the relative difference is 737.3, and same-religion followings of Jews account for a proportion of 8.15% and the relative difference is 392.3. Overall, following someone of the same religion is 646 times as likely as following someone of a different religion based on a macro-average of six denominations, if we estimate the following likelihood with the relative proportion obtained by dividing the raw proportion by the expected value.

Figure 6 plots the proportions of same-religion relations of different types. We compute the average proportions (per user) of being-followed-by, retweet and mention in the same way as we compute that of following. The expected proportions of connections are the same as we have described in the previous section. The same-religion linkage preference exists in all types of connections across all the religious groups. However, because our analysis is conducted on the users who self-reported their religious affiliations, it is probably biased toward very religious users, and for the other religious users who do not disclose their religion/belief on Twitter, the ingroup linkage preference may not be as strong.

From Figure 6 we observe such preference is stronger in the friend network than in the follower network for many religious groups such as Muslim, Jew, and Hindu. Note that this is at first sight paradoxical as when A follows B of the same religion this means that B is followed A by the same religion.[16] In order to explain this phenomenon, we plot the follower-friend ratio against the same-religion follower-friend ratio of the users in each group in Figure 7. It shows that the same-religion linkage preference of follower network is diluted by the out-group followers of the users who have more followers than friends. The ratio of same-religion followers of a local priest (e.g., placing at the bottom-left area in the coordination) may be higher than that of the same-religion friends, while the pope (e.g., placing at the top-right area in the coordination) may have many out-group followers that dilutes the ratio of same-religion followers. When the users in the top-right area contribute more to the overall proportions, the average ratio of same-religion friends is higher than that of the same-religion followers, otherwise, it is lower.

7   Conclusion

In this paper we used data from more than 250k U.S. users to describe the religious landscape on Twitter. We showed the distribution of Twitter users with a self-declared religious affiliation is a reasonable match to the distribution of religious believers according to surveys. We then characterized how different religions differ in terms of the content of the tweets and who they follow, and show that for the task of telling random users from religious users, a user's friends list provides more effective features than the content of their tweets. We find and quantify proof of within-group linkage preference for following, being-followed, mentioning and retweeting across all of our religions.

The ultimate goal of studies such as ours is not to study religion *on Twitter*, but to study religion *per se*, and arguably it will be more and more feasible in the future. Because more and more communication happens online, also more religious communication is likely to happen online. A key feature of religion is replication, and communication is key for such replication. Religion is a replicator – it replicates itself, its dogma, vertically (from generation to generation) and horizontally (across population), and in that sense it relies heavily on transmission media, Twitter being one of them.

---

[16] Some readers might rightly think of the somewhat related "Friendship Paradox" that your friends or followers have more friends and followers than you [17].

Twitter might be replaced by "the next big thing" but religion itself will not disappear in the foreseeable future, though it is continuously evolving along with the cultural context it is embedded in. To ensure a broader relevance of studies using online data it is therefore important to validate findings through separate channels and to ground research in existing literature. Online data can serve well to form hypotheses related to group formation, emotional stability or demographic correlates such as race or income and to guide follow-up studies looking at more holistic data and root causes.

There are several limitations and at the same time directions for future research. For example, in our current analysis, we only used the content of tweets to discover and describe discriminative tokens. No efforts were made to detect differences in dimensions such as sentiment or mood or other linguistic dimensions. In future work, we hope to gain clues as to what makes a religion stand out, e.g., when it comes to providing emotional stability or dealing with personal setbacks.

References

1. Barro, R., McCleary, R.: Religion and economic growth across countries. American Sociological Review 68(5), 760–781 (2003)
2. Busch, L.: To come to a correct understanding of buddhism: A case study on spiritualizing technology, religious authority, and the boundaries of orthodoxy and identity in a buddhist web forum. New Media & Society 13(1), 58–74 (2011)
3. Campbell, A., Converse, P.E., Rodgers, W.L.: The quality of American life: perceptions, evaluations, and satisfactions. Russell Sage Foundation, New York (1976)
4. Campbell, H.: Spiritualising the internet. uncovering discourses and narratives of religious internet usage. Online-Heidelberg Journal of Religions on the Internet 1(1) (2005)
5. Campbell, H.A.: Religion and the internet: A microcosm for studying internet trends and implications. New Media & Society 15(5), 680–694 (2013)
6. Simon Wiesenthal Center: iReport – online terror and hate: The first decade (2008), http://www.wiesenthal.com/atf/cf/
7. Chen, L., Wang, W., Sheth, A.P.: Are twitter users equal in predicting elections? A study of user groups in predicting 2012 U.S. Republican presidential primaries. In: Aberer, K., Flache, A., Jager, W., Liu, L., Tang, J., Guéret, C. (eds.) SocInfo 2012. LNCS, vol. 7710, pp. 379–392. Springer, Heidelberg (2012)
8. Cox, T.: Food-poisoning tracked through twitter with foodborne chicago app. DNAinfo Chicago (2014)
9. Dawkins, R.: Militant atheism. Ted Talk (2002), http://www.ted.com/talks/richard_dawkins_on_militant_atheism
10. Day, N.E.: Religion in the workplace: Correlates and consequences of individual behavior. Journal of Management, Spirituality & Religion 2(1), 104–135 (2005)
11. De Choudhury, M., Sundaram, H., John, A., Seligmann, D.D., Kelliher, A.: " birds of a feather": Does user homophily impact information diffusion in social media? arXiv preprint arXiv:1006.1702 (2010)
12. Dodds, P.S., Danforth, C.M.: Measuring the happiness of large-scale written expression: Songs, blogs, and presidents. Journal of Happiness Studies 11(4), 441–456 (2010)

13. Fan, R.E., Chang, K.W., Hsieh, C.J., Wang, X.R., Lin, C.J.: Liblinear: A library for large linear classification. The Journal of Machine Learning Research 9, 1871–1874 (2008)
14. Golder, S.A., Macy, M.W.: Diurnal and seasonal mood vary with work, sleep, and daylength across diverse cultures. Science 333(6051), 1878–1881 (2011)
15. Hall, D.L., Matz, D.C., Wood, W.: Why don't we practice what we preach? a meta-analytic review of religious racism. Personality and Social Psychology Review 14(1), 126–139 (2010)
16. Ho, S.S., Lee, W., Hameed, S.S.: Muslim surfers on the internet: Using the theory of planned behaviour to examine the factors influencing engagement in online religious activities. New Media & Society 10(1), 93–113 (2008)
17. Hodas, N.O., Kooti, F., Lerman, K.: Friendship paradox redux: Your friends are more interesting than you. In: Proceedings of ICWSM, pp. 1–8 (2013)
18. Iannaccone, L.R.: Introduction to the economics of religion. Journal of Economic Literature 36, 65–1496 (1998)
19. Kang, J.H., Lerman, K.: Using lists to measure homophily on twitter. In: AAAI Workshops (2012)
20. Kwak, H., Lee, C., Park, H., Moon, S.: What is twitter, a social network or a news media? In: Proceedings of the 19th International Conference on World Wide Web, pp. 591–600. ACM (2010)
21. Lieberman, M.A., Winzelberg, A.: The relationship between religious expression and outcomes in online support groups: A partial replication. Computers in Human Behavior 25(3), 690–694 (2009)
22. McKenna, K.Y., West, K.J.: Give me that online-time religion: The role of the internet in spiritual life. Computers in Human Behavior 23(2), 942–954 (2007)
23. Miller, G.: Social scientists wade into the tweet stream. Science 333(6051), 1814–1815 (2011)
24. Mislove, A., Lehmann, S., Ahn, Y.Y., Onnela, J.P., Rosenquist, J.N.: Understanding the demographics of twitter users. In: Proceedings of ICWSM (2011)
25. Newport, F.: Seven in 10 americans are very or moderately religious (2012), http://www.gallup.com/poll/159050/seven-americans-moderately-religious.aspx#1
26. Nguyen, M.T., Lim, E.P.: On predicting religion labels in microblogging networks. In: Proceedings of the 37th International ACM SIGIR Conference on Research & Development in Information Retrieval, pp. 1211–1214 (2014)
27. Okulicz-Kozaryn, A.: Religiosity and life satisfaction across nations. Mental Health, Religion & Culture 13(2), 155–169 (2010)
28. Okulicz-Kozaryn, A.: Does religious diversity make us unhappy? Mental Health, Religion & Culture 14(10), 1063–1076 (2011)
29. Ritter, R.S., Preston, J.L., Hernandez, I.: Happy tweets: Christians are happier, more socially connected, and less analytical than atheists on twitter. Social Psychological and Personality Science 5(2), 243–249 (2014)
30. Sedikides, C.: Why does religiosity persist? Personality and Social Psychology Review 14(1), 3–6 (2010)
31. Sosis, R.: Does religion promote trust? the role of signaling, reputation, and punishment. Interdisciplinary Journal of Research on Religion 1(7), 1–30 (2005)
32. Wagner, C., Asur, S., Hailpern, J.: Religious politicians and creative photographers: Automatic user categorization in twitter. In: Proceedings of SocialCom, pp. 303–310. IEEE (2013)
33. Weber, I., Garimella, V.R.K., Batayneh, A.: Secular vs. islamist polarization in egypt on twitter. In: ASONAM, pp. 290–297 (2013)

34. Woodruff, A., Augustin, S., Foucault, B.: Sabbath day home automation: it's like mixing technology and religion. In: Proceedings of SIGCHI, pp. 527–536. ACM (2007)
35. Wyche, S.P., Aoki, P.M., Grinter, R.E.: Re-placing faith: reconsidering the secular-religious use divide in the united states and kenya. In: Proceedings of SIGCHI, pp. 11–20. ACM (2008)
36. Wyche, S.P., Caine, K.E., Davison, B., Arteaga, M., Grinter, R.E.: Sun dial: exploring techno-spiritual design through a mobile islamic call to prayer application. In: Proceedings of SIGCHI, pp. 3411–3416. ACM (2008)
37. Wyche, S.P., Griner, R.E.: Extraordinary computing: religion as a lens for reconsidering the home. In: Proceedings of SIGCHI, pp. 749–758. ACM (2009)
38. Wyche, S.P., Hayes, G.R., Harvel, L.D., Grinter, R.E.: Technology in spiritual formation: an exploratory study of computer mediated religious communications. In: Proceedings of CSCW, pp. 199–208. ACM (2006)

Who Are My Audiences? A Study of the Evolution of Target Audiences in Microblogs

Ruth García-Gavilanes[1], Andreas Kaltenbrunner[2], Diego Sáez-Trumper[3],
Ricardo Baeza-Yates[3], Pablo Aragón[2], and David Laniado[2]

[1] Universitat Pompeu Fabra, Barcelona, Spain
[2] Fundación Barcelona Media, Barcelona, Spain
[3] Yahoo Labs, Barcelona, Spain

Abstract. User behavior in online social media is not static, it evolves through the years. In Twitter, we have witnessed a maturation of its platform and its users due to endogenous and exogenous reasons. While the research using Twitter data has expanded rapidly, little work has studied the change/evolution in the Twitter ecosystem itself. In this paper, we use a taxonomy of the types of tweets posted by around 4M users during 10 weeks in 2011 and 2013. We classify users according to their tweeting behavior, and find 5 clusters for which we can associate a different dominant tweeting type. Furthermore, we observe the evolution of users across groups between 2011 and 2013 and find interesting insights such as the decrease in conversations and increase in URLs sharing. Our findings suggest that mature users evolve to adopt Twitter as a *news media* rather than a social network.

1 Introduction

Online social networks like Twitter have become extremely popular. Twitter has grown from thousands of users in 2007 over millions in 2009 to hundreds of millions in 2013. Through the years, users have learned to use Twitter following certain conventions in their messages, limited to 140 characters. In certain occasions, these conventions help users to imagine a target audience or set a topic that goes along with what the community is talking about. For example, the use of the symbol @ (at) before a user name to mark a dyadic interaction between two users and the use of *re-tweets* for spreading the content of a tweet posted by someone else. Likewise, the use of URLs (often shortened) to share external information, etc.

As a consequence, Twitter is used in several contexts, for different audiences and with different purposes. In fact, scholars have argued that Twitter is used as an hybrid between a communication media and an online social network [6,17]. Additionally, user behavior is not static, it changes through the years, the way the first Twitter users interacted with the platform when it started may differ from how they interact now. While the set of research using Twitter data has expanded rapidly, little work has studied the change/evolution in Twitter ecosystem itself.

L.M. Aiello and D. McFarland (Eds.): SocInfo 2014, LNCS 8851, pp. 561–572, 2014.

In this paper, we propose a step towards understanding the evolution of user behavior focusing on *how* people tweet and their audiences. To this end, we carry out a longitudinal study of tweets posted during 10 weeks in 2011 and 10 weeks in 2013 by more than 4M users who have been active in Twitter in both of these periods.

First, we propose a taxonomy of messages based on Twitter conventions (mentions, links, re-tweets). In doing so we obtained 6 tweet formats. To identify models of behavior, we cluster users based on these types of tweets and study how users change their behavior in time. To present our results, we organize the paper as follows. Section 2 provides related work. Section 3 describes the data. In Section 4 we explain our methodology and the taxonomy given to the types of tweets. In Section 5 we report how user behavior changes in 2013 with respect to 2011. We finish with conclusions and next steps in Section 6.

2   Related Work

The goal of this work is to study the variation of tweeting behavior across time based on a taxonomy of tweet types and audiences. In a similar way, researchers have already analyzed how a variety of aspects change across time in Twitter and other online platforms. They have studied the following aspects:

The Nature of Twitter. While most messages on Twitter are conversations and chatter, people also use it to share relevant information and to report news [4]. In fact, scholars have concluded that from the highly skewed nature of the distribution of followers and the low rate of reciprocated ties, Twitter more closely resembles an information sharing network than a social network [6].

Evolution of Users and Behavior. Liu *et al.* [8] studied the evolution of Twitter users and their behavior by using a large set of tweets between 2006 and 2013. They quantify a number of trends, including the spread of Twitter across the globe, the shift from a primarily-desktop to a primarily-mobile system, the rise of malicious behavior, and the changes in tweeting behavior. The main part of this study is based on the accumulative number of tweets. We address, instead, the evolution based on individual users' behavior.

Audiences. Marwick and boyd [10,13] claim that users in Twitter *imagine* their target audiences since they do not know "which few" will read their tweets. They find that users do not have a fixed target audience and that having one would be a synonym of "inauthenticity".

Behavior and Clusters. Naaman *et al.* [11] find 4 relevant categories of tweets based on the content of the messages. For each one of these categories, they cluster users and find two types of users: Meformers (talking about one self) and Informers (sharing news). Luo *et al.* [9] classify tweets based on language and syntactic structure and Huang *et al.* [3] show that tagging behavior (hashtags) has a conversational, rather than organizational nature.

Many attempts have been done to classify users according to their audiences and tweet content. However, most of these studies are language-dependent and need manual labeling. In this work, we categorize audiences and tweet types using a language-independent approach.

3 Data Set

For the results we present here, we crawled the profile information of users who posted tweets with the hashtag *#followfriday* or *#ff* on the first Friday of March, 2011 as in [2].

From this set, we randomly selected 55K users with a number of followers and followees in the range of [100, 1000] and crawled their corresponding followee network (for a user u, it contains all users who u is following).

We then proceeded to collect all of the tweets posted in English by the original 55K users as well as their followees during 10 weeks starting from the second half of March 2011. By crawling the information of the followees, we attempt to target the typical accounts twitterers like to follow. It is mostly on these users and the 55K seed set that all our results are concerned.

In total, we obtained 8M users who tweeted around 2.4B tweets. We then crawled Twitter during 10 weeks between October and December 2013 looking for the same users and found that around 4.3M users tweeted at least once also in 2013. After the end of the crawling period, we identify the language in which tweets are written. We then proceed to classify as *active users* those who tweeted at least 55 and less or equal than 1540 tweets in English during 10 weeks to exclude inactive or hyperactive users and bots. In total we found around 538K users tweeting within this range in both years. We chose this range as to set a threshold of 1 tweet per working day (5 per week) and a maximum of 22 per day. The maximum limit was chosen based on a marketing study by Zarrella [19], which argues that most users tweet an average of 22 times a day. With this we attempt to include users likely to be engaged with the platform excluding those with an abnormal activity (i.e, advertisers or bots). Appendix A describes details about the crawling process and Table A1 presents the summary of the dataset used for the experiments.

4 Methodology

As previously discussed in the related work section, some researchers argue that everybody has an *imagined audience* in a communicative act even if that act involves social media [10]. Given the various ways people consume and spread tweets, it is virtually impossible for Twitter users to account for their potential audience, although we often find users tweeting as if these audiences were bounded. For instance, the use of the @ sign before a user login name allows to "poke" that user which may trigger a reply and start dyadic conversations (through mentions) which are visible at the same time to others as well. In fact, Marwick and boyd [10] found, through interviews to twitterers, that sometimes users are "conscious of potential overlap among their audiences (i.e, friends, family, co-workers, etc)." The authors report cases where users tweet to themselves, to fans, to fellow nerds, to super users, etc.

We propose a language-independent taxonomy of tweet types. The proposed types are based on the conventions established by Twitter such as the mention

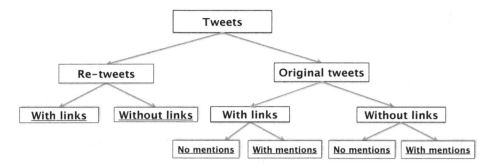

Fig. 1. This classification tree represents the tweet formats used to classify users in different groups. The top groups include the tweets in the subsequent levels. The underlined nodes (leaves of the tree) are used in the clustering process (6 types).

symbol @, the retweet flag and the URLs, *imagining* an audience through the combination of these symbols. Figure 1 shows these categories.

We start by classifying two main groups of tweets: retweets (RT) and original tweets (OT). Retweets refer to those tweets forwarded from other users. We hypothesize that a retweet targets the user who created the forwarded tweet and the followers of the user forwarding the tweet. Next, original tweets refer to tweets posted by users themselves and the audience could vary between the followers and the users themselves. For the RT and OT sets, we make two other distinctions: tweets with URLs and without URLs. We hypothesize that URLs target audiences who are willing to obtain information from the links posted and generally interested in exogenous stimuli. For tweets without URLS, users want to transmit a self-contained idea in maximum 140 characters. For the OT set we make yet another distinction, for the tweets with URLs and without URLs we divide them between tweets containing a mention (conversational) and those without a mention (textual). A OT containing a link with a mention implies that a user calls the attention of another user to open the link shared in the tweet. We do not make this last distinction (mention and link) for the RT set given than all retweets already refer to another user. In this study, we focus on the tweet types at the deepest level of each branch (6 in total): a) re-tweets with links, b) re-tweets without links, c) original tweets with links and no mentions, d) original tweets with links and mentions, e) original tweets without links and no mentions and finally f) original tweets without links and mentions.

Based on this scheme, we classify the tweets of the *active* set of users (538 K) in 2011 and 2013 and find a slight increase in tweets with URLs in 2013 (from 14.62% to 18.74%). Table B1 of Appendix B has the percentage of tweets in each category for *active* users.

Furthermore, for each *active* pair (user, year) we calculate the percentages of tweets belonging to each of the tweet types. Each pair (user, year) is represented by a 6-dimensional vector, *6* being the number of all numerical features (the percentages) used to describe the objects to be clustered. We use the well-known

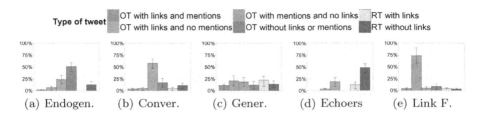

Fig. 2. Clustering based on 6 tweet types posted by *active* users during 10 weeks in 2011 and 2013. The clusters appear from left to right according to their size in descending order. Each bar shows the average percentage of that tweet type. Error bars represent the interquartile range. Clusters (a) and (d) do not contain tweets of all types.

k-means algorithm for clustering. To decide the k points in that vector space, we used the so called *elbow method*. This is a visual standard method [12] that runs the k-means algorithm with different numbers of clusters and shows the results of the sum of the squared error. The value of k is chosen by starting with $k = 2$ and increasing it by 1 until the gain of the solution drops dramatically, which will be the bend or elbow of the graph. This is the k value we want and is chosen visually. We found that the *bend* lingered between 4 and 5 (see Figure B1 in Appendix B). We analyzed both cases and chose $k = 5$ because we observed that it best encapsulates interesting and distinctive patterns of tweeting behavior.

5   Results

We now proceed to the results and study how users have changed their tweeting behavior through time. Figure 2 shows the average composition of tweet type vectors in the clusters. The clusters are ordered by size and the bars indicate the interquartile range for each case. Note that we have abbreviated some of the names in the captions due to space concerns. We observe that each cluster has a dominant tweet type except for the third cluster (*Generalists*) that reports a balance among the tweet types.

We discuss now each of the identified patterns of tweeting behavior and relate them to the concept of the *imagined audiences* discussed in the previous section.

Endogenous: Users in this cluster mostly post and forward messages not linked to external information. Users in this cluster are supposed to use Twitter more as a social network than as a news media. The dominant type of tweets are self-contained posts created by the user herself without mentioning other users such as quotes, thoughts or even futile information. In second place we observe original tweets with mentions which is a sign of conversation with other users.

Conversationalists: Users following this pattern are characterized mostly by tweets containing mentions with no links. Similarly to the *Endogenous* type, users in this cluster are also supposed to use Twitter more as a social network but with an emphasis on interacting with other users more than sharing self-contained ideas.

Generalists: This cluster groups users who use Twitter without a distinctive tweet type. It is interesting to notice that in this cluster, retweets with links and original tweets with links are slightly above the rest which may suggest an inclination to audiences interested in obtaining external information.

Echoers: These are users characterized by forwarding other people's tweets with no links. These users are mostly inclined to read what others have to say, indicating in a way that they make part of the audience of other users's original ideas (being these informative or not). An example of such users are those who follow accounts posting jokes, positive thinking, quotes, etc. The second dominant category in this cluster involves tweets with mentions, which most likely mean that users reply or chat with others.

Link Feeders: This cluster involves all those accounts that mostly tweet messages containing external links. In 2011 [18] found that around 50% of URLs posted in tweets came from media producers. We expect then that the owners of these accounts are mainly news media, journalists, link builders, SEO specialists, etc. Since these are tweets that contain no mentions, the expected target audience is then a general public that aims to obtain information through these accounts (i.e, followers of news papers).

The clustering process was based on the tweets of active users in both 2011 and 2013. Figure B2 in Appendix B shows the number of users falling in one of the clusters for each year.

5.1   Change in Tweeting Behavior

Here we study how users have changed their tweeting behavior in 2013 with respect to 2011. Based on the active users only (those who remained active in 2011 and 2013), we plot these groups into a Sankey diagram in Figure 3 to observe the proportion of users moving from one cluster to another.

We observe that in general around half of these active users remain in the same cluster in both periods, except for the Echoers. On the other hand, we observe an increase in 2013 of the *Generalists* and *Link Feeders* cluster with respect to 2011. The increase in the *Generalists* cluster is expected since our dataset contains users who have remained in Twitter for more than two years. These users have matured with the platform and most likely learned to use it for multiple reasons (chat, share information, retweets, etc). Moreover, the increase in the *Link Feeders* cluster goes along with Table B1, which also shows an increase in the percentage of tweets with URLs. Nowadays, Twitter automatically shortens URLs using the t.co service [1] which makes it easier for users to share links without the need to visit other URL shortener sites. This was not the case in 2011. Additionally, an increasing number of external sites allow to automatically post on Twitter with their link included. It is expected then that by 2013 users share more URLs than before.

On the other hand, we see a decrease in 2013 of the *Conversationalists* type. It seems that some users who used to chat a lot are evolving to chat less and be more *Endogenous* (posting their own tweets with no links or mentions) and *Generalists*. Mature users would have quickly realized that it was hard to

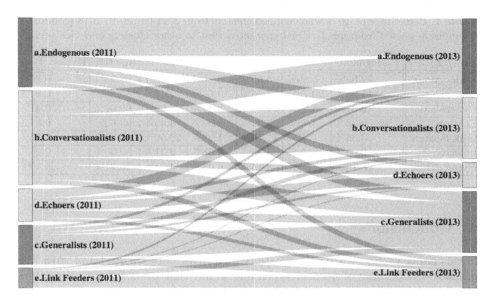

Fig. 3. The Sankey shows how active users have changed the way they tweet in 2013 with regard to 2011

continue conversations once the chat channel has passed in Twitter. On top of that, cross-platform instant messaging services more oriented to conversation purposes (i.e.,WhatsApp) have become increasingly popular. Neverthless, in 2013 Twitter made it easier to follow conversations in the timeline [5]. Perhaps, we will witness an increase in conversations after 2013.

Finally, the decrease in the *Echoers* cluster from 2011 to 2013 shows that users who tend to forward other people's ideas most of the time have evolved to generate more content themselves, moving to the *Endogenous* or *Generalist* clusters.

For a better readability of the evolution of active users' behavior, we did not include in the Sankey diagram the proportion of users who were filtered out of the active set in 2011 and moved to any of the clusters in 2013. We include this information in Table 1 in percentages (of around 4.3 M users) and show in Table B2 (Appendix B) the corresponding absolute values. We observe that the majority of users from any cluster in 2011 become inactive in 2013. Similarly, inactive users tend to remain as such even two years later. Interestingly, the majority of hyperactive users move to one of the clusters but we also observe a significant percentage (26.71%) becoming inactive in 2013.

These findings go along with Liu *et al.* [8], who found a massive percentage of inactive accounts by the end of 2013. As Twitter users mature, many also choose to move to other platforms and to be less active.

Table 1. Percentage of users who changed clusters from 2011 (rows) to 2013 (columns). Some users passed from inactive or hyperactive/bot to other clusters and vice versa.

2011/ 2013	Endogenous	Conver.	Gener.	Echoers	Link F.	Inactive	Hyper./Bots
Endogenous	22.38%	5.89%	5.96%	3.56%	3.02%	58.33%	0.86%
Conver.	11.33%	20.79%	7.26%	3.54%	2.41%	53.80%	0.87%
Generalists	2.67%	3.88%	21.78%	2.17%	7.02%	62.07%	0.41%
Echoers	9.93%	3.72%	8.31%	9.93%	3.65%	63.62%	0.84%
Link Feeders	3.38%	1.47%	11.11%	1.25%	22.59%	59.45%	0.75%
Inactive	6.64%	3.30%	3.12%	2.38%	2.31%	82.00%	0.26%
Hyper./Bots	28.13%	17.42%	8.15%	6.48%	4.91%	26.71%	8.19%

6   Conclusions

In this paper we have carried out a study in Twitter between 2011 and 2013. We propose a taxonomy of 6 tweet types and found that users fall into 5 clusters of behavior: Endogenous (those who mostly tweet without links or mentions), Conversationalists (those who mostly converse with others), Generalists (those who post different type of tweets), Echoers (those who re-tweet more) and Link Feeders (those who share URLs most of the time). We then observed the evolution of users across clusters between these years and noticed a general tendency to become inactive or maintain the same type of behavior over years, with the exception of *echoers* who show to be active in a year full of controversial events. We also observed a decrease of *conversationalists*, likely due to the maturation of users, the emergence of instant message services and the difficulty of chatting in Twitter before 2013. We also found more Link Feeders and Generalists in 2013. In the past, Twitter has been described as hybrid platform, being a social network and a news media at the same time [6]; our results, with the increase in news feeders and decrease in conversationalists, suggest that the main usage of the service by mature users is shifting towards the latter: a news media.

After completing this study, there are several complementary projects ahead. For instance, we plan to look closely at the behavior of the inactive and hyperactive users and bots. We also plan to study the lexical variation in dyadic conversations across time. Furthermore, it would be interesting to analyze if users tweeting in several languages differ in tweeting behavior for each language. Finally, we plan to compare this evolution to the change in user popularity.

References

1. Twitter Help Center. Posting links in a tweet,
   https://support.twitter.com/entries/78124-how-to-shorten-links-URLs
2. García-Gavilanes, R., O'Hare, N., Aiello, L.M., Jaimes, A.: Follow my friends this friday! an analysis of human-generated friendship recommendations. In: The 5th International Conference on Social Informatics, SOCINFO (2013)

3. Huang, J., Thornton, K., Efthimiadis, E.: Conversational tagging in twitter. In: Proceedings of the 21st ACM Conf. on Hypertext and Hypermedia (2010)
4. Java, A., Song, X., Finin, T., Tseng, B.: Why We Twitter: Understanding Microblogging Usage and Communities. In: Procedings of the Joint 9th WEBKDD and 1st SNA-KDD Workshop (2007)
5. Kamdar, J.: Keep up with conversations on twitter, https://blog.twitter.com/2013/keep-up-with-conversations-on-twitter
6. Kwak, H., Lee, C., Park, H., Moon, S.: What is Twitter, a social network or a news media. In: Proceedings of the 19th International Conference Companion on World Wide Web (2010)
7. Lee, K., Eoff, B., Caverlee, J.: Seven months with the devils: A long-term study of content polluters on twitter. In: International AAAI Conference on Weblogs and Social Media, ICWSM (2011)
8. Liu, Y., Kliman-Silver, C., Mislove, A.: The tweets they are a-changin': Evolution of twitter users and behavior. In: International AAAI Conference on Weblogs and Social Media, ICWSM (2014)
9. Luo, Z., Osborne, M., Petrovic, S., Wang, T.: Improving twitter retrieval by exploiting structural information. In: Proceedings of the Twenty-Sixth AAAI Conference on Artificial Intelligence, JToronto. AAAI (2012)
10. Marwick, A., Boyd, D.: I tweet honestly, i tweet passionately: Twitter users, context collapse, and the imagined audience. New Media and Society (September 2010)
11. Naaman, M., Boase, J., Lai, C.-H.: Is it really about me?: Message content in social awareness streams. In: Proceedings of the 13th ACM Conference on Computer Supported Cooperative Work and Social Computing, CSCW 2010 (2010)
12. Ng, A.: Machine learning (2014), https://www.coursera.org
13. Papacharissi, Z.: The presentation of self in virtual life: Characteristics of personal home page. Journalism and Mass Communication Quarterly (2002)
14. Petrović, S., Osborne, M., Lavrenko, V.: Rt to win! predicting message propagation in twitter. In: International AAAI Conference on Weblogs and Social Media, ICWSM (2011)
15. Suh, B., Hong, L., Pirolli, P., Chi, E.H.: Want to be retweeted? large scale analytics on factors impacting retweet in twitter network. In: The 2nd IEEE International Conference on Social Computing (2010)
16. Thomas, K., Grier, C., Song, D., Paxson, V.: Suspended accounts in retrospect: an analysis of twitter spam. In: ACM Proceedings of the 2011 ACM SIGCOMM on the Internet Measurement Conference, Berlin, Germany (2011)
17. Wu, S., Hofman, J., Mason, W., Watts, D.: Who says what to whom on Twitter. In: Proceedings of the 20th International Conference companion on World Wide Web (2011)
18. Wu, S., Hofman, J.M., Mason, W.A., Watts, D.J.: Who says what to whom on twitter. In: Proceedings of the 20th International Conference companion on World Wide Web, WWW 2010 (2011)
19. Zarrella, D.: The science of timing (2011), http://www.slideshare.net/HubSpot/the-science-of-timing

A   Appendix : Detailed Dataset Description

The Follow Friday hashtag emerged in 2009 as a spontaneous convention from the Twitter user base: users post tweets with the *#followfriday* (or *#ff*) hashtag, and include the usernames of the users they wish to recommend on Fridays. Back in 2010 and 2011, this hashtag was one of the most used in Twitter [15,14] and so we hypothesized that engaged twitter users would likely adopt this hashtag because they care about recommending users to follow.

We crawled 55K users with number of followers and followees in the range of [100, 1000] not exceed the limit of the API calls at that time. It also has the added benefit of filtering out less legitimate (e.g., spam) users, since, according to Lee *et al.* [7], the majority of spam users tend to have out-degree and in-degree outside the range of [100; 1000]. Also Kurt *et al.* [16] showed that 89% of users following spam accounts have fewer than 10 followers. So, while we cannot guarantee that our dataset does not contain spammers, previous studies indicate that our sample will indeed have a higher probability of containing mostly legitimate users.

The information collected includes the user id, the screen name, the information in the location field of the profile, the date stamp of the tweet, the number of followers and followees, the *id* and the text of the tweet. We continue by finding the geolocation of each user via the location field entered in their profiles and we kept those geolocated users as to add one additional anti-spam filter. We believed that users who specified a valid geolocation are less likely to be spammers.

There is a higher proportion of active users among those users who tweeted in *both* 2011 *and* 2013 (the 5th row of the 3rd and 4th column) than those who tweeted in 2011 but not necessarily in 2013 (the 5th row of the 2nd column).

Table A1. The second column shows the full data crawled in 2011. The 3rd and 4th column show information of users who tweeted in *both* 2011 and 2013. Rows 3 to 5 contains information about active and inactive users. Rows 7 to 9 contain information of the active users only. Active users are those considered to have tweeted in English more than 55 and less than 1540 times.

	Active and inactive set		
	Full Data Set 2011	**Users active in 2011 & 2013**	
	2011	**2011**	**2013**
Users	8,092,891	4,350,583	4,350,583
Tweets	2,280,707,094	1,527,675,950	679,507,450
English Tweets	1,086,233,182	768,940,902	369,452,361
	Active set		
Active Users	1,868,150	1,315,313	1,125,968
Tweets	1,248,300,919	880,889,333	375,741,789
English Tweets	562,134,366	406,719,99	256,330,241

B   Appendix : Complementary Material

Table B1. Tweets from active users in 2011 and 2013, and the corresponding percentage of tweets that belong to each type

	Full DS 2011	2011	2013
	Tweets	Tweets	Tweets
Original tweets	77.30%	76.94%	74.77%
With URLs	14.93%	14.62%	18.74%
with mentions	6.39%	3.46%	4.16%
without mentions	11.36%	11.16%	14.58%
Without URLs	62.37%	62.32%	56.03%
with mentions	35.18%	35.36%	27.44%
without mentions	27.19%	26.96%	28.59%
Retweets	22.70%	23.06%	25.23%
With URLs	6.29%	6.75%	8.6%
Without URLs	16.41%	16.31%	16.63%

Table B2. The absolute number of users who moved across clusters from 2011 (rows) to 2013 (columns). Some users passed from inactive or hyperactive/bot to the other clusters and vice versa.

2011/2013	Endogenous	Conver.	Gener.	Echoers	Link F.	Inactive	Hyper./Bots
Endogenous	79,472	20,900	21,159	12,657	10,705	207,108	3,036
Conver.	49,832	91,429	31,945	15,570	10,616	236,624	3,807
Generalists	5,886	8,542	47,997	4,784	15,479	136,813	903
Echoers	19,308	7,235	16,149	19,306	7,105	123,704	1,640
Link Feeders	3,573	1,548	11,736	1,315	23,855	62,781	794
Inactive	194,636	96,641	91,391	69,684	67,769	2,403,596	7,481
Hyper./Bots	29,275	18,131	8,484	6,745	5,109	27,803	8,529

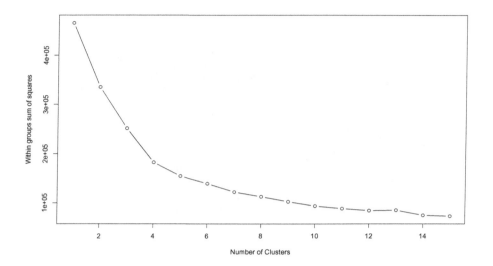

Fig. B1. Elbow method for clustering : the *bend* lingers between 4 and 5

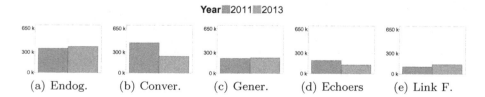

(a) Endog.   (b) Conver.   (c) Gener.   (d) Echoers   (e) Link F.

Fig. B2. Number of active users in each cluster for 2011 and 2013